24th Summer School and International Symposium on the Physics of Ionized Gases 2008

Journal of Physics: Conference Series Volume 133

Novi Sad, Serbia
25-29 August 2008

ISBN: 978-1-61567-162-5

Printed from e-media with permission by:

Curran Associates, Inc.
57 Morehouse Lane
Red Hook, NY 12571
www.proceedings.com

Some format issues inherent in the e-media version may also appear in this print version.

Printed by Curran Associates, Inc. (2009)

Institute of Physics
Dirac House, Temple Back
Bristol BS1 6BE UK

Phone: 44 1 17 929 7481
Fax: 44 1 17 920 0979

techtracking@iop.org

TABLE OF CONTENTS

Low Energy Lepton Scattering: Recent Results for Electron and Positron Interactions 1
Stephen J Buckman, Todd Maddern, Jessica Francis-Staite, Leigh Hargreaves, Michael J Brunger, Gustavo Garcia, Julian C Lower, Subhendu Mondal, James P Sullivan, Adric Jones, Peter Caradonna, Daniel Slaughter, Casten Mackochekanwa, Robert P McEachran

Energy Deposition Model Based on Electron Scattering Cross Section Data from Water Molecules 9
A Muñoz, J C Oiler, F Blanco, J D Gorfinkiel, P Limão-Vieira, A Maira-Vidal, M J G Borge, O Tengblad, C Huerga, M Téllez, G García

Modelling of Anomalous Doppler Broadened Lines, Thermalization of Electrons and the Role of Radicals in Discharges at High E/N 22
Zoran Lj Petrovic, Željka Nikitovic, Vladimir Stojanovic

Low Energy Positron Scattering from Helium 33
J P Sullivan, C Makochekanwa, A Jones, P Caradonna, S J Buckman

A Multi Term Boltzmann Equation Analysis of Non-conservative Electron Transport in Time-dependent Electric and Magnetic Fields 40
S Dujko, R D White

The Expert Systems for Analysis of Electron Energy-loss Spectra 49
Ivan Petrovic, Valerij Bocvarski, Violeta Petrovic

Development, Diagnostic and Applications of Radio-frequency Plasma Reactor 57
N Puac

Modeling of Townsend Discharges at High E/N and Low Pressure 65
V D Stojanovic

Theoretical Study of Excited Electronic States at Surfaces, Link with Photo-emission and Photo-desorption Experiments 73
J-P Gauyacq, A G Borisov, A K Kazansky

Plasma-surface Interactions 85
M J Goeckner, C T Nelson, S P Sant, A K Jindal, E A Joseph, B S Zhou, G Padron-Wells, B Jarvis, R Pierce, L J Overzet

Interactions of Ions with Carbon Nano-structures 95
Z L Mišković

Etching of Bacillus Atrophaeus by Oxygen Atoms, Molecules and Argon Ions 106
J Benedikt, C Flötgen, G Kussel, V Raball, A von Keudell

Surface Analysis with Grazing Incidence Fast Atom Diffraction (GIFAD) 115
Patrick Rousseau, Hocine Khemliche, Nenad Bundaleski, Pierre Soulisse, Anouchah Momeni, Philippe Roncin

Influence of Inelastic Processes on Fast-atom-surface Diffraction 124
F Aigner, N Simonovic, B Solleder, L Wirtz, J Burgdörfer

Channeling of Protons Through Carbon Nanotubes 132
D Borka, D J Mowbray, Z L Mišković, S Petrovic, N Nešković

Collisions of Ions with Insulator Surfaces: Charging and Discharging Dynamics 140
N Bundaleski, H Khemliche, P Rousseau, A Cassimi, L Maunoury, P Roncin

Measurement of Voltage and Current in Continuous and Pulsed RF and DC Glow Discharges 148
V Hoffmann, V V Efimova, M V Voronov, P Šmid, E B M Steers, J Eckert

Advances in the Optical Diagnostics of Kinetic Processes in Atmospheric Pressure Dielectric Barrier Discharges 160
G Dilecce, P F Ambrico, S De Benedictis

Smart Coating Technology by Gas Tunnel Type Plasma Spraying 170
Akira Kobayashi

The Effects of Traces of Molecular Gases (H_2, N_2 & O_2) in Glow Discharges in Noble Gases 181
E B M Steers, P Šmid, V Hoffmann, Z Weiss

Supershort Avalanche Electron Beams and X-ray in High-pressure Nanosecond Discharges 191
V F Tarasenko, E H Baksht, I D Kostyrya, M I Lomaev, D V Rybka

Spectroscopic Study of Hydrogen Rotational, Vibrational and Translational Temperatures in a Hollow Cathode Glow Discharge 203
G Lj Majstorovic

Modeling of a Dielectric Barrier Discharge Used As a Flowing Chemical Reactor 212
D Petrovic, T Martens, J van Dijk, W J M Brok, A Bogaerts

Collisionless Kelvin-helmholtz Instability and Vortex-induced Reconnection in the External Region of the Earth Magnetotail 220
F Pegoraro, M Faganello, F Califano

Simulation Science for Fusion Plasmas 231
S Sudo, M M Škoric, T-H Watanabe, Y Todo, A Ishizawa, H Miura, R Ishizaki, A Ito, H Ohtani, S Usami, H Nakamura, Atsushi Ito, S Ishiguro, Y Tomita, A Takayama, M Sato, T Yamamoto, M Den, H Sakagami, R Horiuchi, S Okamura, N Nakajima

Multilayer Photoionization Models of the Nuclear and Circumnuclear Regions of Type 2 AGN 243
P Rafanelli, L Vaona, S Ciroi, F Di Mille

Spectroscopy Methods Applied to the Research in Plasmas at Atmospheric Pressure 257
M D Calzada

Kinematical Parameters in the Coronal and Post-coronal Regions of the Oe Stars 266
A Antoniou, E Danezis, E Lyratzi, L C. Popovic, M S Dimitrijevic

Processes with Neutral Hydrogen and Deuterium Molecules Relevant to Edge Plasma in Tokamaks 274
Iztok Cadež, Sabina Markelj, Zdravko Rupnik, Primoz Pelicon

Variability of GRB Light Curve: Shock Wave Model Modification 284
Saša Simic, Luka C. Popovic

Confinement and Anisotropy of Ultrahigh-energy Cosmic Rays in Isotropic Plasma Wave Turbulence 292
Miroslava Vukcevic

γ-radiation from the Galactic Center: Dark Matter Annihilation Or More Conservative Astrophysical Models? 299
Alexander F Zakharov, Achille A Nucita, Francesco De Paolis, Gabriele Ingrosso

Author Index

24th Summer School and International Symposium on the Physics of Ionized Gases

Gordana Malović *et al* 2008 *J. Phys.: Conf. Ser.* **133** 011001 (1p) doi: 10.1088/1742-6596/133/1/011001 Help

Gordana Malović, Luka Č Popović and Milan S Dimitrijević

This volume of the *Journal of Physics: Conference Series* contains the Invited lectures, Topical invited lectures and Progress reports presented at the 24th Summer School and International Symposium on the Physics of Ionized Gases – SPIG 2008. The conference was held in Novi Sad, Serbia, 25–29 August 2008.

Throughout the history of scientific discovery, one can see repeatedly how fundamental sciences have solved basic questions and opened new frontiers. In the field of physics, there are many key discoveries, resulting in their useful applications for the benefit of the mankind. It is very important to have meetings to discuss actual problems in particular fields of physics. This Conference provided a forum for 160 active researchers from 25 countries to discuss current advances in the physics of ionized gases and related fields.

The Conference has a long tradition. Let us remember that the first SPIG was organized in 1968. The decay of former Yugoslavia in 1991, caused a disturbance in SPIG meetings, but fortunately, in 1993, SPIG meetings were successfully revitalized. During recent years we have met successively in Belgrade, Kotor, Zlatibor, Soko Banja, Tara, Kopaonik and finally this time in Novi Sad.

The structure of the papers in this Proceedings is as follows: Atomic Collision Processes, Particle and Laser Beam Interactions with Solids, Low Temperature Plasmas and General Plasmas. We hope that this Proceedings will be an important source of information, first of all to students, and also to plasma physics scientists.

First of all, we would like to thank to the invited speakers for participating at the SPIG 2008 and for their efforts writing contributions for this Proceedings. We also express our gratitude to the members of the Scientific and Organizing committees for their efforts in organizing the Conference. Especially we would like to thank the Ministry of Science and Technological Development of the Republic of Serbia for financial support. Also, this Conference was a conference sponsored by the European Physical Society (EPS). And finally we are grateful to all participants for useful contributions and useful discussions.

Gordana Malović, Luka Č Popović and Milan S Dimitrijević

Peer review statement

2008 *J. Phys.: Conf. Ser.* **133** doi: 10.1088/1742-6596/133/1/011002 Help

All papers published in this volume of *Journal of Physics: Conference Series* have been peer reviewed through processes administered by the proceedings Editors. Reviews were conducted by expert referees to the professional and scientific standards expected of a proceedings journal published by IOP Publishing.

Low energy lepton scattering: recent results for electron and positron interactions

Stephen J Buckman[1], Todd Maddern[2], Jessica Francis-Staite[2], Leigh Hargreaves[2], Michael J Brunger[2], Gustavo Garcia[3], Julian C. Lower[1], Subhendu Mondal[1], James P Sullivan[1], Adric Jones[1], Peter Caradonna[1], Daniel Slaughter[1], Casten Mackochekanwa[1], Robert P McEachran[1]

[1]ARC Centre for Antimatter-Matter Studies, Research School of Physical Sciences and Engineering, College of Science, Australian National University, Canberra, ACT, Australia

[2]ARC Centre for Antimatter-Matter Studies SOCPES, Flinders University, GPO Box 2100, Adelaide, SA, Australia

[3]Instituto de Física Fundamental, CSIC, Serrano 113-bis, 28006, Madrid, Spain

Stephen.buckman@anu.edu.au

Abstract. This report will review several new experimental developments for low energy (1-50 eV) scattering of both electrons and positrons from atoms and molecules. Experiments include an apparatus for studies of scattering from unstable molecular radicals, a position- and time-sensitive time-of-flight spectrometer for studies of near threshold electronic excitation of atoms and molecules and a trap-based positron scattering experiment for measurements of low energy positron scattering from rare gas atoms. Where possible, comparisons are also made with other experimental determinations and with contemporary scattering theory.

1. Introduction

Electron and positron interactions with atoms, molecules and materials can provide important insights into reaction dynamics, atomic and molecular structure and a wealth of both realized and potential applications in technology, biomedical science and the environment. At low energies, electron interactions are characterized by, amongst other things, strong scattering resonances involving the temporary trapping of the projectile in the field of the atom or molecule. These resonances often lead to profound effects on the scattering cross sections and, in some cases, can enhance scattering rates for vibrational excitation, for example, by orders of magnitude. For molecular systems they also lead to a very important process known as dissociative attachment whereby the negative ion complex decays via dissociation, resulting in many cases in the production of reactive neutral and negative ion species. This process is known to be important in areas as diverse as radiation damage of DNA and its constituent molecules [1] and the production of reactive etch products in fluorocarbon feedstock gases used in plasma processing [2].

Measurements of positron interactions, on the other hand, are not so extensive, or advanced, as those for their matter counterpart. Positron scattering experiments have until recently lacked the specificity and accuracy of electron scattering measurements, mainly as a consequence of low beam intensities and/or relatively poor energy resolution. A significant advance in recent years has been the advent of trap-based positron beams which, when combined with new techniques developed for charged particle scattering in large magnetic fields, have enabled many new measurements. For example, the improved energy resolution and sensitivity offered by these techniques have enabled the first absolute scattering measurements for excitation processes in atoms and molecules, including the direct measurement of positronium formation (see [3] for a recent review of positron interactions). These techniques promise much new fundamental information on positron interactions with atoms and molecules, including measurements on biologically important molecules which may have relevance for a deeper understanding of the interactions that underpin diagnostic techniques such as Positron Emission Tomography (PET).

For both electron and positron scattering, another important rationale for the measurements lies in the comparison with state of the art quantum scattering theory. In order for theory to develop, particularly in the case of complex molecular targets, it is imperative that accurate 'benchmark' experimental measurements are available for comparison.

In the following sections we discuss several new experimental initiatives in electron and positron physics. First we describe a new apparatus and initial results for electron interactions with molecular radical species. We then describe a technique which, in principle, can provide benchmark measurements for near-threshold electron impact electronic excitation of atoms and molecules. Finally we consider the development of a new positron trap and scattering experiment and results that have been obtained with it for positron scattering from the rare gases.

2. Scattering of electrons by molecular radicals

There is an increasing realization that molecular radicals play an important role in a range of electron-induced processes, from radiation damage in tissue, to plasma processing and deposition technologies. In the latter case, fluorocarbon feedstock gases are commonly used as the etching gas but in the discharge environment, electron-induced dissociation readily results in the production of CF_x radicals which in turn undergo a range of collision processes with other discharge constituents, including the electrons. In the present study we have used a pyrolytic source in combination with a pulsed supersonic nozzle to create a beam of CF_2 molecular radicals, from which we have studied elastic electron scattering. Some of these results have been reported previously [4].

The apparatus used to perform the CF_2 cross section measurements has been described in a recent publication [5] and we will not repeat that detail here. Briefly, the experiment is a crossed-beam electron scattering experiment comprising four differentially pumped chambers. The first chamber contains the solenoid valve, the pyrolytic nozzle, and the skimmer, and is pumped by a 10 in. diffusion pump. The second chamber, separated from the first by the skimmer, houses 9 fixed-angle electron detectors, spanning an angular range of 20–135 deg., and is pumped by a 156 l/s turbomolecular pump. The third chamber contains the electron monochromator, and is differentially pumped by a 50 l/s turbomolecular pump. The fourth chamber, separated from the second by a 30-mm aperture, houses a time-of-flight mass spectrometer (TOFMS) and is evacuated by a 6 inch diffusion pump.

Following pyrolysis, the heated products in the beam undergo cooling through supersonic expansion. A skimmer, placed downstream from the pyrolytic nozzle, serves to collimate the beam for passage through the scattering chamber. The electron monochromator produces a monoenergetic electron beam which is scattered from the molecular beam, and an array of 9 fixed-angle detectors is used to determine the scattering intensities for those angles. The composition of the beam is monitored by the TOFMS with 118-nm radiation used to ionize species through single-photon ionization. The temperature of the pyrolysis nozzle is measured with an optical pyrometer. Differential cross section (DCS) data were placed on an absolute scale by using a modified form of the skimmed supersonic relative density method (SSRDM) [6]. A schematic of the experimental layout is shown in Figure 1.

Examples of the first such results for low energy electron scattering from CF_2 are given in Figure 2. Here we compare the present absolute differential cross sections with several theoretical predictions at incident electron energies of 30 and 40 eV. The level of agreement is remarkably good, given the difficulty of both the experimental measurements and the theoretical calculations.

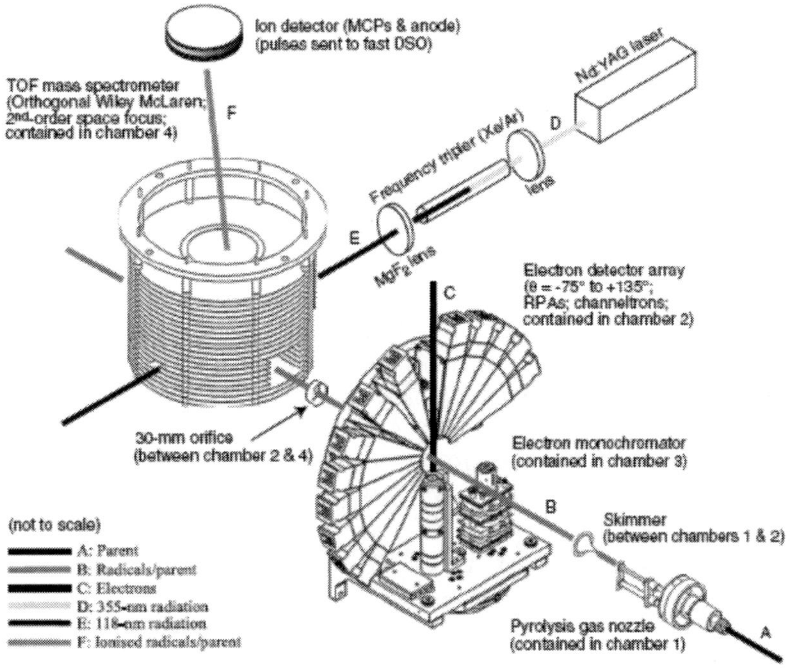

Figure 1: Schematic layout of the molecular radicals scattering experiment.

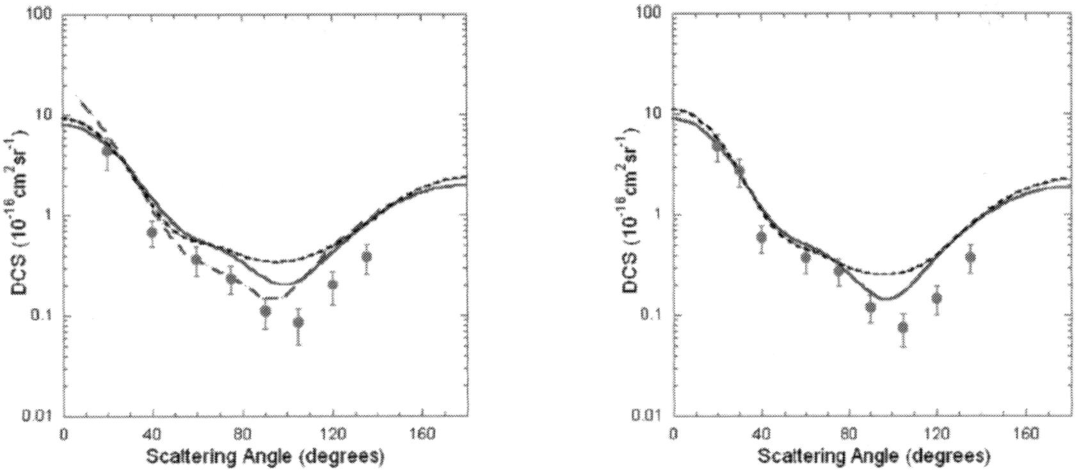

Figure 2: Absolute differential cross sections for elastic scattering from the CF_2 radical at energies of 30 eV (left) and 40 eV (right). The solid line is a Schwinger variational calculation at the static exchange level, the short dashed line the same calculation with polarization included [4] and the long dashed line the ISM-DWA calculation of [7].

3. Near-threshold electronic excitation of atoms and molecules

Electronic excitation can lead to the production of photons, metastable species or dissociative products, including free radicals, and as such it is an important process which is widely exploited in a range of technological devices. At near-threshold energies the cross sections for these processes are often enhanced by the presence of resonances. Somewhat paradoxically perhaps, the accuracy of much of the experimental data available in the literature for such processes is, at times, rather poor, if one takes the level of agreement between various experiments as a gauge. In large part this is due to experimental and/or analysis difficulties.

This energy region within a few electron volts of threshold represents a challenge experimentally as measurements typically require high-resolution electron monochromators and analysers and some means for establishing the absolute magnitude of the scattering cross sections. The most commonly used approach has been to measure the flux of inelastically scattered electrons, for a particular scattering channel and scattering angle, relative to that for the elastic scattering intensity at the same incident energy and scattering angle. If the transmission of the energy analysing device is known, the inelastic scattering cross section can be readily determined from the ratio of scattered electron intensities, providing the elastic cross section is well known. The main experimental issue that has to be resolved is usually the determination of the transmission of the energy analysing device (e.g., a hemispherical analyser) as a function of the scattered electron energy. In the present work we have employed a time-of-flight technique to ensure uniform transmission of the scattered electrons.

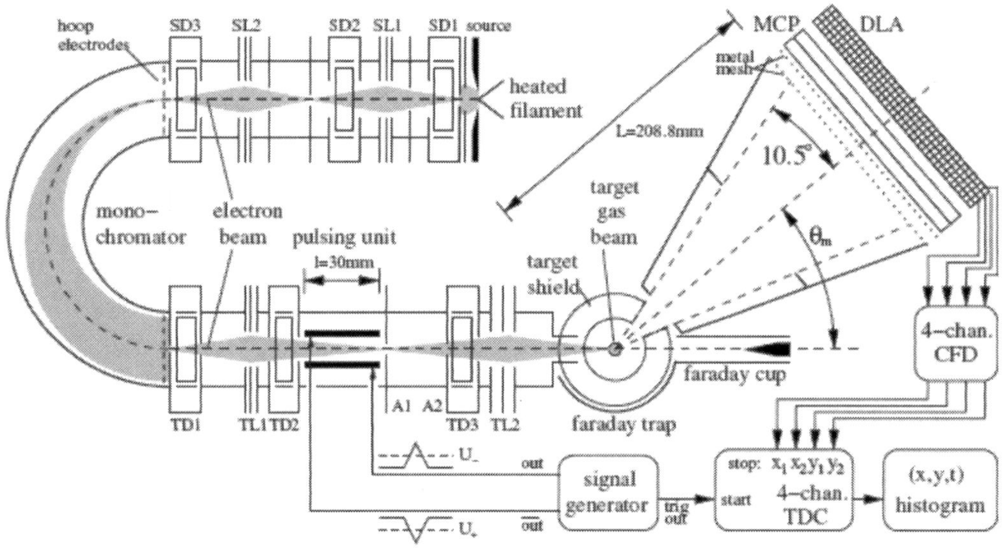

Figure 3: Schematic diagram of ToF apparatus.

The experimental arrangement consists of a crossed electron-molecular beam configuration, with the incident electron beam produced from an electrostatic monochromator. The target beam is formed by effusive flow of helium or N_2 through a single capillary tube. The energy of the incident beam is readily tuneable and calibrated, in the present case, by measuring well-known resonance features in metastable atom/molecule excitation functions. These measurements also indicate that the energy resolution of the incident beam is typically around 100 meV, with a CW beam current of several nA. We use a time-of-flight (ToF) energy analyser as the dispersive element for the scattered electron analysis. Scattered electrons are allowed to drift for 20 cm in a field free environment before being accelerated and striking a large area position sensitive detector comprising an 80 mm diameter channel plate and delay-line anode (Röentdek DLD-80). Scattered electrons can be simultaneously detected over a range of scattering angles of about 22° and, by measuring their time-of-flight, their energy can also be determined. The mean scattering angle is varied by changing the position of the electron monochromator, which is mounted on a rotating turntable. As the experiment involves the measurement of electron flight times, a pulsed beam of several ns duration (FWHM) is formed by sweeping the beam, in a controlled fashion, across a small defining aperture. Space does not permit a full description of the apparatus here but further details can be found elsewhere [8,9]. A schematic layout of the experiment is given in Figure 3.

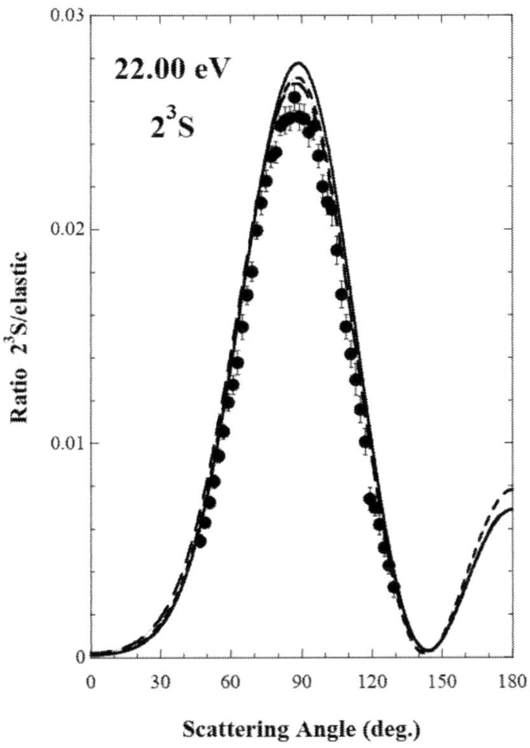

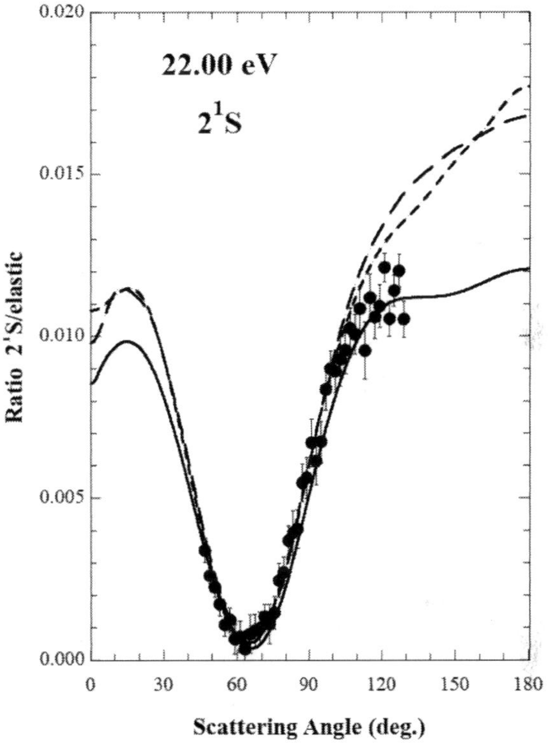

Figure 4: Near threshold excitation of Helium n=2 levels by 22 eV electrons. The solid line is the convergent close coupling calculation [13], the short dashed line the RMPS calculation [14], and the long dashed line the B-spline R-Matrix theory [15]

We have measured differential scattering cross sections for near-threshold energies for several excited states in He and N_2, using this new apparatus. Examples of these measurements are provided in Figures 4-6. The measurements for helium provide a good test of the bona fides of the apparatus and technique as helium is one system where contemporary scattering theory has achieved an extremely high level of accuracy for a range of elastic, inelastic and ionizing events following electron impact. This is demonstrated in Figure 4 where for the first two n=2 states of helium we see excellent agreement between the present measurements for the ratio of the elastic to inelastic scattering probabilities, and the results from three theoretical calculations.

In the case of N_2 (Figures 5 and 6) there are serious discrepancies in the literature for the values of the near-threshold cross sections for a number of low-lying excited levels and for incident electron energies within about 10 eV of threshold (see [10] for the most recent discussion of these issues). Our preliminary results indicate some differences in the experimental cross sections with those of [10] at 10 eV incident energy, but significantly better agreement is found at 15 eV. For example, at 10 eV the total inelastic scattering intensity measured in the present experiment (Figure 5) is somewhat higher than that of [10], although many of the individual excited state cross sections which are unfolded from this data show good agreement (see for example Figure 6, left panel). At 15 eV the agreement in both the total inelastic scattering intensity, and the individual excited state components of this, is considerably better. In general the present results at 15 eV are in better agreement with the recent experiment of [10] than with older determinations of the individual excited state cross sections [eg. 11,12].

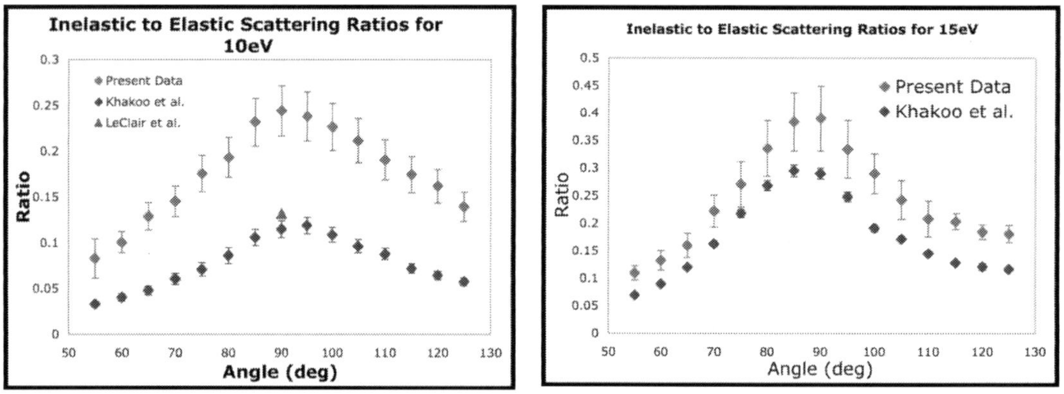

Figure 5: Ratio of total inelastic to elastic scattering intensity for N_2 at incident energies of 10 and 15 eV. The present data are compared with the measurements of [10].

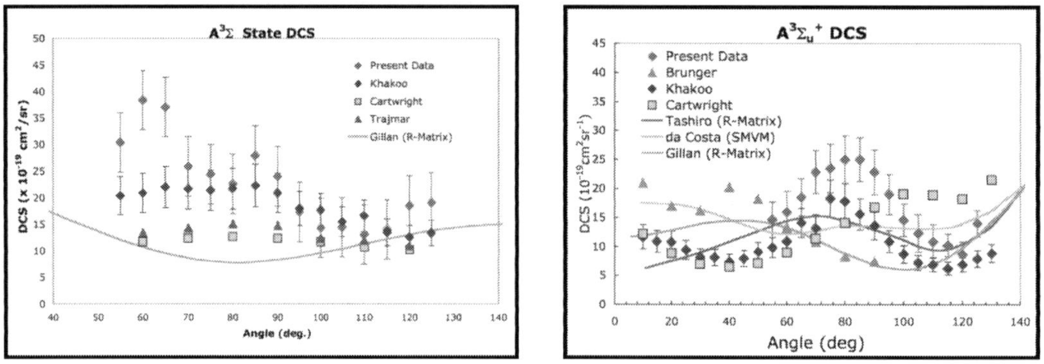

Figure 6: Differential cross sections for the excitation of the $A^3\Sigma_u^+$ state of N_2 at incident energies of 10 and 15 eV.

4. Low energy positron scattering

The advent of high-resolution, trapped-based beams of positrons [16] has revealed many new opportunities for fundamental studies of positron interactions with atoms and molecules. Now it is possible to produce relatively intense, pulsed, variable energy beams of positrons with energy widths considerably less than 100 meV. Such positron sources have enabled energy resolved measurements of elastic and inelastic scattering cross sections, including processes such as vibrational and electronic excitation [17].

We have developed a new, trap-based beam of positrons based on the approach pioneered by Surko and collaborators at the University of California, San Diego. This beam uses positrons from a ~25 mCi ^{22}Na source, which are magnetically guided, trapped and collisionally cooled in a Surko trap and then released in a pulsed beam with an energy width of ~ 60 meV and with a controlled, but readily varied, energy. The pulsed beam is incident on a gas cell containing the target of interest and the energy of the positrons after the scattering cell is analysed with a simple, but highly effective, retarding potential analyzer, before they are detected by a channel plate detector. This technique of scattering in a high magnetic field has been described in considerable detail elsewhere [17], and we do not repeat that detail here. It offers some significant advantages in sensitivity over conventional techniques and enables direct measurements of important quantities such as total, total elastic, total inelastic and total positronium formation cross sections.

An example of preliminary data from this apparatus is shown in Figure 7, where we illustrate measurements of the total positron scattering cross section for neon and compare it with one previous experimental determination and a recent theoretical calculation. The calculation [18] uses the polarized orbital technique. The agreement in the region of the Ramsauer-Townsend minimum is very good between all three cross sections, while at higher energies, discrepancies of the order of 20% exist between all three, up to the positronium formation threshold at 14.76 eV, where the cross section shows a marked increase due to the opening of this important inelastic channel.

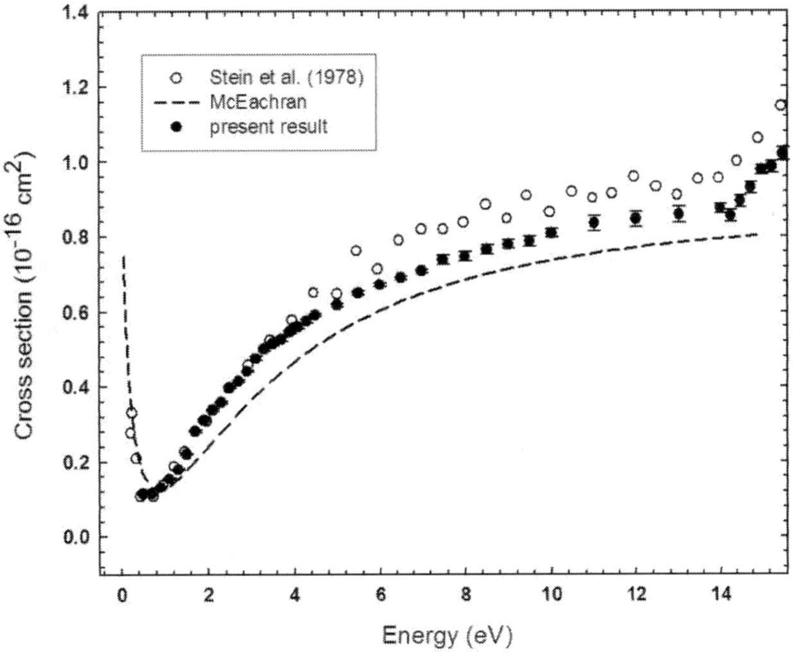

Figure 7: Total scattering cross section for positrons interacting with neon atoms.

5. Conclusions and future directions

This review of recent cross section measurements for low energy lepton scattering has attempted to highlight a number of significant and recent experimental advances in the field. These have led to new, exciting results that have the potential to establish as series of 'benchmarks' for electron collisions with targets which could not be studied previously, such as molecular radicals, and processes for which there exist substantial discrepancies in the literature, such as near-threshold electronic excitation. In addition, new techniques developed for low energy positron scattering are providing similar opportunities to benchmark these collision processes. Future work in these areas will involve an extension of studies to other molecular radicals, potentially CF_3 and CH_2, the extension of the ToF technique to more complex polyatomic molecules such as H_2O and CO_2, and the application of the new positron facilities to a broad range of collision measurements on atoms and molecules.

Acknowledgements
It is a pleasure to acknowledge the support of the Australian Government through the Australian Research Council's Centre of Excellence program and the Department of Innovation, Industry, Science and Research's International Science Linkages Program.

References

[1] Boudaiffa B, Cloutier P, Hunting D, Huels M A and Sanche L 2000 Science **287** 1658
[2] National Research Council I 1996 *Database Needs for Modelling and Simulation of Plasma Processing* (New York: National Academy Press)
[3] Surko C M, Gribakin G F and Buckman S J 2005 J. Phys. B: At. Mol. Opt. Phys. **38** R57
[4] Maddern T M, Hargreaves L R, Francis-Staite J R, Brunger M J, Buckman S J, Winstead C and McKoy V 2008 Phys. Rev. Lett. **100** 063202
[5] Maddern T M, Hargreaves L R, Bolarizadeh M, Brunger M J and Buckman S J 2008 Meas. Sci. Technol. **19** 085801
[6] Hargreaves L R, Francis-Staite J R, Maddern T M, Brunger M J and Buckman S J 2007 Meas. Sci. Technol. **18** 2783
[7] Lee M-T, Machado L E, Brescansin L M, Castro E A and de Souza G L 2006 Phys Rev. A**74** 052716
[8] Lange M, Matsumoto J, Lower J, Buckman S, Zatsarinny O, Bartschat K, Bray I and Fursa D 2006 J. Phys. B: At. Mol. Opt. Phys. **39** 4179
[9] Lange M, Matsumoto J, Setiawan A, Panajotovic R, Harrison J, Lower J C, Newman D S, Mondal S and Buckman S J 2008 Rev. Sci. Instrum. **79** 043105
[10] Khakoo M A, Johnson P V, Ozkay I, Yan P, Trajmar S and Kanik I 2005 Phys Rev. A**71** 062703
[11] Cartwright D C, Chutjian A, Trajmar S, and Williams W 1977 Phys. Rev. A **16**, 1013
[12] Brunger M J and Teubner P J O 1990 Phys. Rev A**41** 1413
[13] Fursa D and Bray I 1995 Phys. Rev A**52** 1279 and see reference [8]
[14] Bartschat K, Hudson E T, Scott M P, Burke P G and Burke M V 1996 Phys. Rev A**54** R998 and see reference [8]
[15] Zatsarinny O and Bartschat K 2004 J. Phys. B: At. Mol. Opt. Phys. **37** 2173 and see reference [8]
[16] Gilbert S J, Kurz C, Greaves R G and Surko C M 1997 Appl. Phys. Lett. **70** 1944
[17] Sullivan J P, Gilbert S J, Marler J P, Greaves R G, Buckman S J and Surko C M 2002 Phys. Rev. A**66** 042708
[18] McEachran R P and Stauffer A D 2008 Private Communication

Energy deposition model based on electron scattering cross section data from water molecules

A.Muñoz[1], J. C. Oller[1], F. Blanco[2], J. D. Gorfinkiel[3], P. Limão-Vieira[4], A. Maira-Vidal[5], M. J. G. Borge[5], O. Tengblad[5], C. Huerga[6], M. Téllez[6] and G. García[7]

[1]Centro de Investigaciones Energéticas, Medioambientales y Tecnológicas (CIEMAT), Avenida Complutense 22, 28040 Madrid, Spain.
[2]Departamento de Física Atómica, Molecular y Nuclear, Universidad Complutense de Madrid, Avenida Complutense s.n., 28040 Madrid, Spain
[3]Department of Physiscs and Astronomy, The Open University, Walton Hall ,Milton Keynes MK7 6AA, UK
[4]Departamento de Física, Universidade Nova de Lisboa, 2829-516 Caparica, Portugal
[5] Instituto de Estructura de la Materia, Consejo Superior de Investigaciones Científicas (CSIC), Serrano 113-bis, 28006 Madrid, Spain
[6]Hospital Universitario La Paz, paseo de la Castellana 261, 28046 Madrid, Spain
[7] Instituto de Matemáticas y Física Fundamental, Consejo Superior de Investigaciones Científicas (CSIC), Serrano 113-bis, 28006 Madrid, Spain
E-mail g.garcia@imaff.cfmac.csic.es

Abstract. A complete set of electrons scattering cross sections by water molecules over a broad energy range, from the meV to the MeV ranges, is presented in this study. These data have been obtained by combining experiments and calculations and cover most relevant processes, both elastic and inelastic, which can take place in the considered energy range. A new Monte Carlo simulation programme has been developed using as input parameter these cross sectional data as well as experimental energy loss spectra. The simulation procedure has been applied to obtain electron tracks and energy deposition plots in water when irradiated by a Ru-106 plaque as those used for brachyteraphy of ocular tumours. Finally, the low energy electron tracks provided by the present model have been compared with those obtained with other codes available in the literature

1. Introduction

Important medical applications of radiation for therapy, diagnostics or patient protection require energy deposition models with increasing resolution. When molecular details are important, as it is the case of DNA damage studies, spatial resolution should be within the order of magnitude of nanometre. For this level of description atomic and molecular properties of the target need to be taken into account. High energy radiation produces abundant secondary electrons which are the main responsible of the energy transfer map and the radiation damage. Pioneering studies of H. Huels, L. Sanche and co-workers [1-3] show that even sub-ionising electrons could produce damage, in terms of strand breaks and molecular dissociations, more efficiently than traditional ways based on ionisation of the medium. Extensive experimental [4-16] and theoretical [17-21] studies have been carried out in recent years to corroborate and quantify these early evidences. However, most medical procedures are based on radiation dose. This means that the magnitude of reference is the energy absorbed per mass unit and therefore only a macroscopic description of the procedure can be achieved. Medical protocols, or codes of practice [22], recommend using ionization chambers calibrated in terms of absorbed dose to water to characterise the radiation beams for radiotherapy. Using these devices for dose assignment implies that the equivalence charge-energy is a well known constant, which does not apply for low energy, and obviously ignore any effect due to subionising radiation. In order to increase the accuracy of these devices by improving their spatial resolution, in the last few years a great effort has been paid to develop small ionisation based detectors as proportional counters. Dose measurement procedures based on reduced area detectors have been developed in the field of microdosimetry [23]. These techniques increased notably the definition of energy transfer maps but its magnitude of reference is still the absorbed dose, no molecular effects are generally included in such models. For this reason, using energy deposition simulation methods as a complement of dose measurements is strongly recommended for radiation planning in electron, photon and ion beam radiotherapy [24-25]. Numerous Monte Carlo simulation codes have been developed for this purpose, i.e. PARTRAC [26-27], MCNPX [28], PENELOPE [29-31], GEANT-4 [32-33], etc... Most of these codes are sophisticated programmes managing radiation interaction data and empirical laws applied to representative radiation particles (electron, photons, ions) interacting with a great variety of materials. However, input parameters and mathematical relations used by these codes are generally valid for high energy particles, principally for energies above 100 eV. Most of them are based on the first Born approximation which, as we showed in previous studies [34-37], overestimates the interaction probabilities by about 30% at 10 keV even in favourable cases as electron scattered by light molecules (N_2, O_2, CH_4, CO_2).

These considerations motivated the present study. We are presenting here a new Monte Carlo simulation programme for electron interactions with molecular media from high incident energies, in the MeV range, slowing down to a few eV by successive collisions with the atoms and molecules constituting the medium. The input parameters for this programme are the electron scattering cross sections, both elastic and inelastic, which has been previously measured or calculated depending on the considered process and energy range. To illustrate this report, we have chosen a medium formed by water vapour. Water is one of the main constituent of living tissue so it is strongly relevant for medical applications. As biomolecular systems are normally surrounded by liquid water, present results will be generalised to liquid conditions by changing the target density. At this stage, no physical properties of liquids, as collective effects, are included in the present model and they will be the subject of further studies.

2. Cross sectional data

In order to define the trajectories of each particle related to the radiation beam, interaction probabilities, or cross sections, for all the accessible interaction processes are in principle needed. When a collision event takes place an amount of the incident energy can be transferred to the target as internal energy (inelastic scattering) or not, simply transferring linear momentum (elastic scattering). In both cases, the energy and direction of emerging particles should be known. In other words, we need to know both differential and integral cross section for all the possible collisions, elastic and inelastic, that can take place in the considered energy range. In order to organise data requirements, the energy range considered in this study has been divided into three regions: a high energy region from 10 keV up to several MeV, an intermediate-high energy region from 10 eV to 10 keV and finally a low energy region for energies below 10 eV. As far as the electron molecule interactions are concerned, this study has been mainly focused on obtaining data at intermediate and high energies. So we are firstly describing the experiments and calculations which have been carried out in this energy range and then going to the high and low energy domains, respectively.

2.1 Intermediate-high energies (10-10000 eV):

2.1.1 *Experiment:* Measurements have been devoted to obtain total electron scattering cross sections (TCS), ionisation cross sections (ICS) and electron energy loss (EEL) spectra for incident energies between 50 and 5000 eV. Details on the experimental techniques as well as a complete discussion of the results can be found in Ref. [38], thus only a brief summary is given here: TCS have been measured with a transmission beam technique [39] which also allows measurements of energy loss spectra for small scattering angles from 0 to 10 deg. In order to record EEL spectra for higher scattering angles, the scattering chamber was replaced by an effusive molecular beam which allowed scattering angle selection by rotating the electron gun with respect to the energy spectrometer. Ionisation cross sections in this range were derived from the simultaneous measurement of electron and ion currents by applying a synchronised pulsing system to extract the ions from the collision chamber. Present experimental results for TCS and ICS, with estimated errors of 5% and 7% respectively, are shown in Figure 1. Following the discussion of Ref. [38], some previous experimental results [40-42] are also shown in this figure for comparison. A complementary discussion of data on electron scattering from water molecules can be found in Refs. [43,44]. Present ionisation cross section results are also shown in Figure 1 together with previous measurements of Ref. [45]. Finally, by averaging energy loss spectra for these energies as a function of the scattering angle, a unique energy loss distribution function with statistical uncertainties within 15% has been derived (see Figure 2).

2.1.2 *Theory:* Calculations were carried out with the optical potential procedure described in previous papers [46-48]. Basically, it is an independent atom representation which includes screening corrections [49] to derive molecular differential and integral elastic as well as

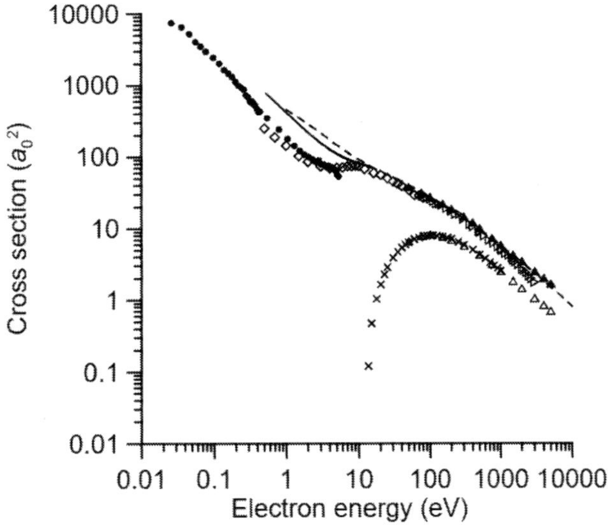

Figure 1. Total electron scattering cross sections: ▲, present measurements;◇experimental data of Ref. [40];▷, experimental data from Ref. [41]; ●, measurements of Ref. [42]; ----, present calculation; —, theoretical data from Ref.[51]. Ionisation cross sections: △, present measurements; ×, experimental results from Ref. [45].

integral inelastic cross sections. As an example, total cross sections calculated with this procedure are shown in Figure 1. Following the suggestion of Ref. [50] these calculations include an approximate additional term to take into account rotational excitations. As shown in this figure the agreement between the present TCS measurements and calculation is perfect between 10 and 5000 eV.

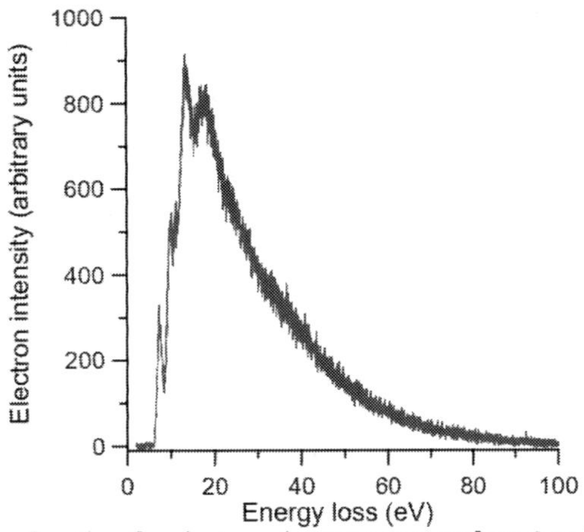

Figure 2. Energy loss distribution function for electrons in water vapour from 0 to 100 eV

2.2 Low energy region (<10 eV)

Below 10 eV the situation is not so ideal. Our calculations, including the rotational excitation term agree reasonably with low energy R-matrix calculations from Ref. [51]. However, both differ from representative experimental data by a factor of 2 below 5 eV. The origin of this discrepancy has been discussed previously [38] and it seems to be connected to the target rotational distribution and the acceptance angle of detectors used in the experiments. Since this point has not been completely clarify at the moment, we will take the experimental results of [40] and [42] as recommendable low energy data for our model.

2.3 High energy region (>10000 eV)

For really high incident energy, molecules can be represented as a sum of atoms and both incident and scattered electron can be considered, in framework of the first Born approximation [52, 53], as plane waves. Under these assumptions integral elastic and inelastic cross section calculations can be represented by simple energy dependent formulae [54]. In order to check the validity of this approximation, Figure 3 is a Fano plot (TCS multiplied by energy versus the logarithm of energy) where the convergence of our measurements and calculations with those approximate values for energies above 10 eV is shown. Due to systematic errors discussed elsewhere [34, 35, 54], previous measurements [41] and calculations [55] diverge from this asymptotic behaviour

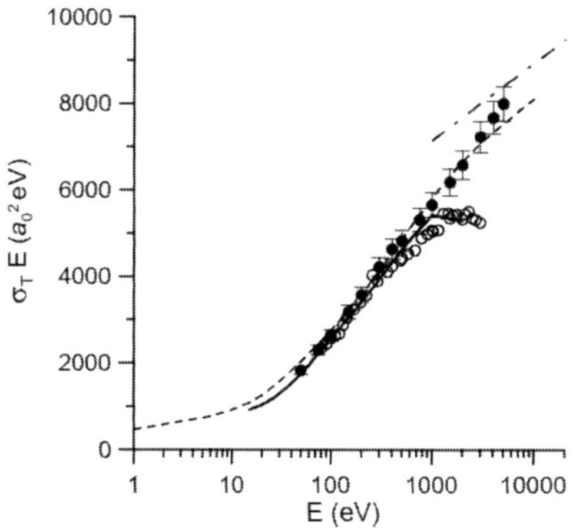

Figure 3. Fano plot (total cross section multiplied by electron energy as a function of energy in a logarithmic scale): ●, present experimental data; ○, experimental data from Ref. [41]; ---, present optical potential calculation; —, calculation of Ref. [55]. ·-· Born-Bethe calculation.

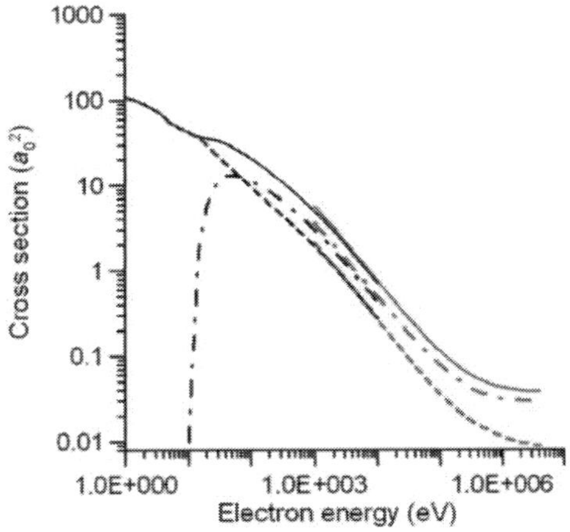

Figure 4. electron scattering integral cross sections from H_2O: —, Optical potential total cross sections; -.- , Optical potential integral inelastic cross sections;---, Optical potential integral elastic cross sections; —, Born-Bethe total cross sections; -.-, Born-Bethe integral inelastic cross sections; ---, Born-Bethe integral elastic cross sections.

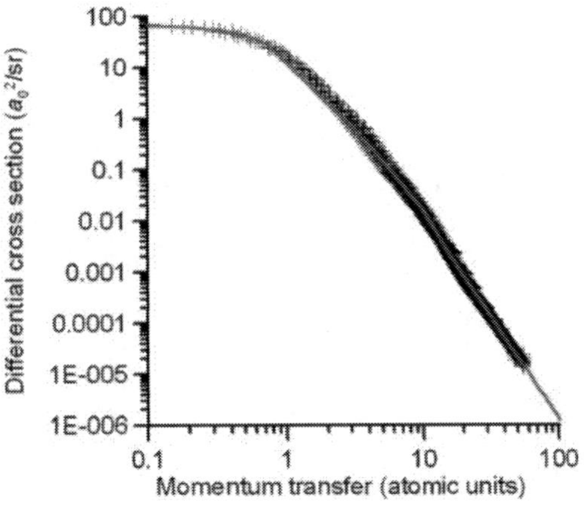

Figure 5. Differential cross section for electron scattering from water molecules for energies above 1 keV as a function of the momentum transfer: +, optical potential calculations; —, Analytical fit for high-energy extrapolation.

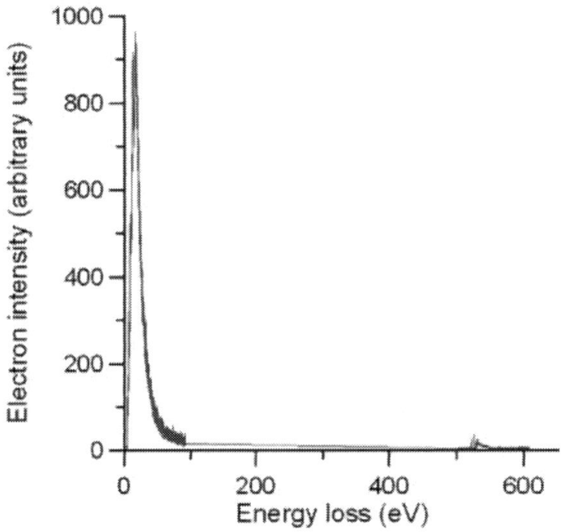

Figure 6. Energy loss distribution function from 0 to 600 eV.

predicted by the Born approximation. Integral cross sections calculated by this method from 1 keV to 3.5 MeV are plotted in Figure 4 together with the model potential calculations described in the previous section. As shown in this figure, at 10 keV there is an excellent agreement between both calculations. In order to obtain differential cross section for high energy electrons, Figure 5 shows a plot of all the differential cross section data we calculated with the above model potential method for energies between 1 and 10 keV as a function of the momentum transfer. As predicted by the Born approximation, these differential values follow a simple exponential dependence on the momentum transfer which allows easily the extrapolation of data toward higher energies. For energies above 500 eV the energy los distribution function of Figure 6 has been used. This function includes excitation and ionization of the atomic oxygen inner shell.

As a combination of data presented in the above three sub-sections, a complete set of electron scattering cross sections has been obtained to be used as input parameters of the simulation procedure described in section IV. Total electron scattering cross sections are experimental with total uncertainties of about 5%. Integral elastic cross sections are calculated with numerical uncertainties within 10%. Integral inelastic cross sections are the result of subtracting the elastic ones from the total scattering cross section with an estimated error of 12%. The ionisation cross sections used were experimental with uncertainties of about 7%. Subtracting the ionisation from the integral inelastic cross section, the remainder curve presents two maxima which can be attributed to electronic excitation and neutral dissociation processes. Identifying the excitation with the lowest energy maximum allows a deconvolution of both components. The estimated error for these reminder inelastic channels is about 25% .Further studies would be needed in order to reduce their assigned uncertainty.

3. Source emission spectra

For monoenergetic electron beams, as those produced by accelerators we would have all required data to feed our simulation procedure. However, interesting brachytherapy applications in ophthalmology are using beta emitters as radiation sources [56] since 1966. Properties of Ru-106 radionuclide for high-dose local irradiation were described by Freundlich [57]. Ocular melanoma is currently treated with Ru-106 plaque brachytherapy and extensive statistical studies has been carried out in order to correlate dosage

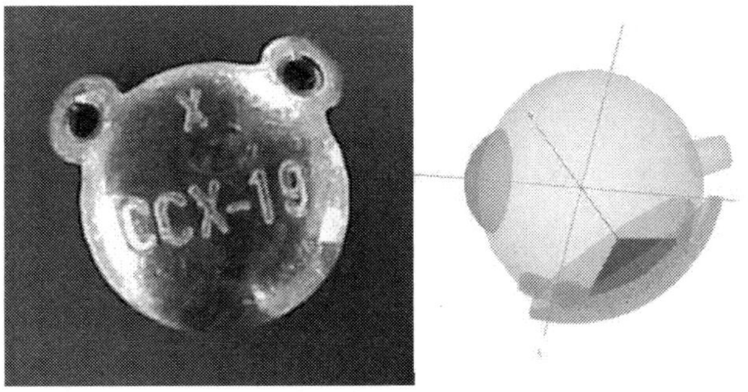

Figure 7. Geometrical aspects of Ru-106 plaque to be used for ocular tumor brachytherapy. Left, picture of the plaque; right, plaque placed on the ocular globe.

with patient evolution [58, 59] as well as comparisons with other radiotherapy techniques [60, 61]. As these studies show, critical out of target region and optic nerve irradiations can affect to visual acuity [62]. This is therefore a clear example in which high accuracy energy deposition models are required. As the ocular medium can easily be assimilated to water, it constitutes an excellent system to check the reliability of the present model. Geometrical aspects of some Ru-106 plaques used in brachytherapy are shown in Figure 7.

3.1 *Beta emission spectra:* Ru-106 decays to Rd-106, with 373.59 day half-life period, by beta emission of 39.4 keV maximum energy. This quickly decays (30 second half-life period) by 5 different beta emission ways to Pd-106, which is stable, with a maximum energy of 3.541 MeV. In these conditions, the Ru-106 decay series reaches secular equilibrium and its electron emission spectrum is a combination of continuum intensity distributions which extends from relatively low energies up to 3.541 MeV. To cover such as broad energy range, a telescopic mounting of three silicon (SiPAD) detectors, of 1500 micron thick and 50x50 mm^2 surface each, has been installed and calibrated with standard beta and alpha sources. By software analysis of coincidence spectra from each detector, the whole electron emission spectrum can be reconstructed (see Figure 8).

3.2 *Photon emission spectra:* As a consequence of the Ru106/Rh106 beta decay there is a subsequent gamma and X photon emission which requires to be considered by the model. Standard calibrated solid state spectrometers have been used to determine the energy and intensity of emitted photons. Representative spectra are shown in Figure 9 for (a) X ray emission and (b) gamma ray emission.

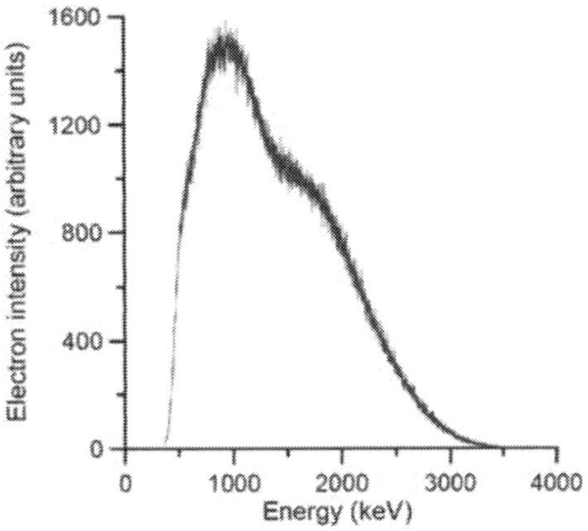

Figure 8. Reconstructed beta emission spectrum of Ru-106 plaque.

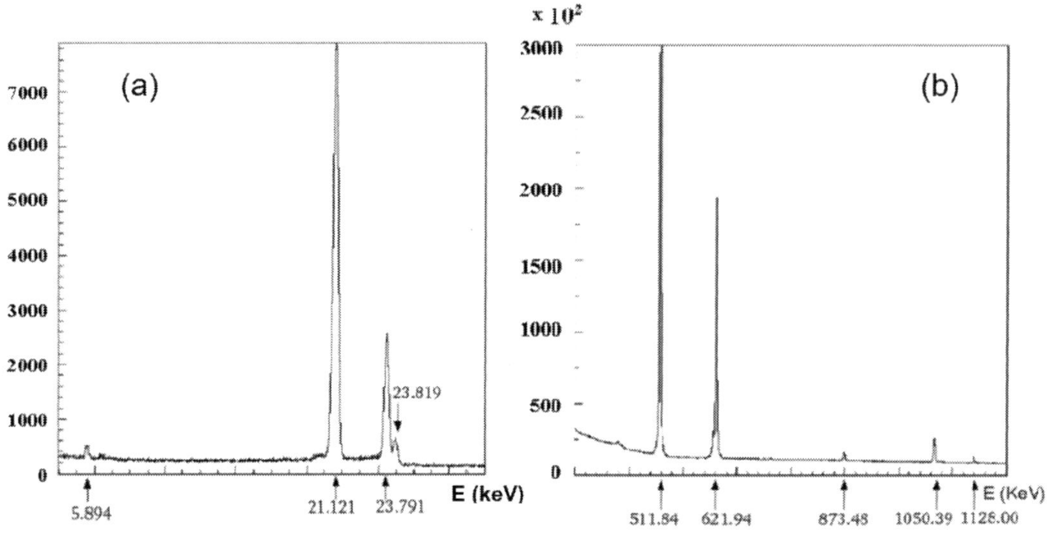

Figure 9. Photon emission spectra of Ru-106 plaque:(a), X-ray spectrum;(b), gamma ray spectrum.

4. Monte carlo simulation procedure

As we have already mentioned, the final goal of this study is to develop a model to simulate electron tracks in water to provide information about the energy deposition and the molecular interactions taking place during the energy degradation procedure. The simulation programme we developed for that was based on the GEANT-4 code [32]. We only used from GEANT-4 the general Monte Carlo routines available on its tool kit but all the physics related to electron molecule interactions constitute a new C++ programme, developed by us, which uses as input parameter the electron scattering cross sections we previously measured or calculated and the experimental electron energy loss spectra. Technical details on this programme can be found in previous publications [63, 64]. Photons are also included in the procedure but according to available photon interaction library [65], their main effect is to produce high energy secondary electrons which enter directly in the electron simulation process.

Track simulation of electrons in water emitted by a plane plaque of Ru-106, with the initial energy distribution given by the spectrum of Figure 8, is shown in Figure 10. We assumed the target density corresponding to liquid water but using the cross sectional data of isolated water molecules. Therefore, collective liquid properties are not taken into account in this model and will be the subject of further investigations. An energy deposition diagram as a function of the penetration of the electron beam in water is given in Figure 11.

In order to show details of the information provide by the model, Figures 12 and 13 represent single tracks for 2 keV electrons in water vapour (200 Torr pressure). Figure 12 shows the energy degradation procedure by successive collisions with target molecules. Different interactions taking place along the tracks are shown in Figure 13.

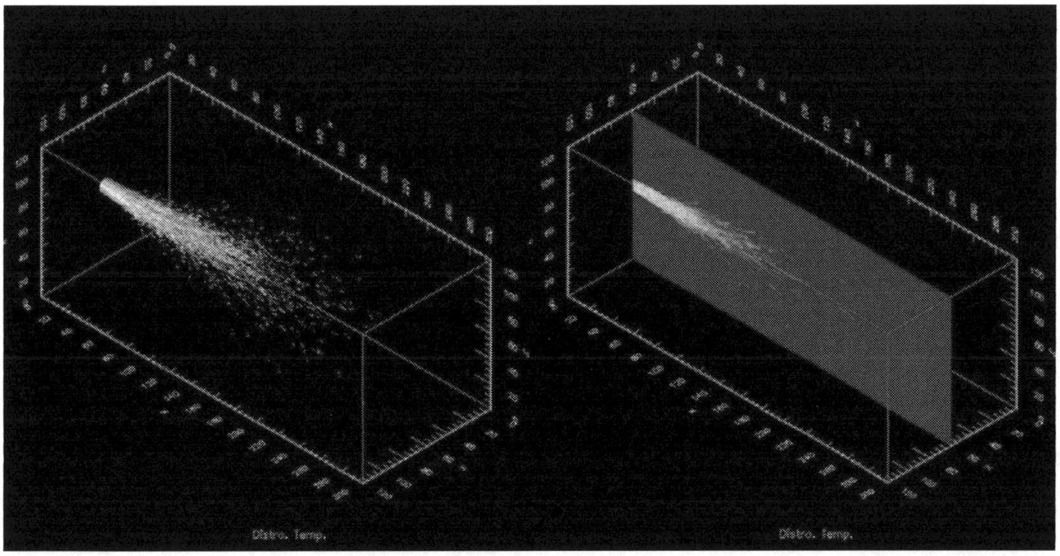

Figure 10. Track simulation of electrons emitted by Ru-106 plaque: Left, tridimensional plot; right, vertical section.

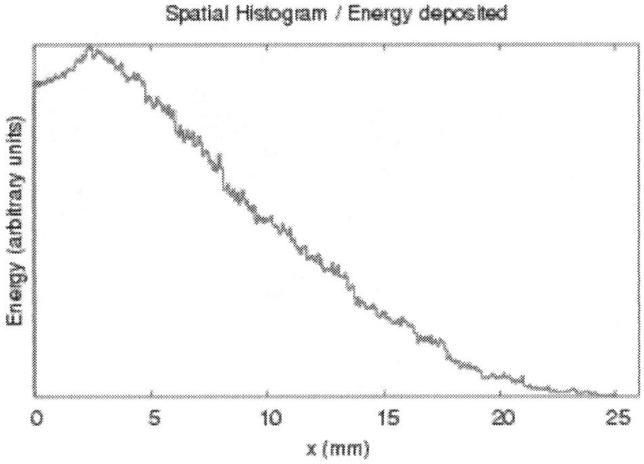

Figure 11. Energy deposition of Ru-106 plaque emission in liquid water as a function of penetration depth (X).

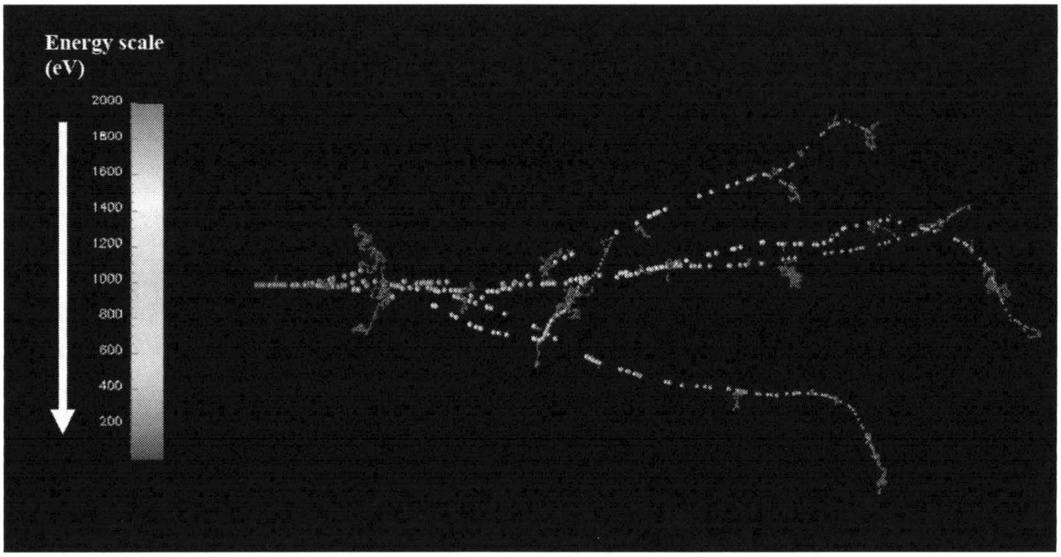

Figure 12. Modeling 2 keV electron tracks in 200 Torr of water vapour. Energy degradation procedure by successive collisions is represented by the colour code shown on the left.

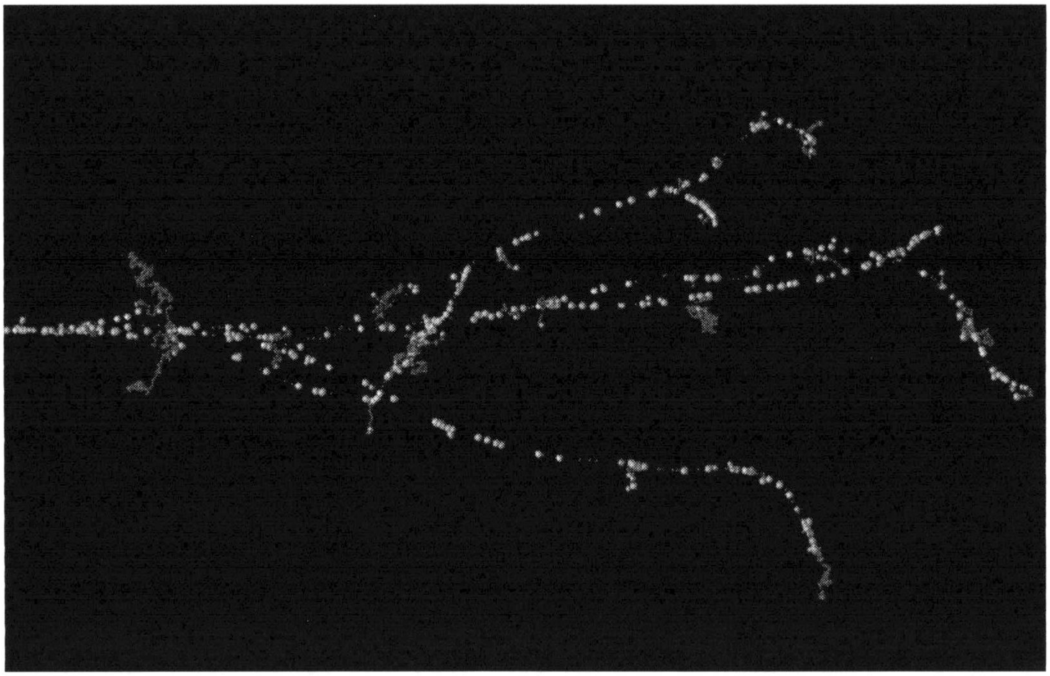

Figure 13. Modeling 2 keV electron tracks in 200 Torr of water vapour. Colliding processes are shown: •, Auger electron generation; •, ionisation; , electronic excitation, •,neutral dissociat

Finally, a comparison between the present simulation and that obtained with the PENELOPE code [29-31] is shown in Figure 14. Both simulations predict a maximum electron penetration in the medium of about 2 mm and therefore the absorbed dose in a 2x2x2 mm^3 volume would be the same. However, single electron tracks and energy deposition events given by each model are completely different. The number of interactions predicted by PENELOPE is less than the present one being higher the energy transferred in each individual collision. Also scattering angles given by PENELOPE simulation tend to be higher than ours. These differences can be explained looking at the energy loss distribution function (see Figure 15). Our energy loss distribution function is a continuous line with a threshold at 6.5 eV which follows the observed electron energy loss spectra in water while that of PENELOPE is an arbitrary delta function set placed on strategic energies to give the conventional value of the electron stopping power in water [66]. This could justify the macroscopic agreement between both models being so different their microscopic descriptions.

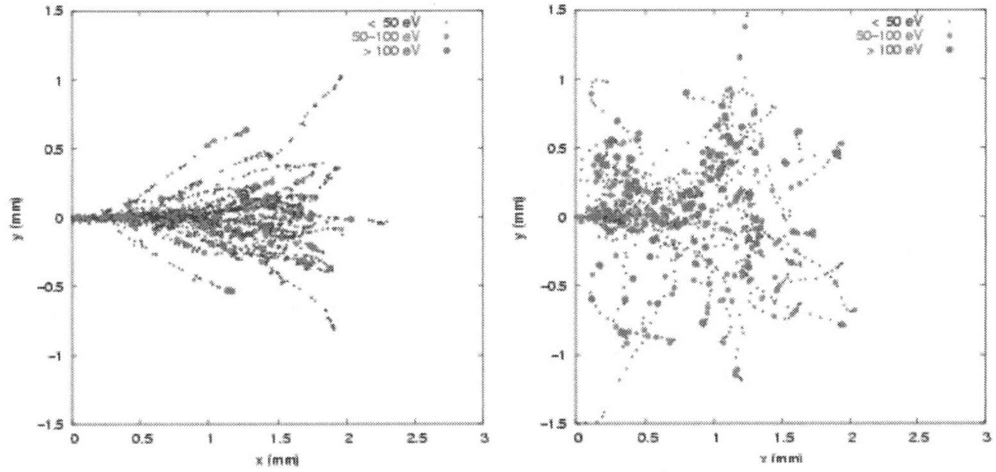

Figure 14. Simulation of 5 keV electron tracks in 1 atm of water vapor. Left, present Monte Carlo code; right, PENELOPE code.

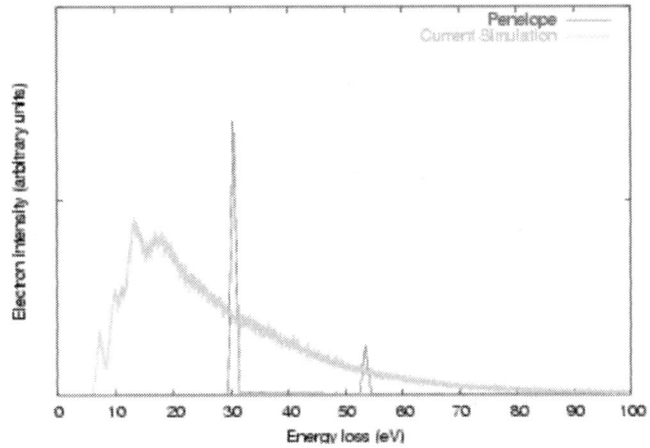

Figure 15. Comparison of (0-100 eV) energy loss distribution functions: ----, present simulation code; —, PENELOPE code.

5. Conclusions

A comprehensive set of electron interaction cross sections both elastic and inelastic with water molecules has been provided in this study by combining experimental results and theoretical calculations, covering a broad range from the MeV down to the meV domains. Consistency of data has been proved for energies above 10 eV. Below this value, important discrepancies appear between theory and experiment which will require further studies in order to be clarified. Using these cross sectional data and experimental energy loss spectra as input parameters, a simulation Monte Carlo program has been developed providing detail of electron tracks, energy deposition and interaction processes art the molecular levels. While macroscopic results of these simulations, in terms of absorbed dose, are in agreement with those given by available simulation programs they give a completely different microscopic description.

Finally the model has been applied to determine the energy deposition in water of Ru-106 beta emitter nuclides (Ru-106/Rd-106) which are commonly used in brachytherapy. The efficiency of these models in medical practice should be checked.

Acknowledgement

This study has been partially supported by the following research projects and institutions: Ministerio de Educación y Ciencia (Plan Nacional de Física, Project FIS2006-00702), Consejo de Seguridad Nuclear (CSN), European Science Foundation (COST Action P9 and EIPAM Project), Acciones Integradas Hispano-Portuguesas (Project HP2006-0042). Hospital Universitario La Paz (Comunidad de Madrid). We acknowledge N. C. Jones and D. Field for providing their unpublished cross section data.

References

[1] Boudaïffa B, Cloutier P, Hunting D, Huels M A and Sanche L 2000 *Science* **287** 1658.
[2] Boudaïffa B, Cloutier P, Hunting D, Huels M A and Sanche L 2002 *Radiat. Res.* **157**, 227.
[3] Huels M A, Boudaïffa B, P. Cloutier, D. Hunting and L. Sanche, *J. Am. Chem. Soc.* **125**, 4467 (2003).
[4] Denifl S, Ptasinska S, Cingel M, Matejcik S , Scheier P and Märk T D 2003 *Chem. Phys. Lett.* **377** 74.
[5] Gohlke S, Abdoul-Carime H, and Illenberger E 2003 *Chem. Phys. Lett.* **380** 595.
[6] Hanel G, Gstir B, Denifl S, Scheier P, Probst M, Farizon B, Farizon M, Illenberger E, and Märk T D 2003 *Phys. Rev. Lett.* **90** 188104.
[7] Caron L G and Sanche L 2003 *Phys. Rev. Lett.* ***91*** 113201.
[8] Martin F, Burrow P D, Cai Z, Cloutier P, Hunting D, and Sanche L 2004 *Phys. Rev. Lett.* **93** 068101.
[9] Scheer A M, Aflatooni K, Gallup G A and Burrow P D 2004 *Phys. Rev. Lett.* **92** 068102.
[10] Abdoul-Carime H, Gohlke S and Illenberger E 2004 *Phys. Rev. Lett.* **92** 168103.
[11] Denifl S, Ptasińska S, Probst M, Hrušak J, Scheier P and Märk T D 2004 *J. Phys. Chem. A* **108** 6562 .
[12] Denifl S, Ptasińska S, Hanel G, Gstir B, Probst M, Scheier P and Märk T D 2004 *J. Chem. Phys.* **120** 6557.
[13] Feil S, Gluch K, Matt-Leubner S, Scheier P, Limtrakul J, Probst M, Deutsch H, Becker K, Stamatovic A and Märk T D 2004 *J. Phys. B* **37** 3013.
[14] Abdoul-Carime H, Langer J, Huels M A and Illenberger E *Eur. Phys. J. D* **35**, 399 (2005).
[15] Ptasińska S, Denifl S, Mróz B, Probst M, Grill V, Illenberger E, Scheier P and Märk T D 2005 *J. Chem. Phys.* **123** 124302.
[16] Denifl S, Zappa F, Mähr I, Lecointre J, Probst M, Märk T D and Scheier P 2006 *Phys. Rev. Lett.* **97**, 043201.
[17] Pan X and Sanche 2005 L *Phys. Rev. Lett.* **94** 198104.
[18] Winstead C and McKoy V 2006 *J. Chem. Phys.* **125**, 074302.
[19] Winstead C and McKoy V 2006, *J. Chem. Phys.* **125** 174304.
[20] Winstead C and McKoy V 2006, *J. Chem. Phys.* **125**, 244302.
[21] Winstead C and McKoy V 20067 *J. Chem. Phys.* **127**, 085105.
[22] Andreo P, Burns D T, Hohlfeld K, Andreo P, Burns D T, Hohlfeld K, Huq M S, Kanai T, Laitano F, Smyth V and Vynckier S 2000 IAEA International Atomic Energy Agency, *Absorbed Dose Determination in External Beam Radiotherapy. An International Code of Practice for Dosimetry Based on Standards of Absorbed Dose to Water*, Technical Report Series no. 398, Vienna
[23] Waker A J 2006 *Radiat. Prot. Dosim.* **122** 369.
[24] Both J A and Pawlizki T 2004 *Int. J. Med. Sci.* **1**, 63.
[25] Chow J C L 2007 *Med. Phys.* **34**, 175.
[26] Friedland W, Jacob P, Paretzke H G and Stork T 1998 *Radiat. Res.* **150**, 170.

[27] Friedland W, Jacob P, Bernhardt P, Paretzke H G and Dingfeldr M 2003 *Radiat. Res.* **159**, 401.
[28] Hendrick J S, McKinney G W, Fensin M L, James M R, Johns R C, Durkee J W, Finch J P, Pelowitz D B, Waters L S and Gallmeier F X 2007 *MCNPX version 26D*, Los Alamos National Laboratory report LA-UR-07-4137.
[29] Baro J, Sempau J, Fernandez-Varea J M, and Salvat F 1995 *Nucl. Instrum. Meth. B* **100**, 31.
[30] Sempau J, Acosta E, Baro J, Fernandez-Varea J M and Salvat F 1997 *Nucl. Instrum. Meth. B* **132**, 377.
[31] Sempau J, Fernandez-Varea J M, Acosta E and Salvat F 2003 *Nucl. Instrum. Meth. B* **207**, 107.
[32] Agostinelli S et al. 2003 *Nucl. Instr. and Meth. A* **506**, 250.
[33] Ivanchenko V N and For Geant 4 Collaboration 2004 *Nucl. Instrum. Meth. A* **525**, 402.
[34] Garcia G and Manero F 1996 *Phys. Rev. A* **53**, 250.
[35] Garcia G and Manero F 1998 *Phys. Rev. A* **57**, 1069.
[36] Garcia G and Blanco F 2001 *Phys. Lett. A* **279**, 61.
[37] Garcia G, Blanco F and Williart A 2001 *Chem. Phys. Lett.* **335**, 277.
[38] Muñoz A , Oller J C, Blanco F, Gorfinkiel J D, Limão-Vieira P and García G 2007 *Phys. Rev. A* **76**, 052707.
[39] Limão-Vieira P, Blanco F, Oller J C, Muñoz A, Pérez J M., Vinodkumar M, García G and Mason N. J. 2005 *Phys. Rev. A* **71**, 032720.
[40] Szmytkowski C 1987 *Chem. Phys. Lett.* **136**, 363.
[41] Zecca A, Karwasz G, Oss S, Grisenti R and Brusa R S 1987 *J.Phys. B* **20**, L133.
[42] Čurík R, Ziesel J P, Jones N C, Field T A and Field D 2006 *Phys. Rev. Lett.* **97**, 123202 and private communication.
[43] *Photon and Electron Interactions with Atoms, Molecules and Ions* 2003 Landolt-Börnstein, Group I, Vol. 17, Pt. A (Springer, Berlin/Heidelberg).
[44] Itikawa Y and Mason M 2005 J. Phys. *Chem. Ref. Data* **34**, 1.
[45] Straub H C, Renault P, Lindsay B G, Smith K A and Stebbings R F 1996 *Phys. Rev. A* **54**, 2146.
[46] Blanco F and García G 2003 *Phys. Rev. A* **67**, 022701.
[47] F. Blanco and G. García, *Phys. Lett. A* **317**, 458 (2003).
[48] Blanco F and García G 2004 *Phys. Lett. A* **330**, 230.
[49] Blanco F and García G 2007 *Phys. Lett. A* **360**, 707.
[50] Jain A 1988 *J. Phys. B* **21**, 905.
[51] Faure A, Gorfinkiel J D and J. Tennyson, 2004 *J. Phys. B* **37**, 801.
[52] Inokuti M 1971 *Rev. Mod. Phys.* **43**, 297.
[53] Inokuti M and McDowell M R C 1975 *J. Phys. B* **7**, 2382.
[54] Garcia G and Blanco F 2000 *Phys. Rev. A* **62**, 044702.
[55] Vinodkumar M, Joshipura K N, Limbachiya C and Mason N 2006 *Phys. Rev. A* **74**, 022721.
[56] Stallard H B 1966 *Br. J. Ophthalmol.* **50**, 147.
[57] Freundlich H F 1949 *Nature* **164**, 308.
[58] Lommatzsch P K 1986 *Br. J. Ophthalmol.* **70**, 844 .
[59] Lommatzsch P K, Werschnik C and Schuster E 2000 *Graefe's Arch. Clin. Exp. Ophthalmol.* **238**, 129.
[60] Wilson M W and Hungerford J L 2002 *Ophthalmolgy* **109**, 1838 .
[61] Johansson B, Persson E, Westman G and Persliden J 2003 *Radiotherapy and Oncology* **69**, 107.
[62] Lommatzsch P K, Alberti W, Lommatzsch R and Rohrwacher F 1994 *Graefe's Arch. Clin. Exp. Ophthalmol.* **232**, 482.
[63] Muñoz A, Williart A, Perez J M, Blanco F and Garcia G 2004 *J. Appl. Phys.* 95, 5865.
[64] Muñoz A, Perez J M, Garcia G and Blanco F 2005 *Nucl. Instrum. Meth. A* **536**, 176.
[65] Cullen D, Hubbell J H, Kissel L 1997 EPDL97: *The Evaluated Photon Data Library*, '97 Version, UCRL-50400, vol. 6, Rev. 5.
[66] *International Commission on Radiation Units and Measurement,* 1984 *ICRU Report* no. 37, Bethesda, MD.

Modelling of Anomalous Doppler broadened Lines, Thermalization of Electrons and the role of Radicals in Discharges at high E/N

Zoran Lj. Petrović, Željka Nikitović and Vladimir Stojanović

Institute of Physics Belgrade POB 68 11080 Zemun Serbia

zoran@phy.bg.ac.yu

Abstract. In this paper we show how Monte Carlo technique developed for swarm studies may be applied to model some real discharges or to provide data for realistic plasma models. In all cases discussed here the Monte Carlo technique without self consistent field is driven to a much greater level of complexity than usually found in swarm studies. We discuss three conceptually quite different situations that require exact modelling by Monte Carlo Simulations (MCS). The first is the case of transport in gases where degree of dissociation is large and thereby radicals formed by the discharge may affect the transport coefficients. Based on recent calculations for electron CF_x radical cross sections we calculated how transport coefficients would be affected by the presence of radicals in pure CF_4 and CF_4/Ar mixtures covering range of abundances often found in plasma etching devices. The second case studied here is modelling of thermalization of high energy electrons in the atmospheric gases as an attempt to provide the basis for detection of very high energy cosmic particles. In that case pressure dependent emission efficiency is studied as a function of electron energy and the system is developed to an arbitrary degree of secondary electron production. Finally we give some recent advances and current status of modelling of high E/N discharges operating in Townsend's regime in hydrogen where anomalous broadening was observed and explained by fast neutral excitation in both directions along and against the acceleration due to electric field.

1. Introduction

Non-equilibrium nature of plasmas [1, 2] makes them especially suited for numerous applications ranging from plasma etching of integrated circuits through treatment of materials, light sources to biomedical treatments and large scale modification of the atmosphere. Perhaps the most promising future applications of non-equilibrium plasmas include massively parallel production of nanostructures for nanotechnologies (top down approach) and for medical treatments. Monte Carlo simulations are the best way to model such plasmas as they allow realistic treatment of complex boundary conditions [3], inclusion of recently found complex kinetic phenomena [4] and essentially 3 dimensional treatments of complex geometries. In this paper we show three examples of how extension of standard and exact Monte Carlo procedure as developed for studies of charged particle swarms [5] may be extended by making the model more complex by one step to include new phenomena required to explain realistic discharges in gases.

The first example involves complexity due to presence of radicals created by the discharge itself and now, thanks to recent calculations of cross sections it has become possible to answer the question of just how much do the radicals change the transport in gases. The second example is that of emission caused by very high energy charged particles in the atmosphere where most of the emission is produced by secondary electrons. Complexity added here is due to very large initial energy and the need to follow a large number of secondary electrons produced by the primary particle. Finally we address and even more complicated situation where both the primary electrons are in runaway, most of the processes are initiated by the secondary electrons, the length of the particle becomes comparable to the mean free path so the energies change form one point to another and most of the excitation is produced by fast neutrals which are generated in charge exchange collisions of ions with gas molecules or in neutralization and reflection of fast ions hitting the cathode.

Observations of anomalously broadened hydrogen lines in gas discharges were made in 60s and the explanations that followed were mainly based on dissociative processes. However, some observations, especially those of Benesch and Li [6] could not be explained without evoking the directionality of the motion of ions. First full explanation of the effect and observations of energies exceeding few tens of eV were made on the basis of a swarm experiment carried out by Petrović et al. [7] which revealed the need to include a new effect, that of the reflection of fast neutrals from the cathode with sufficient energies to produce hydrogen radiation. The model that was developed although having a correct physical phenomenology was based on approximate treatment of non-hydrodynamic transport, i.e. essentially on a beam like treatment of energy distribution functions with the mean energy calculated from the energy balance. In the meantime the set of pertinent atomic collisions has become much more complex and we have based the calculations on a Monte Carlo simulation which gives exact energy distributions. We will present the results on all details of kinetics of relevant particles, from electrons to ions, molecules and atoms. We will also stress the need to model more accurately the heavy particle collisions, especially the angular anisotropy of scattering.

Thermalization of electrons produced by cosmic rays from very high energies and the resulting emission of nitrogen radiation has been proposed as a means to detect properties of extremely high energy elementary particles. The process is exactly the opposite to that taking place in modeling of high E/N discharges and may be treated exactly by a similar Monte Carlo code if the applied cross sections are correct.

Finally we will discuss how dissociated fragments affect the transport properties of electrons having in mind applications such as plasma etching of SiO_2 where abundances of radicals of CF_x are of the order of several percent. The principal effect of radicals on electron transport in pure CF_4 or in CF_4/Ar mixtures is through very different attachment rates which lead to orders of magnitude different effective attachment rate for the mixture with radicals as compared to the properties of pristine buffer gas. Thus one may conclude that even the electronegative nature of the plasma may be changed due to the presence of different radicals, especially of CF_2.

1.1. Monte Carlo simulation technique

The MC code has been developed based on null collision technique and has been tested against all known benchmarks [3, 8]. The code has a facility to use energy dependent differential cross sections defined specially for each process and it also has a facility to follow sequentially products such as ions, fast neutrals, secondary and reflected electrons. We typically follow 100 000 electrons through hundreds of thousands of collisions and we have means to reproduce all observables with accuracy limited only by the quality of the cross sections and by statistical uncertainty.

2. The effect of CFx radicals on electron transport in CF_4

Carbon tetrafluoride is commonly used in today's semiconductor industry for etching of dielectric materials, such as SiO_2 [1, 2] and also for deposition of fluorinated polymer films.CF_4 has also an important role in a number of technological applications. CF_4 belongs to the Freon group of gases that unfortunately significantly contribute to the global warming of our planet with a lifetime of 50 000 years.

In industrial applications a large number of radicals are produced in plasmas, and in etching reactors the densities of radicals may reach several percent [9]. In CCP (capacitively coupled plasma) [10] and ICP (inductively coupled plasma) [11] etching reactors in fluorine-poor gas mixtures the dominant reactive radical is CF_2. The effect of these radicals has already been studied in the literature [12].

Recent advances in electron scattering theory [13] made it possible to calculate the cross sections for electron scattering on CFx radicals in the energy domain relevant to non-equilibrium plasmas. Some of the cross sections for electron- Cox radical scattering can be significantly larger than the cross sections for CF_4 itself especially in the region of the Ramsauer Townsend minimum and for the attachment cross section [13]. Until recently [14] no direct experimental measurements of these cross-sections have been performed.

Of all the CFx radicals the CF_2 molecule may be present at the level of several percent [9, 15] in industrially relevant plasmas. As a result it could play a significant role even at the level of electron transport that determines the electron energy distribution function (EEDF) and the basic kinetics.

In this work we have calculated the electron transport coefficients for mixtures consisting of CF_4/CF_2 [16] and of $CF_4/Ar/CF_2$ [17]. Argon was present at 10% while the abundance of the radical was varied from 0.01% to 10% thus covering realistic abundances found in plasma applications. We have also performed calculations for all possible radicals CF, CF_3, F, F_2 but those, while present in significant amounts are never found at densities greater than CF_2. In general we have used a two term code for calculations and verified them by a Monte Carlo simulation (MCS).

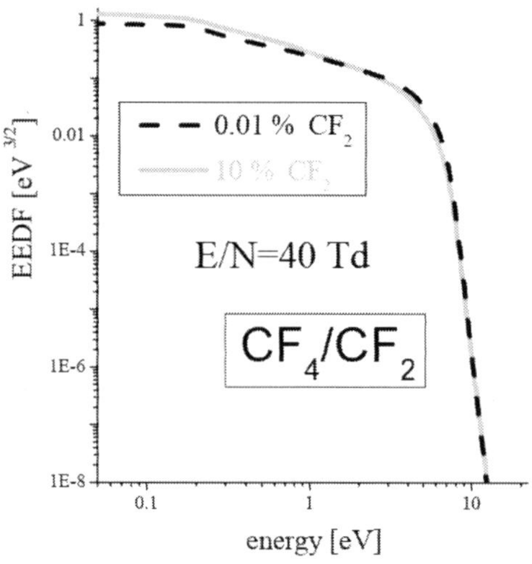

Figure 1. EEDF in CF_4/ CF_2 mixtures [16].

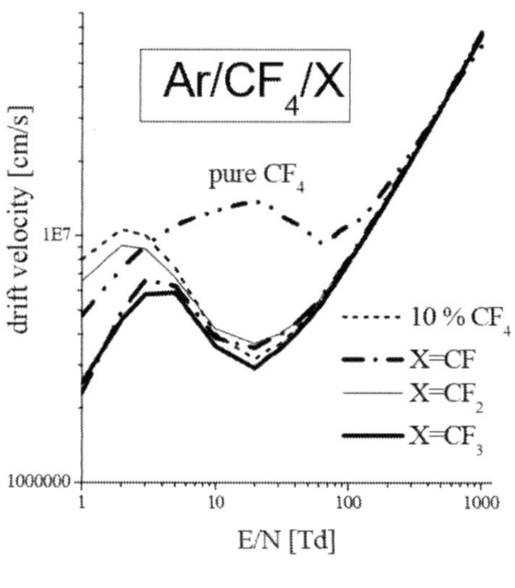

Figure 2. Drift velocities in CF_4/Ar/CFx (x=1-3) [17] mixtures.

2.1. Calculated transport coefficients in mixtures with radicals

Our calculations [16, 17] indicate that the addition of radicals even at the level of 10% does not affect very much the mean energy and the EEDF (see figure 1). On the other hand the effect on the drift velocity is significant and one example is shown for the mixture that includes argon in figure 2. The comparison between the two figures means that while mean energy is not perturbed much by the presence of radical and the shape of EEDF is not affected the elastic cross section for radicals which does not have a minimum fills the cross section in the region of the Ramsauer Townsend minimum in

CF_4 and it changes the optimum conditions for the negative differential conductivity [16, 18]. This is a proof that by selecting a mixture one may control independently the momentum and the energy balance and in particular the number balance.

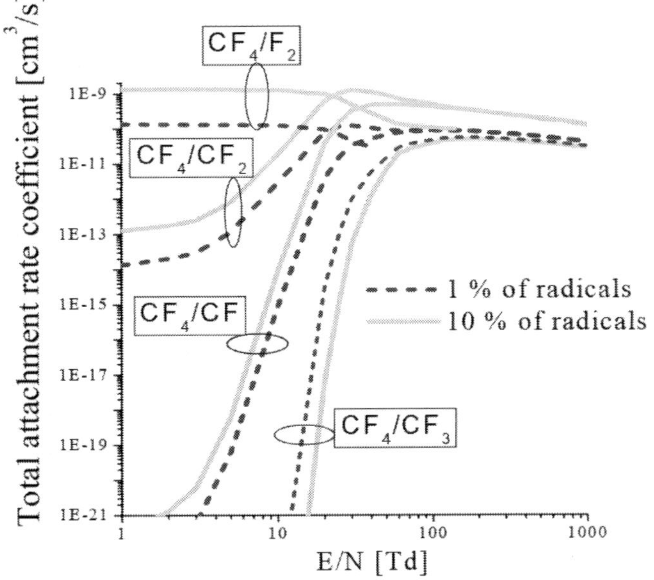

Figure 3. Comparison of total attachment rate coefficient for a) 1 % of radicals and b) 10 % of radicals.

The dominant number changing process in the discharge for energies of practical gas discharges is the attachment. It is a characteristic of the cross sections for electron scattering in CF_4 [19] that attachment has a threshold at a relatively high energy. On the other hand some of the radicals, CF_2 and F_2 in particular have a threshold at lower energies [13, 20]. In figure 3 we show the effective rate coefficients for attachment for two abundances of the radicals one 1% and 10% [16] in pure CF_4. The cross sections for F_2 and F were taken from [20, 21]. Since CF_3 does not have the attachment the curve for 1% of CF_3 is close to that for pure CF_4 while the reduced attachment as the abundance is increased to 10% is due to dilution of the attaching gas in the mixture.

On the basis of our calculations one could conclude that even for the abundances of radicals CF_2 of few percent the effect on the transport coefficients of electrons is significant and requires consideration in plasma modeling. Apart from momentum balance and the presence of the negative differential conductivity the most important effect is on the effective attachment rate. The operating point of plasma will certainly change as the abundance of radicals grows and even the nature will change from almost electropositive plasma to electronegative plasma with all the consequences and changes in the nature and anatomy of the discharge.

3. Thermalization of high energy electrons in nitrogen

The kinetics of excited species at higher altitudes has been often connected to the kinetics of electrons. Emission from the higher layer of the atmosphere has been proposed as a probe of the particles of cosmic origin with extremely high energies [22, 23]. In their papers Blanco and Arqueros [23, 24] have applied a Monte Carlo technique to calculate the emission due to thermalization of very high energy electrons in nitrogen at low pressures that correspond to the conditions in the upper atmosphere. These authors have initially assumed that the cross section may depend on the pressure if one needs to model the low pressure and the high pressure limits. The cross section is a description of a binary collision event and therefore it is independent of the pressure. What was required to fit the experimental results at higher pressures is to model the effective emission/excitation coefficient and the effect of secondary electrons.

These authors have eventually adopted modeling of secondary electrons and effective emission but they confined themselves to a single excitation by each of the secondary electrons and they limited their simulation to a certain escape radius R. In this paper we have performed calculations without such limitations and we based ourselves on a complete set of cross sections and differential cross sections [3] that were tested against standard swarm experiments at high E/N [8]. The MC code used here has representation of the detailed kinetics of primary and secondary electrons, ions and fast neutrals built in together with detailed representation of collisional events (differential cross section, treatment of non-conservative processes). For many years high altitude studies have focused on observation of emission from nitrogen and details of the cross sections from thermal to high energies have been compiled and analyzed [25, 26].

It is important to note that the swarms at very high E/N are a good laboratory equivalent of electrical processes in the upper atmosphere and our codes have been tested against those experiments [27] including the runaway phenomena in the EEDF [28]. Finally it is worth noting that with runaway effects being in the main phenomenology of lightning discharges [29, 30] these models are applicable to study lightings as well as similar atmospheric electrical phenomena such as sprites, elves and blue jays.

In this paper we shall present our results modeling deposition of energy in nitrogen at low pressures that are analogous to and also test the results of Blanco and Arquieros [23]. We have extended the calculations [31] to produce results for spatial profiles of emission that may easily be measured in practice and applied top analyze the effect of cosmic rays in the upper atmosphere.

3.1. *Cross section data*

The cross sections set for electron scattering on N_2 was the same one as in [3]. It consists basically of the data due to Phelps and Pitchford [32] for the total inelastic cross sections, while the total cross section was determined on the basis of their momentum transfer cross section and our collection of differential cross sections. The set was extended by the available data and high energy analytic formulae to 10 keV [3]. Our studies of high E/N data revealed that at high mean energies exceeding several hundred eV it is essential to define very well the differential cross sections as under those conditions of electron runaway it is essential to model properly the energy gain which is not properly calculated by using for example a momentum transfer cross section with isotropic scattering. Therefore our code has an option to use different differential cross sections not only for the elastic cross section which dominates at low energies. Other Monte Carlo codes limit their facility to deal with anisotropic scattering only to elastic scattering. In addition we have employed a number for different differential cross sections for different energy ranges in our code.

3.2. *Thermalization of high energy electrons at low pressures of nitrogen*

The effective macroscopic cross section may be determined from the effective emission. In Ref [23, 24] it is defined as:

$$\varepsilon_{vv'}(\varepsilon) = N\sigma_{vv'}(\varepsilon) + \langle\sigma_{vv'}/\sigma_{inel}\rangle\,(1 - e^{-N\langle\sigma ine\rangle R})\,N\sigma_{ion}(\varepsilon) \quad (1)$$

where cross sections refer to: vv'-transition between two vibrational sublevels of an excited state, inel- total cross section for all inelastic processes, ion- ionization; where e is the electron energy and R is the effective range as introduced in [23]. We may fit the results of simulation or experiment by using (2) and R as a fitting parameter. The effective cross section may be determined from:

$$\varepsilon_{vv'}(\varepsilon) = N\sigma_{vv'}^{apparent}(\varepsilon) \qquad (2)$$

where effective emission is the result of simulations. In our study [31] we sample directly the emission coefficient and therefore we can also sample the effective cross section.

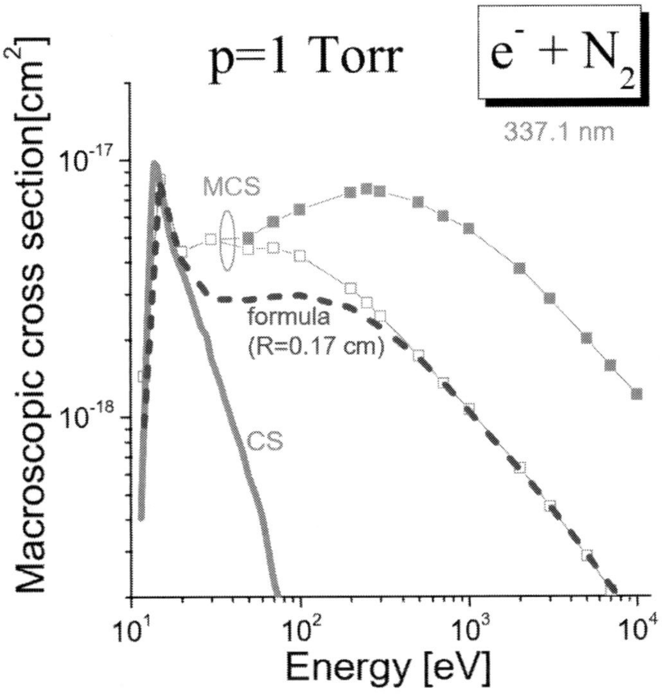

Figure 4 The effective macroscopic cross section for the second positive state of nitrogen. The thick solid line is the microscopic cross section the solid symbols denotes results of our simulations without limitations and the open symbols results with only one excitation collision. The curve by the formula from [23] fitted to our data is separately presented as a thick dashed line.

In figure 4 we show results for the second positive state of nitrogen (337 nm emission). In addition to the basic cross section there is an extension to higher energies due to production of secondaries and subsequent excitation. The secondary electrons have lower energies than the primary beam and therefore they produce emission even at high energies when direct cross section by the primary electron is negligible. We agree with the data from [23] but there is some disagreement just at the maximum of the contribution of the secondaries for energies from 30 eV to 100 eV.

It is hard to separate which effect is due to different assumptions and which due to different cross sections so we have performed a calculation with the assumption of only one generation of secondaries [31]. A relatively good agreement with the data of [23] is also achieved if we fit our results with their formula and with R= 0.17 cm that is close to 0.2 cm used in [23]. At higher energies we see that discrepancy between our data and that from [23] is by a factor of 10 which is due to multiple excitations by secondaries.

The effects for the first negative band and the ionization are smaller but it is still significant to use multiple generations of secondary electrons [31].

3.3. Range of primary and secondary electrons

In addition to spatial emission profiles we may directly sample the range of electrons both primary and secondary. This quantity is well defined as the mean distance traveled by a group of electrons that have some common property. In this case the condition for declaring the end of sampling would be when electron energy drops down below the threshold for excitation. This time may be connected somehow to the time required for thermalization to be achieved but again both depend on the definition of the condition for ending the procedure...

The dependence of the calculated range $<R>$ as a function of the energy of the beam of electrons is shown in figure 5. We are able separate the range defined by the single excitation collision of the primary electrons and the range allowing multiple excitation collisions and the difference is up to a factor of 5. The calculated range was found to scale with pressure (as expected) so the value assumed in [23] is valid only for a narrow range of pressures around 1 Torr. It is also possible to compare the spatial profile of emission due to electrons with and without production of secondaries. While the range is greater if no secondaries are formed, the overall emission with secondaries is larger.

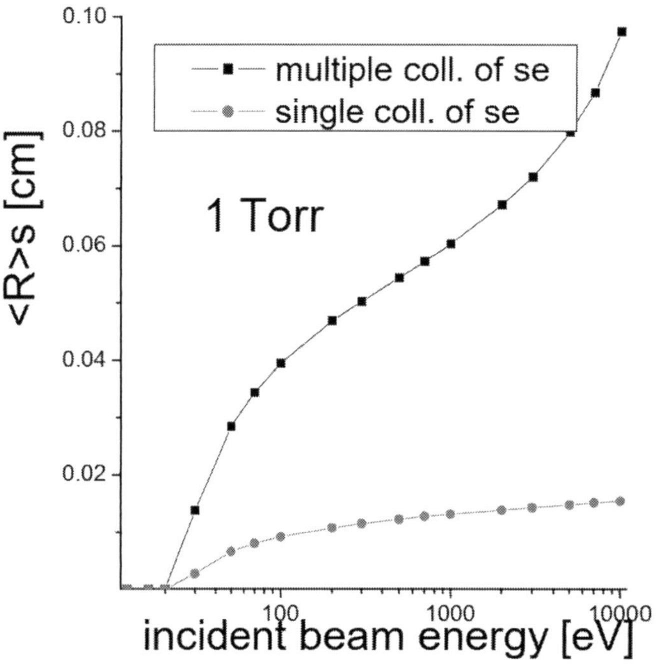

Figure 5 Effective mean range of particles for a single excitation collision (red) and for multiple collisions (black) [31].

3.4 Conclusion

The effect of runaway electrons in high E/N experiments is fully analogous to the thermalization of high energy particles. Thus we propose that the high E/N experiments may be used to benchmark the codes for thermalization. While our results are directly applicable to compare with the results of papers [23, 24] it is also important to note that the upper atmosphere lightning's also involve high energy runaway electrons [29, 30]. In addition our results are relevant to other studies of thermalization of electrons [33] and positrons [34, 35] that in addition to atmospheric phenomena may include biomedical applications and radiation damage.

4. Modelling of Doppler broadened line profiles

The first interpretations of unusually high Doppler broadening of hydrogen lines in non-equilibrium plasmas [6, 7, 36-38] were based on possible dissociative processes, recombination and excitation. Experiments with dc fields revealed a large asymmetric component with energies exceeding the energy available from the repulsive potentials of the dissociating molecules [6, 7]. The explanation of such results was found in a specially designed high E/N swarm experiment [7, 39] which gave a proof that, to the largest degree the excitation is due to collisions of fast neutrals formed in charge transfer collisions of the buffer gas with ions accelerated by the electric field. At high E/N ions gain high energies and consequently create through charge transfer collisions very fast neutrals with comparable energies. In addition it is possible to have reflection and neutralization of ions with reflection as fast neutrals, which leads to different components in blue and red wings of the Doppler profile [7]. Anomalously broadened profiles were observed in hydrogen lines in pure hydrogen and in hydrogen containing molecules such as methane [40] and especially the effects were pronounced in mixtures of such molecules with argon [39]. Here we show the latest results obtained by detailed modeling based on the nest recent cross sections of A.V. Phelps [41] expanded with the analysis of the role of vibrational energy losses and anisotropy of the cross sections on the experimental observables.

4.1. Procedure of simulation

The Monte Carlo code [12] based on null collision technique was used to follow electrons and heavy particles (H^+, H_2^+, H_3^+, fast H and fast H_2) in pure H_2. Processes at cathode and anode surfaces such as secondary electron production, reflection, neutralization, elastic and inelastic collisions were also included based on the available experimental data and models of scattering without any adjustment. Simulations were performer for the conditions of high E/N Townsend experiments. We followed trajectories of all reaction fragments until they neutralized or thermalized below the H_2 excitation energy.

4.2. Results for line and spatial profiles

In figure 6 we show profiles of the Balmer H_2 line and the axial emission profile. A good representation of the profile is obtained with the best available data without any additional adjustments of the cross sections. Only a slightly better fit was obtained if inelastic energy losses due to vibrational excitation are introduced. We have used the suggestion of A.V. Phelps to include vibrational excitation only for H^+ ions colliding with H_2. We have also analyzed the role of anisotropy of scattering on the final results. It is essential to use forward scattering as EEDF will be very different

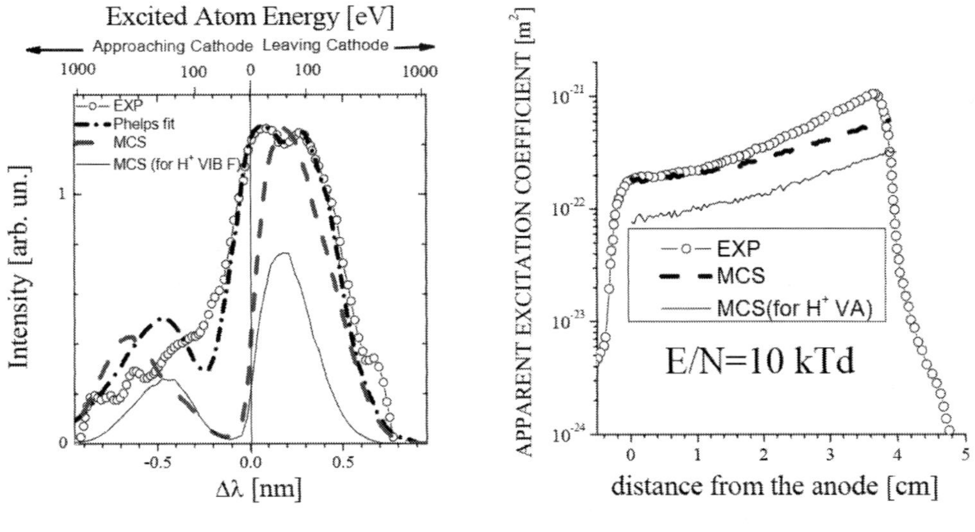

Figure 6 a) Doppler broadened profile, b) spatial distribution of emission, at very high E/N (10 kTd 145 mTorr). Thins line represents the fit with inclusion of vibrational energy losses in inelastic collisions with the buffer gas. "Phelps fit" and "EXP" are results published in [1].

and in disagreement with experiment if one assumes isotropic scattering (even when proper transition from total to momentum transfer cross section is made). However, the details of the differential cross section are open to discussion and may be used to fit the experimental data better.

4.3. Distribution function functions and fluxes

The Monte Carlo code can also provide al the details that are impossible to measure for example the energy distribution functions of fast neutrals at the cathodes (since they are not affected by field the

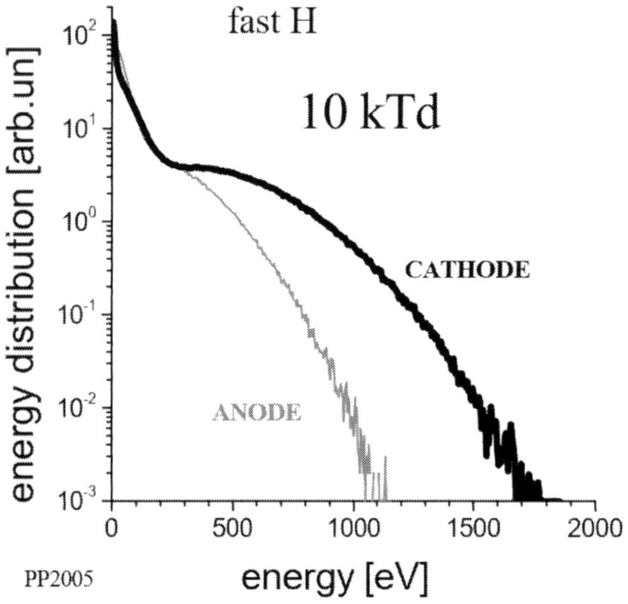

Figure 8 Energy distribution functions of fast H atoms at two electrodes.

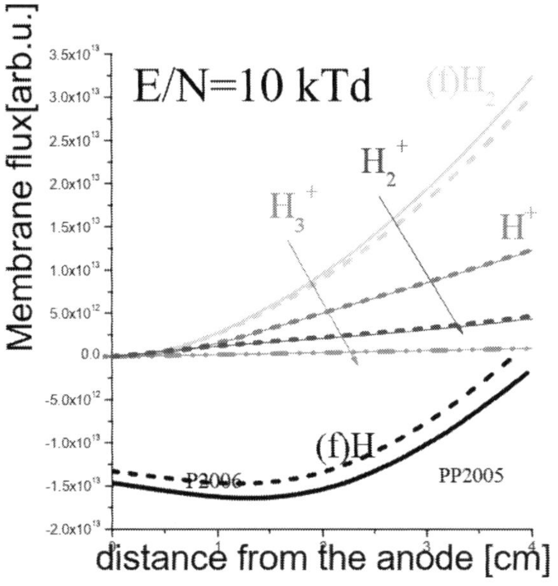

Figure 9 Fluxes of different heavy particles in the gap.

fast neutrals may fly in both directions. In figure 7 we show the distribution function of fast neutrals at the anode and at the cathode and at the figure 8 the axial profile of fluxes of different species in the gap between electrodes. The negative values for fast H mean that more particles flow backwards (as expected from the experimental profile due to reflection with neutralization and with dissociation of molecular ions at the cathode.

Calculations were made for two different yet sets of cross sections.

4.4. Conclusion

A similar degree of success was achieved in modeling of Doppler profiles in rare gas hydrogen (methane) mixtures and in pure methane [40].

One can conclude that the whole phenomenology is well developed and that even without any adjustments good fits of swarm like experiment are achieved. Therefore one may be free to pursue this further either to modify further and achieve a better fit or to proceed to model more complex gas discharges.

In all cases discussed in this text Monte Carlo simulation codes (in one case supplemented by a two term code) provide full description. It also shows a large number of situations where swarm phenomenology as represented by the use of transport coefficients [41] may provide full description.

Acknowledgments

Work at the Institute of Physics is supported by the MNRS, under grant 141025. We are grateful to all of our colleagues who participated to the work presented in this review.

References

[1] Makabe T. and Petrović Z. 2005 *Plasma Electronics: Applications in Microelectronic Device Fabrication*, (New York: Taylor and Francis, CRC Press).
[2] Lieberman M.A., Lichtenberg A.J. 2005 *Principles of Plasma Discharges and Materials Processing* (New York: Wiley).
[3] Stojanović V.D. and Petrović Z.Lj. 1998 *J.Phys.* D **31**, 834.
[4] Petrović Z Lj, Šuvakov M, Nikitović Ž, Dujko S, Šašić O, Jovanović J, Malović G and Stojanović V 2007 *Plasma Sources Sci. Technol.* **16** S1.
[5] Huxley L G H and Crompton R W 1974 *The Diffusion and Drift of Electrons in Gases* (New York: Wiley-Interscience)
[6] Benesch, W. and Li, E. 1984 *Opt. Letters* **9** 338.
[7] Petrović Z Lj, Jelenkovic B M and Phelps A V 1992 *Phys. Rev. Lett.* **68** 325.
[8] Stojanović V.D., Jelenković B.M. and Petrović Z.Lj. 1997 *J.Appl.Phys.* **81** 1601.
[9] Booth J.P. 1999 *Plasma Sources, Sci. Technol.* **8** 249.
[10] Nakano T. and Samukawa S 1999 *J.Vac. Sci. Technol.* A **17** 686.
[11] Hioki K., Hirata K, Nakano N, Petrović Z. Lj. and Makabe T. 2000 *J.Vac. Sci. Technol.* A **18** 864.
[12] Curley G A, Marić D, Booth J-P, Corr C.S, Chabert P and Guillon J 2007 Plasma Sources Sci. Technol. **16** S87.
[13] Rozum I, Limao-Vieira P, Eden S, Tennyson J, and Mason NJ 2006 *J. Phys. Chem. Ref. Data*, **35** 267.
[14] Maddern T M, Hargreaves L R, Buckman S J and Brunger M J 2007 J. Phys. Conf. Series 012006.
[15] Booth J.P, Cunge G, Chabert P and Sadeghi N 1999 *J. Appl. Phys.* 85 3097.
[16] Nikitović Ž, Stojanović V, Booth JP and Petrović Z.Lj. 2008 submitted.
[17] Nikitović Ž, Stojanović V and Petrović Z.Lj. 2008 Acta Physica Polonica submitted.
[18] Vrhovac S.B. and Petrović Z.Lj. 1996 *Phys. Rev.* E **53** 4012.
[19] Kurihara M, Petrović Z. Lj. and Makabe T 2000 *J. Phys. D: Appl. Phys.* **33** 2146.
[20] Morgan W. L. 1992 *Plasma Chem. Plasma Proces* **12** 449.
[21] Gudmundson J. T. 2002 *J. Phys. D: Appl. Phys.* **35** 328.
[22] Nagano M., Kobayakawa K., Sakay N., Ando K. 2003 *Astropart. Phys.* **20** 293.
[23] Blanco F and Arqueros F 2005 *Phys. Lett A* **345** 355.
[24] Arqueros F, Blanco F, Casetellanos A, Ortiz M and.Rosado J 2006 *Astropart. Phys.* **26** 231.
[25] Strickland D.J, Book D.L, Coffey T.P. and. Fedder J.A. 1976 *J. Geophys. Res.* **81** 2755.
[26] Stolarski RS and Green A.E.S. 1967 *J. Geophys. Res.* **72** 3976.
[27] Jelenkovic B.M. and Phelps AV 1987 *Phys.Rev.* A **36** 5310.
[28] Vrhovac S.B, Stojanović V.D, Jelenković B.M. and Petrović Z.Lj. 2001 *J. Appl. Phys.* **90** 5871.
[29] Gurevich A.V. and Zybin K.P. 2005 *Physics Today* May 37.
[30] Pasko V.P. 2007 *Plasma Sources Sci. Technol.* **16** S13.
[31] Petrović Z.Lj, Nikitović Ž and Stojanović V.D. 2008 *Plasma Souces Sci. Technol*, submitted.
[32] Phelps A.V. and Pitchford L.C. 1985 *Phys. Rev.* **31** 2932.
[33] B.L. Tembre, A. Mozumder 1983 Phys. Rev. A 27, M. Kimura and M.Inokuti 1987 *J. Chem.Phys.* **87** 3875.
[34] Šuvakov M, Petrović Z Lj, Marler J.P, Buckman S.J, Robson R.E. and Malović G, 2008 *New J. Phys.* **10** 053034.

[35] Banković A, Marler J.P. Šuvakov M, Malović G, Petrović Z. Lj. 2008 *Nuclear Inst. Methods* B **266/3** 462.
[36] Capelli, A.I., Gottscho, R.A., and Miller, T.A. 1985 *Plasma Chem. Plasma Process* **5** 317.
[37] Barbeau C. and Jolly J. 1990 *J. Phys.* D **23** 1168.
[38] Konjević, N. and Kuraica M. 1992 *Phys. Rev. A: At. Mol. Opt. Phys.* **46** 4429.
[39] Petrović, Z. Lj. and Phelps A.V. 1991 *Proc. of The Internatonal Seminar on Reactive Plasmas*, E-3 351.
[40] Nikitović Ž. D., Strinić A. I., Malović G. N, Stojanović V. D,. Šašić O. M and Petrović Z. Lj. 2006 *Czech. J. Phys.* **56** B 958.
[41] Robson R E, White R D and Petrović Z Lj 2005 *Rev. Mod. Phys.* **77** 1303.

Low energy positron scattering from helium

J P Sullivan, C Makochekanwa, A Jones, P Caradonna and S J Buckman

Centre for Antimatter-Matter Studies
RSPhysSE, Australian National University, Canberra ACT 0200, Australia

E-mail: `jps107@rsphysse.anu.edu.au`

Abstract. Positron scattering from helium has been historically limited to experiments with energy resolution of 0.5 eV or worse. A new apparatus has been developed that allows the investigation of scattering processes with an energy resolution as good as 30 meV, allowing new experimental insight into positron helium interactions. High resolution measurements of low energy scattering from helium will be presented, with a particular emphasis on the development of benchmark cross sections, the investigation of threshold behaviour and the search for positron scattering resonances.

1. Introduction

The advent of buffer gas trapping as a method for creating high resolution positron beams has created the opportunity for the study of positron scattering to be undertaken in unprecedented detail [1, 2]. In particular, there is a need for benchmark cross sections for relatively simple scattering systems, especially above the positronium formation cross section threshold where the theoretical description of scattering becomes significantly more complicated. In addition to this, the investigation of threshold effects with any clarity has been impossible to date, as typical energy resolutions have been 0.5 eV or more. For example, in the case of positron impact ionisation, the extent of the "Wannier region" is not well understood. In this region, the escape of the two charged particle from the residual ion core is described by a simple threshold law. While resonances in positron scattering are far less likely than in the case of electron scattering, any search for resonant features must also rely on a high resolution beam and such searches have been virtually impossible until the development of the Surko trap. The present work describes the first experiments using a new positron trap system based on the San Diego design. Positron scattering from helium has been measured for energies up to 60 eV and the grand total and positronium formation cross sections have been determined. High resolution scans have been undertaken over various regions to both search for resonances and to elucidate threshold behaviour, especially at the positronium threshold where channel coupling effects have previously been reported [3].

2. Experimental apparatus

A detailed explanation of the operation of this apparatus will be provided in the near future [4], so only a brief description is presented here. Positrons are obtained from a 22-Na radioactive source which is located at the end of a two-stage cryogenic cold head, capable of maintaining a temperature of less than 8K. Neon is frozen onto the conical mount containing the 22-Na and

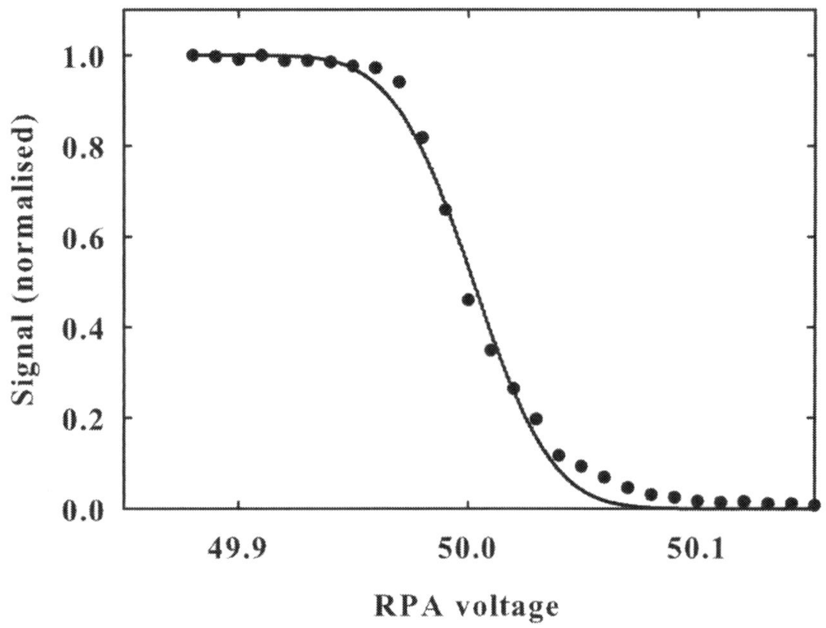

Figure 1. A cutoff curve measured at 50 eV, showing the energy resolution of the positron beam. The fit (solid line) indicates an energy resolution of 65 meV.

acts as a moderator, re-emitting approximately 1% of the high energy (up to 500 keV) positrons in a low energy beam with an energy width of approximately 1.5 eV. This low energy beam is radially confined using a series of solenoids, which provide a magnetic field of approximately 100 gauss. With a 50 mCi source (current source strength is ~30 mCi), a beam of up to 10 million positrons/second is obtained. This positron beam is directed into a Surko trap located in a 530 gauss magnetic field. The trap consists of a series of cylindrical electrodes (with their axis parallel to that of the magnetic field) into which nitrogen and CF_4 buffer gases are admitted. The electrodes form a potential well structure into which the incoming positrons are trapped, losing energy by exciting an electronic transition in the nitrogen molecules. Positrons trapped in the electrode structure continue to collide with the buffer gas, losing energy by exciting electronic, vibrational and rotational transitions in the gas mixture until they are confined in the final stage of the electrode structure and cooled to the gas temperature (room temperature). This reservoir of cooled positrons is then emptied by carefully raising the confining potential so that the positrons are emitted as a pulsed beam, maintaining the energy spread of the cooled positron cloud.

In a strong magnetic field, the energy of the positrons can be considered as having two components, one parallel and one perpedicular to the field lines, $E_{||}$ and $E_{\perp}$ respectively. The $E_{||}$ component can be measured using a retarding potential analyser (RPA), which can simply be a cylindrical electrode with its axis parallel to the magnetic field (in this case, the defocussing effects that are a problem for RPAs in an electrostatic configuration are eliminated). By varying the potential of the RPA until the positrons are no longer transmitted, a "cutoff curve" can be measured and the energy of the positron beam determined. The width of the cutoff step then gives a measure of the energy resolution of the beam. Such a measurement is shown in figure 1, which demonstrates a resolution of 65 meV, typical of the experiments presented in this paper.

Positrons pulses are detected using a microchannel plate combined with a transimpedance amplifier. The signal from the amplifier is proportional to the positron current hitting the plate, and is stored on a computer and analysed to determine the signal strength. The pulses contain 100-500 positrons each and the repetition rate of the pulsed beam is typically 200 Hz.

After their release from the Surko trap, the positrons are directed to a gas cell, also located in a 530 gauss magnetic field. The cell contains the target gas, in this case helium, and the potential on the gas cell (relative to the energy of the positrons released from the trap) defines the interaction energy of the positrons with the gas. Positrons that scatter elastically with a gas atom have energy transferred from the parallel to the perpendicular direction according to the scattering angle with $E_{||} = E_T \sin^2(\theta)$, where E_T is the energy of the collision and θ is the scattering angle. This parallel energy loss can be measured and related to the differential scattering cross section (DCS)[1].

The total cross section can be measured in a similar way, by measuring the proportion of positrons that have lost any parallel energy and then applying Beer's Law, as for a conventional transmission experiment. Unlike a conventional transmission measurement, in this case the angular resolution is not determined by the geometry of the system, but rather by the energy resolution of the beam which will limit the rejection of forward scattered positrons. This problem becomes worse at low energies, as the energy transferred to the parallel direction in a scattering event becomes smaller. Currently, data for the total cross section has had a correction applied below an energy of 1 eV, to correct for the missing forward angle contribution to the total cross section. This correction has used the shape of the theoretical cross sections provided by Bray [5] to estimate the forward angle contribution. As the energy resolution of the experiment is improved, it is anticipated that the impact of this problem on the very low energy measurements can be reduced.

If an inelastic scattering channel is open, it is no longer possible to relate the parallel energy distribution to the differential scattering cross section [1], but the method for measuring the total cross section remains valid. In addition to this, the process of positronium (Ps) formation appears as a loss of beam intensity, as Ps formed in the gas cell annihilates before reaching the MCP detector. This process allows the positronium formation cross section to be measured, in much the same way as the total cross section.

3. Results

Preliminary measurements of the elastic DCS (below the first excited state threshold) have been made at several energies and an example is shown in figure 2, for an incident energy of 5 eV. Data is shown for angles from 10 to 60 degrees, and the error bars in the figure represent the absolute errors in the measurements. While the angular range of the measurements is currently limited, there appears to be good agreement with the theoretical calculations of both McEachran [6] using the polarised orbital approach and Bray [5] who used the convergent close coupling technique. Although the agreement appears to be better with the calculation of McEachran, it is not possible to distinguish between the two calculations with any confidence as there remains some possible systematic errors in the data as the scattering angle increases. Further measurements are planned to both increase the angular range of the measurements and improve the accuracy of the results, with the goal of establishing a benchmark for low energy positron scattering, such as has been achieved in the case of electron scattering from helium [7, 8].

Low energy total elastic cross sections are presented in figure 3. As expected, the measurements agree well with all the latest theoretical calculations [9, 10, 11] as well as the measurements of Mizogawa [12]. High resolution measurements of this cross section have been made at small energy intervals to investigate the claims of resonant features in the Ramsauer-Townsend minimum made by Karwasz et al. [13]. No evidence of these resonant structures was found [14], suggesting that the previous observation was due to some systematic effect in the measurement, as suggested by Zecca [15].

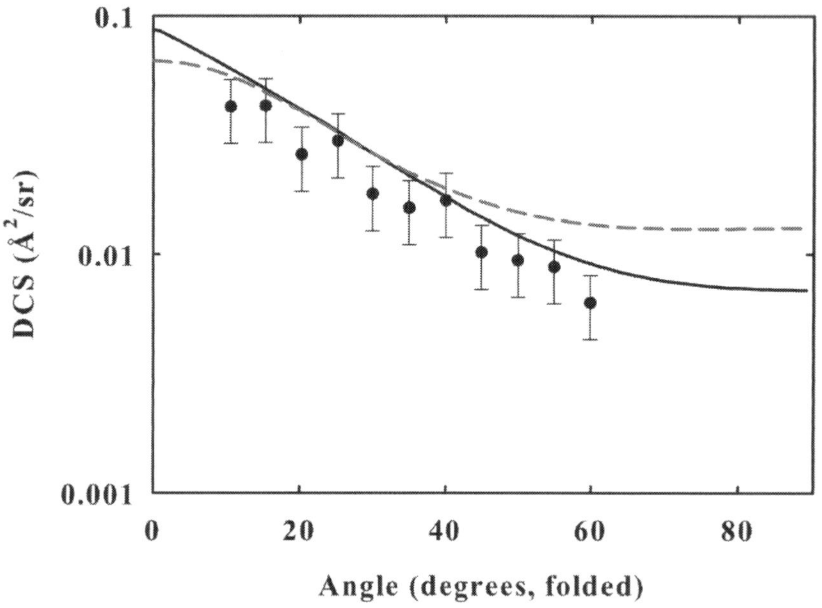

Figure 2. Differential cross section for positron scattering from helium at an energy of 5 eV. Points are the experimental data with the solid line the theory of McEachran [6] and the dashed line the theory of Bray [5].

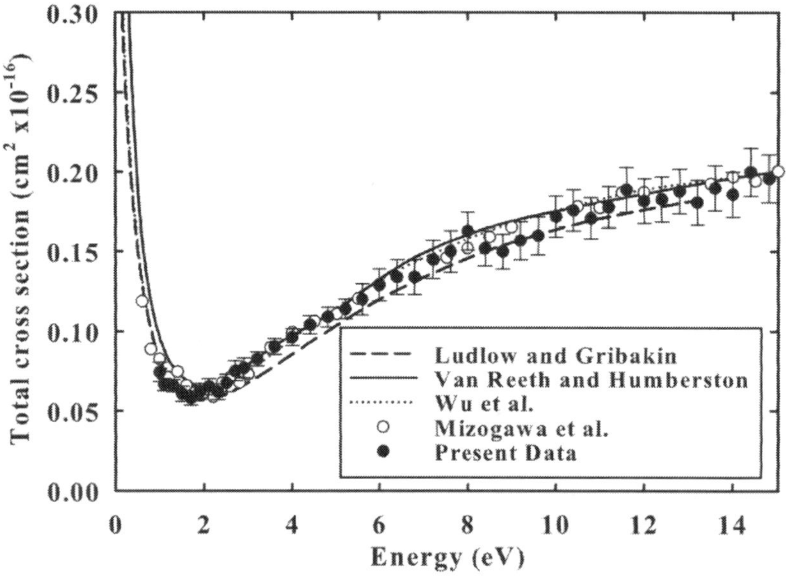

Figure 3. Total elastic cross section for positron scattering from helium.

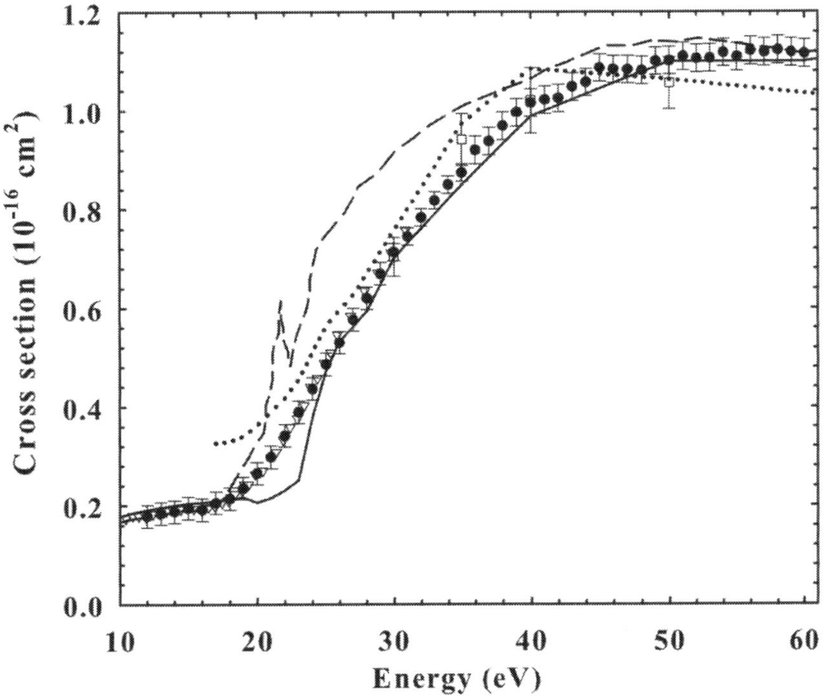

Figure 4. Grand total cross sections for positron scattering from helium. Filled circles are the current measurements. Open triangles and open squares are the measurements of Kauppila et al. [17] and Griffith et al. [16] respectively. Lines are theoretical calculations of the grand total cross section by Campbell et al. [3] (dashed line), Cheng and Zhou [18] (dotted line) and Bray [5] (solid line).

Grand total cross section measurements have been performed up to an energy of 60 eV, with a sharp increase evident in the cross section at the positronium formation threshold, as shown in figure 4, and emphasising the important role that positronium formation plays in the scattering process. The cross section rises smoothly to the maximum energy measured and is in good agreement with previous experimental data [16, 17], although some discrepancies remain with older theoretical calculations [3, 18]. The recent theory of Bray [5] is in good agreement everywhere except in the region between the positronium formation and ionisation thresholds, where difficulties still remain with their theoretical treatment of the positronium formation problem. The challenge of including further open channels in the calculations, especially the Ps formation channel, is one that clearly still needs to be satisfactorily solved.

For energies above the Ps formation cross section, it is also possible to measure the grand total cross section minus the Ps formation cross section. In the Ore gap, where Ps formation is the only inelastic channel available, this amounts to the total elastic cross section. It can be seen from figure 6 that this cross section displays a change in slope or perhaps even a small dip as the energy is increased above the Ps formation threshold, but displays no sign of the sharp cusp predicted by Campeneau [3], in line with the observations of Coleman et al. [19]. This would suggest that there is perhaps some channel coupling effects present, although to nowhere near the extent predicted by Campeneau. In contrast to Coleman, however, we do not the sharp increase in the cross section as it passes through the first excitation threshold at 20.6 eV (for

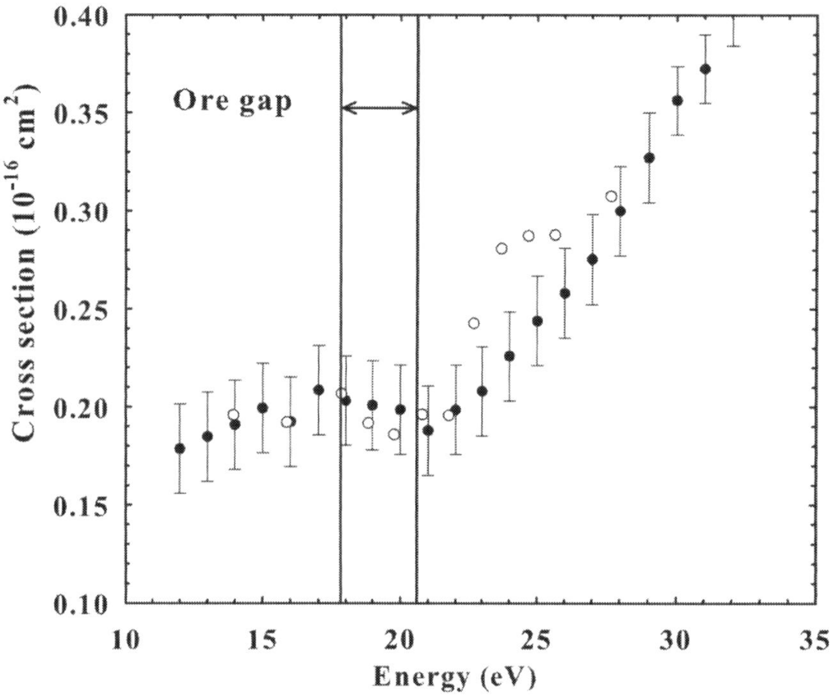

Figure 5. Total scattering cross sections minus Ps formation in the Ore gap. Filled circles are the current measurements. Open circles are the cross sections of Coleman et al. [19].

the 2^1S state of helium). This suggests that the excitation cross section does not turn on as sharply as they suggested, and it is hoped that future measurements may be able to measure this process directly.

4. Conclusion

Results have been presented for positron scattering from helium, giving the grand total, total and differential elastic cross sections over a range of energies. These measurements have been compared to other available data, both experimental and theoretical, with the goal of establishing benchmark values for the processes investigated. Some discrepancies remain, which suggest that further theoretical work is needed to try and resolve the differences. The present data have been taken with the highest resolution available to date for positron scattering from helium, and are able to provide some insight into threshold behaviour. Future measurements will extend the data available for positron scattering from helium, especially with regard to high resolution studies of threshold processes and the measurement of positronium formation, excitation and ionisation cross sections. Once a benchmark experimental data set for positron-helium scattering has been established, it is hoped that this will provide a spur for theoretical calculations to be improved to incorporate the positronium formation process into the theoretical description. In the future, studies of other targets of both fundamental and applied importance will be undertaken using this apparatus.

Acknowledgments
The authors would like to acknowledge the Australian Research Council and the Australian Government's DEST/ISL program for funding this research. In addition, we would like to thank Gleb Gribakin, Igor Bray and Bob McEachran for providing tables of unpublished data and Graeme Cornish for technical support and his work during the construction of this apparatus.

References

[1] Sullivan J P et al. 2002, *Phys. Rev.* A **66** 042708
[2] Surko C M et al. 2005, *J. Phys.* B **38** R57
[3] Campeanu R I et al. 1987, *J. Phys.* B **20** 3557
[4] Sullivan J P et al. 2008a, *under preparation*
[5] Bray I 2008, *private communication*
[6] McEachran R E 2008, *private communication*
[7] Buckman S J and Sullivan J P 2006, *Nucl. Inst. Meth.* B **247** 5
[8] Boesten L and Tanaka H 1992, *At. Data Nucl. Data Tables* **52** 25
[9] van Reeth P and Humberston J W 1999, *J. Phys.* B **32** 3651
[10] Ludlow J and Gribakin G F 2004, *private communication*
[11] Wu H et al. 2004, *J. Phys.* B **37** L1
[12] Mizogawa T et al. 1985, *Phys. Rev.* A **31** 2171
[13] Karwasz G P 2005, *Eur. Phys. J.* B **35** 267
[14] Sullivan J P et al. 2008b, *J. Phys.* B **41** 081001
[15] Zecca A 2006, *Eur. Phys. J.* D **37** 151
[16] Griffith T C et al. 1979 *Appl. Phys.* **19** 431-437
[17] Kauppila W E et al. 1981, *Phys. Rev.* A **24** 725
[18] Cheng Y-J and Zhou Y-J 2007, *Chin. Phys. Lett.* **24** 3408
[19] Coleman P G et al. 1992, *J. Phys.* B **25** L585

A multi term Boltzmann equation analysis of non-conservative electron transport in time-dependent electric and magnetic fields

S Dujko[1,2] **and R D White**[2]

[1] Institute of Physics, University of Belgrade, PO BOX 68, 11080 Zemun, Belgrade, Serbia
[2] ARC Centre for Antimatter-Matter Studies, School of Mathematics, Physics and IT, James Cook University, Townsville 4811, QLD, Australia

E-mail: `sasa.dujko@jcu.edu.au`

Abstract. A multi term technique for solving the Boltzmann equation has been developed to investigate the time-dependent behavior of charged particle swarms in an unbounded neutral gas under the influence of spatially uniform time-dependent electric and magnetic fields. The hierarchy resulting from a spherical harmonic decomposition of the Boltzmann equation in the hydrodynamic regime is solved numerically by representing the speed dependence of the phase-space distribution function in terms of an expansion in Sonine polynomials about a Maxwellian velocity distribution at an internally determined time-dependent temperature. This technique avoids restrictions on the electric and magnetic field amplitudes and frequencies and/or the charged particle to neutral molecule mass ratio traditionally associated with many investigations. The variation of the electron transport coefficients with electric and magnetic field strengths, field frequency, phase difference between the fields and angle between the fields is addressed using physical arguments for certain model and real gases.

1. Introduction

In recent years interest in charged particle transport processes in neutral gases under the conditions of ac electric and magnetic fields has been revived. This interest was motivated by the desire to understand the physics underlying the operation of magnetically controlled/assisted ac plasma discharges in which magnetic field affects both the electron heating mechanisms and charged particle species transport. Large classes of non-equilibrium plasma devices utilize electromagnets, permanent magnets or induced magnetic fields with the goal of enhancing plasma density or improving electron confinement [1]. Hence one of the most challenging areas in plasma modeling is an accurate representation of the effects of a magnetic field on charged particle kinetics.

The history with respect to various aspects of charged particle transport processes in electric and magnetic fields and techniques for solving the Boltzmann equation has been recently reviewed [2, 3]. In contrast to the extensive literature on charged particle swarms in dc electric and magnetic fields, until recently very little has been carried over to charged particle swarms in ac electric and magnetic fields. In order to achieve this goal, a general and accurate theory is required. Such a theory, in ac electric and magnetic fields crossed at arbitrary angle based on the spherical harmonics decomposition of the Boltzmann equation has been published by White *et al* [2]. In this paper we make a further generalization of the theory with respect to the work of White *et al* [2] to consider the explicit effects of non-conservative collisions.

We begin this paper with a brief review of a multi term theory for solving the Boltzmann equation valid for both electrons and ions in time-dependent electric and magnetic fields crossed at arbitrary angles, incorporating the effects of non-conservative collisional processes. We focus on the time-dependent behavior of electron transport properties under conditions which can be generally found in magnetically controlled/assisted rf discharges. We systematically investigate the explicit effects associated with the electric and magnetic fields including field to density ratios, field frequency, field phases and field orientations. In addition we will highlight the explicit modification of transport coefficients by non-conservative collisional processes of attachment and ionization. A multitude of kinetic phenomena were observed that are generally inexplicable through the use of steady-state dc transport theory. We systematically study the origin and mechanisms for such phenomena, their sometimes paradoxical manifestation and possible physical implications which arise from their explicit inclusion into plasma models.

2. Theory

The behavior of charged particles in gases under the influence of electric and magnetic fields is described by the phase-space distribution function $f(\boldsymbol{r}, \boldsymbol{c}, t)$ representing the solution of the Boltzmann equation

$$\frac{\partial f}{\partial t} + \boldsymbol{c} \cdot \frac{\partial f}{\partial \boldsymbol{r}} + \frac{q}{m} \left[\boldsymbol{E}(t) + \boldsymbol{c} \times \boldsymbol{B}(t)\right] \cdot \frac{\partial f}{\partial \boldsymbol{c}} = -J\left(f, f_0\right) . \tag{1}$$

Here r and c denote the position and velocity co-ordinates, q and m are the charge and mass of the swarm particle, t is time while E and B are magnitudes of the electric and magnetic fields, respectively. In what follows, we employ a co-ordinate system in which $\boldsymbol{E}$ defines the z-axis while $\boldsymbol{B}$ lies in the y-z plane, making an angle ψ with respect to the $\boldsymbol{E}$. The right-hand side of Eq. (1) denotes the linear charged particle-neutral molecule collision operator, accounting for elastic, inelastic and nonconservative (e.g. ionizing and attaching) collisions. For elastic collisions we use the original Boltzmann collision operator [4], while for inelastic collisions we prefer the semiclassical generalization of Wang-Chang et al. [5]. The attachment and ionization collision operators employed are detailed in Ref. [6].

Swarm conditions are assumed to apply where the charged particle number density is much less than number density of neutral species and mutual interactions between swarm particles are negligible compared with swarm particle-neutral particle interactions. We assume the neutral molecules (of number density n_0) remain in thermal equilibrium at a temperature T_0 and their internal states are characterized by a Boltzmann distribution. No space charge fields are considered and both the electric and magnetic fields are spatially homogeneous and time-dependent.

2.1. *Time-dependent hydrodynamic regime and definition of transport coefficients*

The connection between experimental and theoretical investigations of swarm behavior is usually made through the equation of continuity

$$\frac{\partial n(\boldsymbol{r}, t)}{\partial t} + \nabla \cdot \boldsymbol{\Gamma}(\boldsymbol{r}, t) = S(\boldsymbol{r}, t) , \tag{2}$$

where

$$n(\boldsymbol{r}, t) = \int f(\boldsymbol{r}, \boldsymbol{c}, t) \mathrm{d}t , \tag{3}$$

while $\boldsymbol{\Gamma}(\boldsymbol{r}, t) = n\langle \boldsymbol{c} \rangle$ is the swarm particle flux and $S(\boldsymbol{r}, t)$ represents the production rate per unit volume per unit time arising from non-conservative collisional processes.

Far from boundaries,sources and sinks, the hydrodynamic regime is assumed to apply. In the time-dependent hydrodynamic regime the space-time dependence is entirely carried out by the density $n(\boldsymbol{r}, t)$ of charged particles and the distribution function has the form

$$f(\boldsymbol{r}, \boldsymbol{c}, t) = \sum_{k=0}^{\infty} f^{(k)}(\boldsymbol{c}, t) \odot (-\nabla)^k \, n(\boldsymbol{r}, t) \,, \tag{4}$$

where $f^{(k)}(\boldsymbol{c}, t)$ are time-dependent tensors of rank k and $\odot$ denotes a k-fold scalar product.

Assuming the functional relationship (4), the flux $\boldsymbol{\Gamma}(\boldsymbol{r}, t)$ and source term $\boldsymbol{S}(\boldsymbol{r}, t)$ in (2) are expanded as follows:

$$\boldsymbol{\Gamma}(\boldsymbol{r}, t) = \boldsymbol{W}^{\star}(t) n(\boldsymbol{r}, t) - \boldsymbol{D}^{\star}(t) \cdot \nabla n(\boldsymbol{r}, t) \,, \tag{5}$$

$$S(\boldsymbol{r}, t) = S^{(0)}(t) n(\boldsymbol{r}, t) - \boldsymbol{S}^{(1)}(t) \cdot \nabla n(\boldsymbol{r}, t) + \boldsymbol{S}^{(2)}(t) : \nabla\nabla n(\boldsymbol{r}, t) \tag{6}$$

where $\boldsymbol{W}^{\star}(t)$ and $\boldsymbol{D}^{\star}(t)$ define, respectively, the *flux* drift velocity and *flux* diffusion tensor. Substitution of expansion (5) and (6) into the continuity equation (2) yields the time-dependent diffusion equation,

$$\frac{\partial n(\boldsymbol{r}, t)}{\partial t} + \boldsymbol{W}(t) \cdot \nabla n - \boldsymbol{D}(t) : \nabla\nabla n = -R_a(t) n \,, \tag{7}$$

which define the *bulk* transport coefficients

$$R_a = -S^{(0)} \qquad \text{(loss rate)} \,, \tag{8}$$

$$\boldsymbol{W} = \boldsymbol{W}^{(\star)} + \boldsymbol{S}^{(1)} \qquad \text{(bulk drift velocity)} \,, \tag{9}$$

$$\boldsymbol{D} = \boldsymbol{D}^{(\star)} + \boldsymbol{S}^{(2)} \qquad \text{(bulk diffusion tensor)} \,. \tag{10}$$

In swarm experiments the bulk transport coefficients are generally measured and tabulated. These transport coefficients are associated with the swarms centre of mass transport. The explicit influence of non-conservative collisional processes on the swarms centre of mass transport is described by the correction terms $S^{(1)}$ and $S^{(2)}$. Obviously, in the absence of non-conservative processes, these two sets of transport coefficients coincide. The distinction between these two sets of transport coefficients was discussed at length in the 1980s, but has been ignored in the majority of previous work in the plasma modeling community. This has lead to a potentially serious mismatch between input swarm data (generally the bulk transport properties) and the parameters (often the flux transport properties) required in many plasma fluid models [9]. Note that only theory, i.e. Boltzmann equation calculations and/or Monte Carlo simulations, can resolve any such mismatch, by providing both flux and bulk transport coefficients.

2.2. Spherical harmonics decomposition of the Boltzmann equation

The angular dependence of the phase-space distribution function in velocity space can be represented in terms of an expansion in spherical harmonics:

$$f(\boldsymbol{r}, \boldsymbol{c}, t) = \sum_{l=0}^{\infty} \sum_{m=-l}^{l} f_m^{(l)}(\boldsymbol{r}, c, t) Y_m^{[l]}(\hat{\boldsymbol{c}}) \,, \tag{11}$$

where $Y_m^{[l]}(\hat{\boldsymbol{c}})$ are spherical harmonics and $\hat{\boldsymbol{c}}$ denotes the angles of $\boldsymbol{c}$. The value of l is incremented until some predefined accuracy criterion is satisfied. This value indicates the deviation of the velocity distribution function from isotropy.

Assuming the time-dependent hydrodynamic regime, the spatial dependence is represented by

$$f_m^{(l)}(\boldsymbol{r}, c, t) = \sum_{s=0}^{\infty} \sum_{\lambda=0}^{s} \sum_{\mu=-\lambda}^{\lambda} f(lm|s\lambda\mu; c, t) G_\mu^{(s\lambda)} n(\boldsymbol{r}, t) , \tag{12}$$

where $G_\mu^{(s\lambda)}$ is the irreducible gradient operator [6]. Truncation at $s = 2$ is necessary to determine transport coefficients up to and including diffusion when non-conservative collisions are operative.

The speed dependence of the coefficients in Eq. (12) is represented by an expansion about a Maxwellian at an arbitrary time-dependent temperature $T_b(t)$, in terms of Sonine polynomials

$$f(lm|s\lambda\mu; c, t) = \omega(\alpha(t), t) \sum_{\nu=0}^{\infty} F(\nu lm|s\lambda\mu; \alpha(t), t) R_{\nu l}(\alpha(t)c) , \tag{13}$$

where

$$R_{\nu l}(\alpha(t)c) = N_{\nu l} \left[\frac{\alpha(t)c}{\sqrt{2}}\right]^l S_{l+1/2}^{(\nu)} \left(\frac{\alpha^2(t)c}{2}\right) , \tag{14}$$

$$\omega(\alpha(t), c) = \left[\frac{\alpha^2(t)}{2\pi}\right]^{3/2} \exp\left[-\frac{\alpha^2(t)c^2}{2}\right] , \tag{15}$$

$$\alpha^2(t) = \frac{m}{kT_b(t)} , \tag{16}$$

$$N_{\nu l}^2 = \frac{2\pi^{3/2}\nu!}{\Gamma(\nu + l + 3/2)} , \tag{17}$$

and $S_{l+1/2}^{(\nu)}\left(\frac{\alpha^2(t)c}{2}\right)$ are Sonine polynomials. Using the appropriate orthogonality relations of the spherical harmonics and modified Sonine polynomials the following system of coupled differential equations for the moments $F(\nu lm|s\lambda\mu; \alpha(t), t)$ is generated:

$$\begin{aligned}
\sum_{\nu'=0}^{\infty} \sum_{l'=0}^{\infty} \sum_{m'=-l'}^{l'} \Bigg\{ & \partial_t \delta_{\nu\nu'} \delta_{ll'} \delta_{mm'} + \omega(000; t) \delta_{\nu\nu'} \delta_{ll'} \delta_{mm'} + n_0 J_{\nu\nu'}^l \delta_{ll'} \delta_{mm'} \\
& + ia(t)(l'm10|lm)\alpha(t) < \nu l || K^{[1]} || \nu' l' > \delta_{mm'} \\
+ \frac{q}{m} B(t) \Bigg\{ & \frac{\sin\psi}{2} \Big[\sqrt{(l-m)(l+m+1)} \delta_{m'm+1} - \sqrt{(l+m)(l-m+1)} \delta_{m'm-1} \\
- im\cos\psi\delta_{mm'} \Big] \Bigg\} & \delta_{\nu\nu'} \delta_{ll'} \Bigg\} F(\nu' l' m' | s\lambda\mu; \alpha(t), t) = X(\nu lm|s\lambda\nu; \alpha(t), t) ,
\end{aligned} \tag{18}$$

where $J_{\nu\nu'}^l$ and $\langle \nu l || K^{[1]} \nu' l' \rangle$ are reduced matrix elements, $\omega(000)$ represents the net creation rate and $(lm|l'm10)$ is a Clebsch-Gordan coefficient. The explicit expressions for the RHS are given in Ref. [7]. Discretising in time using an implicit finite difference scheme converts the system of coupled differential equations into a hierarchy of coupled complex equations. This sparse system of equations is solved using standard sparse inversion routines. The explicit expressions for both bulk and flux transport coefficients including the mean energy, gradient energy vector components and temperature tensor components are given in Refs. [2, 7].

3. Results and discussion

3.1. Preliminaries

In this section we consider the electron transport in rf electric and magnetic fields under hydrodynamic conditions using a multi term theory for solving the Boltzmann equation. Similar studies have been published previously for the Reid ramp model in a crossed field configuration [2, 3]. We try to complement these previous publications by a comprehensive description of electron transport for the most general case of arbitrary field orientations and phase differences between the fields. In addition, in this paper we make a further generalization with respect to the work of White *et al* [2] to consider the explicit effects associated with non-conservative collisions. The ionization model of Lucas and Saelee [8] is employed in this work with the goal of understanding the effects of the ionization processes. Another important aspect of this work is that we investigate the influence of varying field frequency and phase difference between the fields on the electron transport properties in carbon tetrafluoride (CF_4). CF_4 provides an example of a gas which has applications in a wide range of devices where the electron kinetics plays an important role in device behavior. For example, CF_4 has application in rf plasmas mostly realized in capacitively coupled plasma (CCP) [10] and even in ICP for silicon etching [11]. The knowledge of electron transport coefficients and in particular the values of electric and magnetic field strengths for which non-conservative collisions (attachment/ionization) may or may not have a significant effect on the drift and diffusion properties may be important for the operation of these devices. We employ a set of cross sections for CF_4 developed by Kurihara *et al* [12]. For illustrative purposes we chose an applied electric field of 100 Td (1Td $= 10^{-21}$ Vm^{-2}) to ensure the average energy of the swarm is in the range where some interesting kinetic phenomena may be induced. The theory and associated code have been benchmarked against an independent time-resolved Monte Carlo simulation [13].

3.2. The effects of the magnetic field strength

In this section we investigate the influence of the magnetic field strength on the electron transport properties in CF_4. In Figure 1, we display the temporal profiles of the (a) mean energy, (b) longitudinal and (c) transverse drift velocity components and (d) diffusion coefficient along the $\boldsymbol{E}\times\boldsymbol{B}$ direction, as a function of B_0/n_0 in a crossed field configuration for a fixed phase difference of $\pi/3$ rad between the fields. The mean energy is decreased for an increasing B_0/n_0. We may observe that the symmetric profile of the mean energy for the magnetic field free case becomes asymmetric and triangular at high B_0/n_0 with a fast increase and slower decrease.

Both components of the drift velocity lose the symmetry with increasing magnetic field amplitude and at the highest magnetic field strong oscillations are induced due to the cyclotron rotation of the electron swarm. These oscillations represent essentially collision-less gyro-orbits of the swarm particles during this phase of the cycle. Transverse component along $\boldsymbol{E}\times\boldsymbol{B}$ direction is more asymmetric than the longitudinal component, thus resulting in a cycle-averaged value that is non-zero. These figures clearly show how dramatic the influence of the magnetic field can be. The magnetic field and resulting Lorentz force in the $\boldsymbol{E}\times\boldsymbol{B}$ direction produces a macroscopic drift in that direction as well as complex, asymmetric behavior of the longitudinal component of drift velocity.

The transverse diffusion coefficient along the $\boldsymbol{E}\times\boldsymbol{B}$ direction shows some unexpected features. We observe that it has all the features of anomalous diffusion [3] but perhaps the most striking phenomena is the presence of negative transiently diffusivity in the limit of the highest B_0/n_0 of 2000 Hx (1Hx $= 10^{-27}$ Tm^3). These results independently confirm the previous calculations and the reader is referred to [14] for details. Generally speaking, the temporal profiles presented in Fig. 1 (a)-(d) are not predictable from steady-state dc theory. The temporal non-locality in association with the effects of magnetic field result in complex temporal profiles of the electron transport coefficients in rf electric and magnetic fields.

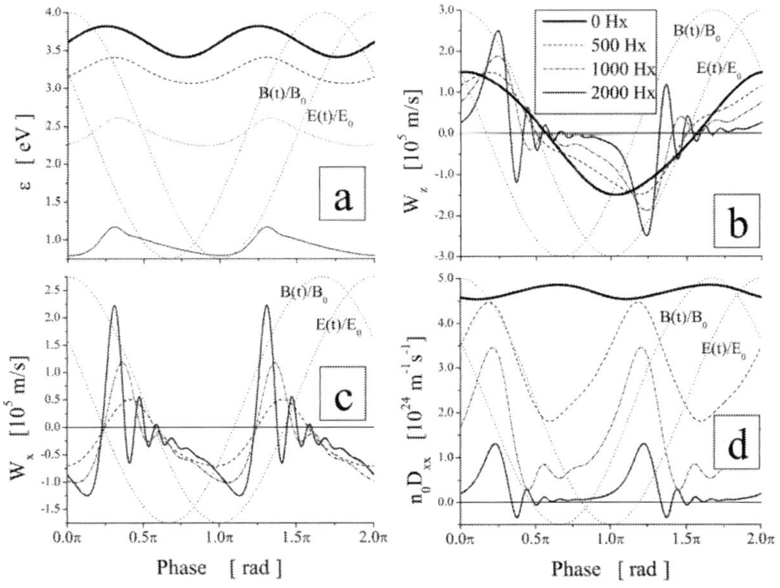

Figure 1. Temporal profiles of (a) mean energy, (b) longitudinal and (c) transverse drift velocity components and (d) diffusion coefficient along the $\boldsymbol{E} \times \boldsymbol{B}$ direction for electrons in CF_4 under the influence of B_0/n_0 for a fixed phase difference of $\pi/3$ rad. The electric field amplitude is set to 100 Td, the field frequency is 100 MHz and $n_0 = 3.54 \times 10^{22}$ m^{-3}.

3.3. The effects of the phase difference and the field orientations

In this section we investigate the influence of varying the phase difference between the fields on the electron transport properties for the ionization model of Lucas and Saelee. In Fig. 2, we display the 3-dimensional plots of the temporal profiles of the mean energy and ionization rate as a function of the phase difference between the fields. In the limit of small phase differences, the magnetic field is large when electric field peaks and electrons cannot gain much energy from the electric field - the magnetic field cools the swarm. Consequently, the mean energy is significantly reduced. As the phase difference increases, the magnetic cooling effects are reduced, particularly in phases where electric field peaks (and magnetic field magnitude is small) and we observe that the modulation amplitude and cycle-averaged value are increased.

In contrast to mean energy, the ionization rate exhibits a specific and unexpected behavior. In the limit of large phase differences the ionization rate is significantly modulated while in the limit of small phase differences this transport quantity is significantly reduced. Note that the mean energy does not follow this trend, which implies that the distribution function (and in particular the tail of the distribution) must be also strongly modulated. Thus the phase difference between the fields has important implications for the efficiency of the ionization processes and other inelastic channels that play vital role in maintaining the plasma.

In Fig. 3 we show the 3-dimensional plots of the temporal profiles of the ionization rate and longitudinal diffusion coefficient as a function of the field orientations. As the angle between the fields increases, both transport quantities monotonically decrease for a fixed phase. While the modulation amplitude of $n_0 D_{zz}$ is increased for an increasing angle between the fields, the ionization rate displays the opposite behavior. In addition, in the limit of perpendicular fields the ionization rate is significantly reduced. One may understand this behavior by considering

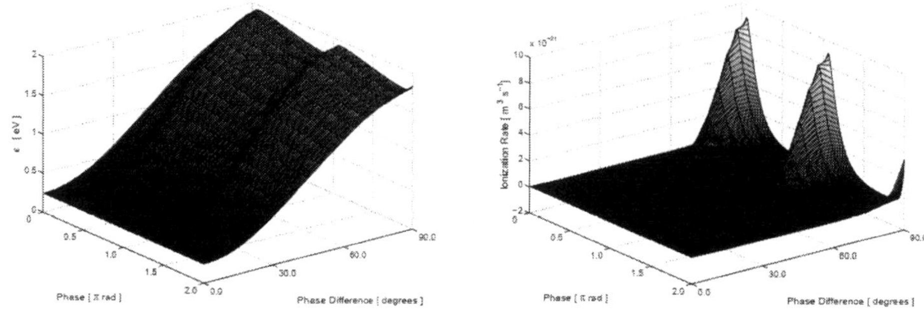

Figure 2. The 3-dimensional plots of the temporal profiles of the mean energy and ionization rate as a function of the phase difference between the fields for the Lucas-Saelee model. The electric and magnetic fields have the following forms $E(t) = 10\cos(\omega t)$ Td and $B(t) = 1000\cos(\omega t + \theta)$ Hx, respectively, where θ is the phase difference while ω/n_0 is the reduced angular frequency of 1×10^{-15} rad m^{-3} s^{-1}.

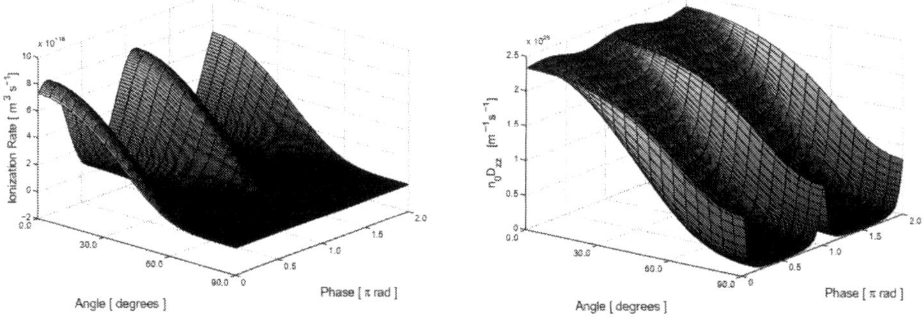

Figure 3. The 3-dimensional plots of the temporal profiles of the mean energy and ionization rate as a function of the angle between the fields for the Lucas-Saelee model. The electric and magnetic fields have the following forms $E(t) = 10\cos(\omega t)$ Td and $B(t) = 1000\sin(\omega t)$ Hx, respectively. The reduced angular frequency ω is set to 1×10^{-15} rad m^{-3} s^{-1}.

the cooling effects associated with the action of an orthogonal component of the magnetic field.

3.4. The effect of the cyclotron resonance on electron transport

One important aspect of this work is that we propose a new additional mechanism for collisional heating in rf electric and magnetic fields caused by the synergism of temporal non-locality and cyclotron resonance effect. The effect of the cyclotron resonance represents the resonant absorption of energy from a radio-frequency or microwave-frequency electromagnetic field by electrons in a uniform dc/rf magnetic field when the frequency of the electromagnetic field equals the cyclotron frequency of the electrons. In Fig. 4 (a) and (b) we demonstrate this effect for the Reid ramp model. The cycle-averaged mean energy is presented as a function of magnetic field amplitude, field frequency and phase difference between the fields. For the field frequency of 50 MHz, the cycle-averaged mean energy is a monotonically decreasing function of magnetic field. However, as the field frequency is increased, the oscillatory-type behavior is clearly evident. The cycle-averaged mean energy is affected by the phase difference between the fields as shown in Fig. 4 (b). For an increasing phase difference, the cycle-averaged mean energy is increased. These figures clearly show that under specific conditions the magnetic field can efficiently pump the energy into the system. The magnetic field amplitude and field frequency can be tuned to exploit/control this behaviour.

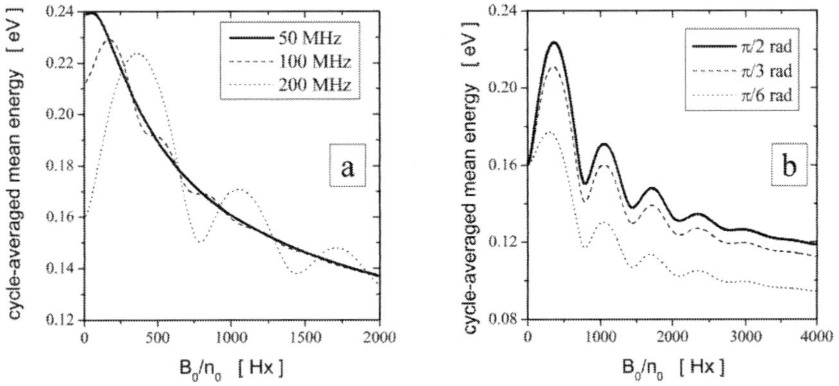

Figure 4. The cycle-averaged mean energy as a function of B/n_0 for various (a) field frequencies, and (b) phase differences for Reid ramp model. The electric field amplitude is set to 14.14 Td, the field frequency is 200 MHz and $n_0 = 3.54 \times 10^{22}$ m^{-3}.

To understand this phenomena we look at the power deposited by the electric field by considering the longitudinal drift velocity shown in Fig. 5. In this figure we can see that increasing magnetic field amplitude decreases the phase delay of the longitudinal drift with respect to the electric field while leaving the drift amplitude relatively unchanged. Consequently the increase in the cycle-averaged mean energy over the range of magnetic field amplitudes 0-400 Hx then follows.

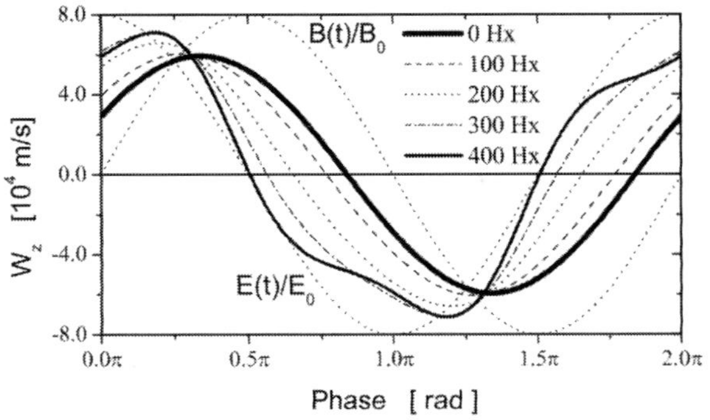

Figure 5. Temporal profiles of the longitudinal drift velocity component as a function of B/n_0 for the Reid ramp model. The electric field amplitude is set to 14.14 Td, the field frequency is 200 MHz and $n_0 = 3.54 \times 10^{22}$ m^{-3}.

4. Conclusion

In this work we have systematically investigated the effects of magnetic field amplitude, field frequency, field orientations and phase difference between the fields on the electron transport properties in rf electric and magnetic fields for a range of model and real gases using a multi term theory for solving the Boltzmann equation. There is a multitude of new phenomena observed which are not predictable from d.c. steady state transport and these results demonstrate the importance of a time-resolved study of electron transport under these conditions. Further to this, and of particular significance in this study, is the ability to suppress/optimize (cycle-averaged) collisional heating under rf electric and magnetic fields through tuning the magnetic field magnitude, frequency and/or phase-difference.

Acknowledgments

The authors would like to thank the support of the Australian Research Council and the Centre for Antimatter-Matter studies. One of the authors (SD) was partly founded by the MNTR project 141025. It is a pleasure to acknowledge helpful discussions with Prof Robert Robson, Dr Kevin Ness and Prof Zoran Petrović.

References

[1] Makabe T and Petrović Z Lj 2006 *Plasma Electronics: Applications in Microelectronic Device Fabrication* (New York: Taylor Francis Group)
[2] White R D, Ness K F and Robson R E 2002 *Appl. Surf. Sci.* **192** 25
[3] Petrović Z Lj, Raspopović Z M, Dujko S and Makabe T 2002 *Appl. Surf. Sci.* **192** 1
[4] Boltzmann L 1872 *Wein. Ber.* **66** 275
[5] Wang-Chang C S, Uhlenbeck G E and DeBoer J 1964 *Studies in Statistical Mechanics* vol 2, ed J DeBoer and G E Uhlenbeck (New York: Wiley) p 241
[6] Robson R E and Ness K F 1986 *Phys. Rev. A* **33** 2068
[7] Dujko S 2008, PhD Thesis, School of Mathematics, Physics and IT, James Cook University
[8] Lucas J and Saelee H T 1975 *J. Phys. D: Appl. Phys.* **8** 640
[9] Robson R E, White R D and Petrović Z Lj 2005 *Rev. Mod. Phys.* **77** 1304
[10] Kitajima T, Takeo T, Petrović Z Lj and Makabe T 2000 *Appl. Phys. Lett* **77** 489
[11] Hioki K, Hirata H, Nakano N, Petrović Z Lj and Makabe T 2000 *J. Vac. Sci. Technol. A* **18** 864
[12] Kurihara M, Petrović Z Lj and Makabe T 2000 *J. Phys. D: Appl. Phys.* **33** 2146
[13] White R D, Dujko S, Ness K F, Robson R E, Raspopović Z Lj and Petrović Z Lj 2007 *Anziam J.* **48** (CTAC2006) C50
[14] White R D, Dujko S, Ness K F, Robson R E, Raspopović Z M and Petrović Z Lj 2008 *J. Phys. D: Appl. Phys.* **41** 025206

The expert systems for analysis of electron energy-loss spectra

Ivan Petrović, Valerij Bočvarski and Violeta Petrović

Department of Physics, Faculty of Science, Kragujevac, Serbia

violeta.petrovickg@gmail.com

Abstract. In this paper is presented development of the expert systems for threshold spectra analysis, starting from firstly developed expert system for atom threshold spectra analysis, up to the expert system for analysis more complex molecule threshold spectra (firstly two-atomic and then tree-atomic molecules). The goal of the expert systems application is to automate, speed up and improve process of spectra analysis. Interest for threshold spectra originates from the fact that, by studding them, it is possible obtained information about processes which happened on energies near reaction threshold. Also, in these spectra can be seen peaks that correspond to forbidden transitions and which can't be seen in the optical spectroscopy. For testing of the expert systems, we used the threshold spectra for the He atom, and N_2 and SO_2 molecules. All these spectra are obtained on threshold spectrometer in Institute of Physics in Zemun (Laboratory for Atomic Collision Processes).

1. Introduction

In process of studying atomic and molecular collision processes, the threshold spectroscopy method takes important place [1]. With this method is performed a detection of electrons with "zero" energy, i.e. electrons which have lost their incident energy in the non-elastic collisions. Because of that, by using a specific configuration of electric field, it is possible to separate aforementioned electrons from the others. With this is accomplished that a process is examined on reaction threshold, hence the name of this method. The importance of this method originates from the fact that it enables a detection of excited as well as vibrational states, along with a recognizing of resonant states.

Expert systems presented in this paper are developed based on the backward chaining methodology and with a goal to analyze spectra obtained by threshold spectroscopy. For their development is used **ESBT** I2+ which in process of conclusion uses the aforementioned methodology that starts with goal and goes toward data. One of advantages of this methodology is that there is no conflict resolution, which can be complicated factor in process of practical exploitation of one expert system.

In the case of threshold spectra analysis, which can be generalized on all kind of spectra in electron spectroscopy, an application of ES enables the following: an *estimation* of usability of experimentally obtained spectra for future analysis, *selective extraction* of physical events from events that originate from noise, *identification of an energy states* of atoms and molecules whose spectra are recorded, and *analysis* of identified states, with an aim to separate the vibrational levels and resonances.

2. Experiment

Characteristic threshold spectra, which are used in process of ES development and testing, are obtained with threshold spectrometer in Institute of Physics in Zemun (Laboratory for Atomic

Collision Processes), by using a modified method of Cvejanović and Read [2]. Position of electrodes in this method is schematically presented on the Fig 1.

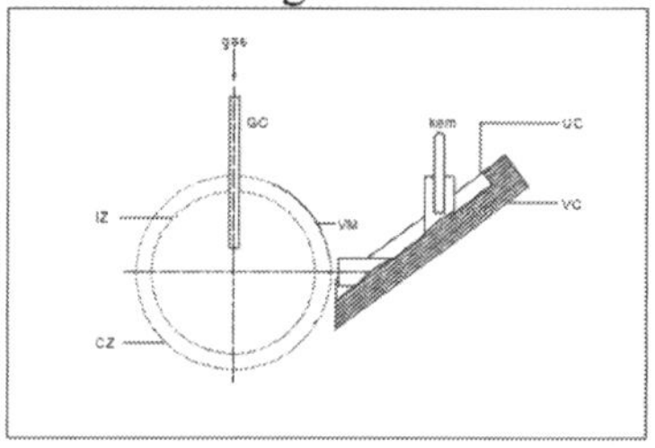

Figure 1: Position of electrodes in modified method of Cvejanović and Read [1]

The essence of this method is a controlled penetration by outer electric field inside the area where the collision is performed between electron and gas beams. This electric field creates, in the center of interaction area, a potential which enables the extradition of electrons with zero energies.

Yield of detected electrons is shown in form of peaks that correspond to excitation of discrete states of atoms and molecules.

Recorded spectra contain the information about channel, i.e. energy, corresponding count, and also information about channel energy value that depends on, chosen in advance, a voltage scanning range on the monochromator plates. In that way, it is possible to assign exact energy value to each channel, so that the spectrum can be characterized as energetic (Fig. 4).

3. Expert system for analysis atom threshold spectra

First step in the process of spectrum analysis is inputting spectra from multi-channel analyzer, and their translation in the tables of appropriate format.

On Figure 2 is shown a characteristic threshold spectrum of He atom [3].

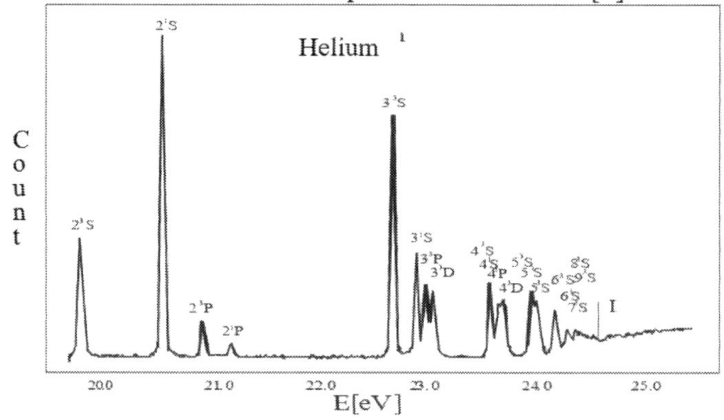

Figure 2: The threshold He spectrum

Beside tables with recorded spectra, it was necessary to form the databases of energy levels of electron, as well as vibrational and resonance states, and also a tables which contain data modified by using Pascal programs.

Identification of peaks of states is first phase in working of ES, which encapsulates the following steps in process of spectra analysis: inputting data, finding and extracting peaks, ordering peaks in non-ascending array, energy scale calibration, resolution estimation, peaks classification and identification, and writing of data obtained by this analysis in the database.

Input of necessary, starting data is performed by interactive communication between ES and user. Those data are primarily used to direct ES on atom i.e. molecule whose spectra will be analyzed. Also, the ES obtains information about energy range on which the spectrum was recorded, and also the names of tables which are necessary for its operation.

After the data input, the program will start with spectrum analysis. First step is extraction of all channels with peaks. However, experimentator's experience says that the big part of extracted peaks originates from noise, so the goal is removing this noise influence in the biggest possible measure. Because of that, the ES sorts all peaks in descending array. Reason for this lies in assumption that the most intensive peak originates from a discrete energy level, and not from noise. Based on average value of exactly determined number of peaks, for example ten biggest and ten smallest in spectrum that is analyzed, ES calculates a ratio between signal and noise. The criterion for peak to be analyzed is that the reciprocal value of relation - ***examined peak/biggest peak*** - is greater then the ratio between signal and noise. Peaks that satisfy this criterion then form a new table with peaks which will be analyzed. In this way the number of peaks that are interesting for analysis is reduced, and because of that, also necessary time for analysis.

Calibration of energy scale is next and very important step in the process of spectrum analysis, because the good calibration is necessary precondition for well-done energy positioning of peaks, and in the further work is a precondition for correct identification. Depending of experimental condition, the calibration can be done on two ways.

First way is based on inputting in system, at the same time, atom whose spectra is to be analyzed, and some characteristic gas. It is important that the gas has clearly separated and well known state from optical spectroscopy which enables an exact determination of his energy position.

Second way of calibration is based on some clearly defined structure (peak) that exists in the analyzed spectrum. Such structure (peak) is marked as "characteristic peak".

After the calibration of the energy sale, next step is an estimation of equipment resolution. Resolution (or the equipment power of decomposition) determines the least energy distance on which two near peaks can be seen separately, i.e. on which two peaks can be separated. If the energy value between two peaks is less then value of resolution, then those two peaks will be presented as one wider peak which will correspond to group of states.

"Characteristic peak" is also used for determination of parameters of referent peak which will have very important role in the further process of analysis.

The ending part of analysis in this phase is classification and identification of peaks based on the experimental data and table of energy level for atom or molecule which is used in experiment. Based on the performed calibration and calculated energy value of a channel, the ES calculates a corresponding energy value for each peak separately. This value is compared with possible energy values for states of given atom (or molecule). If ES determines presence of such a state in the resolution range, the peak obtains its classification number [3], [4], [5]. In order for peak to be classified as a state it is necessary that the energy deviation of examined peak compared with energy of corresponding state lies within range of maximal measurement error, which is determined by experimentator based on his work experience with experimental equipment [6]. The user inputs this value on the starting point of an analyzing process.

In the atom threshold spectra it is possible recognize a resonant influence on certain states. In groups, determined by value of their principal quantum number, the peaks form decreasing series. Resonant increment, which is consequence of emptying of resonance, results in deviation from this dependence (see Fig. 1).

4. Expert system for analysis two – atomic molecule threshold spectra

Although atom and molecule spectra are by their form discrete, the important difference between them originates from presence of vibrational levels in molecule spectra. Analysis is additionally complicated because of peak overlapping on the almost all energy ranges, which leads to forming of "compound structures" and also "shoulders" on the peaks of more intense states.

In the two-atomic molecule spectra, this is especially expressed (characteristic) on higher energies, while in the tri-atomic and poly-atomic molecule spectra this situation is present on whole energy range.

ES which is developed with a goal to perform an analysis of those spectra should also enable an identification of molecule vibrational energy levels. Problem arises in the area of higher energy values, where peaks from vibrational levels (that originate from different states) are overlapping and make aforementioned "compound structures" (Figure 3).

On Fig. 3 is given characteristic N_2 molecule threshold spectrum [7].

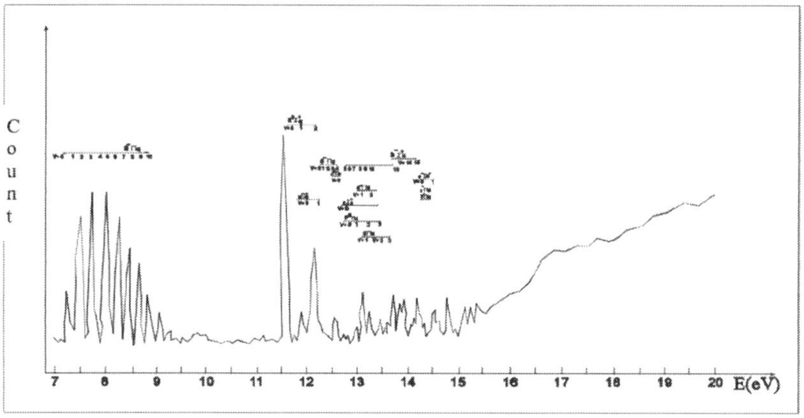

Figure 3: Characteristic N_2 molecule threshold spectrum

Also, in this phase of development of ES, it was necessary to examine is it possible to reproduce a form of real spectrum, based on knowledge of certain parameters.

For testing is chosen a molecule of N_2, because of the practical reasons. Namely, the spectrum of this molecule is the best covered by data. This fact enables us to verify does logic that was used for ES development give satisfactory results. On the Fig. 5 is given a block-diagram of ES which performs an analysis of N_2 threshold spectra.

Upgrading of existing ES was meant to solve the problems of presence of superimposed peaks on the peak that correspond to the energy state or vibrational level. They also cause a disturbance of symmetrical form of peak by creating of "shoulders" that appear on higher energy side of peak.

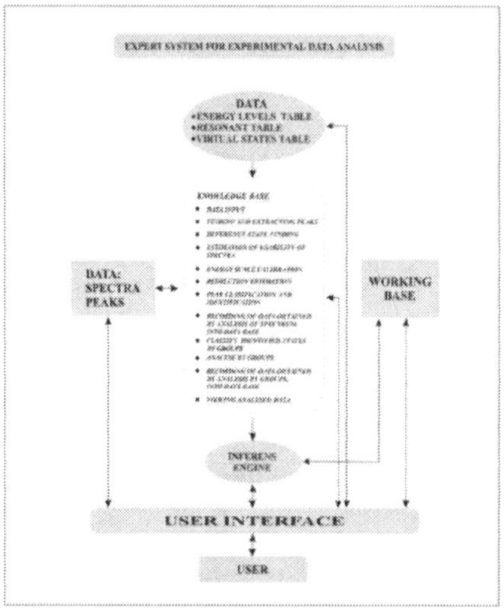

Figure 4: Block diagram of expert system for analysis N_2 molecule threshold spectra

In order for aforementioned two problems to be solved, two assumptions were made: first, that the peaks of states and vibrational levels have normal distribution, i.e. they are in Gaussian form, and second, that vibrational levels of certain state are on almost identical energy distance. As the energy value of channel is known, ΔE, based on the knowledge of characteristic energy value, Ev, it is possible, for each state, to determine a characteristic number n of channels between vibrational levels like a ratio between those two values:

$$n = \frac{E_v}{\Delta E} \qquad (1)$$

For starters, the ES loads a channel on which, in given part of spectrum, is identified a state with lowest energy value. After that, using parameters determined by characteristic peak, ES generates Gaussian for each state (or vibrational level) separately.

The values of counts for corresponding state calculated in this way are put in table with spectrum, inside the field COUNT_G [3]. In the field named DEVIATION, the ES writes the values which are defined by deviation between calculated and recorded values. For further analysis are interesting only the channels whose corresponding count in colon DEVIATION satisfy condition to be classified as peak. ES using, for this purpose made Pascal programs, marks all aforementioned channels and in the field ORIGIN puts the mark P. In this way, ES is concentrated only on those channels. The goal is to find a cause for peak existence on aforementioned channels. Corresponding energy is calculated by following formula:

$$E = E_0 + \Delta E * n, \qquad (2)$$

where E0 is energy of state or vibrational level that has been analyzed, dE is energy value on channel and n is difference between channel on which is a peak interesting for analysis, and channel on which is a state or vibrational level.

When the energy value has been calculated, ES can start with determination of examined peak nature. In this phase of work are necessary the tables of energy levels. By comparing the calculated energy value with energy values within the tables of vibrational levels and resonances, ES determines the origin of peaks, and in the case of positive identification, marks the peaks with an identification number 2 or 3. This number is written in field named ORIGIN. If in no table exists the appropriate energy value, ES concludes that the peak originates from noise, and gives it a classification number 4. Identical procedure is repeated for all peaks which are situated on channels within the range of state peak width, i.e. vibrational level.

After the peaks analysis is completed, ES calculates channel on which the vibrational level should appear, by following formula

$$x = x_0 + k * n, \quad k = 1, \ldots\ldots, v \qquad (3)$$

where x0 is a channel with state, k is integer multiplier that represents the number of vibrational level that belongs to one state, n is characteristically number of channels between two vibrational levels of given state. Quantity x that is calculated determines the serial number of channel on which ES positions itself in order to continue further analysis.

When the value of channel on which is expected appearance of vibrational level is calculated, first step is to check does a peak exist on a channel with serial number that corresponds to the value of variable x. If the result of checking is positive, ES positions itself on that channel, and generates peak in Gaussian form, and the procedure is repeated in the same way as in the case of a previously described state. When ES finishes with analysis of vibrational levels of one state, it starts with next. Aforementioned procedure is repeated until the analysis does not encapsulate all state and vibrational level peaks in the recorded energy range.

5. Expert system for analysis tri – atomic molecule threshold spectra

The threshold spectra of tri-atomic molecules are more complex than spectra of two-atomic molecules. "Compound structures" are characteristic for all energy ranges so that there are no separated and clearly defined structures.

On Fig. 5 is given characteristic part of SO_2 molecule spectrum [5].

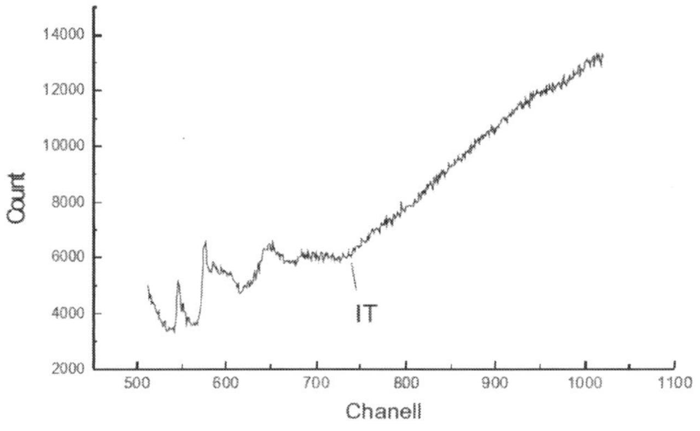

Figure 5: The part of threshold spectrum of SO_2 molecule (9 eV - 12.5 eV)

Problem in the process of analysis arises from the fact that only the parts of spectrum which are well covered by data can be analyzed (if such structures exist at all).

Next phase in developing process of this ES should give an answer on question is the same logic can be applied while analyzing the parts of spectra of polyatomic molecules with, for them, characteristic structures.

In order to obtain an answer to this question it was necessary to upgrade already existing ES with goal to show if ES can generate, based on minimal quantity of data, a spectrum which corresponds to experimentally obtained spectrum.

On the start of this phase, the user must input, in the database, an approximate number of vibrational levels. This data ES will use in process of spectrum generating if the exact number of vibrational levels isn't known for given state. On the beginning ES positions itself on the channel with lowest energy value in the given energy range. This channel is noted as "central". Generating of peaks is performed by I2+ and Pascal programs. When generating of peak of state is finished, ES begins with generating of peaks of vibrational levels on the same principle, moving itself on the channels where vibrational levels are situated (see formula 3).

However, what can be done in the case when the distance between channels isn't exactly known, and how ES can determine a value for moving. There are two ways for solving of this problem: first is moving for number of channel which corresponds to average value of known energy distances between vibrational levels, and second is initial movement for number of channels which correspond to energy difference between state and first vibrational level. More practical is the second way because then is possible, on the beginning, to obtain at least one vibrational level that corresponds to real position. When generating of vibrational levels is finished, the spectrum looks like one on Fig. 6-a.

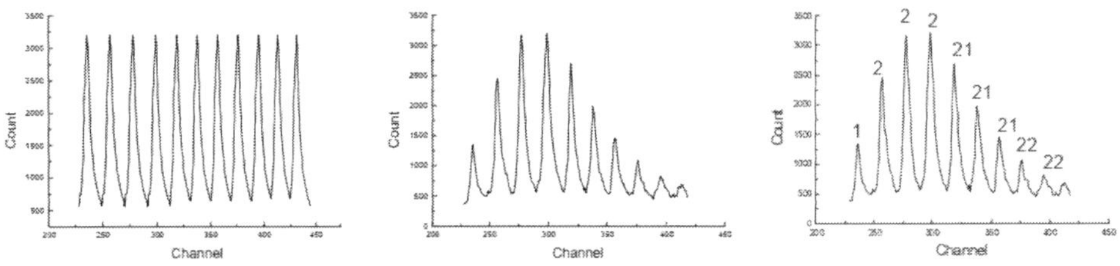

Figure 6: Generating vibrational levels a), b) and c)

Next step is correction of peaks height [4]. After this part of ES work, the spectrum looks like one Fig. 6-b. The spectrum obtained in this way resembles a real spectrum and that was the goal anyway. However, now it is necessary to check a position of peaks in order to see which are on their right place and if needed to make a movement. In this part of ES work, the peaks obtained the following marks: 2 - peaks which are on right place from the start, and 21, 22 and 23 - peaks which will be positioned on right place after first, second or third pass.

Table 1: Table after performed analysis

Channel	Count	Count_L	Deviation	Peak	Peak type	Pass
:	:	:	:	:	:	:
574	6546	6546	1000	0	1	
:	:	:	:	:	:	:
580	5480	4829	1651	0		
581	5535	5535	1000	0	2	22
582	5454	4829	1625	0		
583	5758	5115	1643	0		
584	5859	5859	1000	0	2	2
585	5850	5115	735	0		
586	5673	3754	1919	1	5	
587	5596	4941	345	0		
588	5660	5660	1000	0	2	21
589	5607	4941	666	0		
590	5394	3627	1767	0		
591	5415	3623	1792	1	5	
592	5475	4936	593	0		
593	5654	5654	1000	0	2	2
594	5542	4936	606	0		
595	5546	3623	1923	1	5	
596	5543	4851	692	0		
597	5556	5556	1000	0	2	21
:	:	:	:	:	:	:
733	5992	3938		0	6	
:	:	:	:	:	:	:

On the end of this phase, the obtained spectrum looks like the one on Fig. 6-c.

Also, by inputting of new peaks between existing ones, it need to be shown that is possible to obtain a spectrum which in greatest measure reproduces a look of real spectrum in the areas where there are more complicated structures as is case with tri-atomic molecule spectra, that are a specially interesting for analysis. Goal is to show that complex structures in those spectra can be interpreted like a sum of larger number of peaks which originate from vibrational levels of certain state.

Idea was to put one peak between every two peaks. Channel on which this peak will be positioned is calculated like arithmetic mean of channels between which the aforementioned peak was put.

Count of new peak is calculated by same principle.

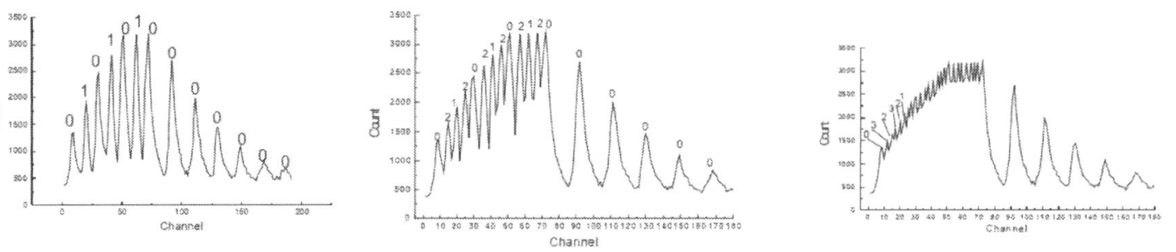

Figure 7: Peaks of first, second and third generation a), b) and c)

Additional peaks - peaks first generation - have mark "1" on Fig. 7-a.

By comparing the spectrum obtained in this phase with the part of tri-atomic molecule spectrum of SO_2 (see Fig. 4), it can be seen that there exists more space for inputting of new peaks. Because of that, ES repeats aforementioned method twice and puts the peaks of second and third generation between existing peaks.

ES developed in this way enables detailed and very exact analysis of electron spectra of tri-atomic and poly-atomic molecules. Testing of aforementioned ES is performed on SO_2 molecule threshold spectra.

6. Conclusion

Application of ES in analysis process of threshold spectra gave expected results. Quantitative and qualitative improvement of analysis process has shown a complete justification of possible future work on the development of ES which would find their usability in processes of analysis of spectra that are obtained in characteristic processes in other areas of physics.

Acknowledgments

We are grateful to the Serbian Ministry of Science and Technological Development for financial support through Project 141011.

References

[1] Jureta J 1978 Phd Thesis University of Belgrade 1-109

[2] Cvejanović S and Read F H 1974 J. Phys. B: Atom. Molec. Phys. **7** 1180-1193

[3] Petrović V and Bočvarski V 2003 Int. J. Mod. Phys. C **14** 433-440

[4] Petrović V and Bočvarski V 2005 Int. J. Mod. Phys. **16** Nb **9** 1395-1407

[5] Petrović V, Petrović I and Bočvarski V 2007 Int. J. Mod. Phys. **18** Nb **7** 1133-1148

[6] Cvjetković V, Bočvarski V and Radenković B 1998 Expert System with Applications **14** 274-282

[7] Bočvarski V 1981 Master Thesis University of Belgrade 1-65

Development, diagnostic and applications of radio-frequency plasma reactor

N. Puač

Institute of Physics, Pregrevica 118, 11080 Belgrade, Serbia

nevena@phy.bg.ac.yu

Abstract. In this paper we present volt-ampere-power characteristics of a plasma needle operating in the flow of helium at the atmospheric pressure. Also, we will present discharge images obtained by using ICCD camera. In addition, we show some examples of how such plasma affects plant tissues. In characterization of the plasma needle, current and voltage waveforms were recorded by two derivative probes. These two probes are similar to the probes previously used in measuring transmitted power in low pressure CCP rf discharge [1]. The instantaneous power was calculated from current and voltage waveforms and *U-I* characteristics of the discharge were determined. Regimes of operation with and without the grounding ring at the tip of the needle were considered. We have chosen two model systems to study the effect of plasma needle on plant cells and tissues: sweet fern gametophyte (prothallus) and calli produced *in vitro*.

1. Introduction

Nonequilibrium plasmas proved to be able to produce chemically reactive species at a low gas temperature while maintaining highly uniform reaction rates over relatively large areas [2, 3]. The fact that vacuum systems are necessary equipment is just one part of the problem as many targets cannot be treated at all under the low pressure conditions. Treatment of organic materials and living tissues falls under that category.

Lately, non-thermal atmospheric pressure plasmas have drawn considerable attention due to their enormous potential for technological applications mostly in surface modifications especially for materials or living tissues that may not allow putting it into vacuum chambers. It was found that plasmas may be easily operated under atmospheric pressure at RF frequencies in a flow of rare gases which mix with the atmospheric gases in the region of the discharge [4]. Apart from this, many different configurations and applications of non-thermal atmospheric plasmas have been developed and studied [5].

The effects of these plasmas on living cells and tissues have been studied on numerous occasions in recent literature. Dependent on plasma conditions, several refined cell responses are induced in mammalian cells [6]. Furthermore, application of plasma treatment in dentistry and deactivation of bacteria have been intensively studied [7]. However, there are few data in the literature about plasma effects on plant cells and tissues [8]. So far, effect of low pressure plasmas on seeds was investigated. It was shown that short duration pretreatments by non equilibrium low temperature air plasma were stimulative in light induced germination of Paulownia tomentosa seeds [1, 9]. As membranes of plants have different properties to those of animals and as they show a wide range of properties, we have tried to survey some of the effects of typical plasma which is envisaged to be used in biotechnological applications on plant cells.

In this paper we will also present properties of our plasma source that is very similar to the plasma needle as first described by Stoffels and colleagues [10]. This source operates at atmospheric pressure and meets all the necessary conditions for treatment of organic materials and living tissues. It is desired that the plasma should be non-aggressive, local, with small penetration depth and, at the same time, that it produces chemically active species at a low gas temperature. Plasma needle meets all those requirements.

For the purpose of successful treatment and comparison of different samples by plasma needle it is necessary to characterize the plasma itself the best way possible. The standard parameters for treatment of samples are duration of treatment, power transmitted to the plasma and distance of the sample to the tip of the needle. For the power measurements the derivative probes were used. The percentage of the total power and mean power distributed to plasma itself was determined from the recorded voltage and current waveforms. ICCD camera was used in order to obtain light emission from the discharge.

2. Experimental set-up

Plasma needle is atmospheric pressure plasma source powered by 13.56 MHz RF generator that operates in a mixture of air and helium. Its construction makes it very useful for treating biological samples like plant tissue, for bloodless surgery, dental cavity antibacterial treatment and many other applications. The needle consist of central tungsten wire (0.5 mm in diameter) placed into a ceramic tube with slightly larger diameter and both placed into the glass tube with 6 mm diameter (see figure 1).

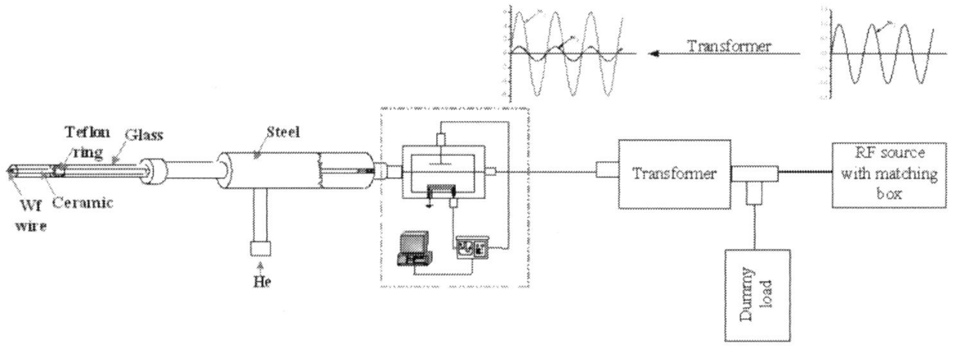

Figure 1. Schematic representation of the plasma needle setup: body of the plasma needle, stainless steel box with derivative probes, home-made voltage transformer, power divider (dummy load) and RF source with matching box.

In addition to the standard configuration (see figure 1) of plasma needle we have used configuration with grounded copper ring placed at the tip of the glass tube. The central wire represents the powered electrode and the grounded electrode is the surface in the vicinity of the plasma needle tip. Helium flows between ceramic and glass tube with typical flow rate of few hundred standard cubic centimeters per minute.

The rest of the electrical circuit is the same that we have already used [8] and is shown in figure 1. Low-temperature RF discharge at 13.56 MHz is generated using a Dressler Caesar 1010 power supply, in combination with Variomatch matching network. In order to increase the peak-to-peak voltage, we have used a home-made transformer and inserted it between the plasma needle and the RF matching network. We have implemented a dummy load to make the discharge more stable and to be sure that only a small part of the power supplied by the RF power supply is transmitted to the discharge. In this configuration the dummy load serves as a 'power divider'.

Instantaneous voltage and current are monitored using two derivative probes developed previously [1, 8] and they are somewhat different from the probes already proposed in the literature. Both probes were placed inside a stainless steel box opposite to each other. The box was placed as close as possible to the plasma needle. The output of the probes was connected to a digital oscilloscope (Tektronix TDS220) with the cables of equal length. All waveforms were collected by the computer for further manipulation.

In order to record emission intensity of the discharge we have used ICCD camera Andor iStar DH720-18U-0. Camera was positioned orthogonally to the needle tip so we could obtain side-on images of plasma. Images were obtained in integrated mode for exposure times of 5 ms and 100 ms. In case of ICCD and U-I diagnostics target was rubber box. Distances between tip of the plasma needle and target were 1 mm.

2.1. Plant material and methods

Plasma treatment of plant tissues is demonstrated on two months old prothalli of Polypodium vulgare L. (fam. Polypodiaceae). The mature fern gametophyte (prothalium) is a multicellular, autotrophic yet small (1-2 mm) structure. As an experimental organism, the fern gametophyte is ideal for a study given that it can be cultured to maturity in a Petri dish and all aspects of its growth and development may be observed and manipulated in a nondestructive way [11].

The plant samples were directly exposed to the plasma needle. The duration of treatment and power of the plasma varied from 15 s to 2 min and 0.01 W to 0.25 W, respectively (of the power supplied to the plasma). Just after plasma treatment, plants were submerged in 1 ml of 0.1% aqueous Evans blue for 1 hour. Evans blue stain was used for microscopic determination of cell death. This led to nonpermeating or exclusion dye leak through ruptured membranes and stained the content of the dead cells [12]. Stained plants were observed using light microscopy.

In order to determine the plasma influence on the fresh weight of the plant tissue, we have used parts of the calli Fritillaria imperialis (fam. Lilliaceae) (about 5 mm diameter in size). It was grown on Murashige and Skoog (MS)-solid medium [13]. Callus is a distinctively organized mass of proliferating cells, with specific morphology and anatomy, and may be obtained from almost any type of plant.

3. Results and discussion

3.1. Electrical measurements

Plasma was generated for a range of powers of the RF power supply from 30W to 100W. Derivative probes were used to measure current and voltage waveforms. After collecting, current and voltage signals were transferred from time domain to frequency domain by using Fast Fourier Transform. While in frequency domain signals were adjusted according to the calibration curves for both probes. Calibration of the probes gives their sensitivity and has been determined by using a network analyzer.

Transition to the frequency domain is performed for several reasons. The first one is that in the frequency domain we can clearly see from the harmonics some basic properties of the discharge. The second reason is that all calculations could be performed more efficiently in the frequency domain. It is important to note that only a single harmonic was found to be significant for both configurations (with and without grounded copper ring) which give an indication of the symmetry of the plasma and its regime of operation.

Current and voltage waveforms were, after adjusting according to the calibration curves, transferred back to the time domain by using an Inverse Fast Fourier Transform. Current peak-to-peak values lie between 0.5-1.2 A and for voltage these values are between 300-600 V (see figure 2). Above the breakdown, the voltage changed very little with increasing the power given by the generator, but the current (and consequently the power consumed by the plasma discharge) changed almost linearly.

One should note that here we have not subtracted the reactive component of the current. The current observed in figure 2 is mainly the reactive component due to the capacitance of the cable between the probes and the tip of the needle (the distance is 30 cm). The reactive component is much greater than the active component (the capacitance of 18 pF corresponds to an impedance of about 670 Ω versus the impedance of the plasma of about 10 kΩ). We have not subtracted the reactive component of the current (although it is doable) because its influence is cancelled out in determination of the integral of the power.

After returning voltage and current signals to the time domain we were able to obtain instantaneous power (see figure 3). Mean values of power were also calculated by using 6T ($T=7.375e10^{-8}$ s) and they are shown in figure 4. By using a 'power divider' we have insured that most of the power feed goes to the dummy load. Less than 10% of power given by RF power supply went to the plasma needle branch and only less than 1% into the discharge. We can see that more power is transmitted to the plasma needle in the case when we added grounded ring at the tip and, also, discharge could be ignited for lower power given by RF power supply.

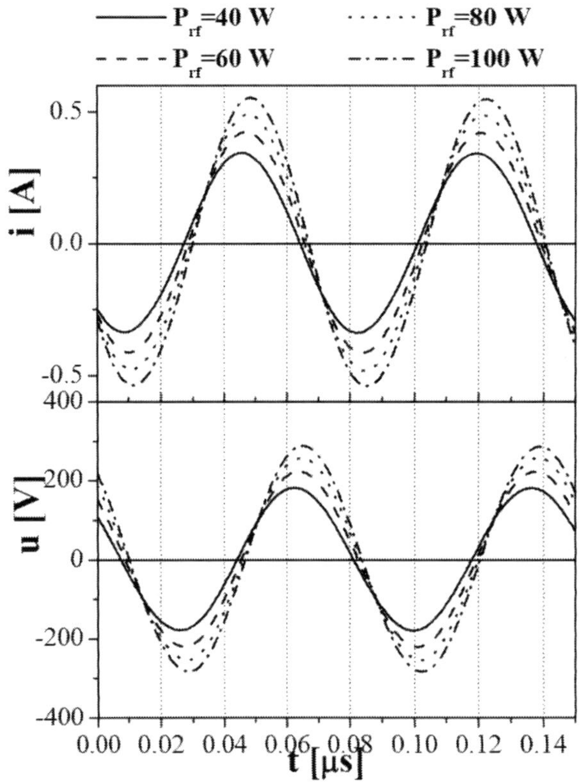

Figure 2. Current and voltage waveforms after processing with FFT procedure and corrected according to the calibration of the probes. Note that the reactive current has not been subtracted from the current waveforms [8].

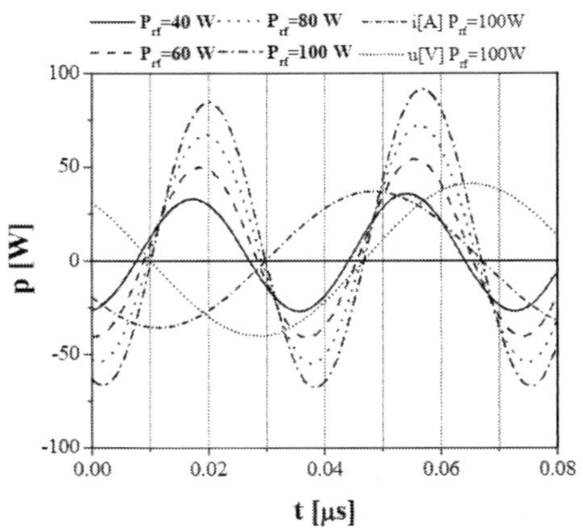

Figure 3. Instantaneous power obtained by multiplying current and voltage waveforms.

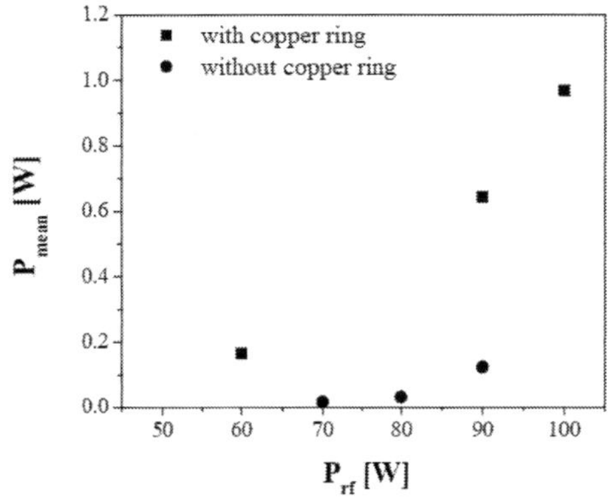

Figure 4. Mean power as function of power given by RF power supply.

In order to treat sensitive samples of biological origin power transmitted to the discharge should be less that 1 W and both configurations, with and without grounded ring, which we have used satisfy this condition. In all our treatments we have used configuration without grounded ring and power transmitted to the discharge did not exceeded 0.5 W.

3.2. Light emission

Plasma needle can operate in two different modes [14, 15]. This was also confirmed in simulations done by Sakiyama and coworkers [16, 17]. In both cases plasma appears as a glow on the tip of the needle. It can work in unipolar mode, when there is no surface in the vicinity of the tip of the needle and discharge is localized at the tip of the needle. If it is brought in the vicinity of surface of the treated sample, plasma switches to a bipolar mode, where the glow spreads over the object's surface.

We have used ICCD camera in order to record integrated light emission from the discharge. In figure 5 light emissions of the discharge for the different powers given by RF power supply are shown. Exposure time was 5 ms, flow of helium was 110 sccm and configuration with the grounded ring was used. Plasma is ignited for the power of 60 W given by RF power supply and intensity of the discharge rises with the increase of the power. When we increase power above 80 W we can see significant increase of the light emission. This increase of the light emission is in correspondence with the increase of the power transmitted to the plasma needle (see figure 5, with grounded ring). Also, sheath is formed at the treated surface and it can be observed with bare eye.

We can see from figure 5 that formed discharge is not completely symmetrical and it has shape of the bean. This indicates that tip of the needle is not completely flat and that even that smallest irregularity of the needle surface influences the shape of the discharge.

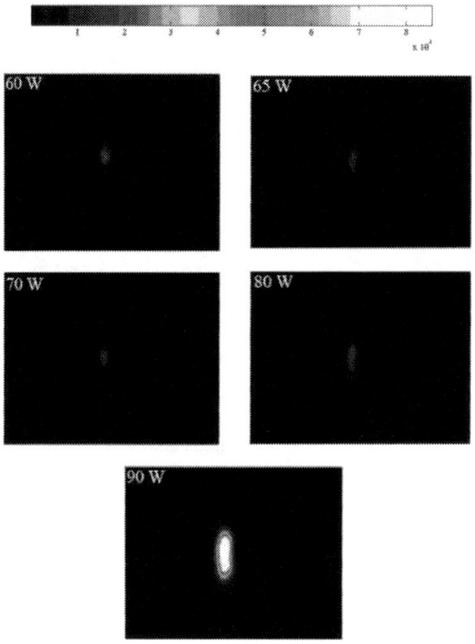

Figure 5. Discharge emission obtained by ICCD camera. Exposure time was 5 ms and distance between needle tip and target was 1 mm.

4. Plasma treatment of plant tissues

We have chosen two model systems to study the effect of plasma needle on plant cells and tissues: sweet fern gametophyte (prothallus) and calli produced *in vitro*. These systems are chosen because their characteristics are determined in great detail. Both systems were treated with plasma needle for different powers (less than 1 W), different treatment times (15 to 120 s) and flow of working gas of 110 sccm. Flow of working gas was kept constant for all treatments.

We have demonstrated that cell death of the Polypodium prothalium occurred after a high dose of plasma (t=120 s). The untreated control plant is shown in figure 6 along with negative control plants treated with absolute ethyl alcohol for 12 h (inserts (*a*) and (*b*)). One should be careful when choosing right conditions for treatment of sensitive samples. In figure 7 samples treated for two different treatment times (P_{rf}=80 W) are shown. One can observe that smaller sample, in not so advanced growth stadium, was completely destroyed while the bigger sample is only partially damaged.

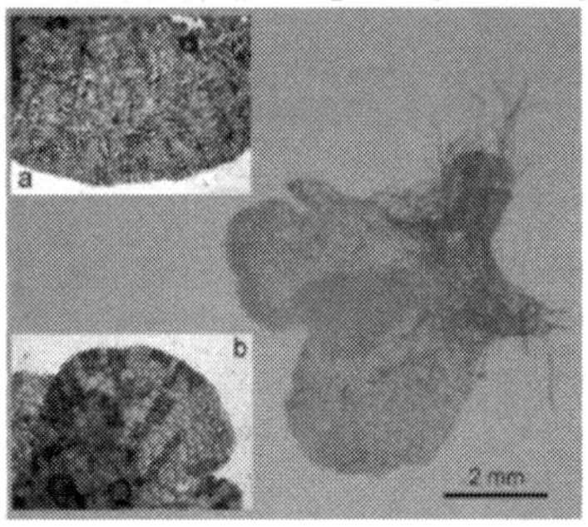

Figure 6. Untreated (control) plant. Inserts (a, b): negative control plants treated with absolute ethyl alcohol for 12 hours. [8]

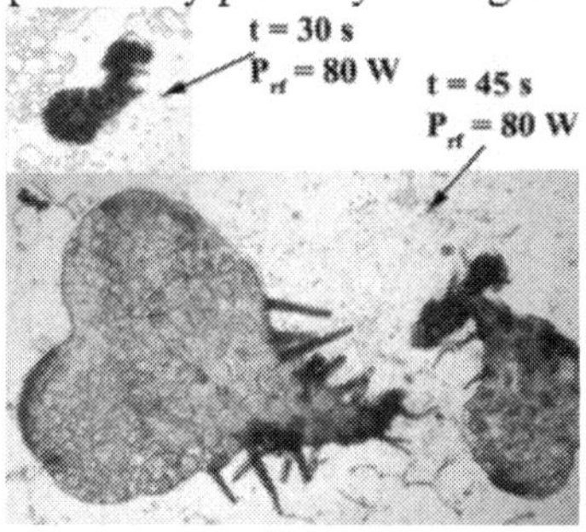

Figure 7. Plasma treatment of sweet fern gametophyte. Smaller picture: one month old gametophyte.

When sweet fern prothalii were exposed over shorter periods to the plasma needle, with appropriate power, cell injury and necrosis were minimized and plants continued their growth. Detailed cytological and histological analysis of the referred gametophyte could elucidate the delicate alteration (if any) induced by the plasma treatment. Again, in order to determine the plasma influence on the fresh weight of the plant tissue, the plant samples were directly exposed to plasma. The time and power of the plasma varied from 15 s to 2 min and 0.1 W to 0.2 W, respectively (of the power supplied to the plasma). After the plasma treatment, calli were transferred to solid medium and fresh weight of the samples was measured every 7 days.

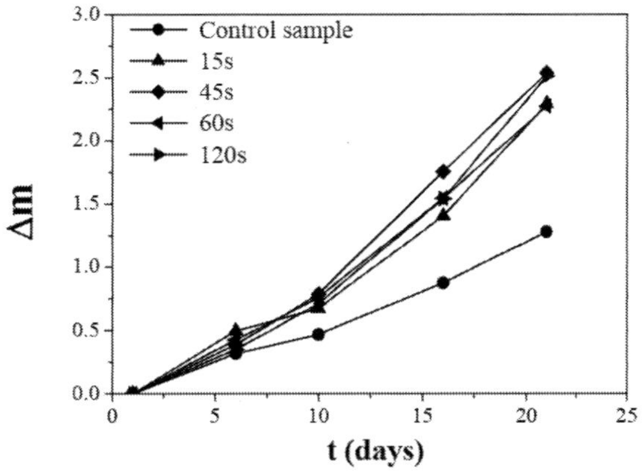

Figure 8. Increase of the fresh weight (FW) of *Fritillaria imperialis* calli as a function of plasma treatment. The calli (about 5 mm diameter in size) were exposed to plasma for the indicated time, and then transferred to the fresh medium. Fresh weight increase was recorded for three weeks and is given by (FW_f–FW_0) / FW_0 ratio. FW_f –fresh weight measured at the indicated time (x-axes); FW_0 – initial fresh weight. [8]

Plasma needle treatment causes an increase in the fresh weight of compact calli (figure 8). Moreover, values of the measured parameter were doubled for all treated calli compared with the untreated sample, even for longer exposure times. Increase in the fresh weight is an obvious implication of calli growth—an irreversible increase in size, accomplished by a combination of cell division and cell enlargement.

5. Conclusion

We have developed a plasma needle similar to that of Stoffels and coworkers [6, 10]. The power supply in our case is somewhat different, with a transformer and a dummy load. In order to obtain voltage-current-power characteristics we have added voltage and current probes. Typically we have achieved a transfer of power below 1 W to the plasma needle which is the condition to reduce thermal transfer to the surrounding gas and target living tissue. We have analyzed operation of plasma needle for two different configurations, with and without grounded copper ring placed at the tip of the needle. It seems that in case with additional ring more power is transmitted to the discharge and discharge is more stable, but for the effect of the treatment the proximity of the target surface will remain as a critical parameter. Apart from derivative probes we have used ICCD camera in order to obtain light emission of the discharge created at the tip of the needle.

Concerning plasma treatment of the plant tissues, we have analyzed interaction of the plasma needle with gametophytes and calli as representatives of small multicellular plant organisms that may be grown and analyzed in a controlled fashion. We have demonstrated that cell death (necrosis) of the

Polypodium prothalium occurred after high doses (power-treatment time) of plasma. As for calli, it was shown that plasma treatment induces the weight gain for monocotyledonous. Further histological and cytological analysis is required to elucidate the mechanisms of interaction, but one may expect that the effects are triggered by abundant radicals and reactive molecules produced by the plasma needle [14].

Acknowledgments
This work was performed as a part of requirement for a PhD thesis which was completed under supervision of Dr. Zoran Lj. Petrović. Also I would like to express gratitude to my co-supervisor Dr. Gordana Malović; Dr. Antonije Đorđević, Saša Lazović and colleagues from Biological institute "Siniša Stanković". This research has been supported by the Ministry of Science Serbia under the contract number 141025 of the project Physics Fundamentals of Applications of Non-Equilibrium Plasmas in Nanotechnologies and Treatment of Materials.

References

[1] Puač N, Petrović Z Lj, Živković S, Giba Z, Grubišić D and Đorđević A 2005 *Plasma Processes and Polymers* ed. R d'Agostino, P Favia, C Oehr, M R Wertheimer (Wiley-VCH:Berlin) p193

[2] Makabe T and Petrović Z Lj 2006 *Plasma Electronics* (Taylor and Francis:New York); Lieberman M A and Lichtenberg A J 2005 *Principles of Plasma Discharge and Materials Processing*, (Wiley:Hoboken);

[3] Roth J R 1995 *Industrial Plasma Engineering* (Institute of Physics:Bristol, UK); Kelly-Wintenberg K, Hodge A, Montie T C, Deleanu L, Sherman D, Roth J R, Tsai P and Wadsworth L 1999 *J. Vac. Sci. Technol. A* **17** p 1539

[4] Park J, Henins I, Herrmann H W, Selwyn G S and Hicks R F 2001 *J. Appl. Phys.* **89** p 15

[5] Kunhart E E 2000 *IEEE Trans. Plasma Sci.* **28** p 189

[6] Stoffels E, Kieft I E, Broers J L V, Ramaekers F C S and Slaaf D W 2003 *Proc. ISPC 16 International Symposium on Plasma Chemistry*, 22-27.jun 2003, Taormina, Book of Abstract, Ed.s R. d'Agostino, P. Favia, F. Fracassi and F. Palumbo, CD [ISPC-884]

[7] Sladek R E J. and Stoffels E 2005 *J. Phys. D: Appl. Phys.*, **38**, 1716

[8] Puač N, Petrović Z Lj, Malović G, Đorđević A, Živković S, Giba Z and Grubišić D 2006 *J. Phys. D: Appl. Phys.* **39** p 3514

[9] Živković S, Puač N, Giba Z, Grubišić D and Petrović Z Lj 2004 *Seed Science and Technology* **32** p 693

[10] Stoffels E, Flikweert A J, Stoffels W W and Kroesen G M W, 2002 *Plasma Sources Sci. Technol.* **11** p 383

[11] Banks J A 1999 *Annual Review of Plant Physiology and Plant Molecular Biology* **50** p 163

[12] Baker C J and Mock N M 1994 *Plant Cell Tissue and Organ Culture*, **39**, p 7

[13] Murashige T and Skoog F 1962 *Physiologia Plantarum* **15** p 473

[14] Stoffels E, Aranda-Gonzalvo Y, Whitmore T D, Seymour D L and Rees J A 2006 *Plasma Sources Sci. Technol.* **15** p 501

[15] Stoffels E, Kieft I E, Sladek R E J, van den Bedem L J M, van der Laan E P, Steinbuch M 2006 *Plasma Sources Sci. Technol.* **15** p S169

[16] Sakiyama Y and Graves D B 2006 *Journal of Physics D: Applied Physics* **39** p 3451

[17] Sakiyama Y, Graves D B and Stoffels E 2008 *Journal of Physics D: Applied Physics* **41** p 095204

Modeling of Townsend discharges at high E/N and low pressure

V. D. Stojanović
Institute of Physics, Pregrevica 118, 11080 Zemun, Serbia
E-mail: stoyanov@phy.bg.ac.yu

Abstract.
This report presents the results of an investigation of complex behavior of electrons and ions in Townsend discharge by a Monte Carlo simulation (MCS) technique. Although recognized as a relatively simple for implementation and at the same time difficult to test and verify, this technique is revealed as a right choice for modelling Townsend discharges. This work begins with selection of input data and benchmarking of codes. We have included description of effects of ions and fast neutrals to kinetics of excitation, description of nonhydrodynamic effects in the vicinity of electrodes and at the high E/N the runaway of electrons and ions. We also present description of impulse discharges and time resolved current impulses under similar conditions. Main presented results were electron energy distribution functions (EEDF), spatial excitation coefficients and arrival time spectra (ATS) obtained for the conditions of Townsend discharges in nitrogen and argon at low pressures and very high E/N. In addition we have compiled cross section sets that may serve for discharge modelling in argon and nitrogen in broad range of conditions and discussed numerous approximations that were recently used in modelling of plasmas.

1. Introduction

Nonequilibrium manifestations in electron and heavy particle transport in gaseous discharges [1] are standard part of all devices that use low pressure discharge. Nonequilibrium effects are related to behavior of these particles in gas that deviates from the hydrodynamic regime. In that sense phenomena presented in this work are part of the swarm physics for which it is characteristic that particles do not interact with particles of the same kind but only with the unperturbed background gas particles. Only partial information is available on elementary collisional processes that are important for modelling of gas discharges. As a result all the research on gas discharges is to a large degree connected to the fundamental studies in atomic and molecular physics.

Phelps and co-workers [2, 3] have performed a series of experiments at moderate to very high electric field to gas number density ratios E/N at low currents and in parallel plate geometry which were the first useful primary experimental data for comparisons with models for high E/N. So far only one dimensional beam models have been developed for such conditions [4]. Such models and the very high E/N experimental data are needed for understanding the nonlocal transport of electrons in the sheaths of the high current discharges [5, 6], microwave breakdown [7] and pseudospark discharges [8] and therefore an independent verification of the models of kinetics at high E/N is justified. In addition, a very interesting feature of the high E/N

transport is the importance of heavy particles in excitation and ionization [2, 4, 9], requiring coupling of electron transport with ion and fast neutral transport models with enough energy for the excitation of electronic transitions. High mean energies, especially in the case of electrons, require reflection of electrons from the anode to be taken into account and low pressures imply that equilibrium is not established throughout the gap.

While some attempts of developing the beam-like models or Boltzmann equation solutions exist, the best method to describe processes that include such complex boundary conditions, spatially dependent transport coefficients of electrons, ion and fast neutral transport, and distribution functions that may include sharp peaks is MCS. We have developed the Monte Carlo (MC) code for electron transport and a more detailed description of that code, together with a comparison with experimental results for the processes dominated by electron excitation, was given in [10].

Basic motivation for this research was found in the fact that existing models of discharges could not explain measurements of Phelps and coworkers (assumes the emission of radiation both spatially and time resolved) as well as our measurements in these conditions, that refer to the left side of the Paschen curve.

According to knowledge of many authors, in these conditions nonhydrodynamic behavior both of electron transport and of ions and fast neutral particles appears and also other effect such as effect of runaway electrons and ions and fast neutral particles, effect of back–scattered electrons, excitation by heavy particles.

Modelling in this work is based on experimental conditions in drift tubes of the Laboratory of Gaseous Electronics in Belgrade or in Phelps's experiments. Discharges are with currents of the order of few μA, voltages of the order of kV and gas pressures of the order of mTorr to few Torr.

2. Monte Carlo technique

MC code was built to include both boundary effects and nonequilibrium effects, and includes also particle scattering anisotropy, nonconservative collisions and electron back–scattering on surfaces.

2.1. Basic principles

MCS comprised running of four MC codes each based on null collision technique [11]: code for electrons [10], codes for ions and fast neutrals [12] while separate code was used for back–scattered electrons.

At first we followed primary(direct) electrons that were released at the cathode. Grand total scattering cross section was used to determine time of the collision. Scattering was anisotropic and differential cross section (DCS) was used to determine scattering angle after collision. Angular distributions were selected in a way to keep agreement with momentum transfer cross section which was determined from total cross section and original source of effective cross sections.

According to the literature data [13] secondary electron share energy with primary electron after the ionizing collision.

MCS for heavy particles transport was based on the codes for ions and fast neutrals which were similar except by electric field value.

In all codes we have assumed stationary target, velocities in LAB system after collision were obtained from the velocities in CMS after the collision, fast neutrals were followed only up to the point where their energy is under the excitation threshold ε_L.

2.2. *Electron backscattering*

Phelps and coworkers [2] have noticed the peak in the spatial profile of the emission closes to the cathode which was attributed to the electrons back scattered from the anode and latter was confirmed by MCS [14]. MCS were also used to obtain effect of back–scattered electrons on multiplication on argon [15].

Backscattered electrons were divided in two groups: reflected and secondary electrons. For both, from the literature data [16, 17, 18] we were able to select initial number, initial energy and direction from electrode surfaces of graphite (gr) and stainless steel (ss).

2.3. *Selection of input data and benchmarks*

Input data required to run MCS in both Ar and N_2 were based on best compilation we found in publications of Phelps and Hayashi. These include both total cross sections and differential scattering cross sections.

For the purpose of running the code for electrons we have prepared complete sets of cross sections of Phelps [19, 20] for molecular nitrogen and of Hayashi [21] for argon atom. Both were completed with differential cross sections for energies below 1 eV and up to 10 keV for argon [12] and molecular nitrogen [19, 10].

The cross sections for heavy particles were not found in great detail as for the electrons. Nevertheless, well validated cross sections of Phelps [22] were used as a base while additional modifications and verifications were performed to properly take into account the ratio of resonant charge transfer collisions to elastic collisions [23, 10].

Data used in plasma modelling are almost exclusively based on the swarm theory and experiments [24]. It was a common practice to apply benchmark calculation to any new theoretical model and numerical code before the code was accepted as accurate. Benchmark calculations were done for all available scattering models relevant for DC electric fields [25, 11, 26, 27] for all our transport codes as well as for common sampling procedures and conditions [28, 29].

3. Modelling of emission in dark Townsend discharges at high E/N

In drift tube experiments the spatially resolved emission obtained side on in large part of the discharge usually exhibits exponential trend or nearly exponential trend in the most of the interelectrode space. Exponential trend is a consequence of the equilibrium electron transport for these electron groups [19]. Thus appears "expected" that perfect agreement of Townsend ionization coefficient (TIC) from the MCS slope of the spatially resolved emission and experiments of Haydon and Williams [30] can be obtained. So if there were no other effects TIC was possible to obtain from the slope of the spatially resolved measurement of emission as was experimentally done for many gases [2, 31, 32, 33, 34, 35, 36, 37].

By increasing reduced electric field spatial emission dependencies with 391.4 nm in large part of the space stay nearly exponential while with some emission bands like 2^+ band (see Fig. 1), drastically alters shape of the spatial dependency with respect to the exponential.

Similar behavior has also appeared in argon where "expected" behavior is observed for 750 nm line [38] while "unexpected" for 811 nm line [32]. It was found in Scott-Phelps experiments [38] that the cause for such spatial excitation dependency is actually contribution of heavy particles to excitation, precisely speaking fast N_2 in the case of nitrogen discharge.

Modelling results of digitized experimental data [2] for the spatial excitation coefficient for the 337.1 nm emission line of N_2 at high E/N are shown in Fig. 1. Clearly three components of spatial dependence of emission are observed obtained by electron excitation by direct (d) and secondary electrons (s) produced in the gas phase, heavy particle excitation (fn) and excitation by backscattered electrons (b) both reflected and secondary. Results of MCS are normalized to the electron excitation coefficient at the anode that corresponds to the normalization of experimental excitation coefficients to the anode current [33]. Emission at the anode was used since the electron current was known at that point. It was actually the spatial profile of emission normalized to the excitation coefficient at the anode. Effect of low electron reflection (graphite) and high electron reflection (stainless steel) surface was successfully modelled by electron backscattering and present data.

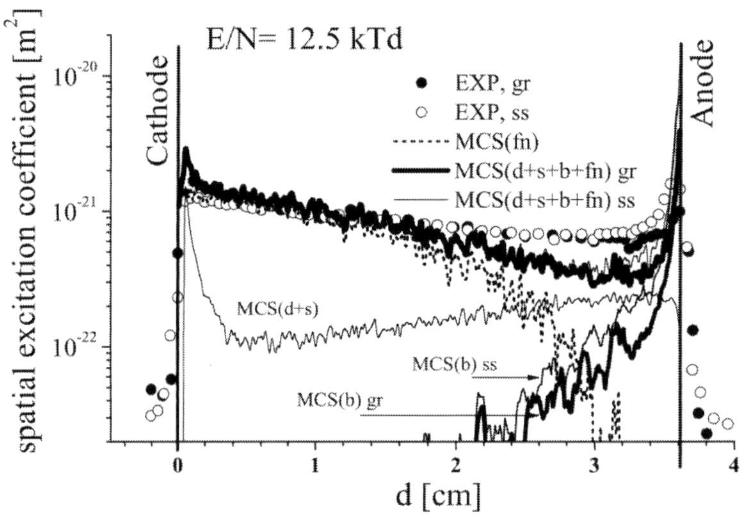

Figure 1. Spatial excitation coefficient for high E/N at pressure p = 61 mTorr (E = 250 V cm^{-1}, d_{AC} =3.61 cm). The results from [2] are shown with EXP.

Excitation coefficients measured in drift tubes provide the basis to normalize the cross sections for specific channels of excitation and thus to avoid the use of the fictitious effective cross sections.

A combination of measurements of the excitation coefficients and the standard sets of the cross section data were also used in order to renormalize cross sections for specific channels [39].

4. Modelling energy distribution functions at the electrodes

One of the fundamental quantities determining the kinetics of the discharge is EEDF, since electrons are responsible for excitation, ionization and dissociation of heavy particles in gas discharges. At low pressure and correspondingly at high discharge voltage and E/N, electrons may not be in local equilibrium with the electric field. Thus, their EEDFs manifest a departure from the hydrodynamic distribution. This departure can be quite substantial, and groups of runaway electrons with beam-like behavior may be formed in the EEDF.

The anode boundary at very high E/N represents significant electron nonequilibrium region due to the very high electron energies and its link to a metal surface.

First our modelling results of the EEDF at the absorbing anode will be presented. They show development of runaway electron groups as a function of increasing electric field i.e. available energy (ε_{max}). EEDF evolution at the anode as a function of E/N, for the Paschen curve [2] conditions at high and very high E/N obtained by MCS is shown in Fig. 2. Main characteristic of this dependence is showing up the peaks at very high energies which characterize beams of runaway electrons. Beam component appeared at moderate E/N has developed over the intensity, as a function of E/N, in such a way that pronounced components of the beam were distinguished at inelastic threshold energy distance. By increasing electron energy beam resolution is less important since largest number of inelastic events happens close to the cathode and largest number of electrons accelerates almost up to the maximum energy. At very high E/N the number of low energy electrons is very low because of their origins in ionization events close to the surface. The anode boundary is characterized by the high energy electrons which may do multiplication in gas or in surface layers of the electrode. In the first case high energy electrons produce reflected electrons and in the second case secondary electrons. Depletion of low energy electrons due to their absorption at the anode is not a dominant characteristic of these EEDFs.

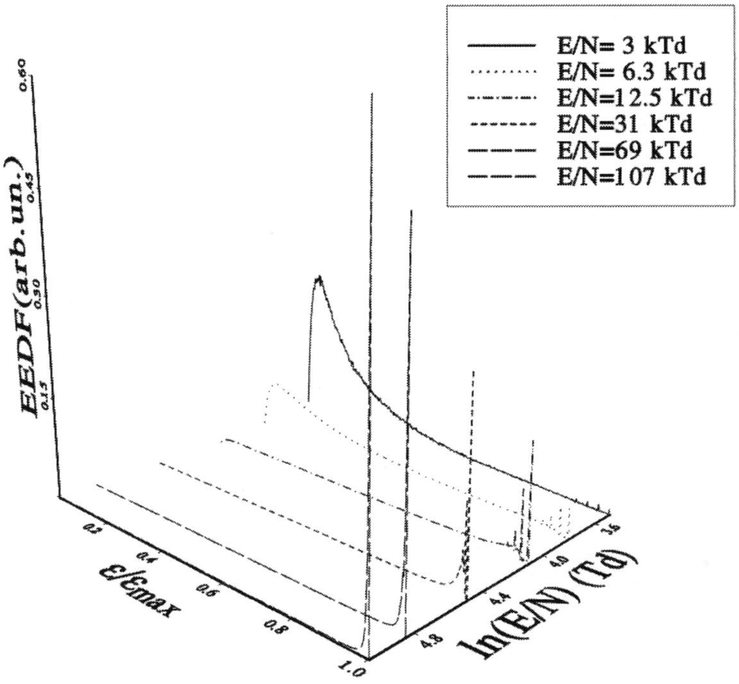

Figure 2. EEDF at the anode in nitrogen obtained by MCS for series of high and very high values of reduced electric field. All EEDF are obtained for distance d =3.61 cm with respect to the cathode, for the conditions of selfsustained discharge [2] between electrodes (sintered graphite). EEDF normalization is with respect to initial number of electrons from the cathode.

As a separate task we did modelling of EEDF measured at the anode by Vrhovac et al. [40] separated from the cathode by d_{AC} =1.85 cm. Electrons with the energies ε arriving at the anode were left to penetrate through the micrometer holes to the other side of the anode where they were collected by the retarding potential analyzer system. By MC modelling we were able to discuss both equilibrium and nonequilibrium parts of the EEDF (Fig. 3). In the modelling, in order to successfully describe measured EEDF it was necessary to include both secondary and reflected electrons.

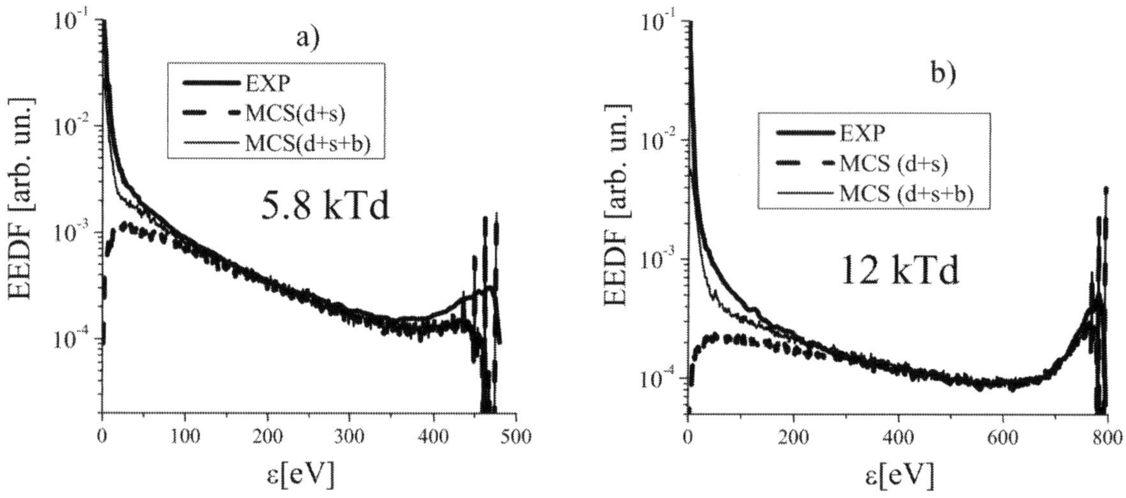

Figure 3. EEDF in nitrogen at a) E/N=5.8 kTd (E =256 V/cm, N =0.438 10^{16} cm^{-3}, M2), b) E/N=12 kTd (E =428V/cm, N =0.354 10^{16} cm^{-3}, M5). EXP-experimental results [40]. We used Maxwellian distribution of secondary electrons with peak at energy of 2 eV (M2) and 5 eV (M5). Small oscillations of MC curves represent the statistical uncertainty of the obtained simulation results. MCS data are normalized to the experimental data at the energy of 200 eV (M2) and 400 eV (M5).

5. Modelling impulse discharge

The studies of impulse discharges showed up as very promising since they may give useful information about the processes relevant to breakdown formation.

In this work our goal was to model Gylys-Phelps experiment (GPE) [3], an impulse discharge operating in prebreakdown conditions which exploits pulsed photoemission technique to initialize discharge. In GPE basic setup consisted of two planparallel electrodes to which high voltage was applied. One of the electrodes was semitransparent (cathode) and was illuminated by the pulsed laser beam (10 Hz). The laser energy was sufficient to induce low energy electron emission at the other side of the cathode. By detecting emission and/or determining the charge passing through the circuit one may figure out about breakdown of such a system.

The analysis of time dependence of accumulated charge at the capacitor connected in series to the electrode system was complicated by displacement currents which were the result of particles transport in discharge. If one wants to determine electron multiplication from the time dependence of the accumulated charge than time interval between the end of the electron ATS and beginning of heavy particles ATS should be larger than time resolution of recording device. Because of that the knowledge about the kind of particles we follow in certain time interval is of crucial importance in determination of the electron multiplication in discharge.

The MC technique we applied to model such a system was similar to the ones described previously, except that now we followed arrival time of the particles to the electrodes.

ATS's of the avalanches reaching the electrodes are shown in Fig. 4. Isolated avalanche at the leading edge of the pulse train was a consequence of direct electrons, having the maximum pulse corresponding to runaway electron group. The continuing shape shows arrival times of secondary electrons until slowest and nearest electrons close to the anode have arrived. The direct electron and ion contribution were separated from the secondary contributions due to ionization by ions and fast neutrals. Excellent agreement was obtained for electron multiplication obtained from MCS and from GPE results for nitrogen.

Generally ATS in Fig. 4 a) are slightly wider than in Fig. 4 b) due to the higher gas density.

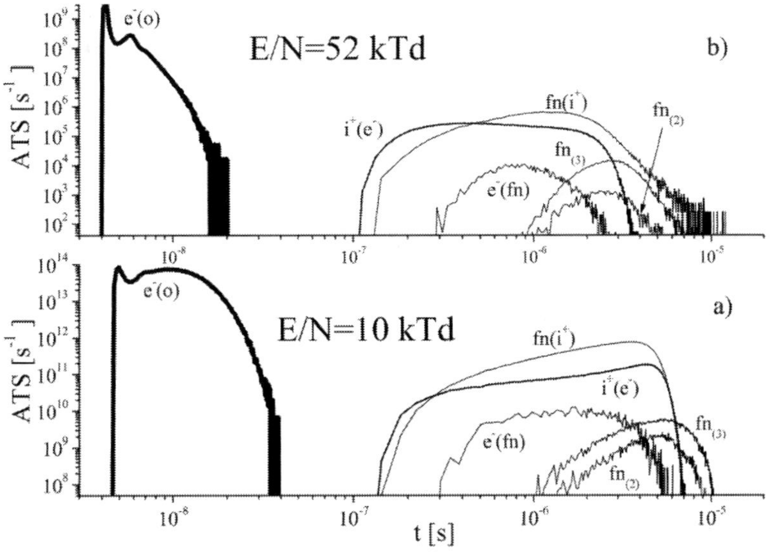

Figure 4. ATS for electrons, N_2^+ ions and fast neutrals obtained by MCS for the conditions of nonselfsustained discharge in GPE: a) E/N= 10 kTd, d_{AC}= 4 cm, Nd_{AC} = 8 10^{15} cm^{-2}, ε_L < 15 eV, b) E/N = 52 kTd, d_{AC} = 4 cm, Nd_{AC} = 2 10^{15} cm^2, ε_L < 10 eV. All ATS are normalized to the initial number of electrons. ATS of electrons arriving at the anode e^-(o) - direct and secondary produced in gas, $e^-(i^+)$ - formed by ion impact ionization of $i^+(e^-)$ avalanche; e^-(fn) - formed by fn transport formed by ion avalanches $i^+(e^-)$. ATS of N_2^+ ions arriving at the cathode: $i^+(e^-)$ - formed in ionizing collisions of e^-(o) electron avalanche with N_2. ATS of fast neutrals to the cathode: fn(i^+) - formed by $i^+(e^+)$, $fn_{(2)}$ - formed by i^+ (not shown) following $e^-(i^+)$ avalanche (not shown) formed by $i^+(e^-)$ ion impact ionization, $fn_{(3)}$ - formed by i^+ (not shown) following fn(i^+) avalanche (not shown) ionizations.

6. Conclusion

We build and tested Monte Carlo codes in order to follow transport of electrons, ions and fast neutrals in homogenous electric field between two planparallel electrodes.

We formed database for scattering of electrons, ions and heavy particles in nitrogen and argon, as well as electron scattering on metal surfaces (graphite, stainless steel).

We precisely determined contributions of electrons, ions and fast neutrals to spatially resolved emission in argon and nitrogen.

By using the precise models we calculated the spatial excitation coefficients at very high E/N and proved that excitations in the cathode boundary were produced by neutral particles.

By MCS we determined EEDF at the graphite anode and obtained good agreement with experimental results in the whole energy range.

The analysis of series of transient measurements of Phelps in nitrogen was performed. We showed the precise time frame in which Gylys-Phelps experiment proceeds from moderate to very high E/N.

Acknowledgments

The work is supported by the MNRS under grants 141025. The author acknowledges all colleagues from the Institute of physics, Belgrade for numerous discussions. Special thanks go to Prof. Z. Lj. Petrović and Dr. Ž. Nikitović for their continuous encouragement and support and to Prof. A. V Phelps for his numerous suggestions and comments.

References

[1] Petrović Z Lj, Šuvakov M, Nikitović Ž, Dujko S, Šašić O, Jovanović J, Malović G and Stojanović V 2007 *Plasma Sources Sci. Technol.* **16** S1

[2] Jelenković B M and Phelps A V 1987 *Phys. Rev. A* **36** 5310

[3] Gylys V T, Jelenković B M and Phelps A V, 1989 *J. Appl. Phys.* **65** 3369

[4] Phelps A V, Jelenković B M and Pitchford L C 1987 *Phys. Rev.* **36** 5327

[5] Ventzek P L G, Hoekstra R J and Kushner M J 1994 J. Vac. Sci. Technol. B **12** 461; Date A, Kitamori K, Sakai Y and Tagashira H 1992 ibid. **25** 442

[6] Den Hartog E A, Doughty D A and Lawler J E 1988 *Phys. Rev. A* **38** 2471

[7] Hays G N, Pitchford L C, Gerardo J B, Verdeyen J T and Li Y M 1987 *Phys. Rev. A* **36** 2031

[8] Boeuf J P and Pitchford L C 1991 *IEEE Trans. Plasma Sci.* **19** 286

[9] Petrović Z Lj, Jelenković B M and Phelps A V 1992 *Phys. Rev. Lett.* **68**, 325

[10] Stojanović V D and Petrović Z Lj 1998 *J. Phys. D: Appl.Phys.* **31** 834

[11] Reid I D 1979 *Aust. J. Phys.* **32** 231; ibid. 1982 **35** 473

[12] Petrović Z Lj and Stojanović V D 1998 *J. Vac. Sci. Technol. A* **16** 329

[13] Opal C B, Peterson W K , and Beaty E C, *J. Chem. Phys.* **55** 4100 1971

[14] Petrović Z Lj, Stojanović and Jelenković B M *J. Appl. Phys. 81* 1601

[15] Petrović Z Lj, Stojanović V and Radmilović M 2000 SPIG Book of contributed papers, Zlatibor Yugoslavia

[16] Woods M E, Hopkins B J, Matthews G F, McCracken G.M, Sewell P M, and Fahrang H 1987 *J. Phys. D* **20** 1136

[17] Thomas E W 1985 in Particle Interaction with Surfaces Oak Ridge National Laboratory, Oak Ridge, Vol. III.

[18] Gergely G, Gruzza B, and Meynard M 1980 *Acta Phys. Acad. Sci. Hung.* **48** 337

[19] Phelps A V and Pitchford L C 1985 *Phys. Rev. A* **31** 2932; 1985 JILA Report No 25

[20] Phelps A V 2004 ftp://jila.colorado.edu/collision data/electronneutral/electron.txt

[21] Hayashi M 1992 personal communication

[22] Phelps A V 1991 *J. Phys. Chem. Ref. Data* **20** 557; ibid. 1992 **21** 883

[23] Phelps A V 1994 *J. Appl. Phys.* **76** 747

[24] Huxley L G H and Crompton R W 1973 *The diffusion and drift of electrons in gases* New York: Willey Interscience

[25] Lucas J and Saelee H T 1975 *J.Phys. D: Appl. Phys.* **8** 640

[26] Nolan A M, Brennan M J, Ness K F and Wedding A B, 1997 *J.Phys. D: Appl.Phys.* **30** 2865

[27] Skullerud H R 1973 *J. Phys. B: Atom. Molec. Phys.* **6** 728

[28] White R D 1997 *J. Phys. D: Appl. Phys.* **30** 810

[29] Sakadžić S and Petrović Z Lj 2000 20th SPIG, Zlatibor, Yugoslavia 127

[30] Haydon S and Williams O 1976 *J. Phys. D* **9** 523

[31] Stojanović V, Božin J, Petrović Z Lj and Jelenković B M 1990 *Phys. Rev.* **42** 4983

[32] Phelps A V and Jelenković B M 1988 *Phys. Rev. A* **38** 2975

[33] Stokić Z, Fraga M M F R, Božin J, Stojanović V, Petrović Z Lj and Jelenković B M 1992 *Phys Rev A* **45** 7463

[34] Šašić O, Malović G, Strinić A, Nikitović Ž and Petrović Z Lj 2004 *New Journal of Physics* **6** 74

[35] Malović G N, Strinić A I, Petrović Z Lj and Sadeghi 2004 N *Spectrochimica Acta part B* **61** 951

[36] Malović G N, Strinić A I, Petrović Z Lj, Božin J V and Manola SS 2000 *Eur. Phys. J. D* **10** 147

[37] Nikitović Ž, Strinić A, Malović G, Stojanović V and Petrović Z Lj 2008 *Acta Chim. Slov.* **55** 219

[38] Scott D A and Phelps A V 1991 *Phys. Rev. A* **43** 3043

[39] Strinić A I, Malović G N, Petrović Z Lj and Sadeghi N 2004 *Plasma Sources Sci. Technol.* **13** 333

[40] Vrhovac S B, Stojanović V D, Jelenković B M and Petrović Z Lj 2001 *J. Appl. Phys.* **90** 5871

[41] Kelly L J and Blevin H 1989 Proc. 19th ICPIG Belgrade Editor J.M. Labat vol. IV 892 Donko Z and Rozsa 1995 *Acta Physica Universitatis Comenianae* **36** 35

Theoretical study of excited electronic states at surfaces, link with photo-emission and photo-desorption experiments

J.-P. Gauyacq[1,2], A. G. Borisov[1,2] and A. K. Kazansky[3,4]

[1]CNRS, Laboratoire des Collisions Atomiques et Moléculaires, UMR 8625, Bât.351, Université Paris-Sud, 91405, Orsay Cedex, France

[2]Université Paris-Sud, Laboratoire des Collisions Atomiques et Moléculaires, UMR 8625, Bât.351, 91405, Orsay Cedex, France

[3]Fock Institute of Physics, State University of Saint-Petersburg, Saint Petersburg, 198504,Russia

[4]Donostia International Physics Centre, E-20018 SanSebastian/Donostia, The Basque country, Spain

jean-pierre.gauyacq@u-psud.fr

Abstract. Excited electronic states at surfaces play a very important role in a variety of surface processes. These excited states have a finite lifetime due to electron-transfer processes and their efficiency as reaction intermediates depends crucially on their lifetime. A review of several physical situations, where an excited electronic state localized on an atom interacting with a metal surface intervenes in a surface process, is presented with an emphasis on the way the metal electronic structure influences the excited state dynamics. Examples are chosen among the alkali/metal systems, stressing how the same transient electronic state can influence different dynamical processes.

1. Introduction

When an atom or molecule is electronically excited near a metal surface, it can evolve into two paths: the excitation can either relax into the metal corresponding to the finite lifetime of the excited state, or trigger an internal evolution of the system opening the way to a variety of processes via electron-heavy particle couplings. The role of excited states as intermediates in surface mechanisms has been invoked in many surface processes: collisional charge transfer, molecular dissociation, desorption and more generally surface reactivity. This important role of excited states prompted many theoretical and experimental studies of excited electronic states at surfaces. Since energy transfer from electronic to heavy particle motion is a slow process, one can expect that the efficiency of a transient state as a reaction intermediate depends crucially on its lifetime. It is thus of paramount importance to understand the various parameters that govern the lifetime of an excited state on a surface and possibly see how it can be optimised to favour a reaction mechanism.

In this paper, we present a review on the dynamics of excited electronic states at metal surfaces. It is not a comprehensive review, we focus on excited states localized on an atom in front of a surface (an adsorbate or an atom colliding with the surface) as examples of the excitation dynamics at surfaces. Much emphasis is put on the influence of the surface electronic structure on the excitation dynamics and more precisely on the excited state lifetime. Various processes are discussed: charge transfer in atom-surface scattering, long-lived states in Alkali/noble metal systems as they can be probed by Two-Photon Photo-Emission (2PPE) experiments, photo-desorption and atomic manipulation. The examples are taken in Alkali/metal systems (see a recent review in [1]), to emphasize how the same excited electronic state participate in different processes.

2. Relaxation of an electronic excitation on a metal surface

When an excited atom is set in front of a metal surface, various de-excitation paths are possible [2]. First it can decay by a one-electron transition: the excited electron can be transferred into a bulk state of the same energy, it is a resonant, i.e. energy conserving transition, and it is only possible if electron states in the bulk exist in the excited state energy range. Secondly, it can decay by two-electron Auger-like transitions: if an excited atom has a vacancy in an inner orbital, an electron from the surface can fill this vacancy while the excess energy is used to eject another electron. Finally, an inelastic interaction ('collision') between the excited electron and a bulk electron can also bring the atom into the ground state. It then leads to the ejection of an electron or to a collective excitation of the surface (a plasmon) [3]. The mono-electronic process, when it is energetically allowed, is in general the most efficient one and it is discussed more in detail below. There are two equivalent ways to look at this one-electron resonant charge transfer (RCT) process. In a first view, the excited atom state is a discrete state localized on the atom, it is degenerate with the continuum of metallic states and coupled with it, so that it acquires a finite lifetime τ inverse of its decay rate, Γ, given by the Fermi golden rule. In a second view, the excited electron is bound in a potential well around the atom; there is a potential well inside the metal too and a potential barrier in between. The excited state then decays by electron tunneling through this barrier. In both views, one expects the electron transfer rate to vary rapidly with Z, the atom-surface distance, and a simple argument based on overlap between atomic orbitals and metal continuum states predicts an exponential behavior with Z at large distance. One can also expect a very fast electron transfer rate at close distances when basically the potential barrier separating the atom and the surface disappears. Actually, atomic levels that are above the surface Fermi level decay by electron transfer to the metal; for levels below the Fermi level, the argument has to be repeated for holes and a hole on the atom decays by transfer into the metal, i.e. by electron transfer from the metal to the atom.

3. Illustration of the atom-surface charge transfer

3.1 *Static case*

The electron transfer process between an atom and a surface is illustrated by the results of recent wave-packet propagation (WPP) calculations. This method consists in solving the one-electron time-dependent Schrödinger equation on a grid of points (see [4,5] for details on the WPP method):

$$i\frac{d}{dt}\Psi = H\Psi = (T+V)\Psi \qquad (1)$$

where Ψ is the electron wave function, T is the kinetic energy operator and V the interaction potential between the excited electron and the atom+surface system. Efficient propagation schemes are now available that allow solving equation (1) starting with a well-chosen initial wave function Φ_0. When the atom is at a fixed finite distance from the surface (static situation), the energies and the lifetimes of the resonant states of the system are extracted from the time dependent wave function Ψ obtained by solving eq.(1). For the resonances associated with excited atomic states interacting with the surface the decay rates thus obtained corresponds to the RCT process [4,5]. The WPP method can also be applied to the RCT in a collisional situation, i.e. to the RCT process when projectile collides with the surface (see below).

One can also extract the wave-function of the resonant state from the solution of equ.(1) and as a typical example, Figure 1 shows the case of a Cs atom interacting with a Cu metal surface. The resonant Cs* state is the lowest lying state localized on the Cs atom, it corresponds to a hybrid of m = 0 symmetry, mixture of 6s and 6p orbitals of the free atom (m is the projection of the electron momentum on an atomic axis normal to the surface). In front of a usual metal surface, the binding energy of this state is lower than the surface work function so that this state is an excited state of the Cs-surface system. It decays by transfer of the electron into the metal and the Cs adsorbate at low coverage is positively charged (see a discussion in [1]). Two different cases are considered: a Cs atom at 10 a_0 from a free-electron metal and a Cs atom at its adsorption distance on a model Cu(111) surface (the two distances are chosen to roughly correspond to the same decay rate of the Cs* state). Let us first consider the case of the free-electron metal (right part of Fig.1). One recognizes a quasi-circular electronic cloud localized around the atom centre that corresponds to the resonant wave function and that is not much perturbed from the atomic 6s orbital. The electron density inside the metal corresponds to the electron flux leaving the resonance state, i.e. to the resonant charge transfer process. Electron tunnelling through the potential barrier that separates the atom and the metal is favoured along the normal to the surface where this barrier is the thinnest and one can see that the outgoing electron flux corresponding to RCT is concentrated around the surface normal. The RCT rate increases roughly exponentially when the atom approaches the surface and the transfer rate becomes very fast, reaching the eV range at typical adsorption distances (lifetime of a fraction of a fs).

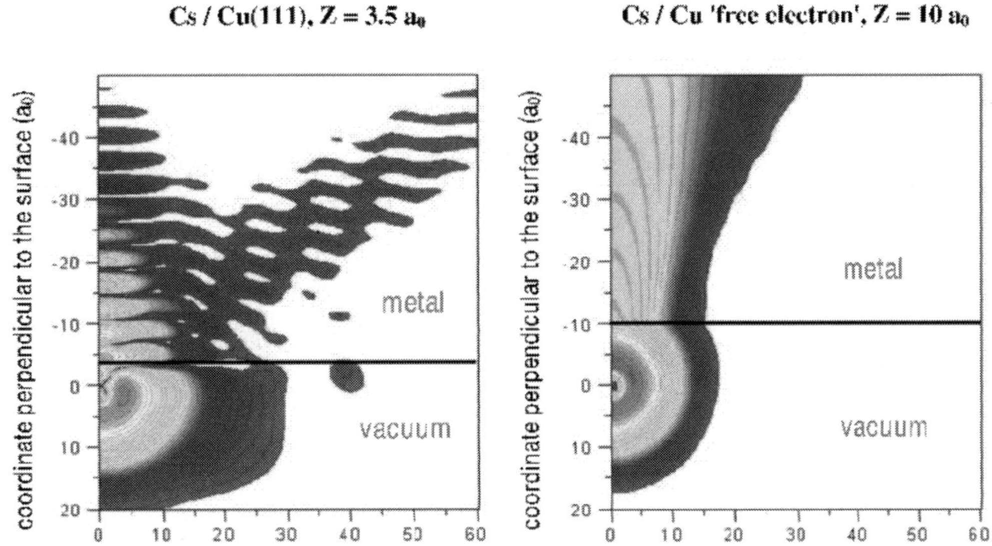

Figure 1: Electron density associated to the resonant Cs-localized state in the case of the Cs atom located (a) at 3.5 a_0 (adsorption distance) from the image plane of a Cu(111) surface and (b) at 10 a_0 from the image plane of a free-electron metal surface. The horizontal black line indicates the position of the image plane of the metal surface and the Cs atom centre is at the origin of coordinates. The logarithm of the electronic density (square modulus of the wave function) is presented in cylindrical coordinates parallel and perpendicular to the surface. z, the coordinate perpendicular to the surface, is positive in vacuum. The electron density decreases when going from red to violet. White corresponds to very small electron densities.

The case of the Cu(111) surface (left part of Fig.1) appears completely different. First, one can notice the difference between the electronic clouds around the atom; this is an effect of hybridisation of 6s and 6p orbitals that occurs for short atom-surface distances (the distance is the Cs adsorption distance on Cu(111)). The excited state discussed here is the hybrid polarized away from the surface. Second and most important, the electronic flux going into the metal is much different from the free-electron case. This is a direct consequence of the electronic structure of the Cu(111) surface, compared to that of a free-electron surface (Figure 2).

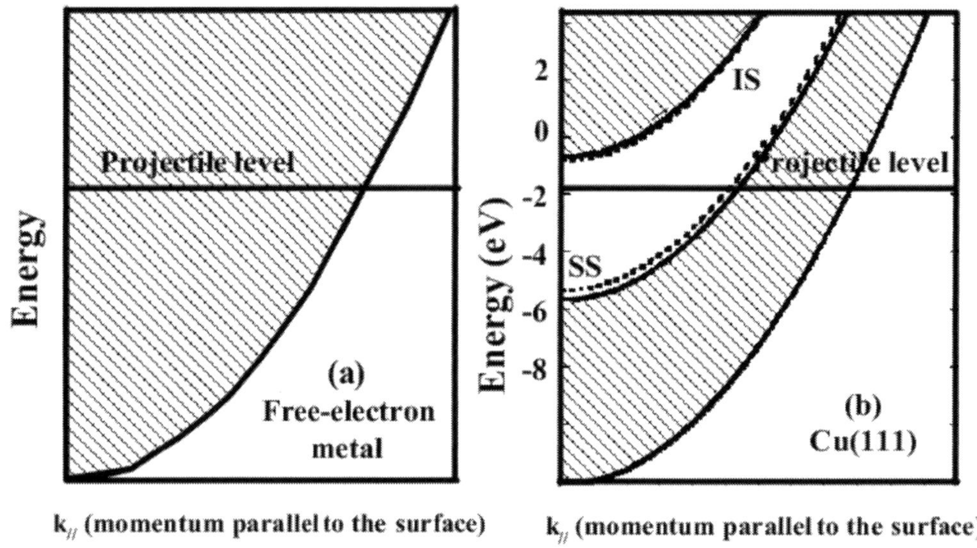

Figure 2: Schematic electronic structure of the Cu(111) surface compared to that of a free-electron metal. The energy of the electronic band of states is shown as hatched areas as functions of $k_{//}$ the electron momentum parallel to the surface. The zero energy is at the vacuum level.

In the free-electron case, the electronic structure is very simple; the energy of the bands simply follows the parabolic dependence of a free electron as a function of $k_{//}$, the electron momentum parallel to the surface (Fig.2a). In a real metal, the 3D-periodicities of the crystal result in the existence of certain directions and energy regions in which electron propagation is impossible. In the case of a surface, these are called surface-projected band gaps and one is visible in the schematic picture in Fig.2b: there is not any electronic state in the [-0.7, -5.8] eV energy range with $k_{//}=0$, i.e. an electron cannot propagate along the surface normal in this energy range. As a consequence, several states can exist in this band gap that are confined to the surface area and that cannot penetrate into the metal (see e.g. in [6]): the surface state (SS in Fig.2b) that is localized close to the surface and the image potential states (IS in Fig.2b) that correspond to an electron bound by its interaction with its own electrical image inside the metal. The difference in the electronic structure between a free-electron metal surface and a surface with a projected band gap has direct consequences on the RCT. On a free-electron metal, we have seen that electron transfer is hugely favoured along the surface normal, where the transparency is the highest. The electron is then preferentially transferred into the $k_{//}=0$ states of the substrate which are available at the Cs* state energy in the free-electron case (Fig.2a). On a Cu(111) surface, we can see in Fig.2b that there is not any state with $k_{//}=0$ at the Cs* state energy in which the excited electron could be transferred. An electron can only be transferred into states with a finite $k_{//}$, either in the surface state continuum or in the 3D-propagating band. The barrier transparency effect is still there and electron transfer occurs mainly as close to the surface normal as possible allowed by the band structure. One can recognize in Fig.1 an outgoing electronic flux going from the Cs adsorbate into the metal, a finite angle away from the surface normal. In the Cs/Cu(111) case, the flux into the surface state continuum is very weak. On the Cu(111) surface, since the states the most efficient for the RCT are missing, the RCT rate is much smaller than on a free-electron metal surface. At the adsorption distance (3.5 a_0), the Cs* decay rate amounts to 900 meV on a free-electron metal and it drops to 7 meV on Cu(111) [7]. There is thus a quasi-blocking of the RCT by the surface-projected band gap. Since one-electron transitions are weak, one should consider the contribution from multi-electron interactions. In the Cs* case at the adsorption distance, the multi-electron decay rate amounts to around 20 meV and thus dominates the electron lifetime [8]. One can then conclude on a very efficient RCT process in the case of a free-electron metal and it is much suppressed in the presence of a surface-projected band gap.

In this section we have seen that the RCT process is very efficient in the case of a free-electron metal and it is strongly influenced by the presence of a surface-projected band gap. The quasi-blocking of the RCT by the surface-projected band gap and the drastic reduction of the global electron transfer rate is of paramount importance for a series of physical situations that are underlined below.

3.2 Collisional charge transfer

During an atom-surface collision, when the atom is close to the surface, an electron can jump from/to the projectile. This process is just the one discussed in the previous section in the static case (fixed atom-surface distance). It is then very tempting to assume that the collisional atom moves on a classical trajectory and that at each time along the trajectory the electron transfer rate is the same as in the corresponding static case. Within this adiabatic approximation, the charge state of the atom along the trajectory is given by a rate equation. In the case of transitions between a neutral atom (population N_0 and a positive ion (population N_+) for example, it is written as:

$$\frac{dN_0}{dt} = -N_0\Gamma_0 + \Gamma_+ N_+ \qquad (2)$$

Depending on the relative energy position of the atomic level and of the Fermi level, the RCT process occurs in one direction or the other. This rate equation approach can be derived from a more general treatment of bound state-continuum transitions, within a set of semi-classical approximations [9]. Correlation effects on the atom (degenerate levels, equivalent electrons) are easily taken into account in this formalism [10], they simply reduce to statistical factors and extra equations. This approach proved to be very successful in a variety of collisional systems, in particular for free-electron metal targets in connection with non-perturbative parameter-free determination of the level energy and of the charge transfer rate [11,12]. In particular, detailed accounts of the cinematically (parallel velocity) assisted resonant collisional electron transfer in grazing collisions could be achieved. In grazing collisions, projectiles are impinging at the surface under very small angles with respect to the surface plane. The motion perpendicular to the surface is slow and the rate equation approach is valid. On the other hand, the motion parallel to the surface is characterized by the large velocity component $v_{//}$. The electron has an energy in the target frame different from that it has in the projectile moving frame, and the collision velocity is thus able to assist the electron transfer, modifying the resonance condition between atomic and target states [12].

One expects the effects from the target band structure (reduction of the RCT rate, role of 2D surface continua) to be directly transferred into the dynamical process, via a rate equation approach. And indeed, joint theoretical-experimental studies confirmed the strong effects induced by surface-projected band gaps on collisional charge transfer in grazing angle collisions. The reduction of the charge transfer rates results in an important shift of the effective charge transfer distance toward the surface and most effective, contributions from the 2D-surface state continuum on Cu(111) deeply modify the cinematically assisted resonant electron transfer processes [13,14].

However, recent studies at higher impact angles i.e. at higher collision velocity perpendicular to the surface revealed that the situation is more complex and that the effect of the target electronic structure is not always as obvious as in the above-mentioned systems [15,16].

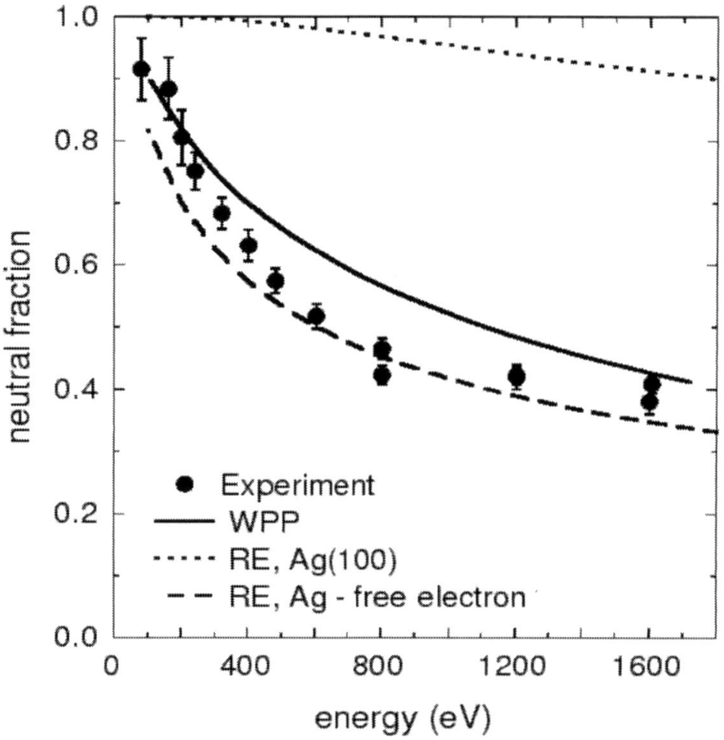

Figure 3: Neutralization probability of Li^+ ions colliding on a Ag(100) surface (scattering angle 135°). Symbols: experimental results as a function of the collision energy. Dotted line: theoretical results using a rate equation approach with an adiabatic Ag(100) description. Dashed line: theoretical results using a rate equation approach with a free-electron metal description. Full line: dynamical WPP results (from Canario et al, [16])

Figure 3 presents the neutralization probability of Li^+ ions colliding on a Ag(100) surface as a function of the projectile energy in the few hundreds eV range [16]. The scattering angle is 135°. Experimental results (circles) are compared with the results of three different theoretical approaches. Ag(100) surface exhibits a surface-projected band gap that strongly affects the energy and width of the Li atom levels interacting with the surface. Two theoretical results were obtained using a rate equation approach (eq.(2)): either with the energy and width corresponding to Ag(100) or with those corresponding to a free-electron metal with the same surface work function. A third set of results (labeled WPP) was obtained by a full dynamical study of electron transfer by wave packet propagation. It consists in solving the time-dependent Schrödinger equation for the active electron in the field of the Ag(100) surface and of the moving atom. The atom is assumed to travel along a classical trajectory and so one has to solve equation (1), in which the potential operator depends parametrically on time via the projectile trajectory. This latter study provides the exact result for the dynamics of the collisional charge transfer (within the choice of potential modeling). In particular all dynamical effects are included. It appears in Fig.3 that the full dynamical WPP result quantitatively accounts for the experimental observations. In addition, the dynamical behavior is very far from the adiabatic rate equation prediction for Ag(100), though very close to the free-electron result. Surprisingly, the Ag(100) surface behaves almost like a free-electron metal and its electronic structure does not seem to influence the outcome of a collision, although it influences the static situation. This is due to the existence of non-adiabatic transitions induced by the projectile motion that are not taken into account by the rate equation approach and that tend to cancel the effect of the target electronic structure. This subtle effect can be understood in the following way. It is linked to the finite time character of a collisional process and has been well illustrated in model calculations on the H^--Cu(111) system [5]. When the electron is in an atomic orbital localized on the projectile, it does no 'know' about the electronic structure of the surface and of a possible surface-projected band gap. Tunneling from the atom to the metal is thus simply determined by the potential barrier separating atom and target, which is practically the same for a free-electron surface and for a surface with a surface-projected band gap. Later, when the electron wave packet penetrates the metal, it is partially reflected by the various atomic planes. These reflections and the corresponding interferences are at the origin of the impossibility of electron propagation in certain energy ranges and more generally of the target electronic band structure. Thus, the electron 'needs' a finite time to fully adjust to the target electronic structure; in the mean time, it behaves as in the case of a free-electron metal. This explains the results in Fig.3: for a finite collision time, the system 'does not know' of the target band structure and RCT occurs as on a free-electron metal. The typical minimum time required for the band structure to fully develop can be estimated to be of the order of the time needed by the electron to travel between the atomic planes in the crystal.

4. Transient excited states localized on an adsorbate on a metal surface

When an atom is adsorbed on a metal surface, excited electronic states can be localized on it; these states are exactly the same as the transient states that were invoked in the collisional charge transfer process, discussed above. In the adsorbate case, the heavy particle motion associated to thermal vibration of the adsorbate is very slow as compared to the previous section. An adiabatic view is then well adapted to describe the system dynamics, so that the effect of the substrate electronic band structure is expected to play a key role.

The development of fast lasers in the fs range allows the study of excited states at surfaces directly in the time domain [17]. In a 2-photon photo-emission experiment (2PPE), a substrate electron is excited by a first photon to a transient excited state and is photo-emitted in vacuum by the second photon. The spectrum of photo-emitted electrons thus reflects the spectrum of transient excited states. In addition, the time evolution of the population in the intermediate state can be probed in an experiment with two separated laser pulses with a controlled time delay between the two pulses (time–resolved 2-photon photo-emission, TR-2PPE). If the phase relation between the two pulses is well defined, the non-linear character of the 2PPE process also allows to perform an interferometric study, i.e. the second pulse probes both the population and the phase of the excitation produced by the first laser pulse and so it can probe the coherence time of the transient intermediate state [17]. These are the two key parameters for describing an excited state dynamics and its role as an intermediate in a reaction mechanism. The lifetime of the state determines how long it survives and then can induce transitions toward a reaction channel and the coherence time determines the extent to which it can be involved in a quantum control scheme. At this point, one can stress that the reaction induced by an excited electronic state involves an energy transfer from the electrons to the nuclei; due to the large mass ratio, this energy transfer is in general very weak but it is favoured for long-lived transient states for which a weak process can be active for a long time.

Adsorbates are located very close to the surface, so that resonant charge transfer between an excited state and the metal substrate is generally very fast. As an example, on a free-electron metal at adsorption distances, alkali levels have lifetimes of a fraction of a fs. This makes reaction mechanisms at surface a priori weakly efficient. The situation changes drastically if the adsorbate level lies inside a surface projected band gap. As shown from theoretical results in section 3.1, in the case of Cs/Cu(111) the RCT rate is much affected by the surface-projected band gap, which reduces it by two orders of magnitude. Taking into account the multi-electron decay of the Cs*, the theoretical lifetime of the Cs* state is in the 30 fs range. Experimental studies on the Cs*/Cu(111) have shown a Cs* state located a few eV above Fermi level with a lifetime in the few tens of fs range [18,19]. Extracting a precise lifetime from these TR-2PPE experiments is not straightforward due to the Cs motion induced by the Cs* excitation (this point is further discussed below). Coherence time are shorter and have been measured and computed in the 10 fs range [19].

Various alkali/noble metal surface systems exhibit excited states with significant lifetimes. The actual value of the lifetime depends on various parameters such as the position of the level inside the surface-projected band gap or the polarisability of the adsorbate. Up to now, Cs*/Cu(111) is the longest-lived state that has been observed in this context [1].

5. Atomic motion induced by excitation of a transient adsorbate state

As an example, let us consider again the case of Cs adsorbates on Cu(111). Figure 4 schematises the 2PPE process in this system and the induced Cs motion. The left part of Fig.4 shows the 2PPE process in which a substrate electron is excited by a first photon ($\hbar\omega_1$) to the Cs* state and then photo-emitted by a second photon ($\hbar\omega_2$). The Cs adsorbate is initially at its adsorption equilibrium. In the Cs* intermediate state, the interaction between the adsorbate and the substrate is different from that in the ground state and the adsorption distance is not an equilibrium distance for the excited state. As a result, when the Cs* electronic state is formed, the Cs adsorbate starts to move outward from the surface. This induces a downshift of the Cs* electronic energy. The right part of Fig.4 illustrate this effect: the energy profile (in blue) of the excited state created by the first photon shifts down as the Cs adsorbate moves away from the surface. The spectrum of photo-emitted electrons then depends on the time when the second photon is absorbed: the larger is the time delay between the two photons, the lower is the photo-emitted electron energy. This effect has been very clearly observed experimentally [20]: the spectrum of photo-emitted electrons varies continuously when the Cs-surface distance increases, i.e. when the delay between the two photons is increased. This brings a direct evidence of the photo-induced Cs motion allowing its study directly in real time.

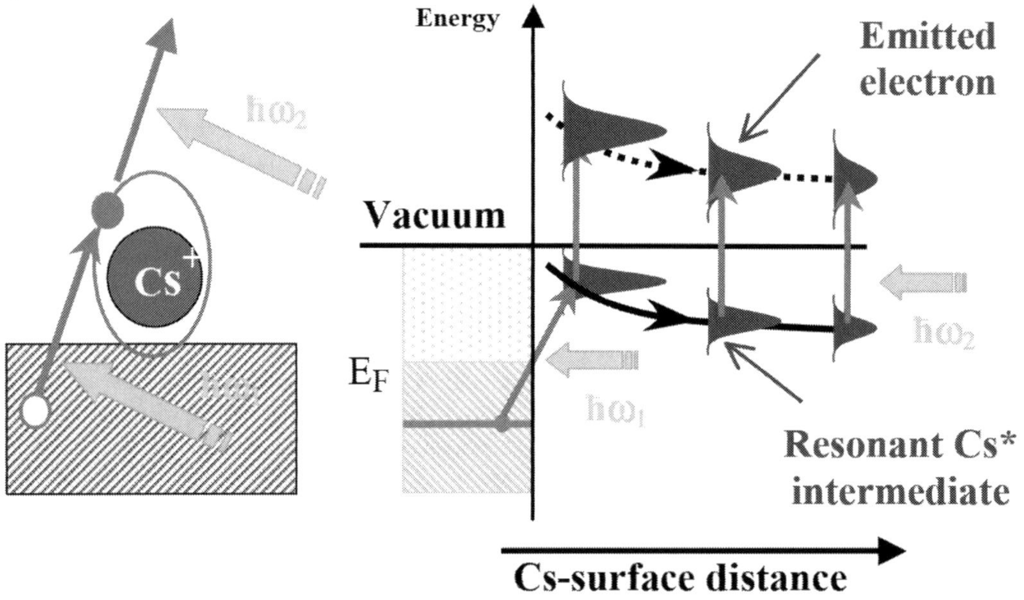

Figure 4: Left part: schematic view of the two-photon photo-emission process in the Cs/Cu(111) system: an electron from the substrate is excited to the Cs* state by a first photon and then photo-emitted by a second photon. Right part: schematic view of the evolution of the Cs*/substrate system. After the first photon absorption, the Cs-surface distance increases, leading to a downshift of the Cs* energy visible in the energy spectrum of electrons emitted by the second photon.

Figure 5 presents the energy shift of the maximum of the energy spectrum of the photo-emitted electrons in the Cs/Cu(111) system as a function of the time delay between the two laser pulses. Experimental data [20] and theoretical results [21,22] obtained from a parameter-free modelling of the photo-emission signal are compared. Results provide a direct view at the way the Cs* energy changes with time: one recognizes a quasi-parabolic behaviour corresponding to the accelerated classical motion on the Cs* potential energy curve of the Cs adsorbate initially at rest. One can stress that the energy change is small during the state lifetime ($\tau \cong 30$fs). Large energy shifts of the Cs* state, i.e. large energy transfers from electrons to heavy particle motion only occur for very late times when the survival probability in the Cs* state is very small and so these are unlikely. The energy shift varies quadratically with τ, confirming the paramount importance of the state lifetime for the efficiency of the evolution induced by the photo-excitation.

Figures 4 and 5 show that the Cs* energy and thus the energy of the photo-emitted electron change with the time delay between the two laser pulses. This demonstrates well the power of the TR-2PPE method in revealing the dynamics of an adsorbate-substrate system directly in real time. In several systems, it allowed the unveiling of the evolution of the system and the characterisation of the different steps in the rearrangement triggered by the photo-excitation: solvation at surfaces [23], small polaron [24] and 2D-polaron [25] formation. The time evolution of the excited state energy makes the quantitative analysis of TR-2PPE signals non-straightforward. Indeed, looking at the photo-emitted signal at a given energy (to isolate a given transient state) yields a signal that depends both on the state lifetime and on the rate of change of the transient state energy (the studied state moves out of the detection window). As a consequence, direct analysis of the TR-2PPE signal at fixed energy yields effective population decay times that a priori depend on the experimental conditions and in particular on the properties of the exciting laser [22,26]. Another possibility is to define 2PPE signals that follow the transient state change with time, but this requires a precise identification of the various contributions in the spectrum and possibly non-straightforward time-dependent deconvolutions.

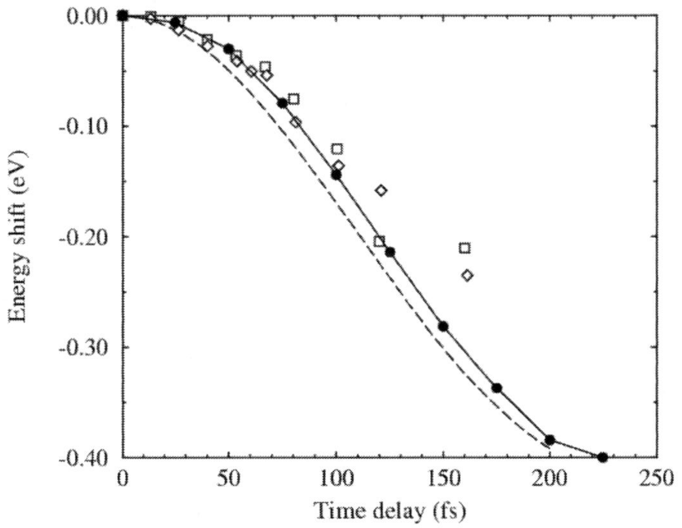

Figure 5: Maximum of the energy spectrum of the two-photon photo-emission from the Cs/Cu(111) system as a function of the time delay between the two photon pulses. Open diamonds and squares: experimental results from [20]; dashed line: theoretical results from [21] and full lines with circles: theoretical results from [22].

6. Desorption induced by electronic excitation

In the previous section, we have seen that the Cs* state excitation in the Cs/Cu(111) system induces an outward motion of the Cs adsorbate from the surface. This corresponds to the beginning of a desorption process and actually, if the energy transfer between the electrons and the nuclei induced by the excitation is large enough the Cs adsorbate will eventually desorb. This is an example of the well-known MGR (Menzel-Gomer-Redhead) [27] mechanism for photo-desorption and many photo-induced or electron-induced desorption processes have been interpreted along this mechanism (see a recent detailed review in [28]). In this mechanism, the excitation of a transient state triggers the desorbing motion of the adsorbate; if the lifetime of the transient state is long, enough energy can be transferred from the electrons to the nuclei to overcome the desorption barrier. In addition, depending on the charge state of the adsorbate at the time the energy transferred becomes large enough, ionic or neutral desorption can occur. If the energy transferred is smaller than needed to overcome the desorption barrier, it results in vibrational excitation of the adsorbate in its adsorption well.

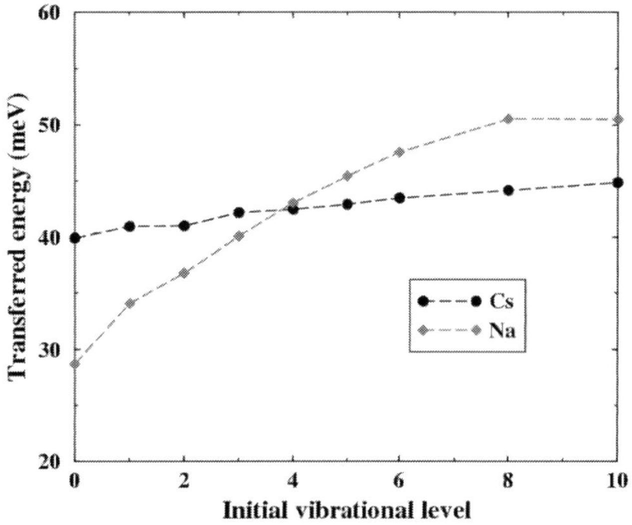

Figure 6: Mean energy transferred to the adsorbate motion following the laser excitation of the Alk* transient state in the Alk/Cu(111) systems (Alk=Na,Cs). The transferred energy (in meV) is given as a function of the initial vibrational level of the Alk adsorbate in its adsorption well (from reference [29]).

The mean energy transferred from electrons to nuclei in a photo-excitation process is presented in Figure 6 in the case of a laser photo-excitation of the Alk* state in the Alk/Cu(111) system (Alk=Na,Cs) [29]. It has been computed using the description discussed above of the Alk/Cu(111) systems. The adsorbate is initially in one of the vibrational levels in the adsorption potential energy well and the transferred energy is seen to weakly vary with the initial motion of the adsorbate. It appears that the energy transferred is of the same order of magnitude in the two systems. The effect of the Na lifetime, which is much shorter than that of Cs, is roughly balanced by the smaller Na mass and by the steeper Na* potential energy curve. Thus, photo-excitation of the Alk* state results in the vibrational heating of the adsorbate in its adsorption well. For both adsorbates, only a small fraction of the energy brought by the photon (of the order of 3eV) is transferred to the adsorbate motion and most of this energy ends up in exciting bulk electrons. The transferred energy is also much smaller than the desorption barrier which is of the order of 1.9 eV in both systems. As a consequence, one can expect desorption to be a weak process, corresponding to the very high-energy wing of the transferred energy distribution.

Photo-induced desorption following the excitation of the transient Cs* state has been observed indirectly in the Cs/Cu(111) systems: a variation of the surface work-function was recorded following the surface irradiation but no direct measurement of the outgoing Cs flux could be performed and only an upper limit of the desorption cross section was obtained. Figure 7 presents the desorption probability per absorbed photon in the Alk/Cu(111) system as a function of the initial energy of the adsorbate in the adsorption well. The desorption probabilities appear to be extremely small if the adsorbate is initially in its adsorption ground state and this is consistent with the experimental observations.

It also appears in figure 7 that an initial vibrational excitation of the adsorbate in the adsorption well results in a drastic increase of the desorption probability. Two effects are playing a role: first, an initial vibrational excitation of the adsorbate means that less energy is required to desorb and so the process is easier. Second and more important, an initial vibrational excitation means a larger velocity of the adsorbate along its trajectory, so that it travels further away before decaying. This effect is the same as the one met in the resonant dissociative attachment process in electron-molecule collisions. In this process, an electron colliding on a molecule is captured to form a transient negative molecular ion; the capture triggers an energy transfer from the electrons to the heavy particle motion, which, if large enough, leads to dissociation. If the transient state is short lived the process is weakly efficient and is strongly favored by an initial vibrational excitation of the target molecule (see e.g. for the H_2 molecule [30,31]).

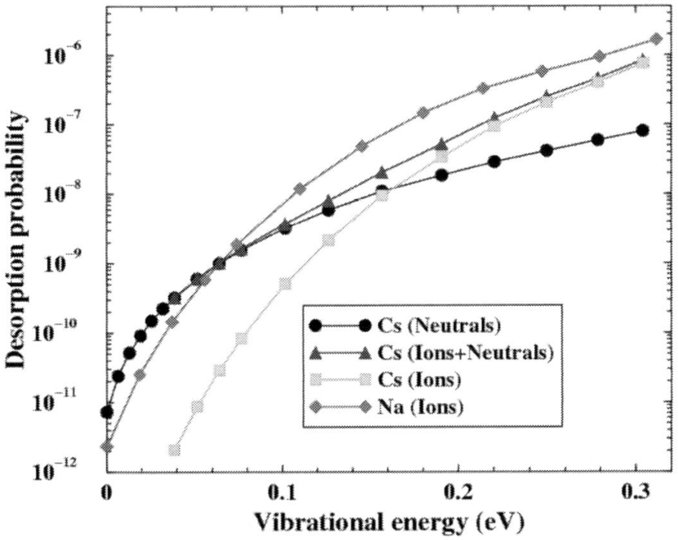

Figure 7: Desorption probability per absorbed photon in the Alk/Cu(111) systems (Alk=Na,Cs), as a function of the initial vibrational energy of the Alk adsorbate in its adsorption well. The desorption probability for the Cs adsorbates are separated according to the desorbing particle charge state. For Na, only ions desorb significantly (from reference [29]).

Finally, one can mention that excited electronic states localized on adsorbates can also be excited by electrons injected through an STM tip and lead to various reactions or atomic manipulations. As an example, one can mention that recently, such a process has been evidenced in the Li/Ag(100) system similar to the Alk/metal systems discussed at length here. An efficient adsorbate diffusion was found to be induced by electron transfer from the STM tip into an unoccupied state on the Li adsorbate, located around 3.6 eV above Fermi level [32]. A good candidate for this active intermediate is the Li* state similar to the ones studied presently (see a discussion of the various Alkali/Metal systems in [1]). Transferring an electron from the STM tip into this state could excite the adsorbate motion perpendicular to the surface in a way similar to the photo-excitation discussed above. Mode mixing, possibly induced by the STM bias, can then also result in the excitation of the adsorbate motion parallel to the surface, leading to the adsorbate manipulation process observed in [32].

7. Conclusions

Excited electronic states at surfaces are often invoked as reaction intermediates in various surface reaction mechanisms. We have reviewed a series of experimental situations where transient states play a substantial role: collisional charge transfer, photo-excitation of transient states localized on adsorbates, vibrational heating of adsorbates, photo-desorption, atomic manipulation by STM. Examples were chosen in the same class of systems, alkali interacting with a metal. It is remarkable that the same quasi-stationary state, corresponding to the transient capture of an electron by an alkali ion, is the active intermediate in all these reaction processes.

Long excited state lifetimes are of paramount importance for the efficiency of the induced processes. On a free-electron metal surface, resonant electron transfer is usually very fast and this makes excited states very short-lived and so weakly effective as reaction intermediates. In contrast, the presence of a surface-projected band gap strongly alters the resonant electron transfer process leading to the possibility of long-lived states. Alkalis on certain noble metal surfaces present striking examples of these stabilization effects.

References

[1] Gauyacq J P, Borisov A G, and Bauer M 2007 *Prog. Surf. Sci.* **82**, 244

[2] Geerlings J J C and Los J 1990 *Phys. Rep.* **190** 133

[3] Lorente N and Monreal R 1997 *Surf.Sci.* **370** 324
Baragolia R A and Dukes C A 1996 *Phys.Rev.Lett.* **76** 2547

[4] Ermoshin V A and Kazansky A K 1996 *Phys.Lett.* A **218** 99
[5] Borisov A G, Kazansky A K and Gauyacq J P 1999 *Phys. Rev.* B **59** 10935
[6] Desjonquères MC and Spanjaard D 1993 *Concepts in Surface Science, Springer-Verlag Series in Surface Science* 40
[7] Borisov A G, Kazansky A K and Gauyacq J P 1999 *Surf. Sci.* **430** 165
[8] Borisov A G, Gauyacq J P, Kazansky A K, Chulkov E V, Silkin V M, Echenique P M 2001 *Phys. Rev. Lett.* **86**, 488
[9] Geerlings J J C, Los J, Gauyacq J P and Temme N M 1986 *Surf.Sci.* **172** 257
Langreth D and Nordlander P 1991 *Phys.Rev.* B **43** 2541
[10] Zimny R 1990 *Surf.Sci.* **233** 333
Bahrim B, Teillet-Billy D and Gauyacq J P 1994 *Phys.Rev.* B **50** 7860
[11] Borisov A G, Teillet-Billy D and Gauyacq J P 1992 *Phys.Rev.Lett.* **68** 257
Borisov A G, Teillet-Billy D, Gauyacq J P, Winter H and Dierkes G 1996 *Phys.Rev.* B **54** 17166
Lorente N, Merino J, Flores F and Gusev M Yu 1997 *Nucl.Inst.Meth.* B **125**, 277
Maazouz M, Borisov A G, Esaulov V A, Gauyacq J P, Guillemot L, Lacombe S and Teillet-Billy D 1997 *Phys. Rev.* B **55** 13869
Hill S B, Haich C B, Zhou Z, Nordlander P and Dunning F B 2000 *Phys.Rev.Lett.* **85** 5444
Niedfeldt K, Carter E A and Nordlander P 2004 *J.Chem. Phys.* **121** 3751
[12] Van Wunnick J N M, Brako R, Makoshi K and Newns D M 1983 *Surf. Sci.* **126** 618
Winter H 2002 *Phys.Rep.* **367** 387
[13] Hecht T, Winter H, Borisov A G, Gauyacq J P and Kazansky A K 2000 *Phys.Rev.Lett.* 84 2517
[14] Borisov A G, Mertens A, Wethekam S and Winter H 2003 *Phys.Rev.* A **68** 012901
[15] Guillemot L and Esaulov V A 1999 *Phys. Rev. Lett.* **82** 4552
[16] Canario A R, Borisov A G, Gauyacq J P and Esaulov V A 2005 *Phys.Rev.* B **71** 121401(R)
[17] Petek H and Ogawa S 1997 *Prog. in Surf. Sci.* **56** 239
[18] Bauer M, Pawlik S and Aeschlimann M 1997 *Phys. Rev.* B **55**, 10040
Bauer M, Pawlik S, Burgermeister R and Aeschlimann M 1998 *Surf. Sci.* **402** 62
Bauer M, Pawlik S and Aeschlimann M 1999 *Phys. Rev.* B **60** 5016
[19] Ogawa S, Nagano H and Petek H 1999 *Phys. Rev. Lett.* **82** 1931
Petek H, Nagano H, Weida M J, and Ogawa S 2001 *J. Phys. Chem.* B **105** 6767
[20] Petek H, Weida M J, Nagano H and Ogawa S 2000 *Science* **288** 1402
[21] Borisov A G, Kazansky A K and Gauyacq J P 2001 *Phys. Rev.* B RC **64**, 201105
[22] Gauyacq J P and Kazansky A K 2005 *Phys. Rev.* B **72** 045418
[23] Gahl C, Bovensiepen U, Frischkorn C and Wolf M 2002 *Phys.Rev. Lett.* **89** 107402
[24] Ge N H, Wong C M, Lingle R L, McNeill J D, Gaffney K J and Harris C B 1998 *Science* **279** 202
[25] Muntwiler M and Xu X Y 2007 *Phys.Rev.Lett* **98** 246801
[26] Bauer M, Wessendorf M, Hoffmann D, Wiemann C, Mönnich A and Aeschlimann M 2005 *Appl.Phys.* A **80** 987
[27] Menzel D and Gomer R 1964 *J.Chem.Phys.* **41** 3311
Redhead P A 1964 *Can.J.Phys.* 42 886
[28] Saalfrank P 2006 *Chem. Rev.* **106** 4116
[29] Gauyacq J P and Kazansky A K 2007 *Surf.Sci.* **601** 5473
[30] Allan M and S.F.Wong S F 1978 *Phys.Rev.Lett.* **41** 1791.
[31] Gauyacq J P 1985 *J.Phys.B* **18** 1859
[32] Simic-Milosevic V, Heyde M, Nilius N, Nowicki M, Rust H P and Freund H J 2007 *Phys.Rev.* B **75** 195416

Plasma-surface interactions

M.J. Goeckner, C.T. Nelson, S.P. Sant, A.K. Jindal, E.A. Joseph, B.S. Zhou, G. Padron-Wells, B. Jarvis, R. Pierce and L.J. Overzet

Department of Electrical Engineering, University of Texas at Dallas

goeckner@utdallas.edu

Abstract. Materials processing is at a crossroads. Currently a large fraction of industrially viable materials processing is via plasmas. Until recently it has been economical to just examine the influence the plasma properties on the desired surface processes and through this ultimately optimize manufacturing. For example, it is well known that the surface processes (etch or deposition), occur in the top few mono-layers of the surface. Thus, in film growth one requires that molecules from the gas-phase land and bond on the surface. However as processing has reached the nano-scale, development of viable processes has become more and more difficult. In part, this is because of all of the free parameters that exist in plasmas. To overcome this economic issue, tool vendors and semiconductor companies have turned to complex computational models of processing plasmas. For those models to work, one requires a through understanding of all of the gas-phase and surface-phase processes that are exhibited in plasmas. Unfortunately, these processes, particularly those at the surface, are not well understood. In this article we describe a viable model of the surface-phase based on cross sections for processes that occur. While originally developed of fluorocarbon systems, the model also appears to be applicable to hydrocarbon systems.

1. Introduction and Significance

1.1. Historical economic background

Currently a large fraction of industrially viable materials processing is via plasma processes. This is because of the unique ability of plasmas to produce surface conditions that one cannot reach under standard thermal equilibrium conditions. This unique ability is due primarily part to the large number of free, yet intertwined, parameters available to plasma systems. However as this processing has reached the nano-scale, development of industrially viable processes has become more and more difficult. In part this is because of all of the intertwined processes that exist in plasma. Unfortunately, these processes are not well understood. [1,2] As such, simple experimentation is no longer an economically viable method of process tool development.

While there are many possible uses for plasma produced films and devices, only those that can be developed below a certain cost threshold will ever be produced in significant numbers. In a simplistic sense, this cost threshold is set by the number of a particular device produced times the sales price. Under this simplified assumption one finds the maximum development cost shown in **Figure 1**. Thus, in 2007 there were ~300 million CPUs sold worldwide. Assuming a purchase price of \$200 would suggest that one might have ~\$0.5 billion available for product development. Applying this same simple calculation to RFID tags one might find that one could only spend ~\$10 million for development. Here we assume 1¢ for the purchase price with 10 billion produced. (These values are based on the needs of 'Big Box' retailers such as Wal-Mart.) Development cost must be even lower for specialized medical devices. In such cases, one might sell 100 devices at \$10,000 each. This makes the development of such devices uneconomical using standard development techniques.

Because of these cost consideration, one must develop a deeper understanding of the fundamental plasma processes that so greatly affect materials processing. Specifically in this article we will examine what is arguably the critical issue, the surface deposition/etch processes. Understanding those fundamental processes should allow one to develop the necessary processing conditions so as to create new methods for synthesis of nano-particles, 3-D nanostructures, thin films with unique chemical functionality and other systems that one can only dream of presently.

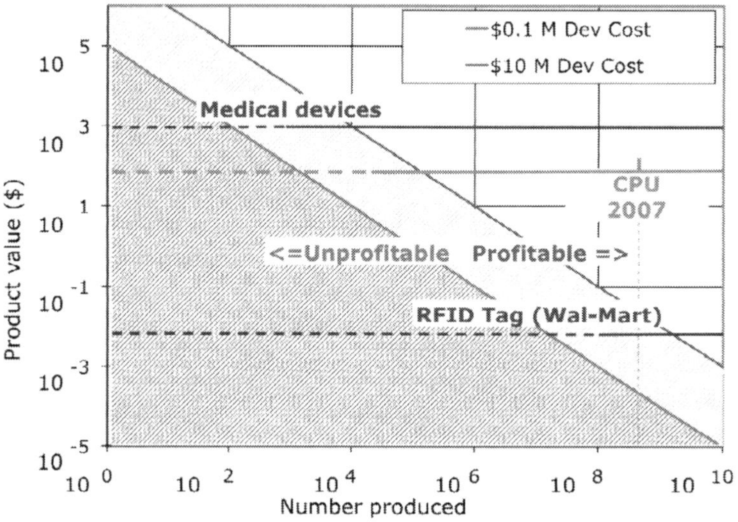

Figure 1: Estimation of possible maximum development cost for various devices.

1.2. Historical scientific background

Plasma processes have been used for semiconductor device fabrication since about 1970. [3] It is important to note that they are particularly complex. A typical oxide etch process can involve multiple steps to achieve the desired critical dimensions (CD), selectivity to the underlying Si substrate and profile control. While not universally the case, oxide etch processes are typically executed using high-density capacitively coupled plasmas. These plasmas are usually low pressure and have a large chamber surface area relative to the wafer area. Etch products are ejected from the wafer at high energies (several eV) [4] and this can lead to their transport from the wafer to opposing surfaces. These surfaces are subsequently coated and etched leading to transport back to the wafer. Etched material can readily deposit on surfaces that are not heavily bombarded by ions (e.g. feature sidewalls) leading to difficulties with profile control or etch stop.

Early studies of such plasma process tools focused on understanding the basic mechanisms. In the book by Chapman [3] (published in 1980) plasma etch (CF_4 or CF_4/O_2 on Si) was described simply by the reaction products (SiF_4 and C_2F_4 or CO, CO_2, COF, COF_2, SiF_4). Critical to this early work was that by Coburn and Winters [5] showing the link between radical flux and ion flux in inducing faster etch. By the mid-80s, considerably more was known about the fluorocarbon plasmas used for Si/SiO_2 etching. [6] At that point, models of the chemistry gave reasonable agreement with experiments. [7,8] Such models made use of large tables of electron impact dissociation and ionization cross-sections. [6] Despite this agreement, major gaps still existed. For example, the cross section tables were and are still incomplete. (Part of this is due to changes in feed-gas chemistry. Part of it is due to not having all of the necessary cross sections.) More importantly, surface processes were at best poorly understood. Specifically, as recently as 2004, when discussing the results from the well known "*Hybrid Plasma Equipment Model*" (HPEM), [9] Vasenkov, Oehrlien and Kushner note that the surface processes where unknown and state "*We acknowledge that the dispositions of surface reactions are critical to the development of a successful reaction mechanism for the low-pressure plasmas of interest.*" [10] Toward this end, numerous experiments and computer simulations have been undertaken to overcome that lack of knowledge. A <u>partial</u> <u>list</u> of those efforts include: Fisher [11,12,13] (interaction of plasma generated radical beams with surfaces), Aydil [14,15,16,17,18,19,20,21] (surface coverage and chemistry in plasma processing tools), Graves [22,23,24,25,26,27,28,29,30,31] (prepared radical and ion beam interactions with surfaces), Oehrlein [32,33,34,35,36] (polymerization and surface chemistry in C_xF_y plasmas), and Economou [37,38,39,40] (numerical simulations of plasma flow and chemistry). Finally there is our work [41,42,43,44,45,46,47] on experimental measurements of a plasma species sticking coefficients. For example, CF_3^+ ions were found to stick at a rate of 2-5%. More data from those studies will be examined below.

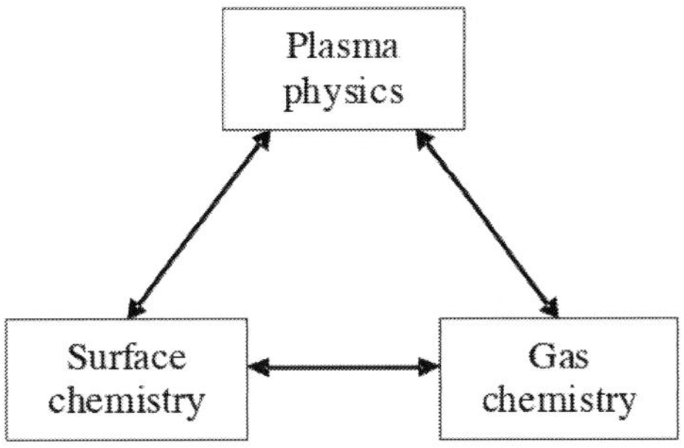

Figure 2: Interaction of various aspects of laboratory plasmas.

To help understand all of the processes that can occur in processing plasmas, we subdivide the system into three main interacting subsystems 1) the surface chemistry, 2) the gas-phase chemistry and 3) the plasma physics. Each subsystem interacts with the others, as is shown in Figure 2. For example, the surface coverage of an adsorbed species in the surface subsystem may depend on the rate of sputtering by the ion flux from the plasma physics modules. The electron temperature in the plasma physics subsystem may depend on the evolution of attaching species from the surface. We know that surface chemistry models and molecular dynamics simulations have been successful in separately addressing bulk plasma and surface processes. [48,49,50,51,52] However, even after almost 40 years of research, there have been few efforts in which studies have self-consistently combined plasma and surface processes to simulate low-pressure systems where wall processes may dominate. As a result, our fundamental understanding of the consequences of plasma surface interactions on the behavior of the bulk plasma would be significantly improved by a comprehensive experimental and computational investigation of these processes. [53,54]

One can understand how these interactions, important to both understanding the fundamental science as well as optimizing manufacturing, ultimately influence the desired plasma properties by examining the surface processes. It is well known [6] that the surface processes (etch or deposition), occur in the top few mono-layers of the surface. For example, growth of a film will require that molecules from the gas phase land and bond on the surface. In order to understand the mechanism of deposition and etch on a surface, we begin with a basic reaction rate density [55,42,43,44,45]

$$R_{particle} = \int_0^\infty \sigma(\varepsilon,T) n_{bonds} v_{particle} n_{particle} f_{particle}(\varepsilon) d\varepsilon \qquad (1)$$

Here $\sigma(\theta,\varepsilon,T)$ is defined as the cross section of a particle of energy ε θ of either filling or creating an available bond site on the a surface of temperature T, n_{bonds} is the density of surface bonds, $v_{particle}$ is the gas particle velocity, $n_{particle}$ is the gas particle density and $f_{particle}(\varepsilon)$ is the gas energy distribution function. Assuming that the fraction of bond sites that are open is Θ, then at any given time the open bond density will be $\Theta\ n_{bonds}$ and the closed bond density is $(1-\Theta)\ n_{bonds}$. Under the assumption that most of these processes occur at open bond sites, our group has arrived at a general formula for the process rate (here applied to C_xF_y etch of SiO_2) of: [43,44]

$$\text{Process Rate} = \left[\begin{array}{l} \sum \frac{K_{ions}}{\rho_{C_xF_y}} \langle \Gamma_{ions} \rangle S.C_{ions} + \sum\limits_{deposit} \frac{K_{C_xF_y}}{\rho_{C_xF_y}} \langle \Gamma_{C_xF_y} \rangle S.C_{C_xF_y} \\ -\sum \frac{K_{ions}}{\rho_{C_xF_y}} \langle \Gamma_{ions,FC} \rangle Y_{ions,FC} - \sum\limits_{etches} \frac{K_{C_xF_y}}{\rho_{C_xF_y}} \langle \Gamma_{C_xF_y} \rangle Y_{C_xF_y} \\ -\sum \frac{K_{C_xF_y}}{\rho_{SiO_2}} \langle \Gamma_{ions,SiO_2} \rangle Y_{ions,SiO_2} - \sum\limits_{etches} \frac{K_{C_xF_y}}{\rho_{SiO_2}} \langle \Gamma_{C_xF_y} \rangle Y_{C_xF_y,SiO_2} \end{array}\right] \quad (2)$$

Here, S.C or Y $\approx \langle R_{particle} \Theta * n_{Surface} \rangle$ is the sticking coefficient or yield, K_X is a species dependent proportionality constant, $\langle \Gamma_X \rangle$ is the particle flux, and ρ_X is the surface material density. Inclusion of closed bond sites results in an additional set of six terms that are identical to those above, except $S.C_{closed}$ or $Y_{closed} \approx \langle R_{particle}(1-\Theta) * n_{Surface} \rangle$.

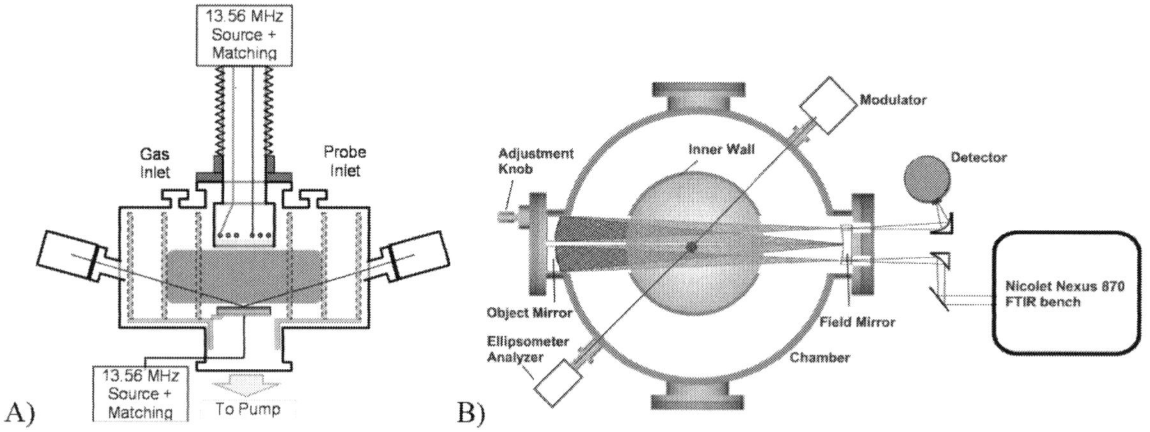

Figure. 3: A) Side and B) top views of the modified Gaseous Electronics Conference (mGEC) system. The system is unique in that effective radius (internal walls) and source to chuck gap can be readily varied. This allows significant variation in process conditions.

From Equation 2 it is now clear how a given interaction, hence process rate, is related to the gas-phase chemistry and the plasma. Toward this end we have developed a reactor system, see **Figure. 3**, that allows 'independent' control of important parameters yet maintains the holistic nature of plasma processes. [41] This 'independent' control is achievable because we can readily change the gas chemistry and/or chamber geometry. The former variable, gas chemistry, allows one to partially decouple the gas/plasma combination from the surface. It is only a partial decoupling as the plasma ions, electron temperature, space potential are all tied to the feed gas chemistry. Fortunately, the later variability, chamber geometry, allows us to decouple the plasma-surface interactions from the plasma-gas interactions as we can use the geometry to independently control ion flux from ion species. For example, simple changes in the reactor geometry can allow us to vary the plasma density near the chuck by over an order of magnitude while maintaining everything else almost constant. [43,44,45] See **Figure 4**. While such a tool might provide industrial advantages, we needed this flexibility to directly determine the influence of the ions and radicals in thin film deposition and etch. The ability to vary the contributions from one component while holding the others constant is crucial for unraveling the processes in thin film etch and deposition because of the complex interrelationships between species in the plasma and on surfaces. This can be contrasted with experiments in typical plasma systems in which if one wanted to change the ion flux, by increasing the plasma power, unavoidable changes also occurred in several other plasma characteristics, such as increased feedstock dissociation. These unavoidable changes cause great difficulty in determining which characteristics - or perhaps the interactions between them - caused observed changes in the deposition/etch. Likewise, unraveling such information using just one species (in a beam experiment for example) is difficult. Plasmas have multiple species that influence the result in a holistic fashion so all species need to be present in appropriate quantities. Changing our reactor geometry in a well-characterized fashion, allows us to unravel some of these effects.

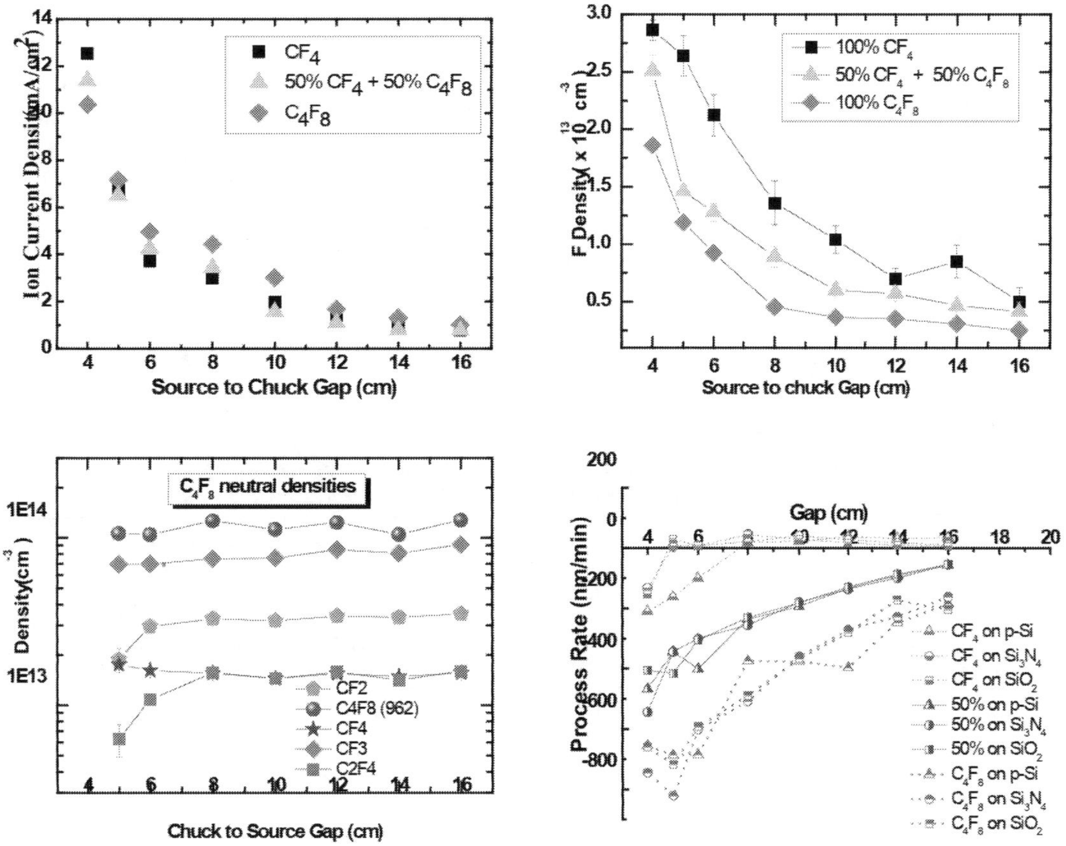

Figure 4: Measured variation in A) Ion current, B) Fluorine Density, C) Other specie density, and D) Process rates. All of these are linked through Equation 2. Data from [44].

Using the mGEC we where able to vary the contact between the plasma and the chuck. In addition to this we used combinations of CF_4 and C_4F_8 to 'independently' control the density of various radicals, see **Figure 4**. Of particular importance is the fact that both the ion current and F density each drop by more than an order of magnitude as the gap is widened. Meanwhile the radical densities are approximately constant, particularly for gaps greater than 8 cm. This variation allows us to look at the process rate at large gaps, effectively eliminating the influence of the ions and the F. From this we are able to obtain upper bound 'sticking coefficients' for the various radicals in the discharge, see **Table 1**. Next we examined the etch rate vs. the normalized flux of ions and F. As seen in **Figure 5**A, at large gaps the rate is only dependent on the ion flux. Thus we can use this to determine the 'sticking coefficients' of the various ions in the system. These values are also given in **Table 1**. Finally, at large gap the ion deposition must balance the etch yield due to F, resulting in an etch yield of ~3.7%. Taking into account these values, we find good fit with independent experiments. Here we again measured the process rate and the densities of the various species. In fact all of our data fit within approximately 20% error bar for our fit values in **Table 1**. See **Figure 5**B as example.

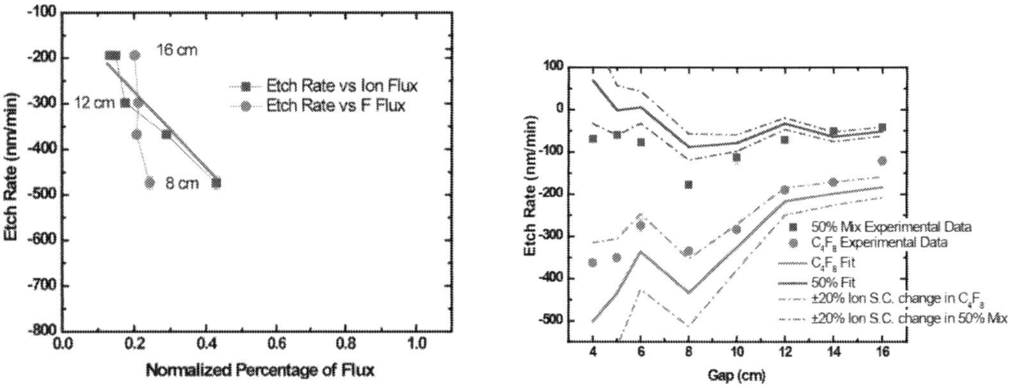

Figure 5: A) Etch rate vs. normalized ion and F flux to the surface. B) Process rate fit sensitivity to a 20% change in the ion-sticking coefficient. Variations in other parameters show similar sensitivity. Data from [44].

Table 1.Upper bound sticking coefficients for radicals in 100% C_4F_8 and 50% C_4F_8 and 50% CF_4 feed gas mixtures and calculated ion sticking coefficients. CF_3^+ sticking coefficient was determined from other data sets. [44].

Molecule/Assumed dominant ion	**Upper Bound - 100% C_4F_8**	**Upper Bound Value - 50% C_4F_8**	**Lowest Upper Bound S.C.**
CF_2	0.57% ±0.07	0.20% ±0.01	0.20% ±0.01
C_2F_4	1.48% ±0.18	1.54% ±0.23	1.48% ±0.18
l-C_4F_8	0.18% ±0.02	0.22% ±0.01	0.18% ±0.02
CF_3	0.45% ±0.05	0.29% ±0.01	0.29% ±0.01
$C_2F_4^+$	15.8% ±0.2		15.8% ±0.2
CF_3^+ / $C_2F_4^+$ mix		15.2% ±0.2	15.2% ±0.2
CF_3^+ (pure CF_4)			1.8-3%

1.3. Future of plasma processing

Plasmas have been used for nearly 40 years for semiconductor processing. At the current technology node, 45 nm, this development is nearing an end. This 'end' must come either at 'single' atom transistors or more likely somewhat before that size. As such, develop will likely end in the next 10 years, unless major changes are made in the material sets used for semiconductor devices.

Plasmas have also been used for years to do other types of processing. Among these are coatings for various applications, ranging from hardening to biological interfaces. It is the latter of these uses that likely has the largest economic potential and yet also is likely the most difficult to achieve. Both the potential and the difficulty are directly related to the potential wide range of uses of specialized films that might be employed. Indeed it is the wide range of possible gas precursors that makes it economically untenable to repeat the development technique used in the semiconductor industry. Thus we need to know if our model of process rates obtained for fluorocarbon discharges can be used for hydrocarbon (or more generally organic) discharges.

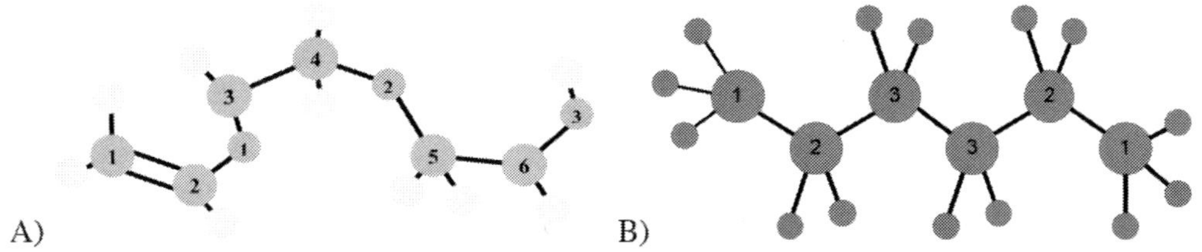

Figure 6 Structures with atoms numbered for A) DEGVE ($C_6O_3H_{10}$) and B) n-Hexane (C_6H_{10}).

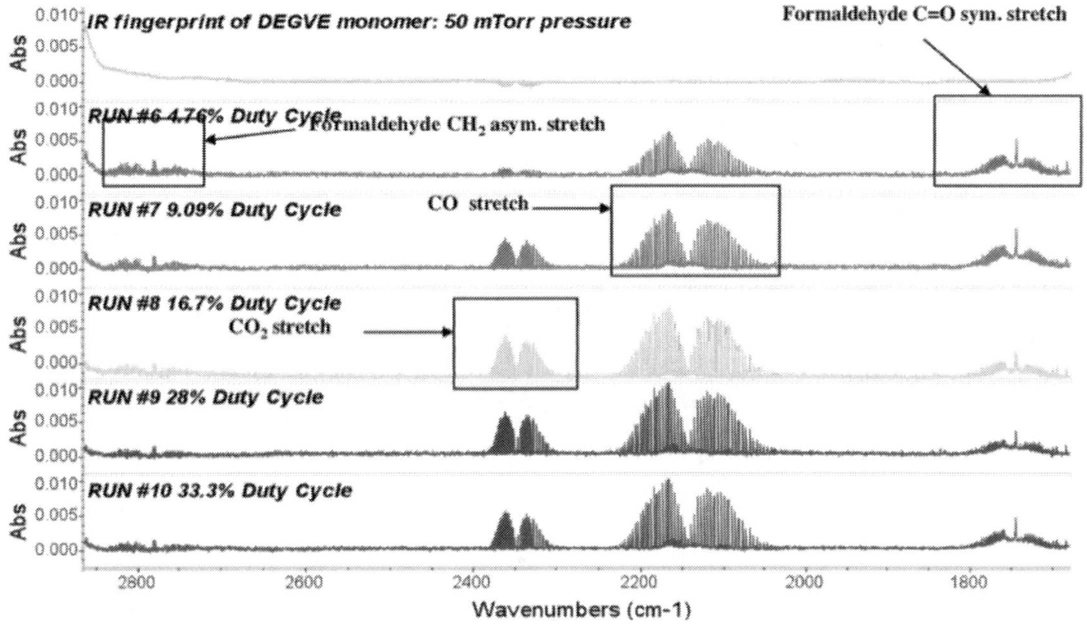

Figure 7 Spectra of various DEGVE discharges showing the production of formaldehyde, CO and CO_2 in the discharge.

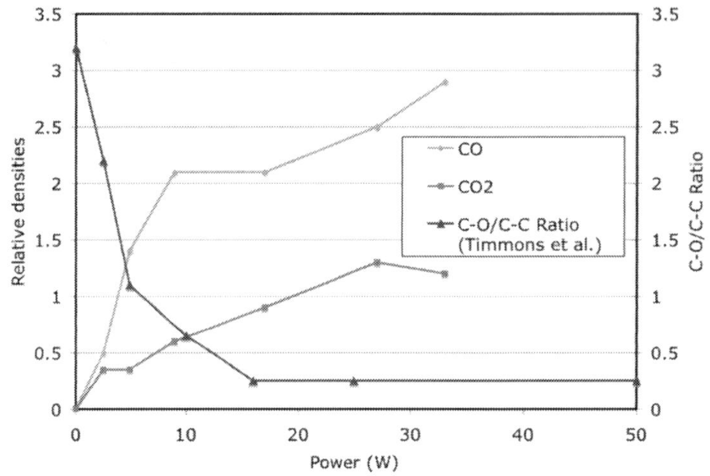

Figure 8 Comparison to relative CO and CO_2 densities to C-O/C-C film ratios as measured by Timmons *et al*. [56]

To determine if the model is appropriate for organic based discharges, we have examined a number of plasmas with organic feedstock. We will consider two in particular here, Di-(ethylene-glycol)-vinyl-ether (DEGVE) pulsed plasmas and n-hexane pulsed plasmas. We find that in the DEGVE discharge the following processes must occur. Scission of the C1-C2, C2-O1-C3 and C5-C6 bond structures shown in Figure 6. The degree of these cleavage reactions depends on the average power delivered to the plasma. In addition we observe a number of abstraction reactions including the following:

$CH_2(g) + \text{Surface} \rightarrow CH_3(g) + \text{Surface} \rightarrow CH_4(g)$

$CH_2(g) + \text{Surface} \rightarrow C_2H_4(g)$

$CH_2{=}CHO\ (g) + \text{Surface} \rightarrow C_2H_4\ (g)$

Such abstraction reactions follow the general form of the $-\sum \frac{K_{ions}}{\rho_{C_xF_y}} \langle \Gamma_{ions,FC} \rangle Y_{ions,FC}$ and $\sum_{deposit} \frac{K_{C_xF_y}}{\rho_{C_xF_y}} \langle \Gamma_{C_xF_y} \rangle S.C_{C_xF_y}$ portions of Equation 2. We find similar results for the n-Hexane discharge. In that case we find the following scission and abstraction reactions:

C1-C2 Scission

$CH_3 + \text{Surface} \rightarrow CH_4$

$C_5H_{11} + \text{Surface} \rightarrow C_5H_{12}$

C2-C3 Scission

$C_2H_5 + \text{Surface} \rightarrow C_2H_4$

$C_2H_5 + \text{Surface} \rightarrow C_2H_2$

$C_4H_9 + \text{Surface} \rightarrow C_4H_{10}$

C3-C3 Scission

$C_3H_7 + \text{Surface} \rightarrow C_3H_8$

2. Conclusions

Based on these measurements and our model of the film-processing rate, we are able to determine upper bounds for the sticking coefficients for a number of the molecular species observed in fluorocarbon discharges. In addition it is important to note that the ions play a major role in the deposition process. It is also observed that hydrocarbon discharges seem to follow the same pattern of film growth. This implies that our process rate equation appears to be fundamental in nature.

References

[1] Kogelschatz M, Cunge G, Joubert O, Vallier L, Sadeghi N, 2004 *Contrib. Plasma Phys.* **44**, 413

[2] Joubert O, Cunge G, Pelissier B, Vallier L, Kogelschatz M, Pargon E, 2004 *J. Vac. Sci. Technol. A* **22**, 553

[3] Brian Chapman Glow Discharge Processes (Wiley, New York, 1980)

[4] Smirnov VV, Stengach AV, Gaynullin KG, Pavlovsky VA, Rauf S, Stout PJ, Ventzek PLG, 2005 *J. Appl. Phys.* **97**, 093302

[5] Coburn JW, Winters HF, 1978 *J. Vac. Sci. Tech.* **15**, 327

[6] Flamm DL in: Plasma Etching an Introduction, edited by DM Manos and DL Flamm, (Academic Press, New York, 1986)

[7] Plumb IC, Ryan KR, 1986 *Plasma Chem. Proc.* **6**, 205

[8] Edelson D, Flamm DL, 1982 *J. Appl. Phys.* **56**, 1522

[9] Kushner MJ, 1982 *J. Appl. Phys.* **53**, 2923

[10] Vasenkov AV, Li X, Oehrlein GS, Kushner MJ, 2004 *J. Vac. Sci. Tech. A* **22**, 511

[11] Martin IT, Fisher ER, 2004 *J. Vac. Sci. Tech. A* **22**, 2168

[12] Zhang JM, Fisher ER, 2004 *J. Physical Chem. B* **108**, 9821

[13] Zhang JM, Fisher ER, 2004 *J Appl. Phys.* **96**, 1094

[14] Agarwal S, Quax GWW, van de Sanden MCM, Maroudas D, Aydil ES, 2004 *J. Vac. Sci. Tech. A* **22**, 71

[15] Sriraman S, Aydil ES, Maroudas D, 2004 *J Appl. Phys.* **95**, 1792

[16] Kim TW, Aydil ES, 2003 *J. Electrochem. Soc.* **150**, G418

[17] Ullal SJ, Kim TW, Vahedi V, Aydil ES, 2003 *J. Vac. Sci. Tech. A* **21**, 589

[18] Marra DC, Kessels WMM, van de Sanden MCM, Kashefizadeh K, Aydil ES, 2003 *Surf. Sci.* **530**, 1

[19] Edelberg EA, Perry A, Benjamin N, Aydil ES, 1999 *J. Vac. Sci. Tech. A* **17**, 506

[20] Ramalingam S, Maroudas D, Aydil ES, Walch SP, 1998 *Surface Science Letters* **418**, L8

[21] Ramalingam S, Maroudas D, Aydil ES, 1998 *J. Appl. Phys.* **84**, 3895

[22] Kimura Y, Coburn JW, Graves DB, 2004 *J. Vac. Sci. Technol. A* **22**, 2508

[23] Humbird D, Graves DB, 2004 *J Appl. Phys.* **96**, 2466

[24] Humbird D, Graves DB, 2004 *J Appl. Phys.* **96**, 65

[25] Humbird D, Graves DB, 2004 *J Appl. Phys.* **96**, 791

[26] Humbird D, Graves DB, Hua XF, Oehrlein GS, 2004 *Appl Phys Lett.* **84**, 1073

[27] Greer F, Fraser D, Coburn, JW, Graves DB, 2003 *J Appl. Phys.* **94**, 7453

[28] Abrams CF, Graves DB, 1999 *J. Appl. Phys.* **86**, 5938

[29] Helmer B, Graves DB, 1999 *J. Vac. Sci. Tech. A* **17**, 2759

[30] Singh H, Coburn JW, Graves DB, 1999 *J. Vac. Sci. Tech. A* **17**, 2447

[31] Kota G, Coburn JW, Graves DB, 1999 *J. Vac. Sci. Tech. A* **17**, 282

[32] Li X, Ling L, Hua XF, Fukasawa M, Oehrlein GS, Wang YC, Vasenkov AV, Kushner MJ, 2004 *J. Vac. Sci. Tech. A* **22**, 500

[33] Vasenkov AV, Li X, Oehrlein GS, Kushner MJ, 2004 *J. Vac. Sci. Tech. A* **22**, 511

[34] Standaert TEFM, Hedlund C, Joseph EA, Oehrlein GS, Dalton TJ, 2004 *J. Vac. Sci. Tech. A* **22**, 53

[35] Li X, Ling L, Hua XF, Fukasawa M, Oehrlein GS, Barela M, Anderson HM, 2003 *J. Vac. Sci. Tech. A* **21**, 284

[36] Schaepkens M, Martini I, Sanjuan EA, Li X, Oehrlein GS, Perry WL, Anderson HW, 2001 *J.*

Vac. Sci. Tech. A **19**, 2946
[37] Nam SK, Economou DJ, 2004 *J Appl. Phys.* **95**, 2272
[38] Kim D, Economou DJ, 2004 *J Appl. Phys.* **95**, 3311
[39] Kim D, Economou DJ, 2003 J. Vac. Sci. Technol. B **21**, 1248
[40] Ramamurthi B, Economou DJ, 2002 *J. Vac. Sci. Tech. A* **20**, 467
[41] Goeckner MJ, Marquis JM, Markham BJ, Jindal AK, Joseph EA, Zhou B-S, 2004 *Rev. Sci. Instr.* **75**, 884
[42] Joseph EA, Zhou B-S, Sant SP, Overzet LJ, Goeckner MJ, 2004 *J. Vac. Sci. Tech. A* **22**, 689
[43] S.P. Sant, PhD Dissertation "Understanding The Complex Chemistry Of Fluorocarbon Plasmas In A Long Residence Time ICP Reactor" University of Texas at Dallas, MJ Goeckner Advisor 2006
[44] Nelson CT, Sant SP, Overzet LJ, Goeckner MJ, 2007 *Plasma Sources Science & Technology* **16**, 813
[45] Joseph EA, Zhou B-S, Sant SP, Overzet LJ, Goeckner MJ, 2008 *J. Vac. Sci. Tech. A* in press.
[46] Zhou B-S, Joseph EA, Overzet LJ, Goeckner MJ, 2006 *J. Vac. Sci. Tech. A* **24**, 114
[47] Zhou B-S, Joseph EA, Sant SP, Overzet LJ, Goeckner MJ, 2005 *J. Vac. Sci. Tech. A* **23**, 1657
[48] Collison WZ, Ni TQ, Barnes MS, 1998 *J. Vac. Sci. Tech. A* **16**, 100
[49] Meeks E, Larson RS, Vosen SR, Shon JW, 1997 *J. Electrochem. Soc.* **144**, 357
[50] Helmer BA, Graves DB, 1998 *J. Vac. Sci. Tech. A* **16**, 3502
[51] Hanson DE, Voter AF and Kress JD, 1997 *J. Appl. Phys.* **82**, 3552
[52] Mantzaris N, Boudouvis A, Gogolides E, 1995 *J. Appl. Phys.* **77**, 6169
[53] Kastenmeier BEE, Matsuo PJ, Oehrlein GS, Langan JG, 1998, *J. Vac. Sci. Tech. A* **16**, 2047
[54] Nishikawa K, Oomori T, Ono K, *J. Vac. Sci. Tech. B* **17**, 127
[55] Schaepkens M, Oehrlein GS, Cook JM. 2000, *J. Vac. Sci. Tech. B* **18**, 856
[56] Wu YJ, Timmons RB, Jen JS, Molock FE, 2000 *Colloids & Surf. B: Biointerfaces*, **18**, 235

Interactions of ions with carbon nano-structures

Z. L. Mišković
Department of Applied Mathematics, University of Waterloo, Waterloo, Ontario, Canada N2L 3G1
E-mail: `zmiskovi@math.uwaterloo.ca`

Abstract. We review some aspects of modeling interactions of charged particles with carbon nano-structures where dynamic response of carbon valence electrons plays prominent roles. Specifically, we evaluate the effects of plasmon excitations on the image and stopping forces on fast ions and dipolar molecules channeled through carbon nanotubes, as well as the effects of single-electron excitations in graphene's π band on the image and stopping forces on slow ions grazingly scattered from a supported graphene.

1. Introduction

Investigation into the properties of fullerene molecules, carbon nanotubes, and the recently contrived graphene has been growing at a relentless rate for a decade or so, promoting carbon nano-structures among the hottest topics in Physics [1]. While interactions with particle beams have been an important part of this research endeavor in the context of various spectroscopic techniques (such as TEM, EELS, ...), the use of energetic electron and ion beams has recently emerged as a novel engineering tool for modification of structure and properties of carbon nano-structures [2]. Namely, while the irradiation of solids by energetic particles is usually perceived to cause undesired and uncontrolled disorder, recent experiments have demonstrated that many properties of carbon nano-structures can be tailored with high precision using particle beams. Moreover, carbon nano-structures exhibit exceptional propensity towards self-organization and self-assembly following such irradiation [2].

Examples of recent experimental demonstrations of the effects of ion irradiation of carbon nano-structures include: doping and chemical functionalization, creation of local defects (requiring typical energy transfer of some 20 eV per carbon atom) which can change electrical and thermal conductance, creation of various kinds of inter-junctions which can affect electrical and especially mechanical properties of carbon nanostructures, particularly for their use in composite materials, etc. [2]. As an illustration, one can mention fabrication of a nanotube-based electronic device, where focused ion beam of 30 keV Ga^+ ions was used to create a pair of defect-rich regions, some 50 nm apart, in a single-wall carbon nanotube lying on a SiO_2 substrate, thus giving rise to a quantum dot with well defined current-voltage characteristics [3].

In the context of using energetic ion beams, it is obviously most important to understand, and make effective use of the energy transfer to carbon nano-structures, especially in view of the traditional concept of separation between the so-called electronic stopping and nuclear stopping [4]. This is particularly important because current experiments employ a very wide range of ion energies, going from several tens of eV to some MeV. However, the concept of electronic stopping power, which is based on averaging over many ion trajectories in solid targets, probably needs to be revised for nano-objects where such statistical approach cannot be applied with confidence. Nevertheless, given that carbon nano-structures are excellent heat and charge conductors, one can assert that structural changes due to ion irradiation must be dominated by knock-on atomic displacements which are most efficient at the low end of the irradiation energy range. Depending on the ion mass, maximum cross section for processes giving rise to defect production in carbon nano-structures occurs at the $10^2 - 10^3$ eV energy range [5].

While the above applications of ion beams to carbon nano-structures require ion impacts with high probability of head-on collision with carbon atoms, the presence of large hollow regions in those structures and a high degree of their ordering and alignment present unique possibilities for achieving the effect of ion channeling [6]. Since this effect requires a sequence of ion collisions on carbon atoms with large impact parameters, one expects strong directional effects during ion passage through regular assemblies of carbon nano-structures. For example, directional effects in interactions of energetic ions with highly-oriented pyrolytic graphite (HOPG) have been recently studied experimentally in several contexts, including ion and molecule implantation in HOPG [7], ion channeling through HOPG [8], and secondary electron emission from HOPG induced by fast ions [9] and molecules [10]. While HOPG can be considered as layered material consisting of stacks of individual graphene sheets separated by about 3.35 Å, there exist other attractive possibilities for ion (and electron) channeling through carbon nano-structures exhibiting well aligned and ordered hollow regions of similar size. Those structures include fullerite, a regular crystalline structure made of C_{60} molecules, and the inter-tube regions inside the so-called ropes (or bundles) of carbon nanotubes forming hexagonal super-lattice. Ion channeling through nanotube ropes has attracted considerable theoretical interest in the past several years [11, 12].

Moreover, channeling through the interior of individual single- and multi-walled carbon nanotubes, possibly held in a matrix of some other material, has also been studied in great detail [11, 14, 15], with a view to possible applications in accelerator physics [13], or to explore processes in nanotubes analogous to those occurring during ion and molecule transport through nano-capillaries in solids [16, 17, 18]. While channeling through carbon nanotubes has been studied mostly theoretically so far, as reviewed in Refs. [11, 14, 15, 12], a few developments have been reported very recently on the experimental side, demonstrating the feasibility of both ion- [19], and electron channeling through carbon nanotubes [20].

Compared to the classical area of ion channeling through single crystals [6], one expects that channeling through nano-structures made of carbon nanotubes offers several advantages, such as, wider channels implying weaker de-channeling and consequently longer channeling distances, broader ion beams that can be used for channeling through nanotube ropes, wider acceptance angles (up to ~ 0.1 rad) and consequently lower minimum required ion energies (< 100 eV), and full three-dimensional (3D) control of beam deflection by bent nanotubes. As regards possible applications of ion channeling through carbon nanotubes, one can think of its use as a diagnostic tool for analyzing nanotube-based structures. Moreover, one should mention creation and transportation of highly focused nano-beams, ion implantation at the nano-scale for the next-generation electronic devices, targeted drug delivery and radiation therapy at the cellular level in medicine, as well as extraction, steering and collimation of ion beams at high-energy particle accelerators.

Going beyond carbon nanotubes, and having in mind that the special scattering geometry under channeling conditions eliminates head-on collision with carbon atoms, it is tempting to further pursue research into interactions of particle beams with graphene under glancing incidence to better understand their electronic response. Besides obvious interest in such processes owed to the fact that graphene is a building block of all carbon nano-structures, one can, and should also explore interesting parallels with the phenomena discovered in the well-known experiments on grazing scattering of ions from solid surfaces [21]. Moreover, one can speculate on the kinds of defect formation in graphene by analogy with the recent studies of nanostructures formed on crystal surfaces due to the impact of highly-charged ions [17].

As regards theoretical modeling and simulation of ion interactions with carbon nano-structures, various methods have been used depending on ion energies and incident trajectories. For ion channeling in the axial direction through structures made of carbon nanotubes at relativistically high energies ($\sim$ 1 GeV), both classical and quantum ion trajectories have been investigated on the basis of Lindhard's continuum approximation for the strings of carbon atoms in the walls of individual carbon nanotubes

[11, 14, 12]. Since the ion passage at such energies is practically instantaneous, electronic response of the target can be neglected, and such processes are perfectly suitable for detecting the atomic arrangements in carbon nano-structures.

For ion impacts at intermediate energies ($\sim$ 1 MeV), where ion speeds match or exceed those of valence electrons in carbon nano-structures, both electron transfer and dynamic response of the target electrons may bring new and interesting effects which do not necessarily play prominent roles at very high or very low ion energies. Those effects become especially prominent under channeling conditions and/or for ion incidence with large impact parameters onto carbon nano-structures, where one can identify effects of the dynamic image force on the incident ions, as well as inelastic processes giving rise to ion energy losses to the collective, or plasmon excitation in carbon nano-structures [15]. Moreover, regularity of ion trajectories under channeling conditions still enables the use of continuum models for interaction potentials at such energies, and consequently restores the concept of stopping power. In this article, we shall review several recent developments in modeling of such phenomena.

Finally, at low ion energies ($\sim$ 1 keV, or less), strong interactions with individual carbon atoms are likely, even under channeling conditions, and one has to obtain ion trajectories using some version of the Molecular Dynamics (MD) simulation [5, 22, 23, 24]. Typically, such simulations solve classical equations of motion for incident ions using various model interaction potentials for ion interaction with individual carbon atoms, depending on ion energy. On the other hand, the use of the so-called Brenner type II reactive bond-order empirical potential for interactions between carbon atoms enables accurate determination of defect formation, its transport across the target, as well as the self-assembly of target after the collision cascade is over. The most difficult question to resolve at low ion energies is the role and significance of electronic processes. In that context, it has been shown recently, by using the Time-dependent Density-functional Theory (TDDFT) for target electrons, that the Born-Oppenheimer approximation and the binary-collision approximation underlying MD simulations are applicable for ion interactions with carbon nano-structures at low energies [4].

2. Particle channeling through a carbon nanotube

As a prototype for modeling collective electron excitations, we consider channeling of a fast, point charge distribution through a straight, infinitely-long single-wall carbon nanotube in vacuum, having the radius a. Since ion trajectories do not interact closely with individual carbon atoms in the target, but rather "integrate" over large numbers of such atoms each time when they "bounce" off the nanotube wall, it is reasonable to assume that the ion-core lattice of the nanotube is described by a two-dimensional, cylindrically shaped continuous distribution of rigid positive charges. Such two-dimensional jellium model is particularly suitable for describing ion channeling through chiral carbon nanotubes, where strings of carbon atoms "swirl" around the nanotube axis.

Furthermore, electron excitations in so described carbon nanotubes may be modeled by assuming that carbon valence electrons form a two-dimensional electron gas, which can be described quantum-mechanically in the so-called random phase approximation (RPA) [25], or semiclassically by using the hydrodynamic model [26]. We note that this model can be applied to all four valence electrons treated as a single fluid [27], or can be developed into a two-fluid model by treating the three σ and one π electron per carbon atom as two separate fluids [28]. Moreover, hydrodynamic modeling of carbon valence electrons is particularly suitable for developing models of multi-walled carbon nanotubes represented as concentric cylinders of interacting two-dimensional electron fluids [29], as well as for taking into account polarization of the dielectric media surrounding the carbon nanotube [30].

We use cylindrical coordinates $\mathbf{R} = \{r, \varphi, z\}$ with the z-axis aligned with the nanotube axis. (In this section, we use atomic units.) One can express the total electric potential $\Phi(\mathbf{R}, t)$ in terms of the external electric potential $\Phi_{\text{ext}}(\mathbf{R}, t)$ by means of the inverse dielectric response function for the nanotube valence electrons, $\epsilon^{-1}(\mathbf{R}, \mathbf{R}', t - t')$, as

$$\Phi(\mathbf{R}, t) = \int d^3\mathbf{R}' \int dt' \, \epsilon^{-1}(\mathbf{R}, \mathbf{R}', t - t') \Phi_{\text{ext}}(\mathbf{R}', t'). \quad (1)$$

We further assume that the electron gas on the nanotube is distributed in an infinitesimally thin layer with number density per unit area $n_0 + n_1(\mathbf{r}, t)$ at the position $\mathbf{r} = \{r = a, \varphi, z\}$ on the surface of the cylinder. Here, n_0 is the equilibrium number density compensating the positive ionic background and $n_1(\mathbf{r}, t)$ is the perturbation of the electron number density due to the external potential. In the linear response regime, one may introduce a polarization function of the 2D electron gas, $\chi(\mathbf{r}, \mathbf{r}', t - t') = \chi(\varphi - \varphi', z - z', t - t')$, and express the first-order perturbation of the electron number density $n_1(\mathbf{r}, t)$ in terms of the external potential impressed on the nanotube surface at points $\mathbf{r}' = \{r' = a, \varphi', z'\}$ and time t', as

$$n_1(\mathbf{r}, t) = \int d\mathbf{r}' \int dt'\, \chi(\mathbf{r}, \mathbf{r}', t - t')\Phi_{\text{ext}}(\mathbf{r}', t'). \tag{2}$$

Invoking the jellium approximation, one can define the Fourier transform for an arbitrary function $A(r, \varphi, z, t)$ as

$$A(r, \varphi, z, t) = \sum_m \iint \frac{dk d\omega}{(2\pi)^3} e^{im\varphi} e^{ikz} e^{-i\omega t} \widetilde{A}(r; m, k, \omega), \tag{3}$$

and rewrite (2) as $\widetilde{n}_1(m, k, \omega) = \widetilde{\chi}(m, k, \omega)\widetilde{\Phi}_{\text{ext}}(a; m, k, \omega)$, with $\widetilde{\chi}(m, k, \omega)$ obtained form the hydrodynamic model.

The total electric potential $\Phi(\mathbf{R}, t)$ is the sum of the external electric potential which satisfies the Poisson equation for the external charge density, $\nabla^2\Phi_{\text{ext}}(\mathbf{R}, t) = -4\pi\rho_{\text{ext}}(\mathbf{R}, t)$, and the induced electric potential $\Phi_{\text{ind}}(\mathbf{R}, t)$ due to polarization of the nanotubes valence electrons, satisfying $\nabla^2\Phi_{\text{ind}}(\mathbf{R}, t) = 4\pi n_1(\mathbf{r}, t)\delta(r - a)$. These equations are easily solved by expressing the Coulomb interaction in cylindrical coordinates as usual,

$$\frac{1}{\|\mathbf{R} - \mathbf{R}'\|} = \sum_m \int \frac{dk}{(2\pi)^2} e^{im(\varphi - \varphi')} e^{ik(z - z')} g(r, r'; m, k) \tag{4}$$

where $g(r, r'; m, k) \equiv 4\pi I_m(|k|r_<)K_m(|k|r_>)$ is the radial Green's function with $r_< = \min\{r, r'\}, r_> = \max\{r, r'\}$, and $I_m(x)$ and $K_m(x)$ being modified Bessel's functions of the first and second kind, respectively. Finally, one can express the Fourier transform of the inverse dielectric function, $\epsilon^{-1}(\mathbf{R}, \mathbf{R}', t - t') = \epsilon^{-1}(r, r', \varphi - \varphi', z - z', t - t')$, as

$$\widetilde{\epsilon}^{\,-1}(r, r'; m, k, \omega) = \frac{\delta(r - r')}{r'} - \delta(r' - a)g(r, a; m, k)\widetilde{\chi}(m, k, \omega). \tag{5}$$

We consider a rigid distribution of external charges with density function $\widehat{\rho}_{\text{ext}}(\mathbf{R})$ which moves inside the nanotube at constant velocity parallel to its axis, $\mathbf{v} = v\,\mathbf{e}_z$, so that its density in the laboratory frame of reference can be written as $\rho_{\text{ext}}(\mathbf{R}, t) \equiv \widehat{\rho}_{\text{ext}}(\mathbf{R} - \mathbf{v}t)$. Therefore, the Fourier transform of the external potential on the nanotube surface can be expressed as

$$\widetilde{\Phi}_{\text{ext}}(a; m, k, \omega) = 2\pi\,\delta(\omega - kv) \int d^3\mathbf{R}\, \widehat{\rho}_{\text{ext}}(\mathbf{R}) f(\mathbf{R}; m, k), \tag{6}$$

where we define $f(\mathbf{R}; m, k) \equiv e^{-i(m\varphi + kz)} g(a, r; m, k)$. The factor $\delta(\omega - kv)$ indicates that the induced potential is stationary in the moving frame of reference attached to the external charge distribution. Denoting this potential by $\widehat{\Phi}_{\text{ind}}(\mathbf{R})$, we note that the induced potential in the laboratory frame is then $\Phi_{\text{ind}}(\mathbf{R}, t) \equiv \widehat{\Phi}_{\text{ind}}(\mathbf{R} - \mathbf{v}t)$.

We further assume that the external charge density $\widehat{\rho}_{\text{ext}}(\mathbf{R})$ is both highly peaked and appreciable only around its center of mass at $\mathbf{R}_0 = \{r_0, \varphi_0, z_0\}$ in the moving frame, giving rise to a well-defined multipole expansion, with total charge $Q = \int d^3\mathbf{R}\, \widehat{\rho}_{\text{ext}}(\mathbf{R})$, and dipole moment $\boldsymbol{\mu} = \int d^3\mathbf{R}\, (\mathbf{R} -$

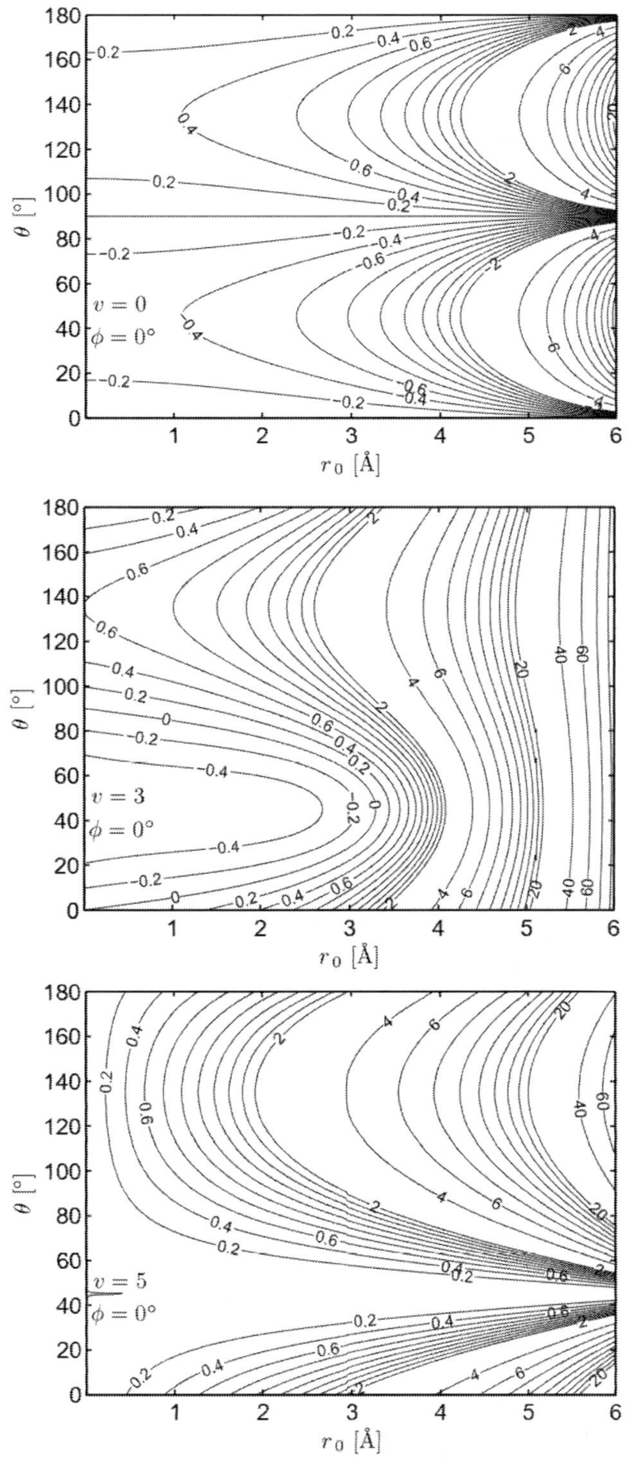

Figure 1. The torque (in meV$/\mu^2$) on a point dipole with $\boldsymbol{\mu}$ oriented in the plane passing through the dipole and the nanotube axis (so that $\phi = 0$), with θ being an angle relative to the direction of motion. The dipole is located at radial position r_0 (in Å) and is moving with speeds $v = 3$ and 5 parallel to the axis of a SWCNT of radius $a = 7$ Å. The contour step-size is $\Delta\tau = 0.2$, 2, and 20 meV$/\mu^2$ beginning at $\tau = 0.4$, -2, and -20 meV$/\mu^2$, respectively.

$\mathbf{R}_0)\,\widehat{\rho}_{\rm ext}(\mathbf{R})$. Consequently, one can approximate the Coulomb structure factor of the external charge in (6) by using the Taylor expansion of $f(\mathbf{R};m,k)$ about $\mathbf{R}_0$, $f(\mathbf{R};m,k)\approx f(\mathbf{R}_0;m,k)+(\mathbf{R}-\mathbf{R}_0)\cdot\vec{\nabla}_{\mathbf{R}_0}f(\mathbf{R}_0;m,k)$. Thus, the induced electric potential in the moving frame of reference can be written, for points $\mathbf{R}$ in the neighborhood of the external charge, as

$$\widehat{\Phi}_{\rm ind}(\mathbf{R})\approx-\sum_m\int\frac{dka}{(2\pi)^2}f^*(\mathbf{R};m,k)\widetilde{\chi}(m,k,\omega)\left[Qf(\mathbf{R}_0;m,k)+\boldsymbol{\mu}\cdot\vec{\nabla}_{\mathbf{R}_0}f(\mathbf{R}_0;m,k)\right], \tag{7}$$

where, from now on, one has to set $\omega\equiv kv$ in $\widetilde{\chi}(m,k,\omega)$. We evaluate the total electrostatic self energy, $\mathcal{E}_{\rm self}$, of the external charge distribution as

$$\mathcal{E}_{\rm self}(r_0)\;=\;\frac{1}{2}\int d^3\mathbf{R}\,\widehat{\rho}_{\rm ext}(\mathbf{R})\widehat{\Phi}_{\rm ind}(\mathbf{R}), \tag{8}$$

$$\approx\;-\frac{1}{2}\sum_m\int\frac{dka}{(2\pi)^2}\widetilde{\chi}(m,k,\omega)\left|Qf(\mathbf{R}_0;m,k)+\boldsymbol{\mu}\cdot\vec{\nabla}_{\mathbf{R}_0}f(\mathbf{R}_0;m,k)\right|^2, \tag{9}$$

where we have used the Taylor expansion of $f^*(\mathbf{R};m,k)$ about $\mathbf{R}_0$.

Similarly, we obtain the dynamic polarization force, $\mathbf{F}_{\rm ind}$, on the external charge density as

$$\mathbf{F}_{\rm ind}(r_0)=\int d^3\mathbf{R}\,\widehat{\rho}_{\rm ext}(\mathbf{R})\widehat{\mathbf{E}}_{\rm ind}(\mathbf{R})\quad\approx Q\widehat{\mathbf{E}}_{\rm ind}(\mathbf{R}_0)+\boldsymbol{\mu}\cdot\vec{\nabla}_{\mathbf{R}}\widehat{\mathbf{E}}_{\rm ind}(\mathbf{R})|_{\mathbf{R}=\mathbf{R}_0}, \tag{10}$$

where $\widehat{\mathbf{E}}_{\rm ind}(\mathbf{R})=-\vec{\nabla}_{\mathbf{R}}\widehat{\Phi}_{\rm ind}(\mathbf{R})$ is the induced electric field in the moving frame of reference. We note that this force consists of a conservative part acting in the radial direction, $-d\mathcal{E}_{\rm self}(r_0)/dr_0$, which corresponds to the classical dynamic image force on external charge distribution, and a dissipative part opposite to the direction of motion, with magnitude $S=-\mathbf{e}_z\cdot\mathbf{F}_{\rm ind}$, corresponding to the standard definition of stopping power.

Finally, the torque $\boldsymbol{\tau}$ on the external charge distribution about its center of mass $\mathbf{R}_0$ in the moving frame may be evaluated as

$$\boldsymbol{\tau}(r_0)\;=\;\int d^3\mathbf{R}\,\widehat{\rho}_{\rm ext}(\mathbf{R})\,(\mathbf{R}-\mathbf{R}_0)\times\widehat{\mathbf{E}}_{\rm ind}(\mathbf{R})\approx\boldsymbol{\mu}\times\widehat{\mathbf{E}}_{\rm ind}(\mathbf{R}_0). \tag{11}$$

As an illustration of the above model, we show the torque on a point dipole moving through a nanotube of radius $a=7$ Å at a distance r_0 from its axis, with $\boldsymbol{\mu}$ lying inside the plane which passes through the dipole and the nanotube axis ($\phi=0$), and having the direction θ relative to the direction of dipole motion. Note that in this case, the torque has only the component perpendicular to the plane passing through the dipole and the nanotube axis, which is shown in Fig. 1 (per μ^2) versus r_0 and θ for dipole speeds $v=0$, 3 and 5 a.u. One notices surprisingly strong variations of the torque on a dipole with its speed, orientation and position inside carbon nanotube. One can assert that these effect are probably the result of an interplay between the image and stopping forces on a point dipole, and they may play important role in the self-alignment of polar molecules moving through constrained geometries [31].

3. Grazing scattering of slow ions from supported graphene

In the previous section we have studied the dynamic screening of fast ions and molecules due to the collective excitations of both π and σ electrons in carbon nanotubes, which can be conveniently described by the hydrodynamic model [30] for particle speeds above the so-called kinematic threshold for plasmon excitation being on the order of Bohr velocity, $v_{\rm B}$. Here, however, we are interested in slow ions, moving parallel to a graphene sheet at the speeds well under Bohr's velocity, so that the dynamic response of graphene is dominated by the single-particle and collective excitations of its π electrons, whereas σ electrons are not excited owing to their having a large band gap. Such low-energy excitations require going beyond the hydrodynamic model, and the dielectric-response theory for graphene, based on the random phase approximation (RPA) using linear approximation for the π electron bands [32, 33] is an adequate, if cumbersome way to proceed.

To make things simpler, we shall use here a semiclassical Vlasov equation for graphene's π electrons in the relaxation-time approximation at finite temperatures, which should perform reasonably well when compared to the more complicated RPA dielectric formalism for extrinsic graphene, which is subject to a sufficiently high gate voltage, V_g [34]. Since such graphene is characterized in equilibrium by a relatively large induced density per unit area of its π electrons, n_g (typically, $|n_g| > 10^{12}$ cm^{-2}), the dynamic polarization of graphene is dominated by the intra-band electron excitations, which are reasonably well described by means of the Vlasov equation for slow ions moving at large distances from graphene. On the other hand, the relative simplicity of the semiclassical approach to dynamic screening in comparison to the RPA method permits an easy treatment of the finite gap between the graphene and a substrate (usually SiO_2), which will be found to exert quite strong influence on the dynamic-polarization forces on the incident ion.

We use a cartesian coordinate system with coordinates $\mathbf{R} = \{\mathbf{r}, z\}$, where $\mathbf{r} = \{x, y\}$, and assume that graphene is placed in the $z = 0$ plane. A substrate with dielectric constant ϵ_{sub} (≈ 3.9 for SiO_2) is assumed to occupy the region $z \leq -h$ underneath the graphene, whereas the region $z > -h$ is assumed to be vacuum or air. In the ground state of such system, the induced excess (or depleted) number density of the π electrons in graphene at temperature T is given by

$$n_g(\mu) = \frac{g_d}{2\pi}\left(\frac{k_B T}{\hbar v_F}\right)^2 \left[\text{dilog}\left(1 + \mathrm{e}^{-\mu/k_B T}\right) - \text{dilog}\left(1 + \mathrm{e}^{\mu/k_B T}\right)\right], \tag{12}$$

where dilog is the standard dilogarithm function, $g_d = 4$ is the degeneracy factor and $v_F \approx c/300$ (c being the speed of light) is the Fermi speed of graphene's π-electron band in the approximation of linear energy dispersion, $\varepsilon_\pm(\mathbf{p}) = \pm v_F p$, with $\mathbf{p}$ being the electron quasi-momentum and p its magnitude. Inverting the relation (12) can be used to deduce the value of chemical potential μ for any value of the externally applied gate voltage, V_g, by using the relation $n_g = \frac{\epsilon_{\text{sub}}}{4\pi e}\frac{V_g}{w_{\text{sub}}}$, where w_{sub} is the thickness of the substrate (typically, $w_{\text{sub}} \gg h$).

When a point charge Q moves parallel to graphene at a fixed distance $z_0 > 0$ above it with a constant velocity $\mathbf{v} = \{v_x, v_y\}$, so that its density can be written as $\rho_{\text{ext}}(\mathbf{r}, z, t) = Q\delta(\mathbf{r} - \mathbf{v}t)\,\delta(z - z_0)$, then the resulting electric potential $\Phi_{\text{ext}}(\mathbf{r}, z, t)$ will perturb the system giving rise to deviations from the ground-state charge densities (per unit area) on both the graphene and substrate surface, which we denote, respectively, by $\sigma_{\text{gr}}(\mathbf{r}, t)$ and $\sigma_{\text{sub}}(\mathbf{r}, t)$.

Next, we resort to the method of Doerr and Yu [35] to treat the effects of substrate on the response of graphene. The same method can be used to treat the effects of a cylindrical cavity in dielectric material on a carbon nanotube grown inside that cavity [30]. Denoting the electric potentials due to the induced charge densities σ_{gr} and σ_{sub}, respectively, by $\Phi_{\text{gr}}(\mathbf{r}, z, t)$ and $\Phi_{\text{sub}}(\mathbf{r}, z, t)$, the total (screened) electric potential in the system, Φ, can be written as the sum $\Phi = \Phi_{\text{ext}} + \Phi_{\text{gr}} + \Phi_{\text{sub}}$. Then, if there are no free charges on the substrate surface (such as, e.g., dopant centers), one can use the continuity of the normal component of the displacement vector,

$$\left.\frac{\partial \Phi}{\partial z}\right|_{z=-h+0} = \epsilon_{\text{sub}} \left.\frac{\partial \Phi}{\partial z}\right|_{z=-h-0}, \tag{13}$$

to obtain σ_{sub} [30, 34, 35]. This is easily implemented by using the Fourier transform (indicated by a tilde above transformed function) with respect to time, $t \rightarrow \omega$, and coordinates in the xy plane, $\mathbf{r} \rightarrow \mathbf{k}$, and by invoking the relation $\widetilde{\sigma}_{\text{gr}}(\mathbf{k}, \omega) = -e\,\chi(\mathbf{k}, \omega)\,\widetilde{\Phi}(\mathbf{k}, z = 0, \omega)$ in terms of the polarization function of graphene, $\chi(\mathbf{k}, \omega)$. As a result, one obtains the Fourier transform of the induced potential at $z > 0$, $\Phi_{\text{ind}} = \Phi_{\text{sub}} + \Phi_{\text{gr}}$, as follows

$$\widetilde{\Phi}_{\text{ind}}(\mathbf{k}, z, \omega) = 2\pi Q\delta(\omega - \mathbf{k}\cdot\mathbf{v})\,\frac{2\pi}{k}\mathrm{e}^{-k(z+z_0)}\left[\frac{1}{\epsilon(\mathbf{k}, \mathbf{k}\cdot\mathbf{v})} - 1\right], \tag{14}$$

where $\epsilon(\mathbf{k},\omega) = \epsilon_0(k) + \frac{2\pi e}{k}\chi(\mathbf{k},\omega)$ is the dielectric function of the supported graphene, with

$$\epsilon_0(k) \equiv \frac{\epsilon_{\rm sub}+1}{2}\,\frac{1+\coth(kh)}{\epsilon_{\rm sub}+\coth(kh)} \tag{15}$$

being the background dielectric function which quantifies effects of the substrate on the response of graphene. We note that, while we assume $\epsilon_{\rm sub}$ to be constant, one can also include the effects of collectiv e modes in the substrate due to, e.g., electron excitations in metallic substrates, or phonon excitations on the surface of an insulator, by assigning some frequency-dependent model to $\epsilon_{\rm sub}$.

To obtain the polarization function of graphene, $\chi(\mathbf{k},\omega)$, one has to solve the linearized Vlasov-type equations for the probability density functions $f_\pm(\mathbf{p},\mathbf{r},t)$ of graphene's π electrons in the conduction (+) and valence (-) bands,

$$\frac{\partial f_\pm}{\partial t} + \mathbf{v}_\pm(\mathbf{p})\cdot\frac{\partial f_\pm}{\partial \mathbf{r}} + e\frac{\partial}{\partial \mathbf{r}}\Phi(\mathbf{r},z=0,t)\cdot\frac{\partial f_\pm}{\partial \mathbf{p}} = -\frac{f_\pm - f_\pm^0}{\tau(\mathbf{p})}, \tag{16}$$

where $\mathbf{v}_\pm(\mathbf{p}) \equiv \partial\varepsilon_\pm(\mathbf{p})/\partial\mathbf{p} = \pm v_F\mathbf{p}/p$ is the electron velocity in the linearized π band, $\Phi = \Phi_{\rm ext} + \Phi_{\rm ind}$ is the total potential, $\tau(\mathbf{p})$ is the relaxation time due to collisions, and $f_\pm^0 \equiv f(\varepsilon_\pm(\mathbf{p}))$ is the Fermi-Dirac distribution. As a result, one finds, in the collisionless limit, $\tau\to\infty$,

$$\epsilon(k,\omega) = \epsilon_0(k) + \frac{k_{\rm TF}}{k}\left[1 - |\omega|\frac{H\left(|\omega| - kv_F\right)}{\sqrt{\omega^2-(kv_F)^2}} + i\omega\frac{H\left(kv_F - |\omega|\right)}{\sqrt{(kv_F)^2-\omega^2}}\right], \tag{17}$$

where $k_{\rm TF} \equiv 2\pi e^2\frac{\partial n_0}{\partial\mu}$, with $n_0(\mu)$ from Eq. 12, is the Thomas-fermi inverse screening length, and H is the Heaviside step function. By using Eq. 17 in Eq. 14, one can evaluate the stopping and the image forces on the moving charge as follows: $F_s = -\frac{Q}{v}\mathbf{v}\cdot\frac{\partial}{\partial\mathbf{r}}\Phi_{\rm ind}\big|_{\mathbf{r}=\mathbf{v}t,z=z_0}$, $F_i = -Q\frac{\partial}{\partial z}\Phi_{\rm ind}\big|_{\mathbf{r}=\mathbf{v}t,z=z_0}$.

As an illustration of the above model, we show in Fig. 2 the velocity dependencies for the stopping and image forces on an ion moving at the distance $z_0 = k_{\rm TF}^{-1}$ parallel to the graphene in the collisionless limit of the dielectric function. Note that all the relevant quantities are defined in terms of the reduced units, where $\overline{F}_s = F_s/\left(Qk_{\rm TF}\right)^2$, $\overline{F}_i = F_i/\left(Qk_{\rm TF}\right)^2$, $\overline{v} = v/v_F$, $\overline{z}_0 = z_0k_{\rm TF}$ and $\overline{h} = hk_{\rm TF}$. One notices almost linear increase in the stopping force at low speeds due to the intra-band single-particle excitations of the π electrons, followed by a massive peak after the speed $\overline{v} = 1$ which is a threshold for energy loss to the collective, or plasmon excitations, and a slow decrease in stopping force at the higher speeds. The image force shows similar resonant features around the speed $\overline{v}$ =1 with the significant tendency of saturation at the lowest and the highest speeds shown. However, the most remarkable features shown in Fig. 1 are the very strong dependencies of both forces on the values of the graphene-substrate gap, with $\overline{h} = \infty$ standing for a free graphene, $\overline{h} = 0$ being the zero-gap approximation usually employed in literature, whereas $\overline{h} = 0.5$ is close to a realistic experimental value [36].

It is of particular interest to obtain analytical expressions for the image and stopping forces in the limit of vanishing projectile speed v describing slow charges moving over graphene. This can be achieved by assuming $\gamma = 0$ and ϵ_0 = constant, so that one obtains a stopping force which is proportional to the ion speed, $\overline{F}_s = -\eta\overline{v}$, with the friction coefficient η given by

$$\eta = \frac{1}{2\epsilon_0^3}\left[\frac{1}{\zeta} + 1 - (2+\zeta)\,\mathrm{e}^{\zeta}\,\mathrm{Ei}(1,\zeta)\right], \tag{18}$$

and the static image force given by

$$\overline{F}_i = \frac{1}{\epsilon_0^3}\left[\frac{1-\epsilon_0}{\zeta^2} - \frac{1}{\zeta} + \mathrm{e}^{\zeta}\,\mathrm{Ei}(1,\zeta)\right], \tag{19}$$

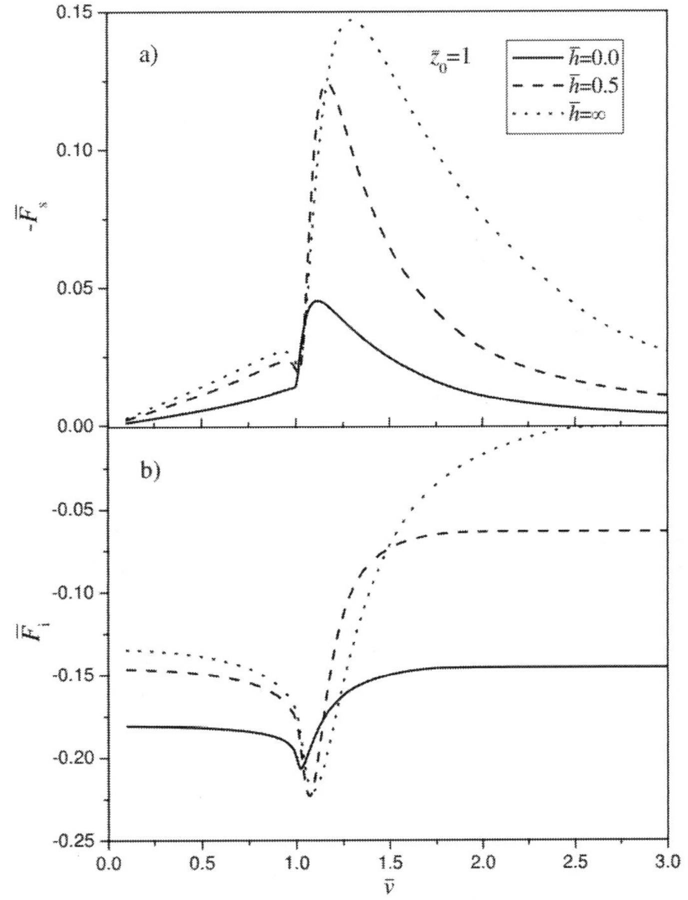

Figure 2. The reduced (a) stopping and (b) image forces as functions of the reduced speed $\bar{v}$ for an ion moving at the reduced distance $\bar{z}_0$ =1 above graphene for three values of the reduced gap between graphene and a SiO_2 substrate: $\bar{h}$ = 0.0 (solid lines), 0.5 (dashed lines), and ∞ (dotted lines), with zero damping, $\bar{\gamma} = 0$.

where $\zeta \equiv 2\bar{z}_0/\epsilon_0$ and $\mathrm{Ei}(1, \zeta)$ is the standard exponential integral. Furthermore, one can easily obtain the leading terms in the asymptotic expansions when $\zeta \approx 4\frac{e^2 z_0}{\hbar v_F}\sqrt{\pi g_d |n_g|} \to \infty$ in the above expressions for graphene in free space ($\epsilon_0 = 1$) which, physically, correspond to large distances z_0 and/or strongly gated graphene with a large equilibrium induced density of charge carriers n_g at low temperatures. So, one obtains $F_i \sim -Q^2/(4z_0^2)$ for the image force, which is typical for a perfectly conducting surface and is consequently independent of n_g, whereas the low-speed stopping force becomes $F_s \sim -\frac{v}{v_B}\frac{Q^2}{16z_0^3}\frac{1}{\sqrt{\pi g_d |n_g|}}$, which decreases as the inverse root of the density $|n_g|$. We note that, in the opposite limit of lightly gated, or intrinsic graphene (when $n_g \to 0$), the linear Vlasov theory for intra-band electron excitations becomes inadequate, as such processes are dominated by the inter-band electron transitions which are properly described by the RPA theory.

4. Concluding remarks

We have reviewed some aspects of modeling ion interactions with carbon nanotubes under channeling conditions and with graphene under grazing incidence, where dynamic polarization of carbon valence electrons plays important roles. Specifically, we have evaluated the dynamic image force on ions and their energy loss to electron excitations. Such calculations were subsequently used to evaluate the role of those forces in the so-called Coulomb explosions of fast H_2 molecules in carbon nanotubes [37], as well as the effects of the dynamic image force on the so-called rainbow effect in angular distributions of fast protons channeled through short carbon nanotubes [38]. By using hydrodynamic model for electron excitations, only plasmon excitations in carbon nanotubes were taken into account while calculating the ion stopping power, which should dominate at high ion speeds. While such mechanism of energy loss was found to have relatively small effect on ion channeling at high speeds [38], we have demonstrated here that the stopping force may play an important role in giving rise to a torque on a dipole which is channeled through a carbon nanotube. On the other hand, the image force on moving charges is adequately described by the hydrodynamic model, as shown in the recent discovery of electron image states around carbon nanotubes [39]. Our exploration of the image force in channeling conditions has demonstrated that the torque on channeled dipoles, Coulomb explosion dynamics of molecules [37], and the rainbow peaks in proton channeling through carbon nanotubes [38] are all strongly affected by the image force.

On the other hand, investigation of interactions of graphene with external particles is still in an early stage. We have, consequently, focused on interactions with slow ions, where peculiar band structure of graphene's π electrons is expected to exhibit the most dramatic effects. While the linear-response treatment of such interactions involving intrinsic graphene is somewhat problematic, we have limited ourselves to the case of gated graphene, where kinetic-equation methods are expected to work well for electron excitations. We have found that both the single-electron, intra-band excitations and the collective plasmon excitations play important roles on the stopping and image forces. However, the most striking effect revealed in our calculations is due to the presence of a gap between the graphene and a polarizable substrate. Surprisingly, all theoretical models of graphene's dynamic response assume a zero gap [32, 33], so we point to a strong need to take into account the finite size of such a gap, which can reach values of ~ 4.2 Å, as documented experimentally [36].

Acknowledgments
This work was supported by NSERC and PREA.

References

[1] B Dumé B, http://physicsweb.org/articles/news/10/5/4/1.
[2] Krasheninnikov A V and Banhart F 2007 *Nature Materials* **6** 723.
[3] Maehashi K *et al.* 2007 *Appl. Phys. Lett.* **90** 023103.
[4] Krasheninnikov A V, Miyamoto Y and Tomanek D 2007 *Phys. Rev. Lett.* **99** 016104.
[5] Krasheninnikov A V and Nordlund K 2004 *Nucl. Instr. Meth.* B **216** 355.
[6] Gemmell D S 1974 *Rev. Mod. Phys.* **46** 129.
[7] Ramos G and Scherzer B M U 1995 *Nucl. Instrum. Methods* B **85** 479; 2001 *ibid.* **174** 329.
[8] Yagi E, Iwata T, Urai T and Ogiwara K 2004 *J. Nucl. Mater.* **334** 9.
[9] Cernusca S, Frsatz M, Winter H P and Aumayr F 2005 *Europhys. Lett.* **70** 768.
[10] Kaneko T, Kudo H, Tomita S and Uchiyama R 2006 *J. Phys. Soc. Jpn.* **75** 034717.
[11] Artru X, Fomin S P, Shulga N F, Ispirian K A and Zhevago N K 2005 *Phys. Rep.* **412** 89.
[12] Petrović S, Borka D and Nešković N (2008 *Adv. Studies Theor. Phys.* **2** 415.
[13] Biryukov V M, Chesnokov Y A and Kotov V I 1997 *Crystal Channeling and Its Application at High-Energy Accelerators*, (Berlin: Springer).
[14] Bellucci S, 2005 textitNucl. Instr. Meth. B **234** 57; 2005 *Phys. Stat. Sol.* (c) **2** 34.
[15] Mišković Z L 2007 *Radiat. Eff. Def. Solids* **162** 185.
[16] Yamazaki Y 2002 *Nucl. Instrum. Meth. Phys. Res.* B **193** 516.
[17] Tókési K, Wirtz L, Lemmel C and Burgdörfer J 2001 *Phys. Rev.* A **64** 042902; Tökési K, Tong X M, Lemmel C and Burgdörfer J 2005 *Phys. Rev.* A **72** 022901.
[18] Arista N R 2001 *Phys. Rev.* A **64** 032901; Arista N R and Fuentes M A 2001 *Phys. Rev.* B **63** 165401.
[19] Zhu Z *et al.* 2005 *Proc. SPIE* 5974.
[20] Chai C *et al.* 2007 *Appl. Phys. Lett.* **91** 103101.

[21] H. Winter, Phys. Rep. 367, 387 (2002).
[22] Krasheninnikov A V and Nordlund K 2005 *Nucl. Instr. Meth.* B **228** 21; *Phys. Rev.* B **71** 245408.
[23] Moura C S and Amaral L 2005 J. *Phys. Chem.* B **109** 13515.
[24] Zhang W, Zhu Z, Xu Z, Wang Z and Zhang F 2005 *Nanotechnology* **16** 2681.
[25] Gumbs G and Balassis A 2005 *Phys. Rev.* B **71** 235410.
[26] Barton G and Eberlein C 1991 *J. Chem. Phys.* **95** 1521.
[27] Wang Y N and Mišković Z L 2004 *Phys. Rev.* A **69** 022901.
[28] Mowbray D J, Mišković Z L, Goodman F O and Y.-N. Wang Y N 2004 *Phys. Rev.* B **70** 195418.
[29] Mowbray D J, Chung S, Mišković Z L, Goodman F O and Y.-N. Wang Y N 2005 *Nucl. Instr. Meth.* B **230** 142.
[30] Mowbray D J, Mišković Z L and Goodman F O 2006 *Phys. Rev.* B **74** 195435.
[31] Mowbray D J, Chung S, Mišković Z L and Goodman F O 2007 *Nanotechnology* **18** 424034.
[32] Wunsch W, Stauber T, Sols F and Guinea F 2006 *New Journal of Physics* **8** 318.
[33] Hwang E H and Das Sarma S 2007 *Phys. Rev.* B **75** 205418.
[34] Radovic I, Hadzievski Lj and Miskovic Z L 2008 *Phys. Rev.* B **77** 075428.
[35] Doerr T P and Yu Y K 2004 *Am. J. Phys.* **72** 190.
[36] Ishigami M, Chen J H, Cullen W G, Fuhrer M S and Williams E D 2007 *Nano Lett.* **7** 1643.
[37] Zhou D P, Song Y H, Wang Y N and Mišković Z L 2006 *Phys. Rev.* A **73** 033202.
[38] Borka D, Mowbray D J, Mišković Z L, Petrović S and Nešković N 2008 *Phys. Rev.* A **77** 032903.
[39] B. E. Granger B E, P. Kral P, H. R. Sadeghpour H R and M. Shapiro M 2002 *Phys. Rev. Lett.* **89** 135506; Zamkov N, Woody N, Shan B, Chakraborty H S, Chang Z, Thumm U and Richard P 2004 *Phys. Rev. Lett.* **93** 156803.

Etching of *Bacillus atrophaeus* by oxygen atoms, molecules and argon ions

J. Benedikt, C. Flötgen, G. Kussel, V. Raballand, A. von Keudell

E-mail: Achim.vonKeudell@rub.de
Research Group Reactive Plasmas, Ruhr-Universität Bochum, 44780 Bochum, Germany

Abstract. The etching of spores of *Bacillus atrophaeus* by oxygen atoms, molecules and argon ions is investigated in a particle beam experiment. Thereby, the conditions occurring in an argon oxygen plasma are mimicked and fundamental inactivation mechanisms are revealed. It is shown that only the combined impact of argon ions and of O atoms or O_2 molecules causes significant etching of the spores. This is explained by the process of chemical sputtering, where an ion induced defect at the surface of the spore reacts with either the incident bi-radical O_2 or with an incident O atom. This leads to the formation of CO, CO_2 and H_2O and thus to erosion. This process is compared to the plasma etching of hydrocarbon thin films as an atomistic model system for the spore coat. It is shown that the etch rate in an inductively coupled argon oxygen plasma is only maximal if both, the electron density and thus the ion flux towards the surface and the O atom flux are optimized simultaneously.

1. Introduction

Sterilization is a key technology in the medical industry[1, 2, 3, 4]. Many sterilization methods are known to be efficient such as heat treatment, gamma radiation or exposure to toxic chemicals (ozone O_3, ethylene oxide C_2H_4O, or hydrogen peroxide H_2O_2). All these methods have in common that they impose a severe stress on the objects to be decontaminated: the wear of surgical instruments is significant and implants made from thermolabile polymers loose their mechanical integrity[5]. One technology that avoids these issues is plasma sterilization. In a plasma, a source gas is dissociated and ionized and reactive species such as electrons, ions, atoms, radicals and UV photons are interacting with the biological system to be inactivated. Typical sterilization plasmas are generated in argon with admixtures of nitrogen, hydrogen, or oxygen[3, 1, 2]. In such a process, the object itself can stay at rather moderate temperatures since the dissociation of the precursor gas is initiated in the plasma state. It has been proven that plasma sterilization can be very efficient in inactivating bacteria or pathogenic biomolecules[6, 7, 8].

The validation of a sterilization technique is usually based on the proof to inactivate endospores of bacteria. They are known to be a very resistant form of a biological system. In the case of plasma sterilization, it is postulated that incident UV photons, ions and chemical active species have a combined effect on endospores[3, 1, 2, 9, 10, 6]. It is generally believed that the inactivation itself is caused by UV photons, which penetrate deep into the cell and cause DNA strand breaks. Sterilization times of the order of seconds for inactivating isolated spores are sufficient at typical photon fluxes in low temperature plasmas[11, 10]. In almost all practical cases, however, the sterilization efficiency is largely reduced, because of shadowing of the UV radiation by multilayered stacks of spores or by biofilms in which the spores are embedded. Therefore, a significant plasma-induced chemical or physical etching of the biological systems is essential. Summarizing, an optimum plasma process needs to provide both, an intense UV-photon flux and a significant etching efficiency.

The etching of biological systems is often performed by adding hydrogen or oxygen to the discharge[11, 12]. In a recent paper, we studied the interaction of hydrogen atoms with endospores[13]. It was shown that the combined impact of ions and H atoms causes significant erosion, which becomes visible as a perforation of the endosporic shell. H atoms alone have no inactivation efficiency. Furthermore, we studied also the ability of oxygen atoms and oxygen molecules [14] to inactivate endospores of *Bacillus atrophaeus*. It is shown, that the impact of O atoms does not cause sterilization at room temperature. However, the simultaneous impact of O_2 molecules or O atoms and of argon ions at around 200 eV causes a very effective etching and perforation of the spore coat. This is due to the process of *chemical sputtering*, where the bi-radical O_2 or the O atom interacts with ion-induced defects at the spore surface leading to the formation of CO, CO_2 and H_2O and thus to etching. This result was deduced from beam experiments where the plasma treatment is mimicked using absolutely quantified beams of oxygen molecules, of oxygen atoms and of argon ions.

In this paper, we transfer the results from the beam experiments in [14] to plasma experiments using an inductively coupled plasma from argon - oxygen mixtures. To assess the efficiency of the chemical sputtering process, we study the plasma etching of thin hydrocarbon films as being a model system for the spore coat.

2. Experiment

The erosion of endospores of *Bacillus atrophaeus* is studied by exposing them to particle beams of oxygen molecules, of oxygen atoms, of argon ions and of combination of these. Thereby, the exposure of a biological system to an argon-oxygen plasma is mimicked. Spores of *Bacillus atrophaeus* (ATCC 51189) were used as biological samples, because they are known to be highly resistant against conventional H_2O_2 based sterilization methods. The spores were analyzed using scanning electron microscopy (SEM). Doped silicon is used as substrate. The substrates were coated with monolayers of endospores using a spraying technique.

A sketch of the particle beam experiment is shown in Fig. 1a. It consists of an exposition chamber containing the particle sources and a load lock. Three atom source and one ion source are implemented: (i) one oxygen source is based on a microwave plasma excited in an Evenson cavity[15]. Oxygen atoms are guided from the microwave cavity to the sample via a quartz tube with a length of 30 cm. The quartz tube has an S-bend shape to avoid direct line-of-sight from the microwave plasma to the samples. Thereby, any UV radiation from the plasma cannot reach the sample, which would otherwise be the dominant inactivation mechanism. (ii) A second oxygen and a hydrogen source are installed consisting of a small iridium/tungsten capillary, which is heated to 1850 K and to 2100 K to dissociate a flow of oxygen or hydrogen molecules, respectively. These sources are not used in the current experiment. (iii) The Ar^+ ion beam is produced using a commercial plasma ion source (type Gen2 by Tectra GmbH)[16]. (iv) In addition, the samples are also exposed to stable oxygen molecules. In that case, a specific pressure of pure O_2 has been adjusted and the arrival rate at the surface has been deduced from simple gas kinetics. Details on the sources and their calibration can be found elsewhere [14].

The etch rate of endospores is also assessed by etching hydrocarbon films in an inductively coupled low pressure plasma from mixtures of argon and oxygen. The plasma reactor consists of a stainless steel vessel with cylindrical shape (see Fig. 1a). A quartz window acts as bottom to separate an ICP antenna

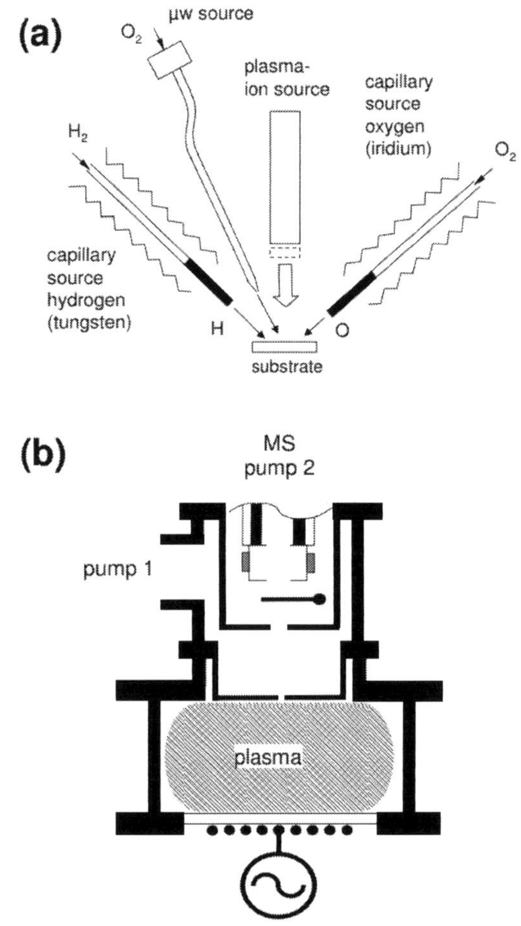

Figure 1. Schematic of the experimental setups: (a) particle beam experiment with three atom sources and one ion source; (b) inductively coupled plasma with a two stage differential pumped mass spectrometer.

from vacuum. The antenna is operated at 13.56 MHz with powers up to 500 W. A ring shower as gas inlet is embedded within the chamber wall. The vessel is equipped with six additional ports to allow for diagnostic, pumping and pressure measurements. The chamber is evacuated with a turbo pump and rough pump, which are connected to the vessel via a butterfly valve. The gas pressure ranges between 0.1 and 20 Pa and gas flows from 1 to 50 sccm can be adjusted. The plasma volume is approximately 5 l. The samples are placed onto a glass carrier which is placed 2 cm above the bottom window. The samples are hard amorphus hydrogenated carbon films deposited in a separate capacitively coupled plasma from acetylene.

The electron density is analyzed using the Langmuir probe system APS3. A tungsten wire with a diameter of 50 μm and a length of 5 mm is used to sample the plasma currents [17]. The probe is inserted into the vessel via a vacuum feed through and is positioned by a belt drive at the center of the discharge. An auxiliary stainless steel probe picks up the floating potential which is used to track an additive offset to the voltage ramp. Passive rf-compensation of the probe characteristics is maintained by a set of band-stop filters in the probe circuit and by capacitive coupling to the auxiliary probe. The current-voltage characteristics are at first numerically filtered as will be presented elsewhere [18]. The resulting second derivative is proportional to the EDF and the Druyvesteyn method is used to calculate electron density and mean electron energy [19].

The oxygen atom density is measured using a Balzers HiQuad Mass spectrometer QMG 700 being mounted inside a two stage differential pumping system on top of the reactor (see Fig. 1b). A mechanical flag is mounted directly in front of the ionizer in the second stage to allow a separate measurement of the background signal without a molecular beam component. This flag is manipulated externally via a mechanical feed through. Details of the mass spectrometry setup can be found in [20, 21].

The quantification of the oxygen atom density is performed using threshold ionization mass spectrometry (TIMS). For this, the electron energy in the ionizer is set to 15 eV and the count rate is measured. This rate is quantified by calibrating the signal using oxygen molecules at a given pressure of 10 Pa. By comparing the cross sections of oxygen molecules[22] and oxygen atoms[23] the count rate of O atoms is converted into a partial pressure of oxygen.

3. Results and Discussion

3.1. Exposure of B. atrophaeus to particle beams

Recently, Raballand et al. [14] studied the exposure of endospores of *B. atrophaeus* (ATCC 51189) to oxygen atoms, oxygen molecules and argon ions and combinations of theses. She showed, that the exposure to an O atom flux ($j_O = 2.4 \times 10^{15}\,\mathrm{cm^{-2}\,s^{-1}}$) and to an oxygen molecules flux ($j_{O_2} = 1.5 \times 10^{17}\,\mathrm{cm^{-2}\,s^{-1}}$) for 240 min does not lead to a biological inactivation of the spores. The inefficiency of inactivation and etching of the spores by O atoms is explained as follows: the endosporic shell of *Bacillus atrophaeus* consists of a peptidoglycan and a protein layer, which can be regarded, on an atomistic level, as a specific type of hydrocarbon compound. The chemical etching of various hydrocarbon compounds by oxygen atoms becomes only effective above a substrate temperature of 400 K, as known from literature [24, 25]. Since the substrate temperature in the experiment stays at 300 K, the absence of any etching is conceivable.

The experiments by Raballand et al. [14] revealed, however, that the oxygen etching can easily be enhanced by an additional ion bombardment. Fig. 2 compares the SEM micrographs of endospores as being treated for 120 min in three different experiments:

- *Argon ions only* (Fig. 2a): endospores exposed to Ar^+ ions at 200 eV ($j_{Ar^+} = 1.8 \times 10^{14}\,\mathrm{cm^{-2}s^{-1}}$) show only a slightly altered surface texture, as can be seen on the SEM micrograph. Apparently, Ar^+ ion bombardment at 200 eV does not cause any significant erosion of the spore coat. This can be explained by the energy dependence of the physical sputtering yield $Y(\mathrm{Ar}^+)$ of hydrocarbon compounds by argon ions. The physical sputtering process becomes only significant for ion energies well above 200 eV[26].
- *Argon ions and oxygen molecules* (Fig. 2b): Endospores exposed to a simultaneous flux of argon ions ($j_{Ar^+} = 1.8 \times 10^{14}\,\mathrm{cm^{-2}s^{-1}}$) and oxygen molecules ($j_{O_2} = 4.5 \times 10^{15}\,\mathrm{cm^{-2}s^{-1}}$) exhibit, however, a very different appearance. Although the absolute size of the spores did not change noticeably, the spore coat became porous, showing deep etch channels. The high sputtering yield is explained by the *simultaneous* impact of ions and oxygen molecules leading to the process of

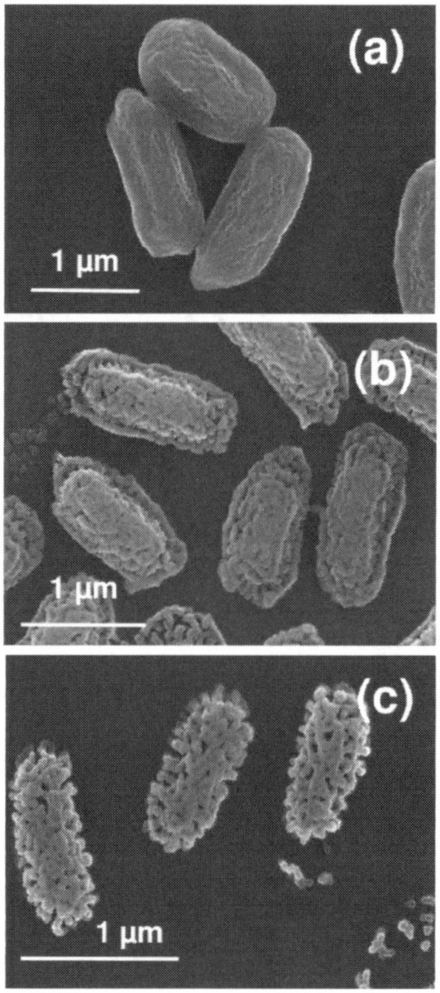

Figure 2. SEM micrographs of *Bacillus atrophaeus* exposed for 120 minutes in the beam experiment to (a) argon ions (j_{Ar^+}= 1.8 × 10^{14} $cm^{-2}s^{-1}$) at 200 eV, to (b) a combined flux of argon ions (j_{Ar^+}= 1.8 × 10^{14} $cm^{-2}s^{-1}$) at 200 eV and O_2 molecules (j_{O_2}= 4.5 × 10^{15} $cm^{-2}s^{-1}$), and to (c) a combined flux of argon ions (j_{Ar^+}= 1.8 × 10^{14} $cm^{-2}s^{-1}$) at 200 eV, O_2 molecules (j_{O_2}= 1.5 × 10^{17} $cm^{-2}s^{-1}$) and O atoms (j_O= 2.4 × 10^{15} $cm^{-2}s^{-1}$). Adapted from [14].

chemical sputtering[27]: impinging energetic ions are able to break bonds in the spore coat within their corresponding collision cascade. The additional flux of oxygen molecules is able to react with these open bonds. The repeated ion-induced bond breaking followed by their reaction with oxygen molecules leads, below the surface, to the formation of presumably CO, CO_2 and H_2O as volatile components. These reaction products diffuse to the surface and desorb[28]. As a result, etching occurs.

At first glance, it is surprising that the impact of stable ground state oxygen *molecules* is able to cause significant etching, if a simultaneous flux of ions impinges on the surface. However, O_2 possesses a triplet ground state with two unpaired electrons, which makes it a so-called bi-radical. The electronic structure of O_2 explains its reactivity: if an ion at the surface creates a defect, namely an open bond with an unpaired electron, the O_2 bi-radical can easily react at this open site. The very same reaction occurs during the slow auto-oxidation of any polymer being exposed to ambient atmosphere[29].

- *Argon ions and oxygen atoms and molecules* (Fig. 2c): The exposure of *Bacillus atrophaeus* to O atoms (j_O= 2.4 × 10^{15} $cm^{-2}s^{-1}$), to O_2 molecules (j_{O_2}= 1.5 × 10^{17} $cm^{-2}s^{-1}$) and to argon ions (j_{Ar^+}= 1.8 × 10^{14} $cm^{-2}s^{-1}$) is performed using the plasma ion source and the microwave source to dissociate 1% of a flow of 5 sccm O_2 molecules. The spores are efficiently etched as indicated by the appearance of deep trenches and pores. The fast etching is again explained by the *simultaneous* impact of ions, of oxygen molecules, and of oxygen atoms leading to the process of *chemical sputtering*.

Summarizing, the experiments using particle beams revealed that a strong etching of the endospores occurs only if argon ions and O atoms or O_2 molecules interact simultaneously with the spores. This is explained by the process of chemical sputtering as the combined impact of neutrals and ions. In the following, we test this observation in a plasma experiment using argon oxygen mixtures.

The quantification of absolute plasma etch rates of endospores is, however, difficult since the etching leads to a perforation of the spore coat rather than to a layer-by-layer removal. Instead, we study the etching of hydrocarbon thin films as being a model system which resembles the stoichiometry of the spore coats on an atomistic level. Such a rough model system is still able to produce comparable results, because *ion-induced etching* in a low pressure plasma does not strongly depend on the exact chemical nature of the materials to be etched. In contrast, a pure chemical erosion process relies very much on the exact nature of the surface groups and the definition of such a simple model system for the spore coat would not be possible.

3.2. Etching of hydrocarbon films in oxygen containing plasmas

An inductively coupled plasma at a pressure of 10 Pa and a power of 200 W is used to etch hydrocarbon films using various mixtures of oxygen and argon. The total flow has been kept at 10 sccm, while varying the oxygen admixture. Fig. 3a shows the electron density, as measured via Langmuir probes, and the oxygen partial pressure, as measured by threshold ionization mass spectrometry. It can be seen that the electron density decreases drastically with increasing oxygen admixture, whereas the oxygen atom concentration slowly increases before it reaches a maximum at an admixture level of 80 %. Such a variation of electron density and atomic oxygen concentration is typical for a plasma experiment where the power is kept constant. At increasing admixture of a molecular gas, more power is dissipated in rotational and vibrational excitation of the molecules and less in ionization. Consequently, the ionization degree and the electron density decreases sharply.

Within the very same parameter range, the etch rate of hydrocarbon films is also quantified, as shown in Fig. 3b. One can clearly see that the etch rate peaks already at a small oxygen admixture of 5%. At higher oxygen admixtures, the etch rate decreases again although the oxygen atom concentration is still increasing. This clearly illustrates that the plasma etch rate of the hydrocarbon films is not only linked to the oxygen atom concentration.

According to the process of *chemical sputtering*, the combined impact of ions and atoms is crucial for an efficient etch rate. Both fluxes, the ion fluxes and the O atom flux are proportional to the bulk densities

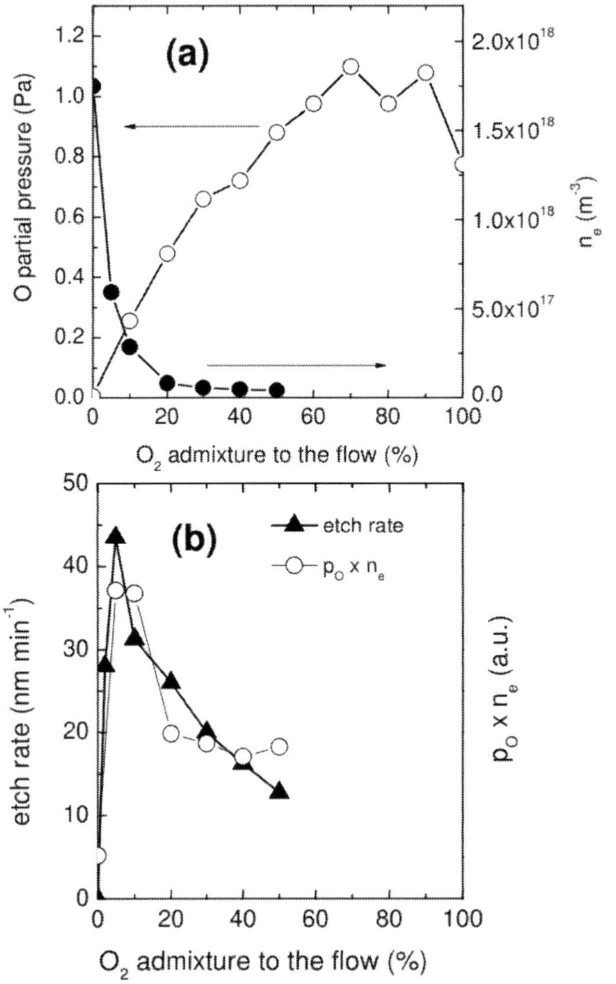

Figure 3. Electron density n_e and oxygen atom partial pressure p_O in dependence of the admixture of oxygen to an argon inductively coupled plasma at a total pressure of 10 Pa, a power of 200 W and a total flow (argon plus oxygen) of 10 sccm. (b) Etch rate of a hydrocarbon film in the ICP plasma at the same process parameters (solid symbols). Product of n_e and p_O (open symbols) using the data in (a). The value for p_O at an admixture of 5% O_2 has been derived by interpolation.

of electrons and O atoms. As a rough hypothesis, one might state that the etch rate is proportional to the product of ion flux and atom flux and thereby also proportional to the product of electron density and O atom density. This hypothesis is tested by plotting the product of oxygen partial pressure p_O and electron density n_e in dependence of the O_2 admixture (The missing value for p_O at an admixture of 5% O_2 has been derived by interpolation). One can clearly see that the overall behavior of the etch rate is very well reproduced by the product $p_O \times n_e$.

The maximum etch rate is observed at an admixture level of 5% O_2. This corresponds to an oxygen atom partial pressure around 0.2 Pa, which converts into a wall collision rate of 7.9×10^{17} $cm^{-2}s^{-1}$ oxygen atoms according to standard gas kinetics assuming a gas temperature of 300 K. The electron density of 6×10^{17} m^{-3} at 5% O_2 admixture converts to an ion flux onto the surface of 1.64×10^{17} $cm^{-2}s^{-1}$, as can be deduced using the Bohm criterion. This results in ratio of the atom to ion arrival rates of 4.8. A ratio significantly above one is usually required for the chemical sputtering process to be efficient, since the ion-induced defects at the surface needs to be quickly saturated by incident neutral species.

Similar observations on the overall etch rate in plasmas from oxygen - noble gas mixtures have been made by several authors: a small admixture of oxygen up to 10% to a noble gas is sufficient to reach the maximum etch rate [30, 31, 32]. Landkammer et al. [31] showed further that the chemical identity of the ions providing the kinetic energy input for *chemical sputtering* is irrelevant: the etch rate is proportional to the number of ion-induced defects within a thin surface layer with a thickness identical to the penetration length of the incident neutrals. Consequently, argon ions are rather efficient at 200 eV for chemical sputtering, because incident argon ion create defects near the surface, where they can easily react with incident neutrals. Lighter ions are in principle less efficient for chemical sputtering, since they create most of the ion-induced defects beyond the penetration length of the incident neutrals leading to a much smaller chemical sputtering rate. But, if the energy of the lighter ions is reduced and their flux is increased to create the same number of defects as argon ions, the same etch rate would result. The chemical reactivity is only dominated by neutral oxygen species [33]. These could be either oxygen atoms or oxygen molecules, depending on their absolute concentrations [34].

Summarizing one can state, that the optimum process parameters to reach an optimum etch rate in an oxygen containing low pressure plasma need to balance the ion density simultaneously with the O partial pressure. Only the combined impact of both species assures an optimum etch rate.

4. Conclusions

Spores of *Bacillus atrophaeus* were exposed to beams of Ar^+ ions, oxygen molecules and oxygen atoms. It is shown, that only the simultaneous impact of O atoms or O_2 molecules and of argon ions at around 200 eV causes a very effective etching and perforation of the spore coat. This is due to the process of *chemical sputtering*, where the bi-radical O_2 or the O atoms interact with ion-induced defects at the spore surface leading to the formation of CO, CO_2 and H_2O and thus to etching. This observation is directly supported in experiments using an inductively coupled plasma from argon - oxygen mixtures. The maximum etch rate is only reached if both the electron density (and thus the ion flux) and the O atom density are optimized simultaneously. The contribution of *chemical sputtering* to the efficiency of the plasma-induced inactivation of bacteria is of great importance for the development and validation of plasma sterilization techniques.

Acknowledgments

This work was supported by the European Commission in the framework of the BIODECON project.

References

[1] Moisan M, Barbeau J, Crevier M, Pelletier J, Philip N and Saoudi B 2002 *Pure Appl. Chem.* **74** 349
[2] Lerouge S, Wertheimer M and Yahia L 2001 *Plasmas and Polymers* **6** 175
[3] Rossi F, Kylian O and Hasiwa N 2006 *Plasma Process. Polym.* **3** 431
[4] Laroussi M 2005 *Plasma Process. Polym.* **2** 391
[5] Lendlein A and Langner R 2002 *Science* **296** 1673

[6] Moisan M, Barbeau J, Moreau S, Pelletier J, Tabrizian M and Yahia L 2001 *Int. J. Pharm* **226** 1
[7] Kylian O, Hasiwa M and Rossi F 2006 *IEEE Trans. Plasma Sci.* **34** 2606
[8] Bernard G, Leduc A, Barbeua J, Saoudi B, Yahia L and de Crescenzo G 2006 *J. Phys. D* **39** 3470
[9] Lerouge S, Fozza A, Wertheimer M, Marchand R and Yahia L 2000 *Plasmas and Polymers* **5** 31
[10] Philip N, Saoudi B, Crevier M, Moisan M, Barbeau J and Pelletier J 2002 *IEEE Trans. Plasma Science* **30** 1429
[11] Halfmann H, Bibinov N, Wunderlich J and Awakowicz P 2007 *J. Phys. D* **40** 4145
[12] Moreau S, Moisan M, Tabrizian T, Barbeau J and Pelletier J 2000 *J. Appl. Phys.* **88** 1166
[13] Opretzka J, Benedikt J, Awakowicz P, Wunderlich J and von Keudell A 2007 *J. Phys. D* **40** 2826
[14] Raballand V, Benedikt J, Wunderlich J and von Keudell A 2008 *J. Phys. D* **41** 115207
[15] Fehsenfeld F, Evenson K and Broida H 1965 *Rev. Sci. Instr.* **36** 294
[16] Anton R, Wiegner T, Naumann W, Liebmann M, Klein C and Bradley C 2000 *Rev. Sci. Instr.* **72** 1177
[17] Scheubert P, Fantz U, Awakowicz P and Paulin H 2001 *J. Appl. Phys.* **90** 587–598
[18] Gälger M, Wenig G, Schulze M, Steland A and Awakowicz P 2007 *unpublished*
[19] Druyvesteyn M 1930 *Zeitschrift für Physik* **64**(11–12) 781–798
[20] Benedikt J, Consoli A, Schulze M and von Keudell A 2007 *Journal Of Physical Chemistry A* **111** 10453–10459
[21] Consoli A, Benedikt J and von Keudell A 2008 *J. Phys. Chem. (accepted)*
[22] Straub H, Renault P, Lindsay B, Smith K and Stebbings R 1996 *Phys. Rev. A* **54** 2146
[23] Thompson W, Sah M and Gilbody H 1995 *Phys. B: At. Mol. Opt. Phys.* **28** 1321
[24] Maruyama K, Jacob W and Roth J 1999 *J. Nucl. Mater.* **264** 56
[25] Landkammer B, von Keudell A and Jacob W 1999 *J. Nucl. Mater.* **264** 48
[26] Eckstein W 1991 *Computer Simulation of Ion Solid Interactions* 1st ed (Berlin and Heidelberg: Springer Series in Materials Science)
[27] von Keudell A and Jacob W 2004 *Prog. Surf. Sci.* **76** 21
[28] Coburn J and Winters H 1979 *J. Appl. Phys.* **50** 3189
[29] Friedrich J, GKühn and Gähde J 1979 *Acta Poylm.* **30** 470
[30] Murillo R, Poncin-Epillard F and Segui Y 2007 *Eur. Phys. J. Appl. Phys.* **299** 37
[31] Landkammer B 1999 *Untersuchung der Erosion von Kohlenwasserstoffschichten in Sauerstoff-Gasentladungen* Ph.D. thesis University of Bayreuth
[32] Hopf C, Jacob W and Rohde V 2008 *J. Nucl. Mater.* **374** 413
[33] Gokan H and Esho S 1984 *J. Electrochem. Soc.* **131** 1105
[34] Collart E, Baggerman J and Visser R 1995 *J. Appl. Phys.* **75** 47

Surface analysis with grazing incidence fast atom diffraction (GIFAD)

Patrick Rousseau[1,2], Hocine Khemliche[1], Nenad Bundaleski[1], Pierre Soulisse[1], Anouchah Momeni[1,3], and Philippe Roncin[1]

[1]Laboratoire des Collisions Atomiques et Moléculaires,
UMR 8625, CNRS -Université Paris-Sud, Orsay, F-91405, France

[2] CIMAP, CEA/CNRS/ENSICaen/Université de Caen-Basse Normandie,
BP 5133, F-14070 Caen Cedex 5, France

[3] LERMA, Université de Cergy-Pontoise & Observatoire de Paris, LERMA, UMR 8112 du CNRS, 95000 Cergy-Pontoise, France

philippe.roncin@u-psud.fr

Abstract. Grazing collisions at surfaces offer rather contrasted conditions. For well ordered flat surfaces, the scattering is spread among several lattice sites, each of which produces only a tiny elementary deflection. If, in addition, the atomic projectile is aligned along a crystallographic direction, the surface appears as made of parallel furrows or as a washboard which act as a diffraction grating for the atomic wave. We will show that the analysis of characteristic diffraction pattern recorded on a position sensitive detector located downstream allows a sensitive measure of the shape of the surface electronic density. A modified Debye Waller factor is proposed to explain the observed diffraction signal.

1. Introduction

Probing surfaces with keV projectile is not a new field. Pushed by applications such as surface modifications or surface analysis, a wide variety of techniques has been developed [1], supported by continuous progress in understanding the underlying physical mechanisms [2]. Diffraction of atoms with thermal energies (TEAS) proved to be very sensitive but comparatively slow and cumbersome. Until recently [3-7], no one however had considered the possibility of Grazing Incidence Fast Atom Diffraction (GIFAD). This paper describes this new diffraction technique where keV atoms or molecules diffract on a single crystal surface and outlines some of the specific advantages. We will show that GIFAD complements very well TEAS just as Reflexion High Energy Electron Diffraction (RHEED) does compared with Low Energy Electron Diffraction (LEED). GIFAD appears as the natural extension of TEAS, however the diffraction of keV atoms was not expected because these have typical de Broglie wavelength below the picometer scale, orders of magnitude smaller than crystal lattice parameters. It was rapidly understood that for incidence angle θ on the order of 1°, diffraction can be understood as that of the slow motion normal to the surface. The relevant energy is therefore reduced by three to four orders of magnitude. At the same time, the multiple collision regime characteristic of grazing incidence strongly reduces the influence of surface thermal fluctuation, allowing larger normal energies than TEAS. These effects and the basic physical conditions prevailing will be briefly recalled bellow. The actual setup and specificity of the technique will be presented with help of selected results.

2. Experimental setup

The general scheme for grazing incidence fast atom diffraction (GIFAD) is displayed in figure 1. The interaction geometry is exactly the same as that of RHEED. Of course both the primary beam and detectors are different but their positions are identical. The production of a neutral beam in the keV energy range is easily achieved by charge exchange of a primary ion beam inside a gas cell. For a moderate gas pressure inside the cell i.e. such that less than 30 % of the beam is neutralized, the energy spread of the neutral beam can be close to that of the primary ion beam. The main limitation is given by the divergence of the beam. Expressed in terms of transverse momentum spread $\Delta k_\perp = k_0.\Delta\theta$ where k_0 is the momentum of projectiles and $\Delta\theta$ is the angular divergence of the beam. If diffraction is to be observed this spread $\Delta k_\perp$ has to be much less than a typical reciprocal lattice vector $G = 2\pi/a$ where a is a lattice vector of the crystal, typically larger than 2 Å. For 100 eV hydrogen projectiles the limit reads $\Delta\theta \ll 0.6$ deg.; but yields $\Delta\theta \ll 0.1$ deg. for 1 keV helium. This low divergence is not common in most ion sources and it is achieved only by passing the beam through small diaphragms. Compared with TEAS, detection of keV atoms is very easy with help of micro channel plates and much more efficient since the whole diffraction pattern is confined into a narrow cone which can be recorded at once without any scan. Compared with RHEED, one or two micro channel plates are needed followed by a position sensitive collector or phosphor screen.

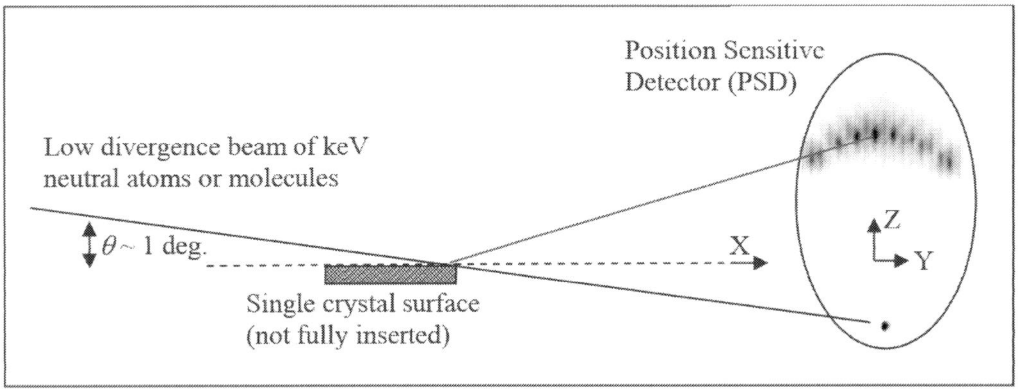

Figure 1. Schematic view of the GIFAD geometry. The primary neutral beam impinges the target at grazing incidence and the diffracted signal is recorded onto a position sensitive detector. The image inside the detector corresponds to diffraction of 400 eV helium on a ZnSe(001) surface along the [1-10] direction.

3. Grazing incidence diffraction of fast atoms.

3.1. Decoupling of fast and slow motions

The naive picture explaining fast atom diffraction is to consider that, viewed from grazing incidence, a crystal surface is not made of atoms but of furrows. More formally, the crystal surface symmetry implies that the momentum **G** exchanged with the projectile has to be a linear combinations of 2D surface reciprocal lattice vectors $\mathbf{G}_1$, $\mathbf{G}_2$. The surface plane is taken to be the XY plane and Z is normal to the surface. The projectile impinges the surface in the XZ plane at grazing incidence, meaning that its momentum k_z is much smaller that k_x, $\tan(\theta) = k_z/k_x$ and $k_y = 0$. At keV energies, $|\mathbf{k}| \gg |\mathbf{G}|$. The idea of structureless furrows aligned parallel to the beam can be established if one shows that only reciprocal lattice vectors G normal to the beam can be exchanged i.e. $\mathbf{k}.\mathbf{G} = 0$ (in the present definition this means that $G_x=0$). Two effects contribute to exclude non zero G_x component; the energy associated is two orders of magnitude larger than that associated with G_y and the coupling responsible for these transitions is negligible, orders of magnitude smaller than for G_y transition. As note in ref. [5], the energy change of the projectile scales as $\sim G^2 + 2.\mathbf{k}.\mathbf{G}$ so that much more energy is required to exchange a G component parallel to the beam. At variance, the coupling strength is almost zero along the beam direction [8]. Along a crystal direction the projectile flies over several target atoms with almost constant z elevation. The associated slowing down and reacceleration compensate each other so that, integrated along the trajectory the overall coupling averages to negligible values (figure 2 may be used to visualize this effect). This is not true in the direction normal to the beam because, if the impact parameter is such that, when the projectile hits an atomic row on the edge all successive atoms of the row will deflect the projectile in the same direction, i.e. all contributions add up with the same sign allowing significant value of the coupling. Both arguments contribute to select transitions with $\mathbf{k}.\mathbf{G} = 0$. As a result, the scattering problem becomes separable in the X direction and YZ plane, transforming the actual surface into a washboard.

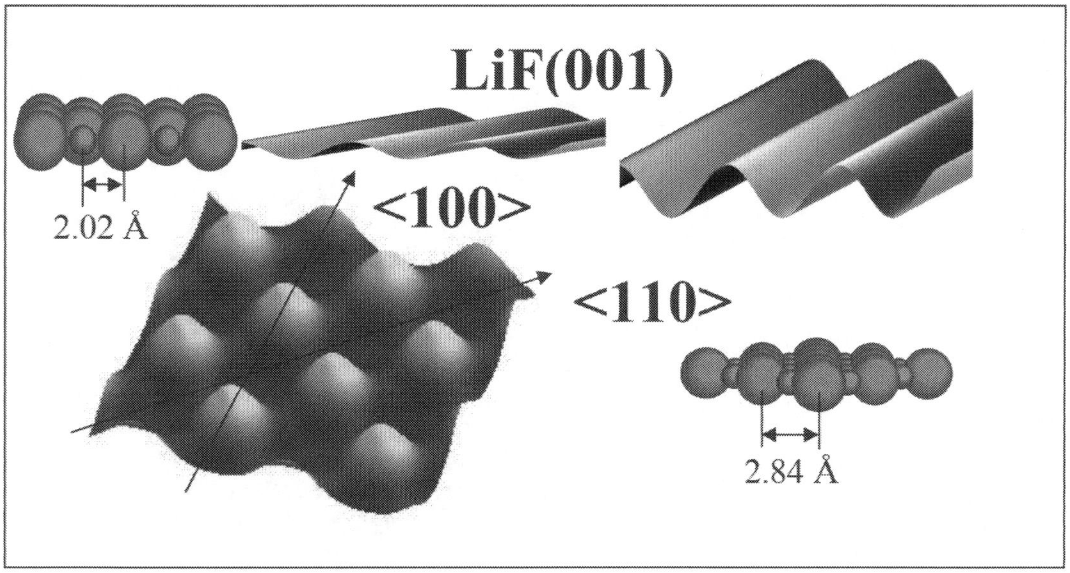

Figure 2. Schematic drawing of the averaging effect intrinsic to GIFAD. Along the <100> direction, the maxima and minima of the egg box compensate for each other resulting in a reduced corrugation. Note that the symmetry of the rows (2.02 Å) is not a translation symmetry of the 2D crystal. Along the <110> direction the averaging effect is reduced.

The shape of the board is given by the average along X of the actual surface. The effective wavelength $\lambda_\perp$ that diffracts on this 1D potential corresponds to the normal energy $E_\perp = E_0 \sin^2\theta$. The periodicity a_y of the washboard is given by modulus of the smallest linear combination normal to the beam $G_y = 2\pi/a_y$. In this YZ plane, writing the initial and final momentum of the projectile k_i and k_f respectively, the energy conservation gives $k_{iz}^2 = k_{fz}^2 + k_{fy}^2$ while Bragg condition is $k_{fy} = n.G_y$. The diffraction pattern is made of spots sitting on a circle of radius k_{iz} and the maximum diffraction order is given by k_{iz}/G_y. The overall validity of this decoupling of the fast and slow direction has been established with help of a test system; helium projectile scattered off a LiF(001) surface along the <100> direction. The relative intensities of the observed diffraction orders are plotted on figure A for normal energies between 20 meV and 500 meV and compared with a parameter free calculation where the atoms surface potential was derived from state of the art Hartree-Fock technique together with a full 3D wave packet treatment of diffraction [5] (full line on the left hand side of fig 3).

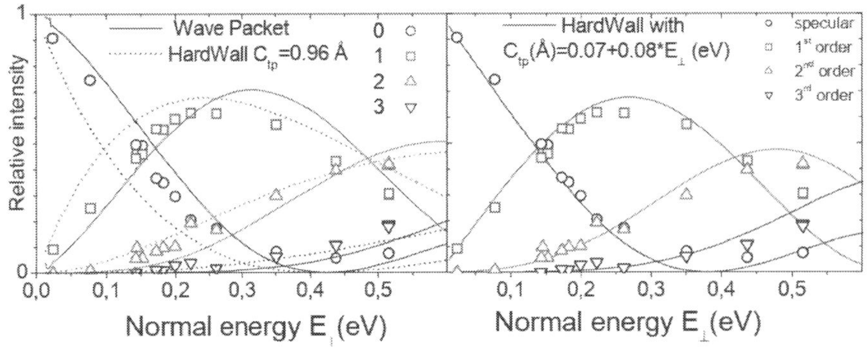

Figure 3. Relative intensities measured for diffraction of 200 eV helium on a LiF(001) surface along the <100> direction are reported as a function of the projectile normal energy $E_\perp$ (symbols). On the left hand side, the data are compared with a parameter free wavepacket calculation [5] (full line) and to a HardWall model (dashed lined) with a trough to peak corrugation C_{tp}=0.096 Å independent of the normal energy [9]. On the right hand side, the HardWall model is used to derive an energy dependent corrugation.

3.2. The HardWall model

The full calculation and the decoupling model suggest that most of the results derived in diffraction of thermal energy atoms (TEAS) should be applicable to GIFAD. In this context, the simplest diffraction model is that of the HardWall which considers that the interaction potential, usually decaying exponentially away from the surface, can be replaced by an infinite potential $Z_t(X,Y)$ located at the classical turning point of the trajectory i.e. such that $V(X,Y,Z_t) = E_\perp$. Applied here, the corrugation function is a 1D function $Z_t(Y)$ defined by $\mathcal{V}(Y,Z_t) = E_\perp$ where $\mathcal{V}(Y,Z) = \langle V(X,Y,Z) \rangle_X$ the bracket indicating the average over X (the beam direction) on a lattice site. The projectile surface interaction potential is now reduced to a 1D function $Z_t(Y)$ describing the shape of the washboard and it transforms into diffraction intensities via a Fourier-like transform [see e.g. ref 5,9]. For a low number of diffraction orders it does not make sense to infer a too complex form for the corrugation function. Indeed, due to the very wave nature of diffraction, sub wavelength details are averaged so that the number of significant values in the potential has to compare with the number of observed diffraction orders. As long as the specular and 1st diffraction order dominate, a single parameter such as C_{tp}, the trough to peak corrugation amplitude is more adequate to track small evolutions trying to infer a shape evolution. The simplest corrugation function is the cosine function $Z_t(Y) = C_{tp}/2 \,.\, \cos(G_y.Y)$, allowing straightforward calculation of the intensity I_m of a diffraction order m. I_m is simply given by the square of a Bessel function of rank m J_m with dimensionless argument, $C_{tp}.\Delta k_z /2 \sim 2\pi.C_{tp}/\lambda_\perp$:

$$\mathbf{I_m = J_m^2(C_{tp}.\Delta k_z /2)} \;,\; \Delta k_z = k_z + (k_z^2 - m^2 G_y^2)^{1/2} \sim 4\pi/\lambda_\perp \text{ (for low m)} \qquad \text{(Eq. 1)}$$

In counterpart, the parameter C_{tp} is measured with "interferometric" accuracy because all intensities are linked to this value. Semi classically, it can be understood as the interference between paths contributing to a given diffraction order. Considering the specular peak (order 0), these paths correspond to reflection in the bottom of the valley or on top of the hill with a path difference of $2.C_{tp}$ and a phase difference $4\pi.C_{tp}/\lambda_\perp$ so that, to first order, both paths interfere destructively each time that $C_{tp}/\lambda_\perp = (2n+1)/4$. The HardWall approach is described in ref [9] and also reported in figure 3 (on the left) where the experimental data are reasonably reproduced by a fixed corrugation of 0.096 Å. Of course it is tempting to allow the corrugation to depend on energy. This is done on figure 3 where the right hand side proposes a linear dependence C_{tp} (Å) = 0.07+0.08*$E_\perp$ (eV). The increased corrugation is probably due to the fact that, in this <100> direction, the valley corresponds to the space in between the rows which is softer that the top of rows. Increasing the normal energy up to tens of eV, the inter-row channel will eventually open. The Bessel formula in Eq.1 indicates that the corrugation is measured in units of the wavelength $\lambda_\perp$ so that even a fixed corrugation produces much more diffraction orders at larger normal energies increasing considerably the shape sensitivity. This is one of the most salient features of GIFAD, large normal energies are possible whereas TEAS is usually limited below 50-100 meV. This aspect related to the Debye-Waller factor is detailed in section 5.

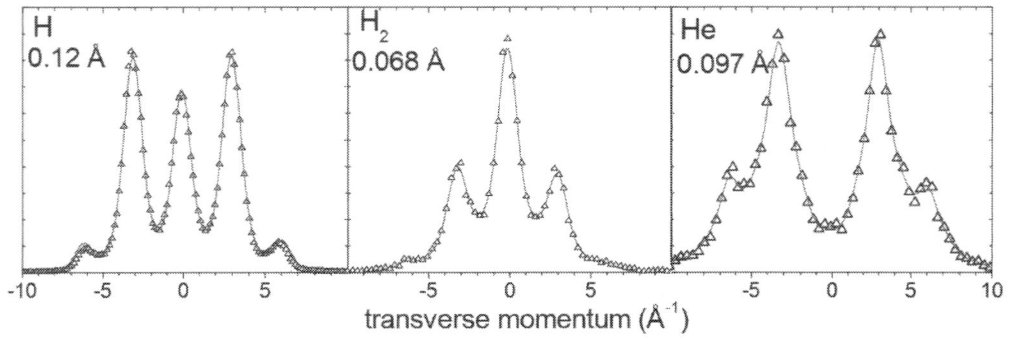

Figure 4. Diffraction spectra recorded on a LiF(001) surface aligned along the <100> direction for H, H_2 and He projectile with a normal energy of 300 meV. The solid line is a fit of the experimental points by Gaussian functions weighted by a Bessel function (see text). The corrugation C_{tp} extracted from the fit in Å is reported on each inset.

4. Case study H , H_2 and He on LiF(001)

4.1. LiF analysed along the <100> direction.

From these encouraging results, it is tempting to extract a corrugation by directly fitting the diffraction spectra. As an illustration, the corrugation measured by H, H_2 and He projectiles on a LiF(001) surface with a beam aligned along the <100> direction are reported on figure 4. The observed G_y peak spacing corresponds to a periodicity of ≈ 2.0 Å i.e. matching that of the spacing between rows. Note that this periodicity does not correspond to translation symmetry of the 2D LiF(100) surface because it would bring a F atom on top of a Li atom (see fig. 2). To measure accurately the corrugation C_{tp}, the experimental data are fitted by Gaussians which intensities are given by the appropriate Bessel function. The quality of the fit justifies *a posteriori* the simple cosine model used for the corrugation function. The amplitude of the corrugation is rather limited as illustrated by the few diffraction orders observed in spite of a comparatively large normal energy $E_\perp$ (therefore small $\lambda_\perp$).

To compare with data from TEAS the 2D interaction potential has to be averaged along the beam direction. The consequences are surprising. LiF(001) has been investigated as early as 1930 by pioneers of quantum mechanics [10]. It had become a test system for TEAS and a 2D potential was proposed consisting of a sum of 2 cosine functions [11,12]. The bumps are located on top of the F^- ions where most of the electron density sits while the troughs are on top of Li^+. This was enough to reproduce the TEAS diffraction data and continuous efforts to observe deviations from this simple form were not fully convincing. Interestingly this sum of cosine function averages exactly to zero along the <100> direction because the crest exactly matches the trough for any impact parameter Y (as can be qualitatively understood in fig 2) therefore only the specular peak should be observed!!! Obviously this is not the case and all three projectiles in figure 4 show well resolved diffraction orders. The corrugation derived is limited but it directly points to the defect of the surface model as a sum of cosine. Along this <100> direction the possible chemical effect are partly washed out because each row is made Li and F ions so that, if present, selective attraction towards Li or F ions does not directly contribute to the corrugation. Of course a general attraction towards the surface is still possible and may explain the fact that a net corrugation exists even at very small normal energy. Without any attraction to the surface, the projectile would be repelled further away from the surface where little corrugation exists.

4.2. LiF analysed along the <100> direction.

Along the <110> direction, the surface can be seen as adjacent rows of lithium and fluorine (Fig. 2). In this direction the sum of cosine averages to a simple cosine with equal corrugation. For H_2 and He, the figure 5 displays the diffraction pattern and the corrugation derived by a fit with eq 1. The corrugation is indeed much larger along <110> than it is along <100>. The other striking feature is that the fit by a cosine potential is rather poor, clearly calling for a better description. For helium, the choice of lower normal energy does slightly improve the cosine fit mainly because diffraction is less sensitive to the shape when probed with a longer wavelength. However, even at a normal energy as low as $E_\perp = 40$ meV, the corrugation remains comparatively large with a C_{tp} of 0.33 ± 0.01 Å. This value compares reasonably well with the value of 0.31±0.01 Å determined around 25 meV by TEAS [11]. It hardly evolves with the normal energy, with a slight tendency to decrease downto 0.32 Å at $E_\perp = 133$ meV as shown in figure 5. However, the cosine shape appears more and more inappropriate so that C_{tp} should be considered only as an effective corrugation. The tendency to decrease could be due to the fact that, along the <110> direction the valley sits on top of a Li^+ row which are probably harder than the F^- ions so that getting closer does not result in a deeper valley. H_2 displays a similar tendency with an apparent corrugation roughly 3 to 4 times larger along the <110> than along the <100> direction.

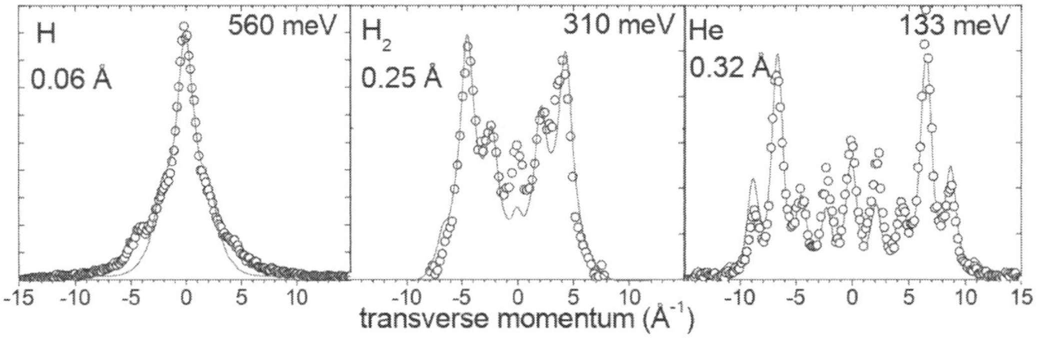

Figure 5. Diffraction spectra recorded on a LiF(001) surface aligned along the <110> direction for H, H_2 and He projectile. The solid line is a fit of the experimental points by Gaussian functions weighted by a Bessel function (see text). The corrugation C_{tp} extracted from the fit in Å is reported on each inset together with the normal energy. Note that the quality of the fit is much less that for the <100> direction indicating a "non cosine" shape of the potential. Note also that in spite of a much larger normal energy the H projectile seems to indicate a weak corrugation along this <110> direction.

The relative surprise is the low apparent corrugation probed by the H atoms. It only amounts to 0.06 Å even at $E_\perp = 0.56$ eV normal energy. This is due to the restriction that the potential is monotonic with only one maximum. Such simplification is not relevant here, as indicated by the side wings in fig.5. The H atoms are strongly attracted to the surface and actually feel directly the Li atoms giving rise to a second maximum inside the unit cell [13]. The apparent reduced corrugation in figure 5 is mainly the wave response to this double well situation. Such "chemical" attraction is of course much reduced with He and H_2 which have a closed shell electronic structure. This aspect opens interesting perspective for probing complex projectile surface potentials in regimes where chemical interaction is important. More work is obviously needed to allow a refined optimization of the shape of the interaction potential to the data.

5. Decoherence effects

5.1. Effect of thermal motion, the Debye-Waller factor

As detailed above, the atoms probe the shape of the interaction potential with the surface. As any diffraction technique, the wavelength is the basic unit for the accuracy, strongly favoring the use of larger normal energies. However, using smaller wavelength increases the sensitivity to decoherence [14]. In TEAS, the atom "bounces on a single atom" so that if the thermal motion of these atoms is on the order of the wavelength of the projectile the diffracted signal becomes negligible. Quantitatively, this random phase shift induced by thermal displacement induces an attenuation of the coherent intensity known as the Debye-Waller factor. The diffracted intensity reduces to $I = I_0 \exp(-\langle(\Delta\mathbf{k}.\mathbf{u})^2\rangle)$ where $\Delta\mathbf{k}$ is the projectile wave vector change and $\mathbf{u}$ is the mean thermal displacement, the brackets denoting thermal averaging. The thermal motion is such that, even at low surface temperatures, 100 meV is considered as an upper energy limit for helium diffraction. In GIFAD, the projectile atoms diffract on the aligned rows rather than on individual atoms so that thermal motion is not so severe. Of course the projectile does not probe the entire row and the averaging is limited to the number of atoms probed in a single grazing trajectory.

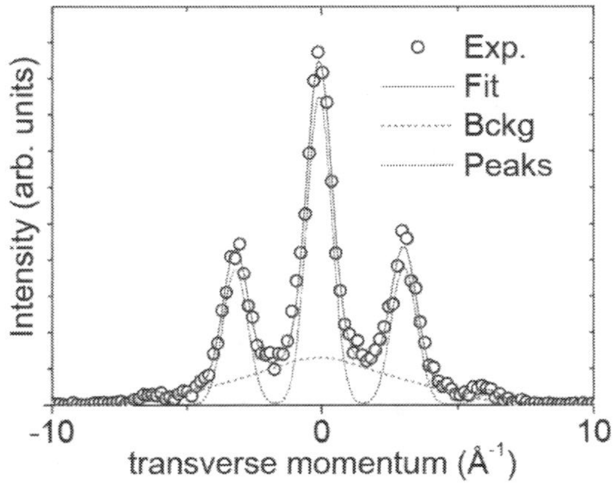

Figure 6. For 200 eV helium on LiF(001) surface along the <100> direction at 1.5 deg. incidence. The intensity I_m given to each diffraction order is splitted in two components; a narrow one of intensity $I_m.R$ plus a broad one of intensity $(1-R).I_m$. The sum of the narrow peaks in dashed blue represents the coherent intensity, the dotted green the sum of the broad peaks is an estimate of the background and the total in solid red is the fit trough the experiment. The ratio R thus determined is the effective Debye-Waller factor, here R=0.65 .

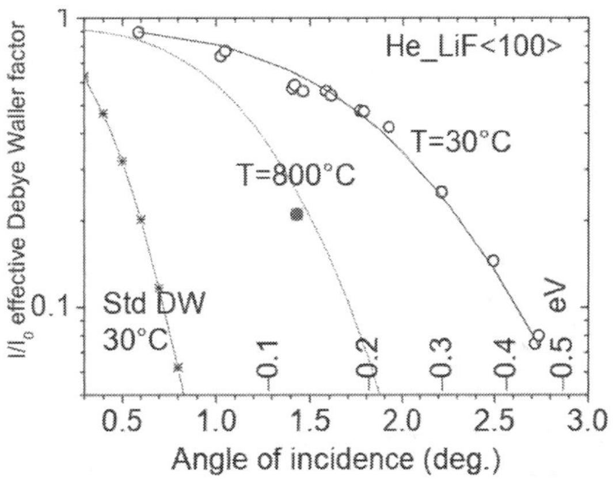

Figure 7. The effective Debye-Waller factor as estimated in fig. 6 is reported vs. the angle of incidence for 200 eV helium atoms and compared with the modified Debye-Waller formula (eq. 2). The blue circles correspond to data taken at room temperature while the red square is for a LiF target heated at 800°C. A surface Debye temperature of 490 K was considered. The estimations are multiplied by 0.92 which is taken as the maximum coherence. The line with crosses stands for the unmodified Debye-Waller factor at room temperature.

For a simple exponential form $V(Z)=V_0 \exp(-\Gamma Z)$ of the projectile surface interaction potential, the effective length of the trajectory can be defined exactly [15,16] as $L_{eff} = 2 \cosh^{-1}(2^{1/4})/(\Gamma.\theta) \sim 2.42 / \Gamma\theta$ where θ is the angle of incidence expressed in radian. The effective number of atoms participating to the scattering is estimated by dividing by the lattice unit a; $N_{eff} = L_{eff}/a$. Accordingly [9], the effective thermal motion of a row u_{eff} is reduced by a factor $N_{eff}^{1/2}$ i.e. $u_{eff} = u.N_{eff}^{-1/2}$. Taking for example typical values for the He-LiF system $\Gamma\sim 1$ Å^{-1} and a = 4 Å yields an effective $N_{eff}\sim 0.6/\theta$ which amount to N_{eff} ~30 at 1 deg. incidence and $u_{eff}\sim u/5$. For the specular peak $\Delta k_z = 2.k_z$ and $k_z = k_0.\theta$ where $k_0 = (2.M_p.E_0)^{1/2}$ is the total projectile momentum. Overall the Debye Waller factor can be rewritten as:

$$I / I_0 = \exp(- 4.k_z^2.u^2/ N_{eff}) = \exp(- 6.67\ \theta\ k_z^2.u^2) = \exp(- 13.3 * M_p E_0\ \theta^3\ .u^2)$$

$$\mathbf{I / I_0 = \exp(- \alpha.M_p E_0\ \theta^3\ .u^2)} \qquad \text{(Eq. 2)}$$

with $\alpha\sim 472$ if M_p is in nuclear mass unit (i.e. 1 for H), E_0 in eV, θ in radian and u in Å.

In grazing scattering, the coherent and incoherent scattered intensity are collected onto the position sensitive detector. The effective Debye Waller factor can be estimated as the experimental I/I_0 ratio i.e. what part of the intensity actually sits under the diffraction peaks. This is estimated by taking a double Gaussian for each peak, a narrow one describing the coherent signal and another one taken as 1.5 time broader than the peak spacing (the reciprocal lattice vector) to reproduce the background. Each peak contributes to the fit as R time the narrow gaussian plus (1-R) time the broad one. The ratio R is the estimate of the effective Debye Waller factor. This is illustrated in figure 6 and the dependence of this effective Debye Waller factor R is compared with the prediction of eq. 2 in figure 7. The procedure is questionable in its details but the figure 7 shows a good qualitative agreement of this modified Debye Waller factor with the estimated coherent fraction of the signal. This modified Debye-Waler factor allows the use of shorter wavelength and normal energies typically an order of magnitude larger and/or the use of comparatively hot surfaces. This mainly explains why diffraction is easily observed at room temperature and with fraction of an eV normal energy. Not only the sensitivity is increased (shorter wavelength) but the distance to the surface is reduced allowing a more sensitive inspection of the surface. Assuming for simplicity that the interaction potential is scaling with the electronic density (this is certainly very bad for reactive species but probably a fair starting point for helium or neon projectiles). Getting closer to the surface means probing larger electronic densities, which are more likely to reflect the chemical reconstruction taking place at the surface.

5.2. Inelastic effects

Mainly the normal energy has been discussed above, while the large parallel velocity seems to be entirely advantageous. It is the source of the decoupling allowing easy tuning of the normal energy by orders of magnitude within a few degrees variation, it has fantastic consequences for detection both in terms of quantum efficiency and of large solid angle and it averages the individual thermal fluctuations. It however also has negative counterparts' namely inelastic process. Electronic excitations of the surface are strongly favored at large collision energy and these will progressively but inevitably lead to decoherence. This aspect was first circumvented with large band gap insulators such as LiF or NaCl but could be dramatic for semi conductors or metals where most applications are foreseen. Fortunately, for the development of GIFAD, diffraction is well observed on semi conductors [19] and even metals [20] allowing bright prospects for applications in surface analysis.

6. Conclusions

Grazing Incidence of Fast Atom Diffraction (GIFAD), a new diffraction technique using keV atoms or molecules has been described in some details. The basic physics is very similar to that of diffraction of thermal energy atoms as discovered in the 1930's. However, it is practically very different because it is a time effective technique, it allows higher surface temperature and leaves the space above the surface free of instrument allowing online monitoring of growth processes with a definite advantage that it is sensitive exclusively to the topmost layer of the surface.

Acknowledgement

We acknowledge fruitfull discussions with AG. Borisov and the financial support from the EEC FP6 network projet 026015 "ITS LEIF".

References

[1] Rabalais JW 2003 "Principles and applications of ion scattering spectrometry" Ed. Wiley ISBN: 978-0-471-20277-6
[2] Winter H 2002 *Physics Reports* **367** 387
[3] Khemliche H, Roncin P, and Rousseau P 2006 Int. Pat. WO/2008/003865, IPC G01N 23/201
[4] Schüller A, Wethekam S, and Winter H 2007 *Phys. Rev. Lett.* **98** 016103
[5] Rousseau P, Khemliche H, Borisov AG, and Roncin P 2007 *Phys. Rev. Lett.* **98** 016104
[6] Schüller A, K.Gärtner K. and Winter H. 2008, EPL, **81** 37007
[7] Schüller A, and Winter H. 2008 *Phys. Rev. Lett.* **100** 097602
[8] Farıas D, Dıaz C, Nieto P, Salin A, and Martın F 2004 *Chem. Phys. Lett.* **390** 250
[9] Manson JR, Khemliche H, and Roncin P Accepted for publication in Physical Review B
[10] Esterman I and Stern O 1930 *Zeitschrift für Physik.* **61** 95
[11] Boato G, Cantini P, and Mattera L 1976 *Surface Science* **55** 141
[12] Garcia N 1976 *Physical Review Letters* **37** 913
[13] Rousseau *et al* In preparation.
[14] Aigner F, Simonovic, Solleder B, Wirtz L and Burforfer J. 2008 In these proccedings
[15] C. Auth, 1996 Ph.D. thesis, Humboldt-Universität Berlin (Shaker Verlag, Aachen,).
[16] Rousseau P, Khemliche H, and Roncin P 2007 *Nucl. Instrum. Meth. B* **258** 13
[17] Villette J 2000 *Thesis Université Paris Sud*, http://tel.archives-ouvertes.fr/tel-00106816
[18] Rousseau P 2006 *Thesis Université Paris Sud* http://tel.archives-ouvertes.fr/tel-00106727
[19] Khemliche *et al* in preparation.
[20] Bundaleski N, Khemliche H, Soulisse P and Roncin P Accepted to *Phys. Rev.Lett.*

Influence of Inelastic Processes on Fast-Atom-Surface Diffraction

F Aigner[1], N Simonović[1,2], B Solleder[1], L Wirtz[3] and J Burgdörfer[1]

[1] Institute for Theoretical Physics, Vienna University of Technology, A-1040 Vienna, Austria, EU

[2] Institute of Physics, P.O. Box 57, 11001 Belgrade, Serbia

[3] IEMN, CNRS, Villeneuve d'Ascq, France, EU

E-mail: `florian.aigner@tuwien.ac.at`

Abstract. Diffraction of fast helium atoms at alkali-halide surfaces under grazing angles of incidence shows intriguing diffraction patterns. The persistence of quantum coherence is remarkably strong, even though high surface temperatures and high (keV) kinetic energies of the incident atoms would strongly suggest the dominance of dissipative and decohering processes. The main source of decoherence is the excitation or absorption of surface vibrations upon impact. The momentum transfer between the surface and the incident helium atom depends on the amplitude of the thermal vibrations of the surface atoms and the energy of the incident particle. We present an ab initio simulation of the quantum diffraction of fast helium beams at a LiF (100) surface in the ⟨110⟩ direction, and compare with recent experimental diffraction data.

1. Introduction

Ever since the pioneering experiments by Estermann and Stern [1], diffraction of massive particles scattered at surfaces has remained one of the most impressive illustrations of the quantum wave nature. Estermann and Stern observed interference patterns of slow ("thermal") helium atoms and H_2 molecules scattered off alkali halide surfaces. In order to investigate the quantum-to-classical crossover, interference of much more massive objects like fullerenes scattered on gratings has been studied successfully [2, 3]. In matter-wave interferometry of large and complex biomolecules, de Broglie wavelengths as small as few picometer ($10^{-12}\,m$) have been reached [3]. The same regime of small de Broglie wavelengths $\lambda_{dB} \to 0$, can also be approached by using lighter particles with a considerabely larger kinetic energy. Recently, sub-picometer wavelengths have been accessed by scattering studies of fast atoms with keV energies at surfaces [4, 5, 6] with λ_{dB} as small as ≈ 200 femtometers. In spite of the remarkably small wavelength, the diffraction patterns obtained by these experiments do not only show clear evidence of coherent wavepacket interference, but even provide detailed structural information about the scattering surface. This is all the more remarkable, as thermal fluctuation amplitudes of the surface atoms are much larger than λ_{dB}. Especially at high surface temperatures, thermal displacement of surface atoms leads to a considerably enhanced momentum transfer between the incident atom and the surface. The collisions of projectiles at keV energies with the surface would strongly suggest the dominance of dissipative and decohering processes.

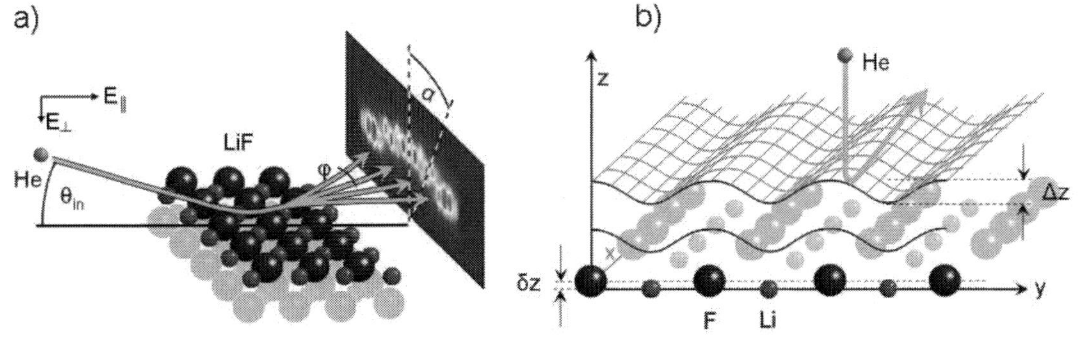

Figure 1. Helium atoms scattered at the LiF-surface at a grazing angle of incidence θ (a). In 100-direction, the He atoms interact with longitudinal chains of either Li or F atoms (b). Averaging the He-LiF potential along the longitudinal direction yields an effective two-dimensional potential.

In this communication we present first examples of ab initio simulations of fast helium atom diffraction (^{4}He) at a LiF(100) surface. We analyze the suppression of decoherence in grazing-incidence scattering within an open quantum system (OQS) approach and quantitatively reconstruct experimental diffraction images [4, 5].

2. Method

Fast atom scattering at surfaces under small grazing angles ($\theta_{in} \approx 1°$) is characterized by widely disparate kinetic energies parallel (Fig. 1a) ($E_\parallel = E_0 \cos^2\theta_{in} \approx E_0$) and normal ($E_\perp = E_0 \sin^2\theta_{in} \approx E_0\,\theta_{in}^2$) to the surface. This leads to vastly different projected de Broglie wavelengths

$$\lambda_{dB\,\perp,\parallel} = 2\pi/k_{\perp,\parallel} = 2\pi\hbar/\sqrt{2ME_{\perp,\parallel}}, \tag{1}$$

where $k_\parallel = k_x$ and $k_\perp = k_z$ are parallel and normal components of the wave vector of the incident beam, respectively. While $\lambda_{dB\perp}$ is comparable to the lattice spacing ($a = 2.01$Å for LiF) $\lambda_{dB\,\parallel}$ is orders of magnitude smaller. The fast motion along the surface leads to an approximate separability of the longitudinal and the transverse degrees of freedom and a weak coupling during the collision between them [7]. By averaging the atom-surface potential along the direction of the incident atom (x-direction), the scattered particle experiences an effective two-dimensional atomic string potential $V_{str}(y, z)$ (Fig. 1b).

Projecting the outgoing wave onto the plane transverse to the beam direction $|\langle k_y k_z|\Psi_{out}\rangle|^2$, two-dimensional diffraction patterns (see Fig. 1a) emerge. The remarkable persistence of quantum coherence in the presence of an energetic collision with keV energies and lattice vibrations at elevated temperatures in the experiment ($T \approx 620$ K [4, 6]) is the key to quantitatively analyzing diffraction patterns and extracting structural and dynamical information.

Starting point of the open quantum system approach (OQS) [2] is the reduction of the Liouville-von Neumann equation

$$i\frac{d\rho(t)}{dt} = [H, \rho(t)] , \tag{2}$$

for the density operator ρ of the entangled system of atomic wavepacket and surface to a Lindblad equation

$$\frac{d\sigma(t)}{dt} = -i\left[H_S, \sigma(t)\right] + R\sigma(t)\,, \tag{3}$$

for the reduced density operator

$$\sigma(t) = Tr_R\left[\rho(t)\right] = \sum_i \{i|\rho(t)|i\}\ , \tag{4}$$

describing the translational motion of the atomic wavepacket where all "reservoir" degrees of freedom (R) representing the LiF surface have been traced out. The partial trace in Eq. (4) includes both the average over thermal excitation of the surface as well as collisional excitations of the surface by the interacting beam, which is the root cause for decoherence represented in Eq. (3) by the so-called relaxation operator R.

Under surface channeling conditions, the longitudinal motion is effectively decoupled from other degrees of freedom (except for cases of resonant coherent excitations [14], i. e. frequency matching of the longitudinal corrugation with internal degrees of freedom of the projectile). We therefore separate the system Hamiltonian in longitudinal and transverse parts.

$$H_S = \frac{P_\perp^2}{2m} + V_{str}(y,z) + \frac{P_\parallel^2}{2m} \tag{5}$$

The Hamiltonian contains the effective transverse interaction potential of atom with the LiF surface, $V_{str}(y,z)$. It is obtained by averaging the full potential $V_{surf}(\vec{r})$

$$V_{str}(y,z) = \frac{1}{d}\int_{x_0}^{x_0+d} dx\, V_{surf}(x,y,z)\,, \tag{6}$$

where d is the space between two adjacent F of Li atoms.

The initial density matrix of the atomic beam prior to entanglement with the solid,

$$\sigma(t=0) = \int d^3p_i\, |\Psi_{\vec{p_i}}\rangle\langle\Psi_{\vec{p_i}}| \tag{7}$$

corresponds to an integral over wavepackets representing the incoherent energy- and angular spread of the initial beam.

In the presence of an environment with a very large number of degrees of freedom, the quantum-to-classical crossover can be understood within the framework of decoherence theory [1], or alternatively, elements beyond standard unitary quantum dynamics, including the frequently involved "collapse" of the wavefunctions [8, 9]. We solve the Lindblad equation by a quantum trajectory Monte Carlo (QTMC) method [10] which is an extension of similar techniques in quantum optics [12]. In the present case of free-particle wavepackets, the QTMC reduces to an ensemble average over solutions of a stochastic linear Schrödinger equation. The evolution of the wavepacket is given by a sequence of collisions and propagation in the effective potential. Each stochastic realization of the wavepacket at the time $t = t_N$ is given by

$$|\Psi_i(t)\rangle = \prod_{j=1}^{N} U_{coll}\left(\vec{q}_j, t_j\right) U_{cont}\left(t_j, t_{j-1}\right)|\Psi_i\rangle\,, \tag{8}$$

where $|\Psi_i\rangle$ is the initial wavepacket at $t_0 = 0$. The evolution operator for an impulsive stochastic collisional momentum exchange $\vec{q}_j$ with the surface is given by

$$U_{coll}\left(\vec{q}_j, t_j\right) = \exp\left(i\vec{q}_j\vec{r}\right)\ , \tag{9}$$

while the continuous evolution in between two "kicks" during the interval $[t_j, t_{j-1}]$ is given by

$$U_{cont}\left(t_j, t_{j-1}\right) = \exp\left(-iH_S\left(t_j - t_{j-1}\right)\right)\ , \tag{10}$$

with the system Hamiltonian H_S.

3. Atom-surface momentum transfer and decoherence

Momentum transfers $\{\vec{q}_j\}$ as well as collision times $\{t_j\}$ are stochastic sequences determined by collisional interactions with the environment. The resulting diffraction pattern in the transverse plane is calculated as an ensemble average over stochastic realizations.

The momentum transfer distribution functions in longitudinal (x-) direction and in the transverse (y, z-) directions entering Eq. 9 are obtained along sequences of binary collisions between helium and lithium and fluorine by averaging over the random displacement of the thermally excited lattice atoms. The mean square displacements of the atoms are given by [11]

$$\langle u_\perp^2 \rangle = \frac{3\hbar T}{M k_B \Theta_D^2}, \tag{11}$$

where Θ_D is the surface Debye temperature, M the mass of the corresponding lattice atom and T the temperature of the surface. The Debye temperature was estimated as $\Theta_D = 530K$. As shown in Fig. 2, the z-component of the momentum transfer (Δk_z) is considerably larger than the momentum transfer in the other two directions. All the momentum transfer distributions become wider as the energy of the incident atom increases, since faster incident atoms lead to larger momentum transfer between the atom and the surface. In the range of $0.5keV$ to $2keV$ however, this increase is only moderate. The widths of the momentum transfer distributions remains in the same order of magnitude throughout this energy range (Fig. 2).

The momentum transfer in the transverse plane (Δk_y, Δk_z) displaces the momentum of each wavepacket. As shown in Fig. 2, the z-component of the momentum transfer (Δk_z) is considerably larger than the momentum transfer in the other two directions. This leads to an elongation of the diffraction peaks in z-direction in the resulting diffraction picture (see Fig. 4). The effect of Δk_x is more subtle: it causes velocity and kinetic energy changes of the parallel motion. The energy exchange with the surface (of the order of $\approx$ eV) results in random phases in the continuous time evolution.

Diffraction images can be successfully reconstructed only when decay of coherence across the beam profile due to encoding of localization information ("which path" information) is accounted for. Neglecting first the initial beam divergence, we consider an incident Gaussian wavepacket (Fig. 3) of the form

$$\Psi_{in}(\vec{r}) = e^{ik_\parallel x}\, \Psi_\perp(y, z) \tag{12a}$$

with

$$\Psi_\perp(x, y) \propto e^{-ik_\perp z} e^{-y^2/2\sigma_y^2} e^{-(z-z_0)^2/2\sigma_z^2} . \tag{12b}$$

$\sigma_{y,z}$ are the transverse coherence length of the incident beam determined by the width of the zeroth-order diffraction peak ($\sigma_y^{in} = \sigma_x^{in} \approx 2L\frac{\lambda_{dB\parallel}}{D}$). The latter is controlled by the collimation slit of width D positioned at a distance L in front of the surface. The transverse coherence length is large compared to the lattice spacing. In fact, Eq. (12) is close to a delocalized plane wave. We expand the initial wavepacket in terms of a coherent superposition of narrow Gaussian wavepackets (Fig. 3a) each of which is propagated according to Eqs. (8 to 10). The random evolution phases acquired due to collisions occuring at different instances in time (i. e. different x coordinates along the string) causes a reduction of the transverse coherence length (Fig. 3b) along the surface normal, σ_z^{out}, of the projected wavepacket, monitored after its center reached distance from the surface $z = z_0$ again.

Two portions of the wave packet which are initially separated by the offset δz hit the surface at different positions on the longitudinal (x-) axis (see Fig. 3). If both trajectories suffer the same amount of momentum transfer δk during the impact on the surface, their phase difference after the impact is $\phi \approx \delta k \delta x$. Only if $\phi \ll 2\pi$, the two paths will interfere constructively. The coherence width of the wavepacket in z-direction (perpendicular to the surface) can therefore be roughly estimated as

$$\delta z = \delta x / \theta_{in} \approx \pi \theta_{in} / \delta k \tag{13}$$

A more detailed numerical analysis of the reduction of the coherence width of the incident wavepacket is shown in Fig. 3. The width of the wavepacket is reduced dramatically due to stochastic momentum transfer with the LiF surface. The coherent width of the wavepacket depends on the energy of the incident He atom and on the surface Debye temperature. As higher energies of the incident atom lead to larger momentum transfer, it also causes stronger decoherence effects and narrower resulting wavepackets. Higher surface Debye temperatures on the other hand correspond to smaller thermal oscillation amplitudes of the surface atoms, which leads to less momentum exchange between the He atom and the surface, and, in turn, to weaker decoherence effects. Although the effect of the energy and the surface Debye temperature on the coherence width of the wavepacket is clearly visible, this effect is only moderate in strength. In a wide parameter range, the resulting wavepackets can be approximated by Eq. 12b, with $\sigma_z \approx 0.5au.$, which translates into a FWHM of $\sqrt{8 \log 2}\sigma_z \approx 1.2au.$.

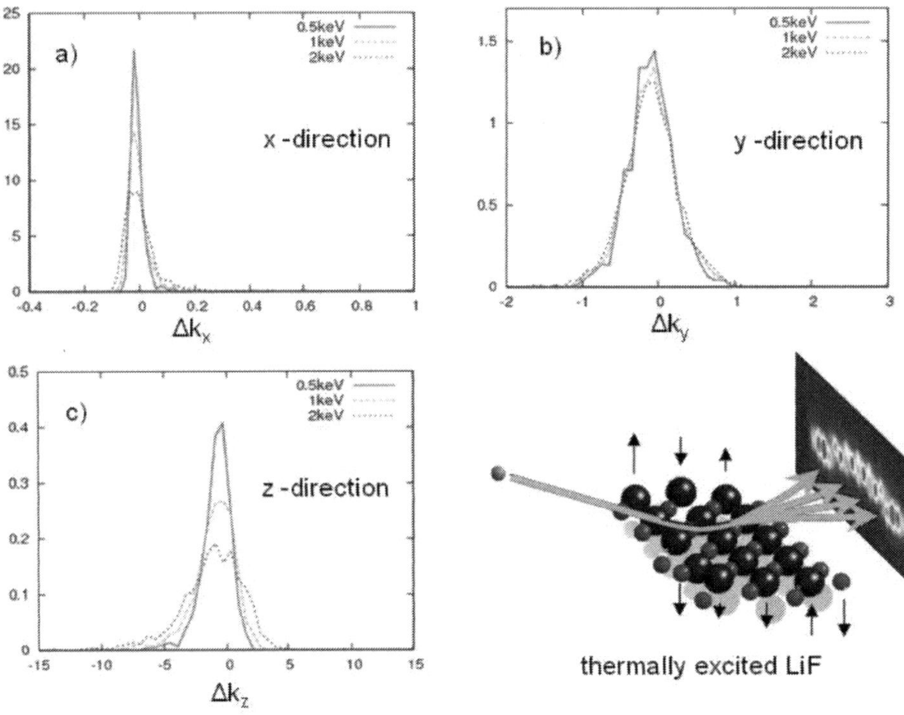

Figure 2. Probability density functions of momentum loss of a He atom hitting a LiF surface with a temperature of $620K$. Momentum transfer distribution functions of the x-component (longitudinal direction, a), the y-component (b) and the z-direction (perpendicular to the LiF surface, c) are shown for different energies of the incident He atoms.

We determine $V_{str}(y, z)$ as the string average along the $\langle 110 \rangle$ direction of the full He-LiF surface potential $V_{surf}(\vec{r})$ calculated for a large LiF cluster on the multi-configuration self consistent field level, using the quantum chemistry code Columbus [15]. The string potential along the $\langle 110 \rangle$ direction is periodic with $V_{str}(y+d,z) = V_{str}(y,z)$, where d is the spacing between adjacent strings of F.

4. Calculating diffraction patterns

Due to the periodicity of the effective two-dimensional atomic string potential ($V_{str}(y,z) = V_{str}(y+d)$, where d is the space between two adjacent F or Li atoms), all the interference maxima correspond to Bragg peaks.

$$k_y^{(n)} = n\frac{2\pi}{d}, \quad n = 1, 2, 3, \ldots \tag{14}$$

A typical diffraction pattern at a well-defined incident particle energy shows modulations of the intensity such that not all the Bragg peaks are simultaneously visible. Whether or not a particular Bragg position $k_y^{(n)}$ features an interference maximum at a certain energy depends on the shape of the effective atomic string potential, in particular on its corrugation amplitude.

A key parameter characterizing the diffractive scattering induced by V_{str} is the corrugation amplitude of the string potential, $(\Delta z)_{str}$ i. e. the maximum variation of the normal (z) coordinate of the equipotential surface $E_\perp = V_{str}(y,z)$. For a wide interval of normal energies (0.5 eV $\leq E_\perp \leq$ 10 eV), within which a converged potential for a large cluster can be generated, $(\Delta z)_{str}$ is remarkably weakly dependent on $E_\perp$ and essentially constant $(\Delta z)_{str} \approx 0.3$Å. As some of the experimental diffraction data are taken at even smaller perpendicular energies for ($E_\perp < 0.5$ eV), where the ab initio potential calculation becomes unreliable, we have extrapolated V_{str} to low-energy surfaces subject to the constraint that $(\Delta z)_{str}$ remains constant. Imposition of such a constraint has proven remarkably successful when compared with experimental data.

Reconstruction of the experimental diffraction pattern requires, in addition to decoherence, incoherent averaging over the geometric angular divergence taken to be rectangular, W_R, and energy spread taken to be approximately, Lorentzian, W_L.

$$\langle I(\theta_{out}, \phi_{out}) \rangle = \int d\theta' \int d\phi \int dE' I(\theta', \phi, E') \times W_L(E' - E) W_R(\theta' - E_{out}) W_R(\theta' - \theta_{out}) \; . \tag{15}$$

We compare the resulting diffraction patterns to experimental results [4] for scattering of He atoms ($E_0 = 1$ keV) on a LiF(001) surface aligned along $\langle 110 \rangle$ direction under a grazing angle of incidence $\phi_{in} = 1.1°$ ($E_\perp \approx 0.38$ eV). We find a near-perfect reconstruction of the experimental diffraction pattern without any adjustable parameter (Fig. 4). The positions of the main maxima are obtained from the interference condition for a periodic potential along the y-direction,

$$\sin \alpha_n = \frac{k_y}{k_\perp} = n\frac{\lambda_{dB\perp}}{d}, \quad n = 0, \pm 1, \pm 2, \ldots . \tag{16}$$

The two outermost maxima are the so-called rainbow peaks. Between the rainbow peaks, there are intricate substructures, formed by additional maxima ("supernumeraries"). These structures are well reproduced by the simulation.

Summarizing, fast keV atomic collisions with insulating alkali-halide surfaces give rise to remarkably sharp and well-resolved diffraction images. Despite de Broglie wavelengths as small as one hundred femtometer, quantum interference persists. Due to the weak coupling of transverse and longitudinal degrees of freedom decoherence is effectively suppressed in grazing incidence collisions. Ab initio quantum trajectory Monte Carlo simulations accurately reproduce the diffraction pattern and allow to extract structural surface information such as buckling at an unprecedented level of detail. This technique also promises application to other surfaces and structures relevant for material sciences.

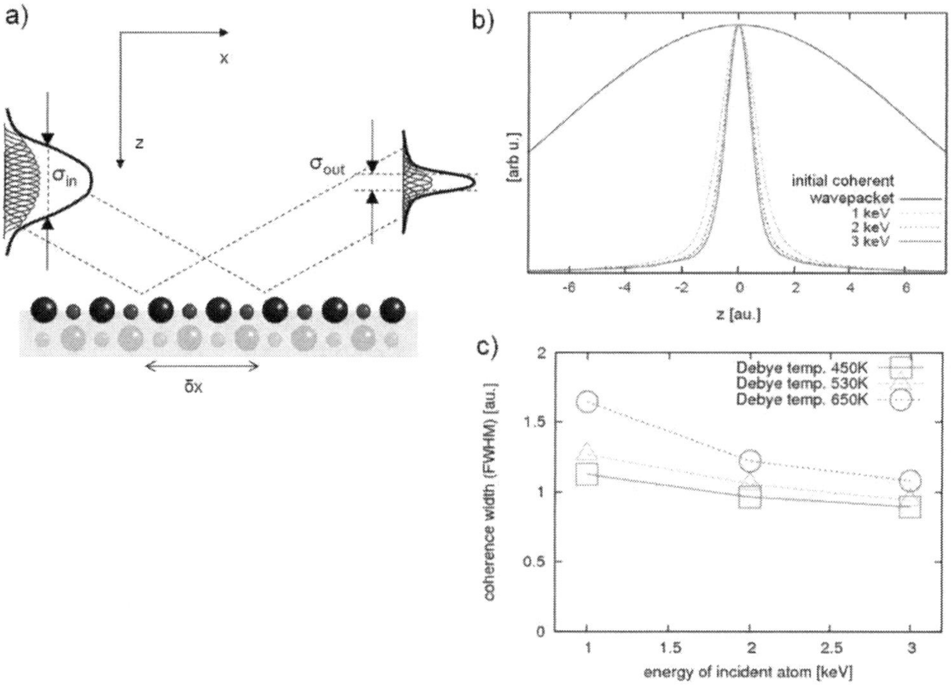

Figure 3. A delocalized wavepacket hits the surface that is considerabely wider than the size of the unit cell. Different components of the wavepacket (indicated in (a) by narrower wavepackets) are subject to different random momentum changes, which lead to a partial loss of coherence. The coherent width of the wavepacket is therefore drastically reduced (b). The effective coherence width of the wavepacket moderately depends on the energy of the incident particle as well as on the surface Debye temperature (c).

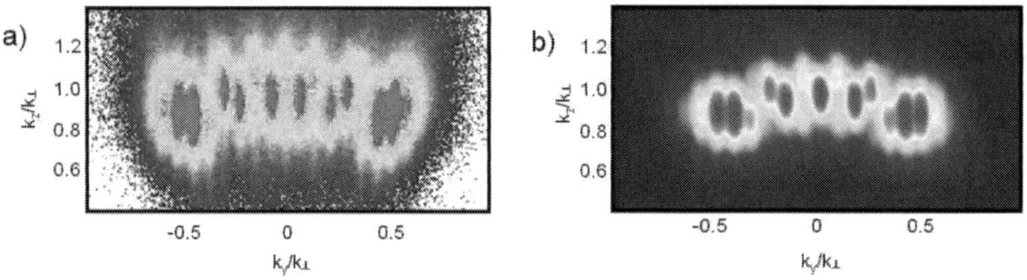

Figure 4. Experimental (from Ref. [4]) (a) and theoretical (b,c) diffraction patterns for scattering of 1keV He from LiF under $\theta = 1.1°$. In b), with surface buckling $\delta z = 0.053$Å, c) without surface buckling $\delta z = 0$. Note the different number of interference maxima.

Acknowledgements

We thank C. O. Reinhold, A. Schüller and H. Winter for helpful discussions. This work was supported by the FWF-Austria, grants no. SFB016 and P17359, and by the EU under contract no. HPRI-CT-2005-026015.

References

[1] Estermann I and Stern O 1930 *Z. Phys.* **61** 95

[2] Arndt M , Nairz O, Voss-Andreae J, Keller C, Van der Zouw G, and Zeilinger A 1999 *Nature* **401**, 680
Hornberger K, Uttenthaler S, Brezger B, Hackermüller L, Arndt M, and Zeilinger A 2003 *Phys. Rev. Lett.* **90**, 160401

[3] Hackermüller L, Uttenthaler S, Hornberger K, Reiger E, Brezger B, Zeilinger A, and Arndt M 2003 *Phys. Rev. Lett.* **91**, 090408

[4] Schüller A, Wethekam S, and Winter H 2007 *Phys. Rev. Lett.* **98**, 016103

[5] Schüller A, and Winter H 2008 *Phys. Rev. Lett.* **100**, 097602

[6] Rousseau P, Khemliche H, Borisov A G, and Roncin P 2007 *Phys. Rev. Lett.* **98**, 016104

[7] J. Burgdörfer, in "Review of Fundamental Processes and Applications of Atoms and Ions", edited by C.D. Lin (World Scientific, Singapore, 1993)

[8] Zurek W H 1991 *Phys. Today* **44**, 36

[9] Ghirardi G C, Rimini A, and Weber T 1986 *Phys. Rev.* D **34**, 470

[10] Minami T, Reinhold C O, and Burgdörfer J 2003 *Phys. Rev.* A **67**, 022902
Seliger M*et al* 2007 *Phys. Rev.* A **75**, 032714 and Refs. therein.

[11] Desjonquères M C, Spanjaard D 1993 Concepts in Surface Physics, Springer, Berlin

[12] Dalibard J, Castin Y, and Mølmer K 1992 *Phys. Rev. Lett.* **68**, 580
Dum R, Zoller P, and Ritsch H 1992 *Phys. Rev.* A **45**, 4879

[13] Mertens A and Winter H 2000 *Phys. Rev. Lett.* **85**, 2825
Pfandzelter R, Mertens A, and Winter H 2001 *Phys. Lett.* A **290**, 145
Roberts J G, Van Hove M A, Somorjai G A 2002 *Surf. Sci.* **518**, 49

[14] Kupfer E, Gabriel H, and Burgdörfer J 1981 *Z. Phys.* A **300**, 35
Kimura K, Ida H, Fritz M and Mannami M 1996 *Phys. Rev. Lett.* **76**, 3850

[15] Lischka H et al. 2006 computer code COLUMBUS, Release 5.9

[16] De Wette F W, Kress W, and Schröder U 1985 *Phys. Rev.* B **32**, 4143

[17] Vogt J, and Weiss H, *Surf. Sci.* **501** 203

Channeling of protons through carbon nanotubes

D. Borka[1], D. J. Mowbray[2], Z. L. Mišković[2], S. Petrović[1] and N. Nešković[1]

[1]Laboratory of Physics (010), Vinča Institute of Nuclear Sciences, P. O. Box 522, 11001 Belgrade, Serbia

[2]Department of Applied Mathematics, University of Waterloo, Waterloo, Ontario, Canada N2L3G1

E-mail: dusborka@vin.bg.ac.yu

Abstract. We investigate how dynamic polarization of carbon valence electrons influences both the angular and spatial distributions of protons channeled in a (11, 9) single-wall carbon nanotube placed in vacuum and in different dielectric media. Proton speeds between 3 and 10 a.u., corresponding to energies of 0.223 and 2.49 MeV, are chosen with the nanotube length varied between 0.1 and 1 μm. In all performed calculations we describe the interaction between proton and carbon atoms on the nanotube wall using the Doyle-Turner potential. The image force on a proton is calculated using a two-dimensional hydrodynamic model for the dynamic response of the nanotube valence electrons and the dielectric media surrounding the nanotube. The angular distributions of channeled protons are generated using a computer simulation method which solves the proton equations of motion in the transverse plane numerically. The best level of ordering and straightening of carbon nanotube arrays is often achieved when they are grown in a dielectric matrix. Consequently, we investigate here how the dynamic polarization of carbon valence electrons in the presence of various surrounding dielectric media affects the angular distributions of protons channeled through (11, 9) single-wall carbon nanotubes. Our analysis shows that the inclusion of the image interaction, gives rise to a number of rainbow maxima in the corresponding angular and spatial distribution. Our analysis shows that the presence of dielectric media surrounding the nanotube influences the positions and appearance of rainbows in the corresponding angular and spatial distributions. In addition, we analyze the possibility of production of nano-sized beams by carbon nanotubes.

1. Introduction

Theoreticians continue to explore possible applications of ion channeling through nanotubes [1, 2, 3], beginning with those based on the *angular* distributions of channeled ions, e.g., for the purpose of deflecting ion beams in accelerators [4] or determining some structural details of the short nanotubes using the rainbow effect [5, 6, 7, 8]. Theoretical studies of the *spatial* distributions of channeled ions have demonstrated the possibility of creating nanosized ion beams, which could find interesting applications in biomedical research and for materials modification [4, 9]. An additional application of ion channeling through nanotubes based on the *spatial* distributions remains to be investigated in the context of extending the classical use of ion channeling for materials analysis [10] to nanotube based materials. Experimental realization of ion channeling through nanotubes is still at the preliminary stage [11, 12]. While the image force has not been found to play any significant role in channels of single crystals [10], its role has been identified clearly in ion-surface scattering [13], and in ion transmission through capillaries in solids [14, 15]. In addition, although the image force plays a minor roles in ion channeling through nanotubes in the GeV and keV energy ranges [3, 2, 4, 16], it has been shown to strongly influence ion trajectories in the MeV energy range. For example, the image force gives rise to the rainbow effect in angular distributions of protons channeled through short chiral single-wall and double wall nanotubes [17].

It has been established that, when the nanotubes are grown in a dielectric medium, one may achieve a very high degree of ordering and straightening. This makes such composite structures very suitable candidates to be used in experiments of ion channeling through nanotubes. Therefore, it is important to study the effects of the surrounding dielectric medium on ion channeling through nanotubes in the MeV energy range. The materials of interest are Al_2O_3 [11], SiO_2 [18] as well as Ni [18] and Pt [12]. It has been shown recently that the image force and the rainbow effect in the angular distributions of protons channeled through nanotubes may be strongly modified by the polarization of the surrounding cylindrical dielectric boundary [19, 20, 21].

2. Theory

The system under investigation is a proton moving through an (11, 9) single-wall carbon nanotube embedded in a dielectric medium. The z axis coincides with the nanotube axis and the origin lies at its transverse entrance plane. The initial proton velocity is taken to be parallel to the z axis. We assume that the nanotube is sufficiently short for the proton energy loss to be neglected. The initial proton speed, v, is varied between 3 and 10 a.u., corresponding to energies between 0.223 and 2.49 MeV, respectively. The nanotube length, L, is varied between 0.1 and 1.0 μm.

We assume that the (repulsive) interaction between the proton and nanotube atoms may be treated classically using the Doyle-Turner expression for the proton-nanotube atom interaction potential averaged axially and azimuthally [23, 24, 25]. The resulting interaction potential may be expressed as

$$U_{rep}(r) = \frac{16\pi Z_1 Z_2 d}{3\sqrt{3}l^2} \sum_{j=1}^{4} a_j b_j^2 I_0(b_j^2 rd) \exp\{-b_j^2[r^2 + (d/2)^2]\}. \quad (1)$$

Here $Z_1 = 1$ and $Z_2 = 6$ are the atomic numbers of the proton and nanotube atom, respectively, d is the nanotube diameter, l is the bond length for nanotube atoms, r is the distance between the proton and nanotube axis, I_0 designates the modified Bessel function of the first kind and the 0th order, and $a_j = (0.115, 0.188, 0.072, 0.020)$ and $b_j = (0.547, 0.989, 1.982, 5.656)$ are fitting parameters [25].

The dynamic polarization of the nanotube and dielectric medium by the proton is treated using a two-dimensional hydrodynamic model for the valence electrons of the nanotube atoms. This is based on a jellium-like description of the nanotube ion cores, extended to include the contribution of the dielectric boundary [20, 21, 26, 27]. This model includes the procedures of axial and azimuthal averaging, used in the repulsive interaction model. It gives the (attractive) interaction potential between the proton and its image. The image interaction potential at the proton position is then $U_{atr} = -\frac{Z_1}{2}\Phi_{ind}(\vec{r}, t)$, where $\Phi_{ind}(\vec{r}, t)$ is the electric potential at the proton position due to polarization charges induced on the nanotube and dielectric boundary by the (screened) proton. This is called the induced potential. A more detailed description of the theoretical hydrodynamic model is given in Refs. [19] and [20].

The total interaction potential between the proton and the nanotube and dielectric medium is $U(\vec{r}, t) = U_{rep}(r) - \frac{Z_1}{2}\Phi_{ind}(\vec{r}, t)$. The angular and spatial distributions of channeled protons in the exit plane are generated using a computer simulation method, which includes the numerical solving of the proton equations of motion in the transverse plane. The rectangular coordinates of the initial proton position, x_0 and y_0, are chosen randomly from a two-dimensional uniform distribution under the condition $r_0 = (x_0^2 + y_0^2)^{\frac{1}{2}} < a - a_{sc}$, where $a_{sc} = [9\pi^2/(128\ Z_2)]^{\frac{1}{3}} a_0$ is the nanotube atom screening radius and a_0 the Bohr radius. We take for the nanotube atom bond length $l = 0.144$ nm [28] and obtain for the nanotube radius $a = 0.689$ nm. The initial number of protons is 3 141 929.

Angular distributions may be analyzed successfully via the corresponding mapping of the impact parameter plane to the scattering angle plane [5, 6, 7, 8, 17, 29]. We also analyze the mapping of the entrance transverse plane to the exit transverse plane. However, since in the case we investigate the total interaction potential is axially symmetric, it is possible to construct the one dimensional deflection function, Θ_x as a function of x_0 for the angular distribution. In the configuration plane we can analyze the mapping of the proton radial axis in the entrance transverse plane x_0 to the proton radial axis in the exit transverse plane x ($x_0 \mapsto x$). The extrema in this mapping are the rainbow maxima or minima, and the corresponding singularities in the angular and spatial distribution of channeled protons are the rainbow singularities.

3. Results and discussion

We use for the nanotube radius a = 13.01 a.u, and for the separation between the nanotube and the dielectric boundaries 3.21 a.u. [22, 30]. Thus, the dielectric boundary radius is $b = a$ + 3.21 a.u. We describe the surrounding SiO_2 using a dielectric constant of 3.9 [31]. The dielectric responses of the surrounding Al_2O_3 and Ni are modeled using the methods described by Arista and Fuentes [15] and Kwei et al. [32], respectively. We have found that for the case of nanotube embedded in a dielectric media a proton moving at a speed below about 3 a.u. does not polarize the dielectric media. This means that the dielectric medium is completely screened by the nanotube and does not contribute to the image force. For a proton speed above about 3 a.u., polarization of the dielectric media occurs. The screening of the dielectric medium by the nanotube is then incomplete and it thus influences the image force.

Figure 1(a) shows two angular distributions of channeled protons along the Θ_x axis for $v = 3$ a.u. and $L = 0.3$ μm, both with and without the inclusion of the dynamical polarization effect. One can see that 10 maxima appear in the angular distribution, located around 10 scattering angles, $\Theta_x \cong \pm$ 0.7 mrad, $\pm$ 1.1 mrad, $\pm$ 1.8 mrad, $\pm$ 2.1 mrad, $\pm$ 2.3 mrad, when the dynamical polarization effect is included. The extremal points are labeled by 1, $2'$, $2''$, $3'$ and $3''$, which result from one (1), two ($2'$ and $2''$) and three ($3'$ and $3''$) deflections of the rainbow trajectories within the image-generated potential well. Figure 1(b) shows two deflection functions that correspond to the angular distributions shown in Figure 1(a). The deflection function with the inclusion of the dynamical polarization effect has 10 extremum points: five minima and five maxima, which are designated by 1, $2'$, $2''$, $3'$ and $3''$. In both cases we notice two additional pairs of extrema near the nanotube wall (for large impact parameters), but these structures are not visible in the angular distributions. Figure 1(c) shows two spatial distributions of channeled protons along the x axis in the exit plane for the proton velocity $v = 3$ a.u. and the nanotube length $L = 0.3$ μm. The nanotube is placed in vacuum. The spatial distributions correspond to the cases when the image interaction potential is neglected and when it is included. In the former case the spatial distribution contains a central maximum and three pairs of peripheral maxima, designated by 1, 2 and 3, located at $x = \pm 7.5$, ± 11.1 and ± 12.1 a.u. In the latter case a central maximum and five pairs of peripheral maxima are seen, designated by 1_i, 2_i, 3_i, 4_i and 5_i, and located at $x = \pm 8.1$, ± 9.9, ± 10.7, ± 11.4 and ± 12.2 a.u. The maxima designated by 2, 3, 4_i and 5_i are very weak. When the image interaction potential is included, the central maximum is much narrower and about three times weaker while the peripheral maxima are much more prominent. Figure 1(d) shows two mappings of the x_0 axis in the entrance plane to the x axis in the exit plane corresponding to the two spatial distributions of channeled protons shown in Fig. 1(c). It is evident that without the image force the mapping has six extrema, three maxima

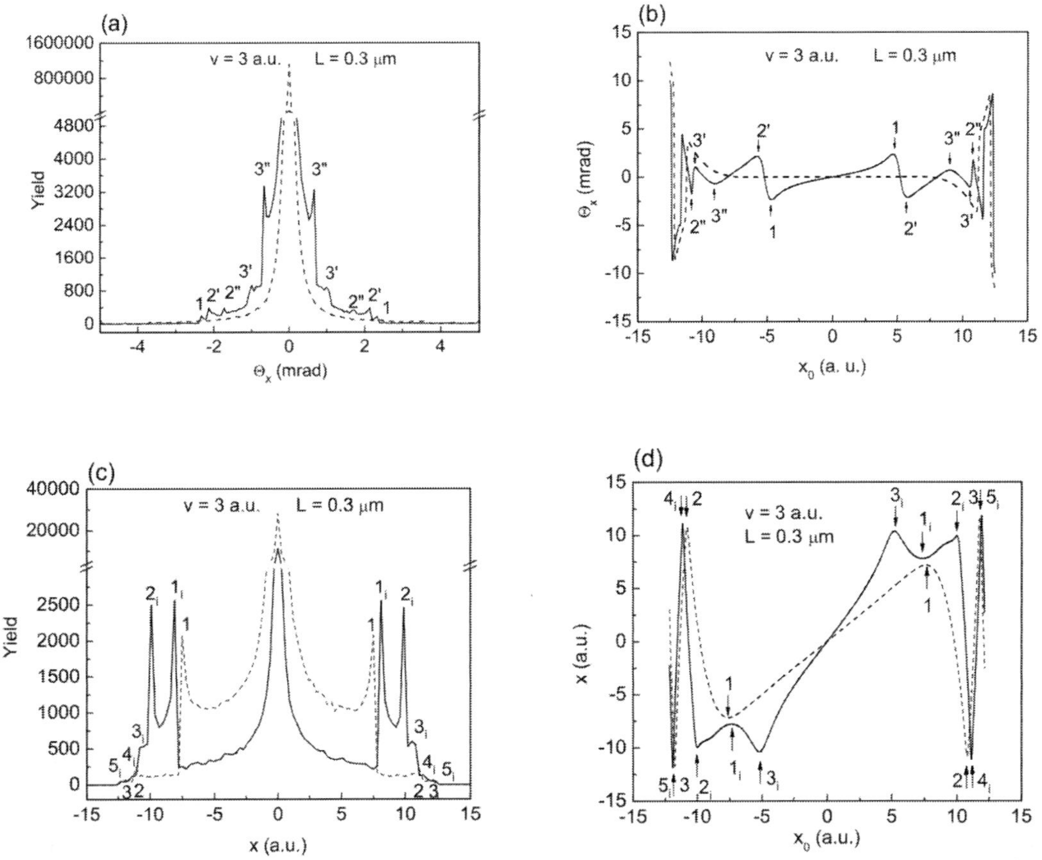

Figure 1. (a) Two angular distributions and (b) two corresponding deflection functions, (c) two spatial distributions and (d) two corresponding mapping $x_0 \mapsto x$ of channeled protons for the proton speed of 3 a.u. and the nanotube length of 0.3 μm. The (11, 9) SWNT is placed in vacuum. The dashed line corresponds to the case without the dynamic polarization effect and the solid line to the case with it.

and three minima. These are designated by 1, 2 and 3. When the image force is included, the mapping has 10 extrema, five maxima and five minima. These are designated by 1_i, 2_i, 3_i, 4_i and 5_i. The maxima and minima designated by 2, 3, 4_i and 5_i are very sharp and lie near the nanotube wall.

Figure 2(a) shows two angular distributions of channeled protons along the Θ_x axis in the scattering plane for the proton speed $v = 5$ a.u. and the nanotube length $L = 0.5$ μm. These correspond to the cases in which the nanotube is placed in vacuum and embedded in SiO_2, and the dynamic polarization effect is included. We notice a significant depletion of the number of rainbow peaks, down to three pairs (1, $2'$ and $2''$) for the nanotube in vacuum, and only one pair (1_d) for surrounding SiO_2. When the nanotube is placed in vacuum the peripheral maxima are located at $\Theta_x = \pm 0.23$, ± 0.43 and ± 0.74 mrad, and when it is embedded in SiO_2 at $\Theta_x = \pm 0.43$ mrad. Figure 2(b) shows two deflection functions that correspond to the angular distributions shown in Figure 2(a). In both cases we notice two additional pairs of extrema near the nanotube wall, but these structures are not visible in the angular distributions. Figure 2(c) shows two spatial distributions of channeled protons along the x axis in the exit plane for the proton speed $v = 5$ a.u. and the nanotube length $L = 0.5$ μm. These correspond to when the nanotube is

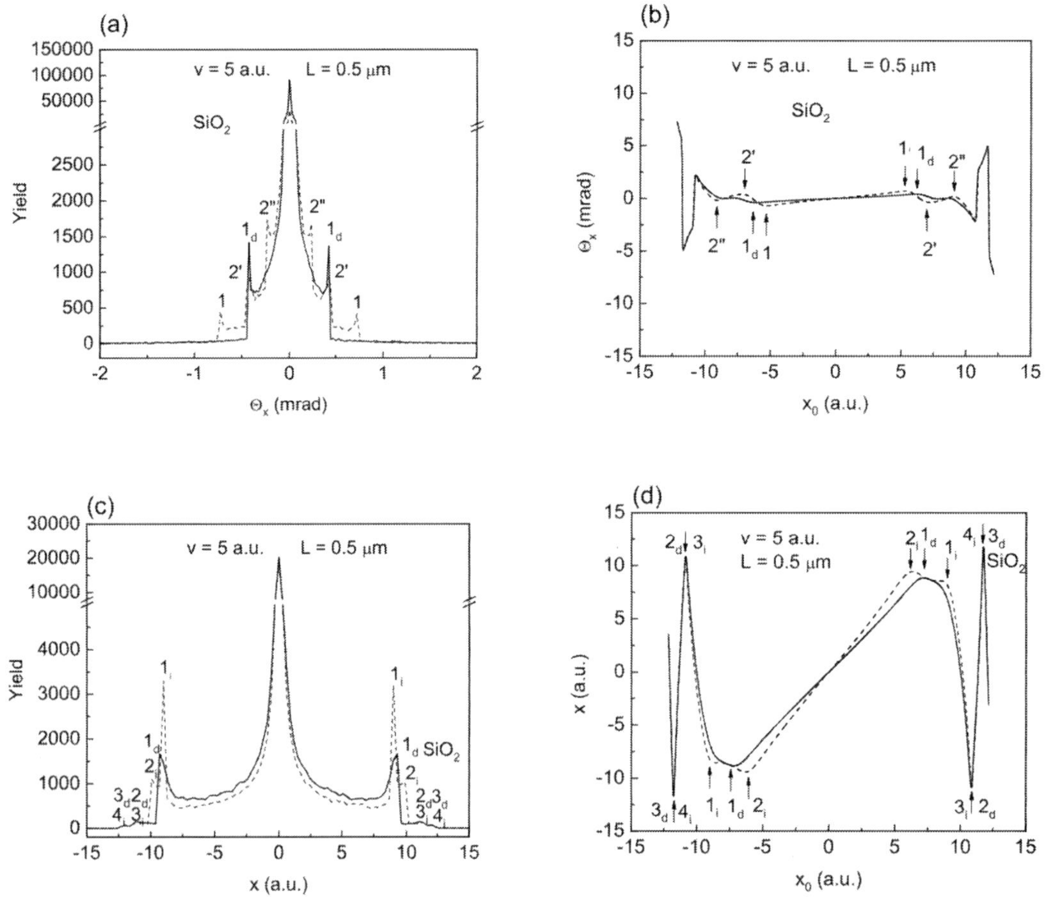

Figure 2. (a) Two angular distributions and (b) two corresponding deflection functions, (c) two spatial distributions and (d) two corresponding mapping $x_0 \mapsto x$ of channeled protons for the proton speed of 5 a.u. and the nanotube length of 0.5 μm. The (11, 9) SWNT is placed in vacuum (dashed line) and in SiO_2 (solid line).

placed in vacuum and embedded in SiO_2, with the dynamic polarization effect included. In the former case the spatial distribution contains a central maximum and four pairs of peripheral maxima are seen, designated by 1_i, 2_i, 3_i and 4_i. In the latter case a central maximum and three pairs of peripheral maxima, designated by 1_d, 2_d and 3_d. The maxima designated by 3_i, 4_i, 2_d and 3_d are very weak. When the nanotube is placed in vacuum the peripheral maxima are located at $x = \pm 9.0$, ± 9.9, ± 11.2 and ± 12.1 a.u., and when it is embedded in SiO_2 at $x = \pm 9.3$, ± 11.2 and ± 12.1 a.u. Figure 2(d) shows two mappings of the x_0 axis in the entrance plane to the x axis in the exit plane corresponding to the two spatial distributions of channeled protons shown in Fig. 2(c). It is evident that when the nanotube is placed in vacuum the mapping has eight extrema, four maxima and four minima. These are designated by 1_i, 2_i, 3_i and 4_i. When the nanotube is embedded in SiO_2 the mapping has six extrema, three maxima and three minima. These are designated by 1_d, 2_d and 3_d.

Figure 3(a) shows four angular distributions of channeled protons along the Θ_x axis in the scattering angle plane for the proton speed $v = 8$ a.u. and the nanotube length $L = 0.8$ μm. These correspond to when the nanotube is placed in vacuum and embedded in SiO_2, Al_2O_3 and Ni, with the image interaction potential included. One notices that the positions of the rainbow extrema, labeled by 1_d, hardly change for different dielectrics. When the nanotube is placed in vacuum there are no peripheral maxima, and when it is embedded in a dielectric medium, periferal maxima are located at $\Theta_x = \pm 0.17$ mrad (in the case of Al_2O_3) and $\Theta_x = \pm 0.21$ mrad (in the cases of SiO_2 and Ni). Figure 3(b) shows four deflection functions that correspond to the angular distributions shown in Fig. 3(a). One finds that in each case of the nanotube embeded in a dielectric the mapping has one pair of extrema, labeled by 1_d. In both cases we notice two additional pairs of extrema near the nanotube wall which are not visible in the angular distributions. Figure 3(c) shows four spatial distributions of channeled protons along the x axis in the exit plane for the proton speed $v = 8$ a.u. and the nanotube length $L = 0.8$ μm. These correspond to the cases in which the nanotube is placed in vacuum and embedded in SiO_2, Al_2O_3 and Ni, with the image interaction potential is included. In each case the spatial distribution contains a strong central maximum and three pairs of peripheral maxima, designated by 1_i, 2_i and 3_i, when the nanotube is placed in vacuum, and by 1_d, 2_d and 3_d, when it is embedded in SiO_2,Al_2O_3 and Ni. It is clear that the spatial distributions when the nanotube is embedded in the three dielectric media are very similar to each other. Also, they do not differ significantly from the spatial distribution when the nanotube is placed in vacuum. The spatial distributions when the nanotube is embedded in SiO_2 and Ni almost coincide. Figure 3(d) shows four mappings of the x_0 axis in the entrance plane to the x axis in the exit plane corresponding to the four spatial distributions of channeled protons shown in Fig. 3(c). One finds that in each case the mapping has six extrema, three maxima and three minima.

All angular and spatial distributions shown are characterized by the same duration of the proton channeling process, i.e., by the same proton dwell time.

The results for the angular distributions for the combination of parameters $L = 0.3$ μm and v= 3 a.u. are displayed in Fig. 1(a), where one can see five pairs of the extremal points, labeled by 1, $2'$, $2''$, $3'$ and $3''$. For the combination of parameters $L = 0.5$ μm, v= 5 a.u. (see Fig. 2(a)), one notices a significant depletion of the number of rainbow peaks, down to three pairs (1, $2'$ and $2''$) for the nanotube in vacuum, and only one pair (1_d) for surrounding SiO_2. The further reduction of the number of rainbow peaks with increasing proton speed is illustrated in Fig. 3(a) for the combination of parameters $L = 0.8$ μm, v= 8 a.u. This is due to the diminished image force on protons at such a high speed. On the other hand, the one rainbow peak (1_d) from Fig. 2(a) has remained in Fig. 3(a) for the case of a nanotube surrounded by SiO_2, although this peak is now very small. This persistence of the rainbow peak for a surrounding dielectric can be explained by the "transparency" of nanotubes at high proton speeds [26], where the image force is dominated by the polarization of the surrounding dielectric.

For $v = 3$ a.u. and $L = 0.3$ μm, the spatial distribution with the nanotube placed in vacuum contains five rainbow maxima [see Fig. 1(c)], as does the spatial distribution with the nanotube placed in SiO_2. For $v = 5$ a.u. and $L = 0.5$ μm the spatial distribution with the nanotube placed in vacuum contains four rainbow maxima, while the spatial distribution with the nanotube placed in SiO_2 three rainbow maxima [see Fig. 2(c)]. For $v = 8$ a.u. and $L = 0.8$ μm the spatial distribution with the nanotube placed in vacuum contains three rainbow maxima, as does the spatial distribution with the nanotube placed in SiO_2 [see Fig. 3(c)]. It is clear that for the same proton dwell time the number of rainbow maxima in the spatial distribution decreases with the proton speed.

4. Conclusions

We have presented theoretical investigations of the effects of dynamic polarization of the nanotube atom valence electrons on the angular and spatial distributions of protons channeled through (11, 9) single-wall carbon nanotubes placed in vacuum and embedded in various dielectric media, for proton speeds between 3 and 10 a.u. and nanotube lengths between

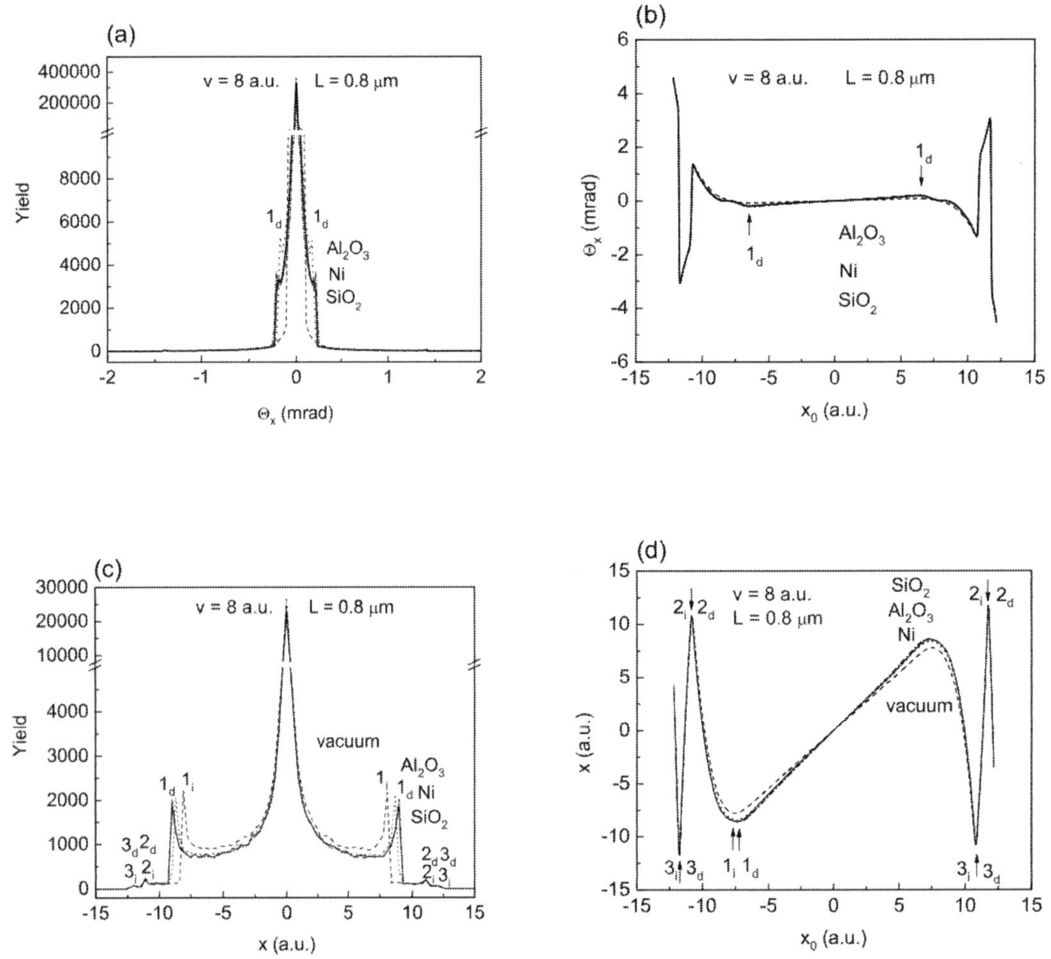

Figure 3. (a) Four angular distributions and (b) four corresponding deflection functions, (c) four spatial distributions and (d) four corresponding mapping $x_0 \mapsto x$ of channeled protons for the proton speed of 8 a.u. and the nanotube length of 0.8 μm. The (11, 9) SWNT is placed in vacuum (dashed line) and in SiO_2, Al_2O_3 and Ni (solid, dotted and dash-dotted lines, respectively).

0.1 and 1.0 μm. We found that the image force gives rise to the rainbow effect, which may be strongly affected by the surrounding dielectric medium for proton speeds above 3 a.u. Our study has revealed that prominent rainbow maxima exist in the angular and spatial distributions of channeled protons. While the rainbows in the angular distributions of channeled ions may be easily measured and used to provide information on the structure and atomic forces inside the nanotubes, the rainbows in the spatial distributions of channeled ions may be employed for detecting and locating the atoms and molecules intercalated in the nanotubes.

We also find prominent peripheral maxima in the spatial distributions at the distances from the nanotube wall on the order of a few tenths of a nanometer. These distances are suitable for applying proton channeling to probe the atoms and molecules adsorbed on the nanotube wall. We have revealed that the image force is responsible for the appearance of additional prominent peripheral maxima as well as for making them more prominent at the expense of the central maximum, as the nanotube length increases in angular and spatial distributions.

All our findings indicate that it is important to carefully consider in future simulations and experiments the role played by dielectric media in ion channeling through carbon nanotubes in the MeV energy range. Such studies may further elucidate dielectric properties of carbon nanotubes in the presence of dielectric media of relevance to nanoelectronics, such as SiO_2. The spatial distribution of channeled protons gives us detailed information about the proton flux within the nanotube. It appears that a careful studying of the speed dependence of the image force as well as of the properties of the nanotube and its surroundings, may help us better understand and perhaps even find a way to induce a spatial redistribution of channeled protons towards the nanotube wall. Redistributing channeled protons might then be used for probing the atoms and molecules intercalated in the nanotube. We also mention the possibility of producing nanosized ion beams with nanotubes embedded in various dielectric media for applications in biomedical research.

References

[1] Mišković Z L 2007 *Radiation Effects and Defects in Solids* **162**, 185

[2] Artru X, Fomin S P, Shulga N F, Ispirian K A and Zhevago N K 2005 *Phys. Rep.* **412** 89

[3] Moura C S, Amaral L 2007 Carbon 45 1802; 2005 *J. Phys. Chem.* B **109** 13515

[4] Biryukov V M and Bellucci S 2005 *Nucl. Instrum. Meth. Phys. Res.* B **230** 619

[5] Petrović S, Borka D and Nešković N 2005 *Eur. Phys. J.* B **44** 41

[6] Petrović S, Borka D and Nešković N 2005 *Nucl. Instrum. Meth. Phys. Res.* B **234** 78

[7] Borka D, Petrović S and Nešković N 2005 *Mat. Sci. For.* **494** 89

[8] Nešković N, Petrović S and Borka D 2005 *Nucl. Instrum. Meth. Phys. Res.* B **230** 106

[9] Bellucci S, Biryukov V M, Chesnokov Y A, Guidi V and Scandale W 2003 *Phys. Rev.* ST AB **6** 033502

[10] Feldman L C, Mayer J W and Picraux S T 1982 *Materials Analysis by Ion Channeling* (Academic Press, New York)

[11] Zhu Z, Zhu D, Lu R, Xu Z, Zhang W and Xia H 2005 *Proc. Of SPIE* Vol. **5974** Bellingham, WA 597-1

[12] Chai G, Heinrich H, Chow L and Schenkel T 2007, *Appl. Phys. Lett.* **91** 103101

[13] Winter H 2002 *Phys. Rep.* **367** 387

[14] Arista N R 2001 *Phys. Rev.* A **64** 32901; Arista N R and Fuentes M A 2001 *Phys. Rev.* B **63** 165401

[15] Tőkési K, Tong X M, Lemell C and Burgdörfer J 2005 *Phys. Rev.* A **72** 022901

[16] Krasheninnikov A V and Nordlund K 2005 *Phys. Rev.* B **71** 245408

[17] Borka D, Petrović S, Nešković N, Mowbray D J and Mišković Z L 2006 *Phys. Rev.* A **73** 062902; 2007 *Nucl. Instrum. Meth. Phys. Res.* B **256** 131

[18] Guerret-Plécourt C, Le Bourar Y, Loiseau A and Pascard H 1994 *Nature* **372** 761

[19] Mowbray D J, Mišković Z L and Goodman F O 2006 Phys. Rev. B **74** 195435

[20] Borka D, Mowbray D J, Mišković Z L, Petrović S and Nešković N 2008 *Phys. Rev.* A **77** 032903

[21] Borka D, Mowbray D J, Mišković Z L, Petrović S and Nešković N 2008 Jour. Phys. Con. Mett., in press.

[22] Hulman M, Kuzmany H, Dubay O, Kresse G, Li L, Tang Z K, Knoll P and Kaindl R 2004 *Carbon* **42** 1071

[23] Lindhard J 1965 *K. Dan. Vidensk. Selsk., Mat.-Fys. Medd.* **34** 1

[24] Zhevago N K and Glebov V I 1998 *Phys. Lett.* A **250** 360; 2000 *J. Exp. Theor. Phys.* **91** 504

[25] Doyle P A and Turner P S 1968 *Acta Crystallogr.* A **24** 390

[26] Doerr T P and Yu Y-K 2004 *Am. J. Phys.* **72** 190

[27] Mowbray D J, Mišković Z L, Goodman F O and Wang Y-N 2004 *Phys. Rev.* B **70** 195418

[28] Saito R, Dresselhaus G, Dresselhaus M S 2001 *Physical Properties of Carbon Nanotubes* (Imperial College Press, London)

[29] Berdinsky A S, Alegaonkar P S, Lee H C, Jung J S, Han J H, Yoo J B Fink D and Chadderton L T 2007 *Nano* **2** 59

[30] Soler J M, Artacho E, Gale J D, García A, Junquera J, Ordejón P and Sánchez-Portal D 2002 *J. Phys. Condens. Matter* **14** 2745

[31] Swart J W, Diniz J A, Doi I and de Moraes M A B 2000 *Nucl. Instrum. Meth. Phys. Res.* B **166-167** 171

[32] Kwei C M, Chen Y F, Tung C J and Wang J P 1993 *Surf. Sci* **293**, **3** 202

Collisions of ions with insulator surfaces: charging and discharging dynamics

N. Bundaleski[a], H. Khemliche[a], P. Rousseau[b], A. Cassimi[b], L. Maunoury[b], and P. Roncin[a]

[a]Laboratoire des Collisions Atomiques et Moléculaires,
UMR 8625, CNRS -Université Paris-Sud, Orsay, F-91405, France

[b]CIMAP, CEA/CNRS/ENSICaen/Université de Caen-Basse Normandie,
BP 5133, F-14070 Caen Cedex 5, France

E-mail: nenad.bundaleski@u-psud.fr

Abstract. The guiding of ions in keV to MeV energy range by insulator microcapillaries opens interesting perspectives such as direct injection of ions inside a living cell, formation of surface nano-structures, etc. We have studied the dynamic behaviour of charge deposition and beam deflection on insulators having well defined planar geometry when the exact number of charges implanted into the surface, as well as the time evolution of the beam deflection, can be monitored. The used materials, borosilicate, fused silica and quartz, are of interest for production of tapered microcapillaries. Depending on the temperature of the sample, surface composition and structure, charging and discharging dynamics varies by orders of magnitude, giving rise to different guiding effects in insulator microcapillaries. The charging process is compared to simple simulations in order to explain the observed behaviour.

1. Introduction

Interaction of slow highly charged ions (HCI) with different materials have interesting perspective in several fields. The physical reason is mainly in their high potential energy, relative to their kinetic energy: during the interaction process the potential energy is locally deposited onto a material. This results in different effects such as potential sputtering, formation of different structures on a nano-scale (nano- dots, hillocks), applications in surface analysis etc. [1]. The potential of slow highly charged ions as tools in nano-fabrication is essentially related to providing intense micro- and nano-beams. The conventional techniques (employing electrical and magnetic fields) are not convenient in this case since they are strongly influenced by the divergence and the energy spread of the initial ion beam. Hence, the discovery of HCI transmission through PET nanocapillaries [2] drew a lot of attention. The main point was that the charge state was kept, meaning that ion transmit without touching the inner walls. By tilting PET foils above 10° some ions were still transmitted which is a clear proof of HCI guiding through nanocapillaries. The observed behaviour was explained by the self-organizing process of inner wall charge up [2], which was confirmed by computer simulations of this process [3].

More recently discovery that a single tapered-glass micro-capillary has been used for guiding and maybe even focusing of HCI beams [4-6] which would have higher technological impact. Since the ion trajectories are parallel to the inner walls (as in PET nanocapillaires), ions are focused due to the taper angle: the observed focusing factor has been estimated close to 10.

In the experiments dedicated to propagation of HCIs through micro-capillaries the macroscopic values observed are mainly time evolution of the intensity of the transmitted beam, its angular distribution, and the charge state analysis [5]. Although important information is gained in these experiments, closer insight into the guiding/focusing effects is hard to achieve without knowing the charge distribution along the capillary including its time evolution. This can be partly overcomed using modelling, for which information on the charging/discharging parameters of a given material are essential. Grazing incidence of ions onto a surface is a good model system for the experimental investigation of charging, discharging and secondary electron emission dynamics of corresponding materials. This experiment would also exclude some 'geometric effects' that influence ion transmission through micro-capillaries.

We report here experimental investigation of the charging/discharging dynamics in the case of grazing incidence highly charged ions onto materials of interest for production of taper-glass micro-capillaries. The influence of the surface temperature, composition and structure, as well as the sample history, is investigated. A simple Monte Carlo simulation of the surface charging evolution by grazing incidence ion beam is also developed. Comparisons between the simulation and the experiments explain some of the observed effects and suggest hints concerning those which are less understood.

2. Experimental

The experimental setup is described in detail elsewhere [7]. A primary beam of highly charged ions is extracted from an ECR ion source. The beam is analyzed using sector magnetic field and collimated before impinging the insulator surface at grazing incidence. The incident angle ψ is in the range 0.5°-1.5°, and the extraction voltages are 1-2.5 kV. A position sensitive detector (PSD) facing the primary ion beam is placed behind the target. When the target is introduced into the beam by translation perpendicular to the primary beam direction, part of the beam hits and charges the target while the rest of the ions fly above the previously deposited charges towards PSD (figure 1). The later are therefore repelled by the charges already deposited at the surface, so their deflection corresponds to surface charging. The charging dynamics is determined by the time evolution of the direct beam deflection. For discharging measurement the beam is left free to discharge with the beam turned off most of the time. It is turned on just for a few seconds in regular time intervals to monitor time evolution of the beam deflection. This method cannot be applied if the discharging is too fast. However, the temperature dependence of the time constant can be determined using the procedure proposed in [8]. It consists of measuring the primary beam rate needed to reach, in equilibrium, a specified direct beam deflection.

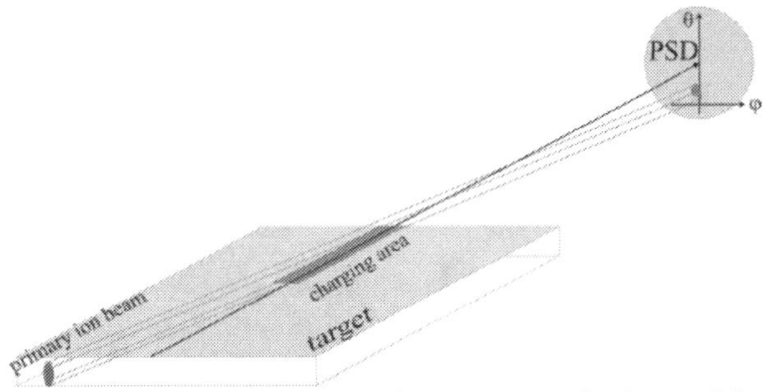

Figure 1. Scheme of the experimental setup; PSD is the position sensitive detector, φ and θ are azimuthal and polar angles respectively

The charging rate is directly proportional to the ion flux hitting the sample surface. This flux can be determined from measuring the intensity of scattered particles if the total reflection coefficient is known. In the case of grazing incidence, the angular distribution of scattered particles is narrow enough to be recorded on the PSD. The number of scattered particles per second is enough to determine the reflection coefficient provided the incident beam intensity is stable. The additional contribution to the surface charging is secondary electron emission which can be also measured by a system of 16 microchannel plates positioned on a hemisphere surrounding the target.

3. Computer simulation

A simple algorithm is adopted to model the surface charging dynamics. The coordinate at which particle hits the surface is the position of a point charge equal to $q + \gamma$, where q is the charge of an ion and γ is the secondary electron yield, both being input parameters. For a specified beam profile, a sequence of ions with appropriate starting conditions is defined using the Monte Carlo procedure. The trajectory calculations are performed by 4th order Runge-Kutta method with adaptive time step. Incident angle and the target distance from the beam axis are input parameters. Whenever an ion hits the target the field distribution is recalculated. Surface discharging is described by exponential time decay (linear model [3]), which corresponds to bulk diffusion of holes. Before proceeding with the calculation of the next ion trajectory, the field is decreased by appropriate exponential factor.

Besides this simulation there is also an additional program used to calculate the distribution of deflected projectiles for a defined charge distribution on the surface. Its execution time is short, and it can give some useful hints for further detailed analysis. This simulation is denoted here as static, while the former is referred to as dynamic simulation. The simulation results will be presented in section 5, in order to support the interpretation of the experimental results.

4. Results

Charging and discharging has been studied on the example of three materials based on SiO_2: fused silica, quartz and borosilicate. Concerning the employed primary beams, no dependence on the ion species was observed. Consequently, we focus to the influence of temperature and material properties on the charging and discharging processes.

4.1. Surface charging

The typical angular distributions obtained by the 50 s data accumulation at the beginning of charging, and for the equilibrium are given in Figure 2. The charging beam is 10 keV Ar^{4+} and the target is fused silica. Before the target is introduced into the beam, there is a single spot on PSD, corresponding to undisturbed direct beam (Fig. 2a). At the beginning of the charging, scattering signal can be observed in the upper half of the PSD and at the same time the direct beam is deflected, magnified and deformed (Fig. 2b)..

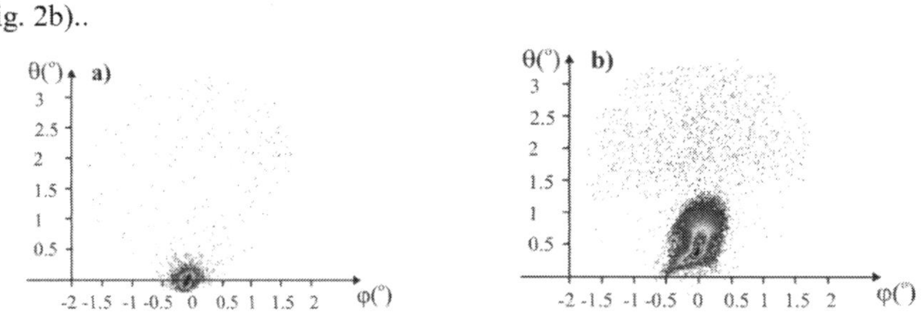

Figure 2. Fused silica charging by 10 keV Ar^{4+} ions, direct beam rate is 1000 ions/s, T = 550 K. The distributions are obtained for a) beginning of the experiment and b) equilibrium beam deflection

The time evolution of the beam deflection and of the number of scattered particles for the same experiment is presented in Figure 3. In general, the beam deflection increases with time and reaches saturation, while the scattering intensity decreases. The time coincidence between the minima of the one and maxima of the other function is a consequence of using the same beam for both surface charging and monitoring the surface charge: whenever the beam is deflected, smaller number of particles hit the surface and vice versa. This is most probably the same negative feedback which provides stability of the transmittance of HCIs through capillaries

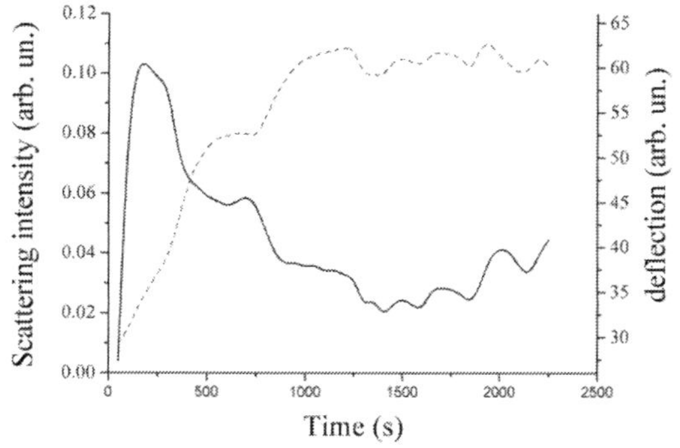

Figure 3. Time evolution of the 10 keV Ar^{4+} direct beam deflection and the intensity of the scattered particles. The surface temperature is T = 550 K. Solid line is the scattering intensity and dashed line is averaged beam deflection

The secondary electron yield depends on the ion type but it disappears quickly after the charge up begins, since the electrons are suppressed by the electrical field of surface charges. Hence, the influence of γ to surface charging is important only at the very beginning of the charging process. This behaviour was observed for all samples. The charging was measured in the temperature range from 440 K to 550 K. Below 400 K the discharging was extremely slow. A strong temperature influence on surface charging is observed. Additionally, the total fluence of ions implanted into the sample also plays a role. Both influences are probably related to the discharging process. Corresponding results will be thus presented in the section 4.2 devoted to discharging, and discussed later.

In the case of single crystal quartz, which has the same composition as the fused silica, the initial charging behavior was quite different: it was not possible to considerably charge up the surface even at room temperature in spite of more than 10 times greater fluxes as compared to those used for fused silica at elevated surfaces temperature. But the real surprise was that the discharge was also very slow, which is going to be discussed later. However, when the target was annealed to about 750 K for about 20 hours, the surface could be charged even with moderate beam flux.

Borosilicate glass, which is of particular interest for production of tapered-glass microcapillaries, shows significantly different behavior in terms of charging as compared to other samples. The typical rates needed for charging was 10-100 times higher as compared to those used for fused silica, at temperatures of 350 K and below. The reason is in extremely efficient discharging of this material. Besides, a number of peculiarities during the charging of this material was observed such as separation of the deflected direct beam into two or more beamlets, sudden appearance of beamlets above the major spot of the direct beam, fast rearranging of beamlet intensities while keeping the same overall intensity of the deflected beam, oscillatory deflection of the directed beam, etc. The time evolution typical angular distribution of the 22.5 keV Kr^{9+} deflected beam(s) is presented on Figure 4. The distributions are obtained by the 50 s data accumulation at three different moments. Two discrete banana-shaped spots clearly appear in fig. 4b. Multiple beams were not observed with other materials.

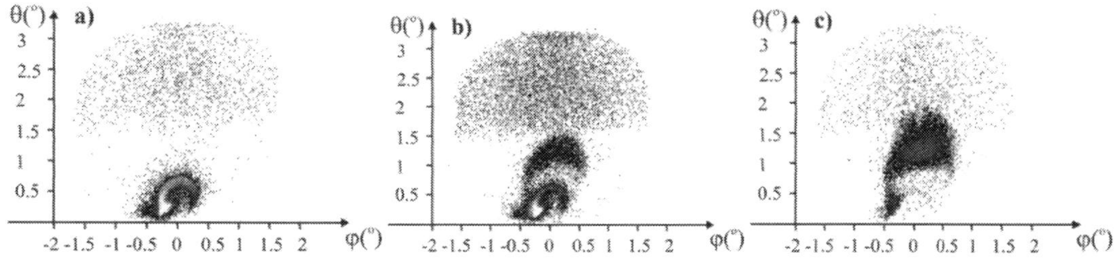

Figure 4. Borosilicate charging by 22.5 keV Kr^{9+} ions, direct beam rate is 16000 ions/s, T = 300 K. The angular distributions are recorded at a) t = 175 s, b) t = 225 s and c) t = 275 s.

4.2. Surface discharging

The time dependence of the direct beam deflection during the discharging of fused silica is presented in Figure 5 and shows a single (5a) or multiple (5b) exponential decay. On the other hand, there are two slopes on figure 5b. The experimental parameters for these two measurements were exactly the same with the exception of the incident angle. The increased incident angle (figure 5a) corresponds to an increase of the surface charge density as the same beam is deposited on a smaller area..The results presented in figure 5 are obtained on the surface which was not previously bombarded. In the discharging measurement for similar experimental parameters as in figure 5a, but on the surface previously bombarded by a fluence of about $5 \cdot 10^{12}$ cm^{-2}, only one slope was observed but the discharge time constant was $\tau = 629$ s which is 2.4 times greater than for a fresh surface.

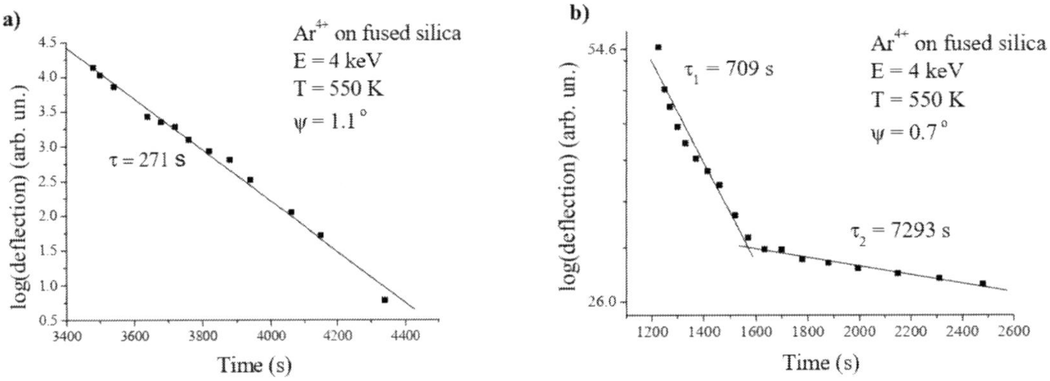

Figure 5. Logarithm of the beam deflection during the fused silica discharging. Primary ion beam is 4 keV Ar^{4+} and T = 550 K. The incidence angles are a) $\psi = 1.1°$ and b) $\psi = 0.7°$

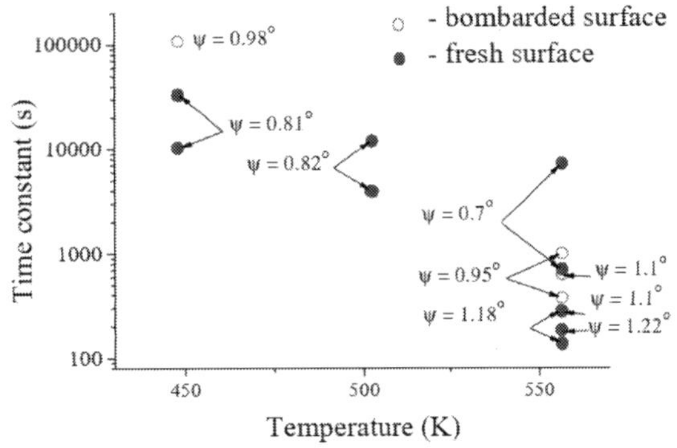

Figure 6. Temperature dependence of the discharge time constant of fused silica; Incident angles and surface condition are denoted in the figure. When two values are reported, they correspond to a two slope discharging dynamics. For identical conditions, a fresh surface (full circles) discharges faster than a previously bombarded (hollow circles). Besides, for identical conditions, discharging is faster as the incident angle increases.

The results of the discharge time constants vs. temperature for fused silica are given in figure 6. The primary beam flux was kept roughly constant. The incident angles and the surface quality (hollow circle for previously bombarded and full circle for fresh surface) are denoted. The observation of two slopes in some cases, influence of the incident angle (decrease of τ with ψ) and of the surface history (increase of τ due to the surface bombardment) makes determining the temperature dependence of τ very hard. Still, it can be concluded that the temperature difference of 100 K produces change of the discharge time constant by two orders of magnitude.

The influence of the surface structure to the discharge dynamics can be clearly seen on the example of quartz which is a single crystal with the same composition as fused silica. Its discharge time constant at room temperature is around 50000 s, while for this temperature discharging of fused silica is too slow to be measured. The strong influence of the composition can be observed on the example of borosilicate which consists mainly of SiO_2 (80.9 %), B_2O_3 (12.7 %) Al_2O_3 (2.3 %) and Na_2O/K_2O (4.1 %). The discharge dynamics could not be measured directly for this material since it discharges too fast. However, using the dynamic equilibrium method explained in [8] where flux is adjusted to reproduce the same deflection, it was possible to observed an Arrhenius behavior suggesting that $\tau \sim \exp(E_a/kT)$, where E_a is activation energy for discharging and k is Boltzmann constant. The measured value of E_a is 0.48 ± 0.04 eV.

5. Discussion

The general physical picture of the surface charging process of glasses is the following. When an ion hits the surface it is efficiently neutralized, so it introduces a charge equal to $(q+\gamma)\cdot e$ in a small area (few nm^2) around the impact point. The impact points are spread along the charging area which is approximately the beam projection to the surface plane. These charges are actually electronic holes initially localized in the nonbonding 2p orbitals of oxygen [9]. They form a field distribution above the surface which deflects the subsequent ions away from the charging area, resulting in beam deflection and in some cases its deformation into a banana (figure 4b) or helmet (figures 2b and 4c) shape, but also in widening of the charged track. The same field is responsible for suppression of the secondary electron yield, which decreases and finally disappears relatively quickly. At the same time the charges migrate along the surface, which further changes their surface distribution, and into the bulk, which is actually discharging. Increase of the surface charge contributes to decreasing the fraction of the primary beam hitting the surface and at some point there is an equilibrium in which charging and discharging are compensated. While the charging is interplay between the surface charge up by ions hitting the surface and the migration of charges, the discharging is solely influenced by the later effect, which makes it simpler to understand. Hence, we shall begin the discussion with the surface discharging.

5.1. Surface discharging

The clear difference between the discharging behavior of fused silica and quartz can be understood in terms of two mechanisms of hole migration. Initially, the discharging takes place by hole hopping between nonbonding 2p orbitals, which is common to both glass and crystal since it depends only on the short range ordering. This migration mechanism, known as intrinsic, should be dominant in the case of quartz. On the other hand, amorphous materials have high concentration of defects which can act as hole traps. Sooner or later, hole gets trapped and its further migration becomes slowed since the potential barrier is much higher than the one corresponding to the intrinsic migration. Above 200 K the transport is thermally activated and the residence time in the hole trap can be described by the Arrhenius behavior [9].

The fact that quartz could not be charged up enough to measure discharging before the annealing seem to be related to small concentration of hole traps: the holes efficiently diffuse by the intrinsic mechanism. Nevertheless, the small concentration of hole traps did not allow fast discharging. The sample annealing probably triggered segregation of hole traps i.e. defects towards the surface (which is thermodynamically favorable), and the surface charging became much more efficient.

In the case of fused silica, the thermal dependence of the discharge time constant qualitatively follows the Arrhenius behavior, although we cannot say much on the nature of this dependence. The influence of the surface bombardment on the discharge time constant measurement is rather surprising knowing that the corresponding ion fluence hitting the surface is very small. Nevertheless, since the surface bombardment increases the time constant, the only explanation would be that additional defects are introduced that way. The observation of two slopes in some cases suggests the existence of two discharge mechanisms.

Two different mechanisms of nanocapillary discharging have been proposed in [3]. The first is the linear model in which the discharge time constant has a constant value. It corresponds to pure diffusion process. The other is a non-linear model in which discharge time constant exponentially decays with the square root of the local electric field. It is convenient for high surface charge concentrations when the local electric field is strong enough to provide electron tunneling by traps. The discharging rate then increases exponentially with the square root of the surface charge density, which was indeed observed in ref. [10]. This physical picture also corresponds qualitatively to here-observed decrease of the τ with ψ. It is hard to say if the existence of two discharge time constants (Fig.3a) could also be explained by this model without the information on the surface charge distribution introduced in every experiment.

The discharging behavior of borosilicate implies that another mechanism takes place besides those proposed above. The temperature dependence of the discharge time constant of borosilicate follows the Arrhenius law. The corresponding activation energy is very close to that for migration of Na^+ ions binded to non-bridging oxygen atoms (NBO) in a borosilicate glass (0.55 eV) [11]. According to this scenario, not the holes but Na^+ ions migrate in order to provide surface discharging. Apparently, weak point of this suggestion is that sodium is preferably binded to BO_2^- defect which has higher activation energy. Hence, Na^+ can be binded to NBO only if the concentration of Na_2O exceeds half of the concentration of B_2O_3 [11] which is not the case unless there is a surface segregation of Na. Indeed, strong surface segregation of Na in borosilicate glass has been measured in ref. [12].

5.2. Surface charging

The basic problem with the mentioned details is that they are not reproducible: not all relevant experimental parameters are well controlled and/or small deviations of some of them yield in significantly different behavior. This particularly stands for the borosilicate sample. We shall concentrate here to possible qualitative explanations of some of the features. Let us emphasize before that the two clear differences in the borosilicate experiment with respect to other samples: a) the employed beam intensity is at least for one order of magnitude higher and b) the discharge time constant is much shorter.

The time evolution of the direct beam deflection is strongly correlated not only to the amount of surface charges but also with its surface distribution. The static simulations show that propagation of the primary beam above the charged surface yields in magnification of the primary beam, and its deformation, as observed in experiment. These simulations reproduce curved shape of comparable radius as the one observed in experiments for the same beam fluence. Depending on the suface charge distribution, the magnification can be either positive (Gaussian charge distribution) or negative (uniform charge distribution). The later can explain sudden appearance of beam spots above the averaged direct beam image on PSD, which is sometimes observed in the borosilicate charging experiments (fig. 4c).

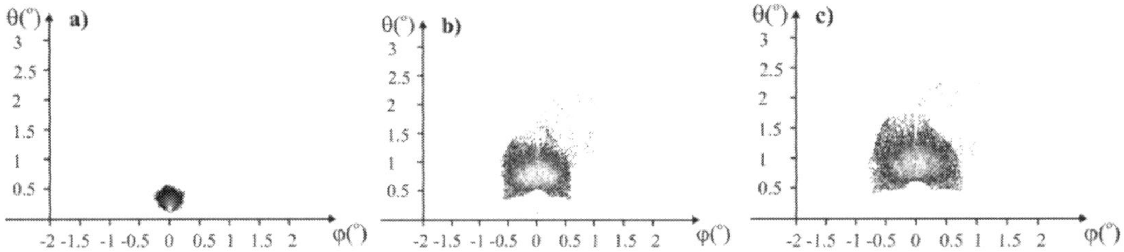

Figure 7. Dynamic simulation of insulator surface charging by 10 keV Ar^{4+} ions, the direct beam rate is 10000 ions/s, γ = 10, and a discharge time constant τ = 1000 s. The angular distributions correspond to snapshots taken at a time a) 0.07 s b) 0.75 s and c) 1.5 s.

One example of the dynamic simulation result is given in Figure 7. The primary ion beam is 10 keV Ar^{4+}, overall number of ions is 156250 (corresponding to overall charge of $1 \cdot 10^{-14}$ C carried by ions), and the discharge time constant is τ = 1000 s, which should correspond to experiments with fused silica at 550 K. Indeed, the magnitude of the direct beam deflection, the spot magnification, including the spot deformation into a helmet shape is comparable to the fused silica experiment and even to that of borosilicate (figures 2b and 4c). Additional feature that usually appears, a sharp well defined spot superimposed to the blurred helmet (figure 2b), the simulation did not succeed to reproduce.

While the simulations have some success in reproducing results on fused silica, it seems to be much less efficient in the case of charging dynamics of borosilicate. Decrease of the discharge time constant down to τ = 10 s, did not yield in observation of multiple bananas. Two bananas were observed in simulation when the surface charge distribution has two maxima. However, it is not easy to imagine physical situation in which this would appear. The oscillations of the beam deflection are observed in the dynamic simulations. The analysis of the surface charge distribution shows that they are correlated with local surface discharging which lowers the beam deflection but at the same time increases the surface charging rate, contributing again to the beam repelling. However, the amplitude of these oscillations are much smaller than those observed experimentally. The possible explanation of the strange behavior of borosilicate could be the interplay between the beam instability and fast discharging. The first results of dynamic simulations with randomly introduced beam shifts give some promising results (fast shifts along the azimuthal direction), but more work is necessary to see if this interpretation is correct.

6. Summary

Discharge time constants of fused silica and quartz were experimentally determined for different physical conditions. Influence of the temperature and previous surface bombardment was observed in the case of fused silica. Charging properties of quartz were modulated by sample annealing, which was explained as a consequence of defect segregation towards the surface. Observation of very fast discharging of borosilicate is of particular interest since taper-glass microcapillaries are made of this material. The determined activation energy of the borosilicate discharge time constant suggests that migration of sodium ions is responsible for discharging rather than that of hole as for other two materials. Influence of the ion species to the studied processes is not observed. Simple simulations of surface charging are introduced, which reproduce in a first approximation, the experimental findings. Still, several effects, particularly in the case of borosilicate, are left unclear. Further work on more refined simulations is in progress.

Acknowledgments

We acknowledge support from the EEC FP6 network project 026015 "ITS LEIF".

References

[1] Schenkel T, Hamza A V, Barnes A V and Schneider D H 1999 *Progr. Surf. Sci* **61** 23
[2] Stolterfoht N *et al.* 2002 *Phys. Rev. Lett.* **88** 133201
[3] Schiessl K *et al.* 2005 *Phys. Rev. A* **72** 062902
[4] Ikeda T *et al.* 2006 *Appl. Phys. Lett.* **89** 163502
[5] Ikeda T *et al.* 2007 *J. Phys:Conf. Ser.* **58** 68
[6] Cassimi A *et al.* 2008 *J. Nanotechnology* **5** 809
[7] Morozov V A, Kalinin A, Szilagyi Z, Barat M and Roncin P 1996 *Rev. Sci. Instr.* **67** 2163
[8] Bundaleski N *et al.* submited
[9] Hughes R C 1976 *Phys. Rev. B* **15** 2012
[10] Stolterfoht N *et al.* 2007 *Phys. Rev. A* **76** 022712
[11] Grandjean A, Malki M and Simmonnet C 2006 *J. Non-Crystal. Solids* **352** 2731
[12] Sumner A L *et al.* 2004 *Phys. Chem. Chem. Phys.* **6** 604

Measurement of voltage and current in continuous and pulsed rf and dc glow discharges

V. Hoffmann[1], V.V. Efimova[1], M.V. Voronov[2], P. Šmíd[3], E. B. M. Steers[3], J. Eckert[1]

[1]Institut für Festkörper- und Werkstoffforschung Dresden, P.O.Box 270116, D-01171 Dresden

[2]AQura GmbH, P.O.Box 915-d115, Rodenbacher Chaussee 4, D-63457 Hanau-Wolfgang, E-mail: VoronovMV@mail.ru

[3]London Metropolitan University, 166-220 Holloway Road, London, N7 8DB, UK, E-mail: p.smid@londonmet.ac.uk, e.steers@londonmet.ac.uk

E-mail: V.Hoffmann@ifw-dresden.de

Abstract. Electrical measurements are an important tool for the characterisation of glow discharges and have proved to be useful for a variety of needs in fundamental studies and as control parameter. Therefore, extensive hardware developments and studies of current-voltage (I-U) characteristics in continuous and pulsed, dc and rf modes have been made [1] and will be presented together with new results.

In continuous dc mode, the I-U curves are non-linear and may be characterised by a threshold voltage U_0 and saturation current I_{max} (both cathode material and pressure dependent). On the other hand P-U curves are to a large extent linear and very similar in the continuous rf mode [2]. The ionic part of time resolved I-U curves of rf discharges however shows almost a linear behaviour and the capacitive component is small. This led to the assumption that gas heating is responsible for the non-linearity between I and U in continuous dc discharges. Consistent with this assumption, a dependence of the I-U curves of pulsed discharges on the duty cycle was found. The comparison of the curves with those at low duty cycle (cold) led to a rough estimation of the gas temperature.

Further investigation and cooperation with modelling groups is needed and planned to explain these results.

1. Introduction

The electrical properties of discharges have always been of interest in discharge physics. Electrical direct current (dc) measurements are easy to perform, and do not disturb the discharge. For dc glow discharges (GD) a lot of documented discharge properties are directly related to the discharge voltage and current [3-5]. Dc discharges used for glow discharge optical emission spectrometry (GD-OES) [6] with the aim to measure the chemical composition of solid samples and layered materials [7], require voltage and current measurements for the regulation of the discharge to constant conditions. A radio frequency (rf) generator is required for the measurement of insulating materials or insulating layers. Reproducible electrical rf measurements are necessary to adjust the discharge to constant conditions or to correct the intensities and sputtering rates in order to determine the chemical composition of all samples with sufficient accuracy. Therefore, extensive hardware developments and studies of current-voltage (I-U) characteristics in continuous and pulsed, dc and rf modes have been made [1] and will be presented together with new results.

For any defined glow discharge source and plasma gas there are three main parameters which characterize the glow discharge: voltage, current and pressure. It is known that there are more parameters existing as gas flow [8], temperature of source and gas [9] or source material [10], which influence the discharge. Because they are difficult to measure or to vary they are mostly treated as boundary conditions.

In many experimental results, these boundary effects cannot be distinguished unequivocally, because e.g. the temperature also has an influence on the secondary electron emission [11, 12]. Also wall losses at constricted discharges were taken into account and change the I-U characteristic by a reduction of the effective cathode area [13].

Payling et al. [6, p. 265] e.g. found an increasing sputtering rate with depth at constant voltage and current and assumed that changes in the temperature could be the reason for this effect. Kasik et al. [14] have mentioned that cathode heating lead to changes of the I-U characteristics in GD-MS, mainly to a decrease of the current at the same voltage. Bogaerts et al. [15] confirmed the same effect in modelling calculations and simulations. In this modelling work a nearly linear relation between the cathode and discharge gas temperature was presented. Furthermore, old fundamental measurements about normal glow discharges did show that more than 50 % of the input energy were found in gas heating and heat transport by the discharge gas to the walls [16].

There are some papers about the experimental determination of the discharge gas temperature. Ferreira et al. [17] used line width measurements for the determination of gas temperatures. Li [18] inserted a "thermometer molecule" $W(CO)_6$ in the discharge gas which loses a defined quantity of CO groups at different gas energy distributions and TOF-MS was used for the detection. Wilken at al. [19] estimated surface temperatures of the sample from crater bending. However, many authors have mentioned the lack of experimental data about sample surface and gas temperatures, e.g. [9].

In analytical glow discharges two electrical parameters (e.g. U and I or P and U) are kept constant to get the same sputtering, ionization and excitation conditions for the standard material and the analyzed sample, which is necessary for quantification [20]. Whereas the measurement of voltage and current in continuous glow discharges is reliable and modelling groups can use these data without difficulties, for the analysis of nonconducting materials by RF-GD, a special rf technique for the measurement of voltage and current was developed and used in this work [21]. For continuous analytical glow discharges there is a good agreement between the experimental results and modelling calculations [dc: 22, rf: 23], but the situation is different for pulsed discharges. The difficulty to get reliable data about voltage and current in those discharges becomes clear in the course of some modelling work. At the beginning Bogaerts [24] used the gas temperature as a parameter to fit the measured current in the first time interval of μs pulsed dc discharges, but later it was found that difficulties in the measurement led to wrong results in the experimental data but also at modelling [25]. Our own measurements confirm very high currents at the beginning of dc discharges, but also the difficulties for accurate measurements in the first microseconds. Therefore, we have not focused our work on this problem, but will show that similar problems can arise in the interpretation of afterglow effects, if no clear data about the discharge are given.

Furthermore, I-U or P-U dependences permit to compare dc and rf discharges, to estimate breakdown voltage and to make some theoretical conclusions about the discharge physics. They can also serve as indicators of processes taking place in the discharge. But the aim of this paper is not to draw those conclusions but to present data of measurement together with experimental details, which are necessary to understand and evaluate these results.

2. Experimental

2.1. Direct current continuous and pulsed discharge

For dc and pulsed dc measurements a 4-mm Grimm type source from Spectruma (GDA-750) was used. The discharge pressure was measured in the anode body by a Pirani gauge, calibrated by a baratron manometer (MKS Baratron Type 127; range 0–100 mbar). A high voltage generator (FUG MCN 350-2000) was used to power continuous GD. We used 10 mm thick pure Cu samples (Ø 30 mm) for all electrical measurements and 60 s presputtering time.

The pulsed GD was excited either with the high voltage pulse generator (RUP 3-3a), synchronized by an external TTL pulse generator (HP 33120A) or with a high voltage pulse generator (IRCO M3kS-20N). For evaluation, discharge voltage and current of the pulsed discharge were transferred via a digital oscilloscope (Tektronix 11201) to the computer.

Before the measurement of I-U characteristics of dc pulsed GD, these two pulse generators were compared. The main reason for the comparison of these two pulsers is the significant difference in the electronic circuit. Mainly, the RUP 3-3a, unlike the IRCO M3kS-20N, has an additional high voltage switch, which discharges the load after the termination of the pulse (figure 1).

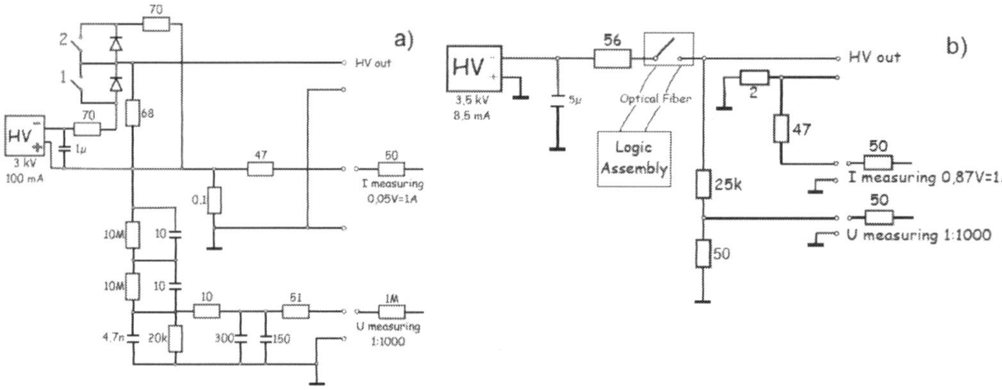

Figure 1. Electronic scheme of (a) RUP 3-3a, (b) IRCO M3kS-20N.

The differences in the electronic circuit lead to differences in the current and voltage signal behaviour (figure 2). The discharge voltage signals were measured by the generators itself, using internal voltage dividers. For the electrical current measurements the electronic schemes of the generators provide special outputs where a voltage proportional to the current is formed (figure 1).

With the RUP 3-3a during the pulse, switch 1 is on and 2 is off (figure 1 a). To terminate the pulse the first switch is turned off, but at the same moment the second is turned on and discharges the residual load current. Therefore there are no voltage and current signals after the termination of the pulse. To know this feature is very important for example at the investigation of temporal light emission of pulsed GD. Therefore at the research of the pulsed discharge, especially of the afterglow, it is very important to pay attention to the electronic circuit of pulse generator. The influence of this different voltage and current characteristic on the light emission is shown by Efimova et al. [26]. For the subsequent experiments only RUP 3-3a was used.

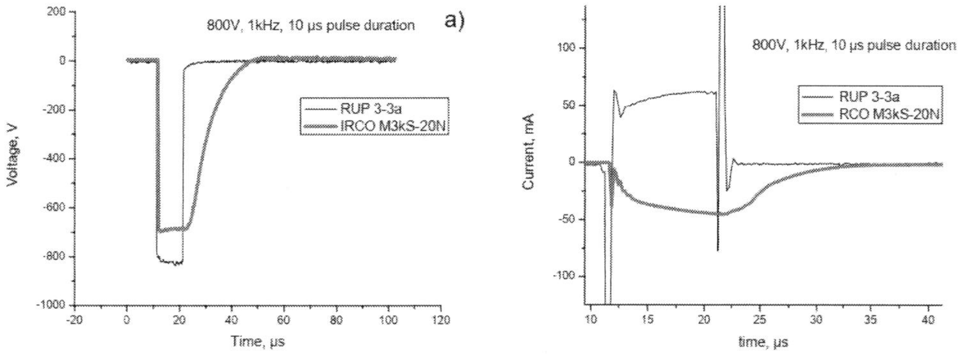

Figure 2. Comparison of voltage (a) and current (b) signals produced by different power supplies.

2.2. Radio frequency continuous and pulsed discharge

For rf and pulsed rf measurements a standard free standing Grimm type glow discharge source with 4 mm anode tube was used. The glow discharge was excited with an rf generator (Forschungstechnik IFW 3.37 MHz) working in free running regime, i.e. the glow discharge source was a part of the oscillation circuit powered by a dc power supply [1].

For the measurement of the instantaneous signals of voltage and current the glow discharge source was equipped with integrated voltage and current probes. The voltage probe uses a capacitive voltage divider which means that the recorded voltage signal does not contain a dc bias component. The dc bias was calculated (see [1]). The integrated current probe is a current-voltage transformer which converts the measured current into a secondary current. This current is passed through a resistor and the voltage drop over this resistor is measured. Both voltage and current probes are mounted between the cathode plate and the body of the glow discharge source which ensures high mechanical stability and low displacement current distorting the measured current. Nevertheless there is some displacement current which was subtracted numerically.

The signals from the voltage and current probes were fed into a high speed oscilloscope (LC584A, LeCroy). The recorded signals were subsequently transferred into a PC memory where further evaluation was done.

3. Results and Discussion

3.1. Continuous direct current discharge

The discharge in most analytical devices is operated in the abnormal regime. In this mode all cathode surface is fully covered by the discharge and an increase of the current leads to an increase of the current density on the cathode surface and therefore to an increase of the voltage. The plasma behaves electrically rather similar to a resistor. In abnormal mode the operating voltage U can be described by the equation

$$U = U_0 + I \cdot R \qquad (1)$$

where U_0 is threshold voltage, I the current and R the differential resistance of the plasma [6, p. 193]. R is considered to be constant, which is not the case in reality (figure 3).

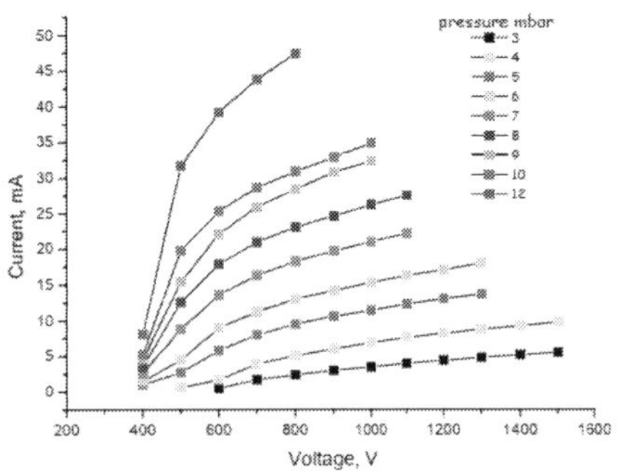

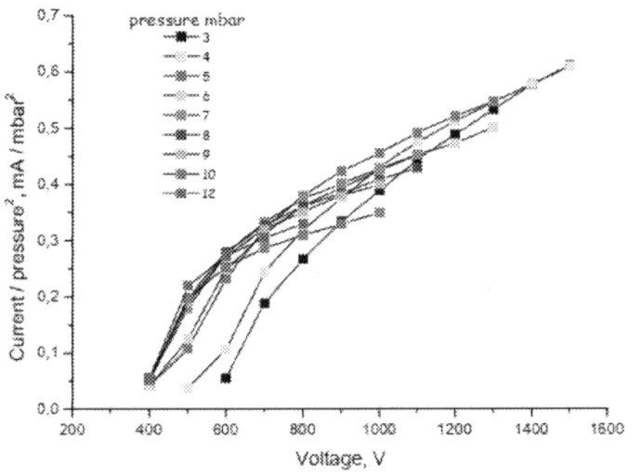

Figure 3. I-U characteristics of dc GD.

Figure 4. I/p^2-U characteristics of dc GD.

For the low pressures of 3-4 mbar the linear range is wider than at high pressures. Here the convex non linear behaviour becomes more and more pronounced with increasing pressure and at high voltages (and corresponding powers *P*) a saturation of the current can be expected. Even if the determination of the threshold voltage is difficult in this presentation it becomes clear that the threshold voltage decreases with higher pressure according to the law of Paschen [27]. The low voltage and current data are not very reproducible and the extrapolation to zero current limit is not possible.

In order to see if the similarity law $I_1/p_1^2=I_2/p_2^2$ is fulfilled, figure 4 shows this presentation, which has also been used for a long time in analytical glow discharges [28].

At the low pressures of 3 and 4 mbar the changed threshold voltages have a strong influence and with increasing pressures there are deviations form this law at high voltages and powers. Most probably, these are caused by sample and gas heating, which is known to exist at abnormal discharges [29]. Nevertheless, the similarity law is valid over a wide range of pressures and voltages also for the used abnormal discharge in a Grimm type source and deviations from this law can be used to characterize the discharge.

Unlike the I-U plots P-U plots are nearly linear. The power of the dc discharge can be evaluated as

$$P = U \cdot I = I_{\max}(p, matrix) \cdot \left[U - U_0(p, matrix)\right] \quad (2)$$

$$I = I_{\max} \cdot \frac{U - U_0}{U} \quad (3)$$

where I_{max} is a saturation current, p is a pressure. The meaning of the slope as saturation current becomes clear in equation 3. At very high voltages (e.g. 10 times $U_0 \approx 3000$ V) the quotient becomes nearly one and the current converges to the limiting value I_{max}.

There is no physical background for this presentation, but the difficulty to measure the current in an rf discharge led to this empirical approach. Even if the rf power measured by a reflectometer may be not very accurate and e.g. a subtraction method mostly is applied additionally, these curves can be obtained quite easily. As proved in figure 5, these curves are linear over a wide range and extrapolation to zero power/current limit is possible. Of course also the curves for the reduced power in dependence on voltage in figure 6 obey the similarity law to the same extent as the reduced current (figure 4), but the reduced power curves remain nearly linear.

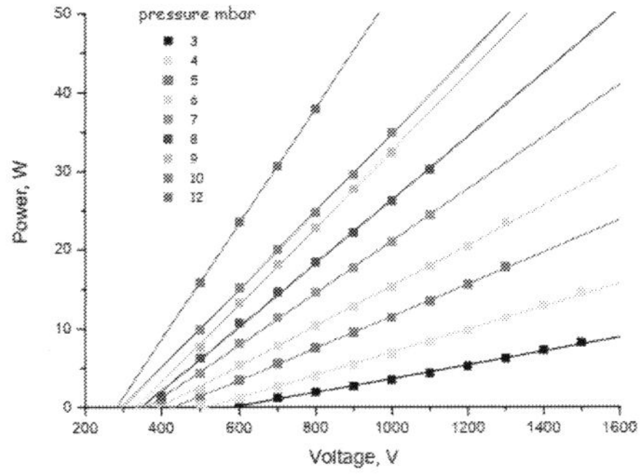

Figure 5. P-U characteristics of dc GD.

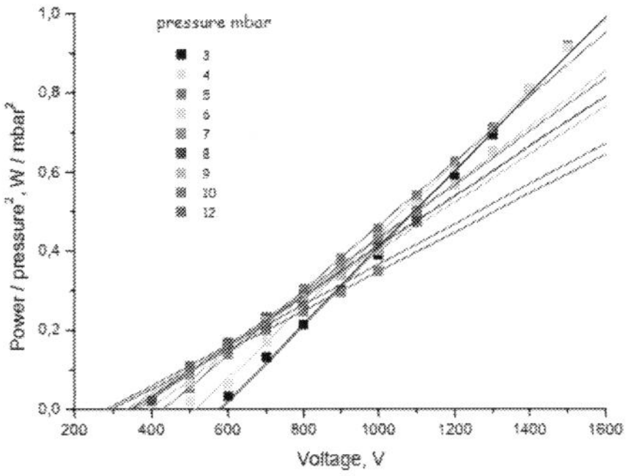

Figure 6. P/p^2-U characteristics of dc GD.

The threshold voltage can be determined from the P-U plots by a linear extrapolation to P = 0, and delivers objective and reproducible data in the case of the Grimm type sources (figure 7). It becomes clear that the source is operated at the left side of the Paschen curve and a reduction of pressure in the ring slit means a constriction of the plasma inside the anode with direct contact to the wall.

figure 8 shows that the saturation current depends linearly on pressure. For lower pressures no extrapolation can be done, because the discharge is not stable at these conditions. Of course, threshold voltage and saturation current depend strongly on the applied matrix material and can be used to characterise e.g. the secondary electron emission yield.

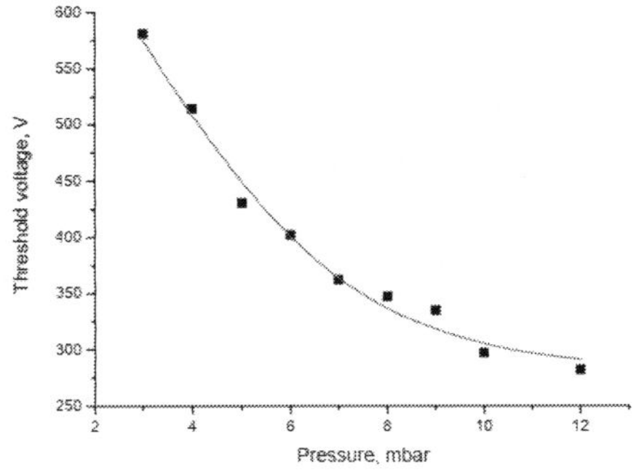

Figure 7. Threshold voltage in dependence on pressure from figure 5.

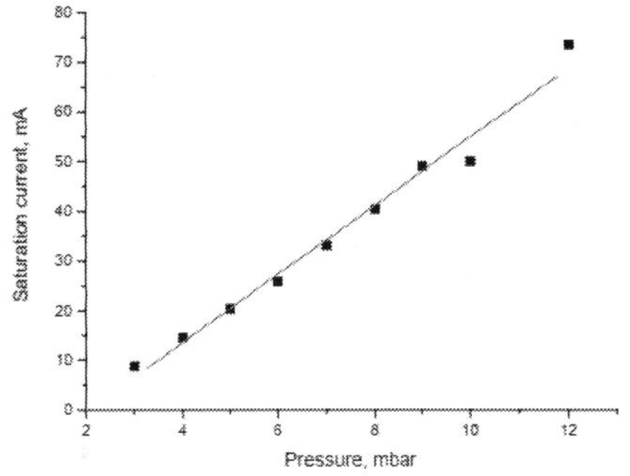

Figure 8. Saturation current in dependence on pressure from figure 5.

3.2. Continuous radio frequency discharge

The measurement system developed at IFW Dresden allows to measure instantaneous voltage U(t) and current I(t) waveforms of rf discharge. These time dependent curves can be converted into current voltage characteristics. In this paper we mostly present the ionic part of the I-U curves, which can be compared to the dc ones. As in dc mode the impedance of the plasma decreases with increasing pressure. The ionic part of the I-U characteristics of the rf discharge is quite linear and shows hysteresis (figure 9). The hysteresis is caused by the capacity of the sheath at the powered electrode, which is determined by the number of positive charges in the sheath at the powered electrode per volt.

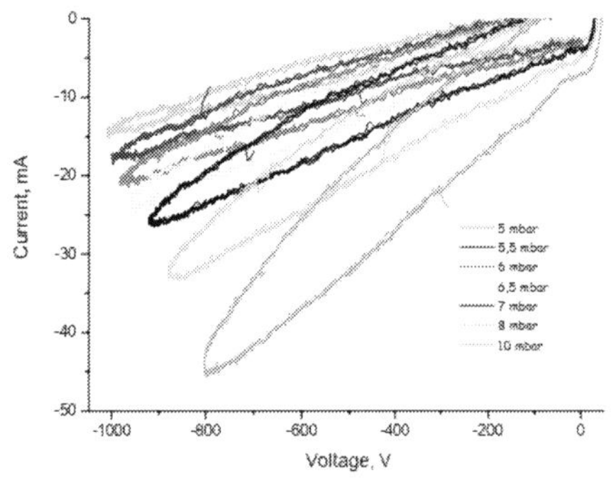

Figure 9. I-U characteristics of rf GD at constant anode voltage of 1100 V.

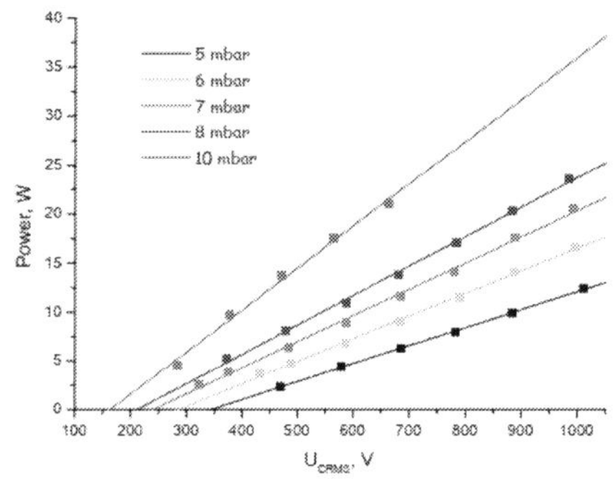

Figure 10. P-U plots of rf GD.

With increasing pressure also the hysteresis is increasing which means an increase of the cathode sheath capacitance too. The ionic part of time resolved I-U curves of rf discharges shows almost a linear behaviour. This led to the assumption that gas heating is responsible for the non-linearity between U and I in continuous dc discharges.

For the characterization of the rf discharge and also for quantification it is important to reduce the number of data. Voltage and current depend on time and even if harmonics in the voltage are suppressed, the rf voltage is still superimposed by a dc part, called dc bias. If a sinusoidal rf voltage is applied, the rf current is non linear because of the I-U characteristics of the plasma. Therefore, the cycle root mean square value (CRMS) is used to characterise the rf voltage,

$$U_{CRMS} = \sqrt{\frac{1}{t_p}\int_o^{t_p}(U_{bias} + U_p(t))^2\,dt} \tag{4}$$

where t_p is a time of one or several periods. The CRMS value of the rf voltage is also called the effective voltage U_{eff}. Marshall et al. reported for the first time, that the effective voltage in rf mode is comparable to the voltage in dc mode. Sputtering rates as well as emission yields become comparable [30].

The power waveform $P(t)$ is calculated as the product of the plasma voltage waveform $U_p(t)$ and the plasma current waveform $I_p(t)$:

$$P(t) = U_p(t)I_p(t) \tag{5}$$

The mean power is the definite integral of the power waveform:

$$P = \frac{1}{t_p}\int_0^{t_p} P_t(t)dt \tag{6}$$

At the variation of voltage at constant pressure the impedance of the plasma remains nearly constant in the ionic part of the I-U characteristics also in pulsed rf mode. The P-U plots of rf discharge (figure 10) show a nearly linear behaviour like in dc mode (figure 5). Both curves are also quantitatively very similar, which proves the similar meaning of dc and effective voltage in dc and rf mode, respectively. The reduced power in dependence on voltage (not shown) obeys the similarity law as in figure 6.

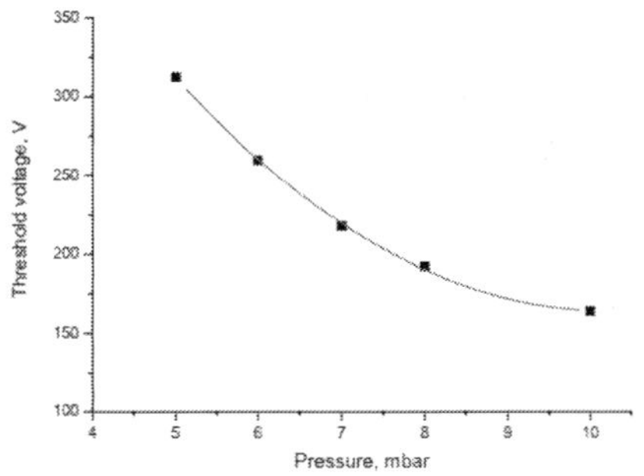

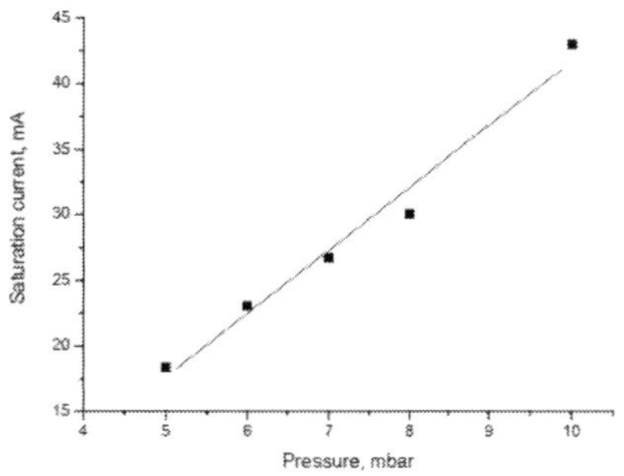

Figure 11. Threshold voltage in dependence on pressure from figure 9.

Figure 12. Saturation current in dependence on pressure from figure 9.

The threshold voltage shown in figure 11 obeys also the typical behaviour of the Paschen curve at the left side of the minimum. The values are lower than in figure 7, which could be caused to a certain extent by differences in the source geometry, because the threshold voltage is a function of p·d. The distance between anode and cathode is only about 0.15 mm and varies from source to source. On the other hand this result agrees with the fact that rf sources can operate at lower powers than dc sources. figure 12 shows that also in the rf case the saturation current depends linearly on pressure and the absolute values are similar.

3.3. Pulsed direct current discharge

In the case of pulsed discharges besides current, voltage and pressure the pulse duration τ and frequency f can also be varied. Additionally, the evaluation of the current value is not as simple as in dc, as the current is not constant. In figure 13 there is a comparison of current signal shapes at constant and different duty cycles. For figure 13 a) the frequency was kept constant and the pulse length was varied resulting in a varying duty cycle as well. For figure 13 b) both frequency and pulse length were varied so that the duty cycle of 1% kept constant. At the beginning and end of each pulse there are artificial signals caused by high currents at ground potential. The data presented in figure 13 were measured with an 8 mm anode and therefore the absolute values cannot be compared with data from the other figures.

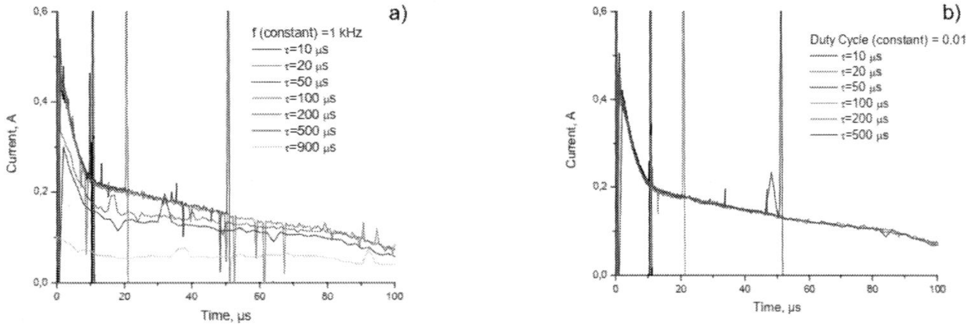

Figure 13. Comparison of current signal shapes at different (a) and constant (b) duty cycles at 1000 V and 6.7 mbar.

At a constant duty cycle of 1 % (figure 13 b) the currents have the same time dependence and only stop at different times according to their pulse length. But with increasing duty cycle the current decreases at constant frequency 1 kHz (a). This effect is very similar to the decrease of current under higher voltages in the dc case and can also be caused by the discharge gas heating. In order to investigate the influence of pulse length and frequency at the variation of the duty cycle, measurements were done at constant pulse duration of 50 µs, varying the frequency and at constant frequency of 1 kHz, varying the pulse length (figure 13).

For the evaluation of these curves the current of pulses with any length is averaged within a time interval skipping the artificial signals. This algorithm provides the same magnitude of the electrical current for all pulses having the same time dependence, even if the durations of the pulses are different. This algorithm was used to reflect the changes of the current at the first and last 5 µs, which represent changes in the I-U characteristics in this time. The voltage was almost constant during the pulse (see figure 2 a).

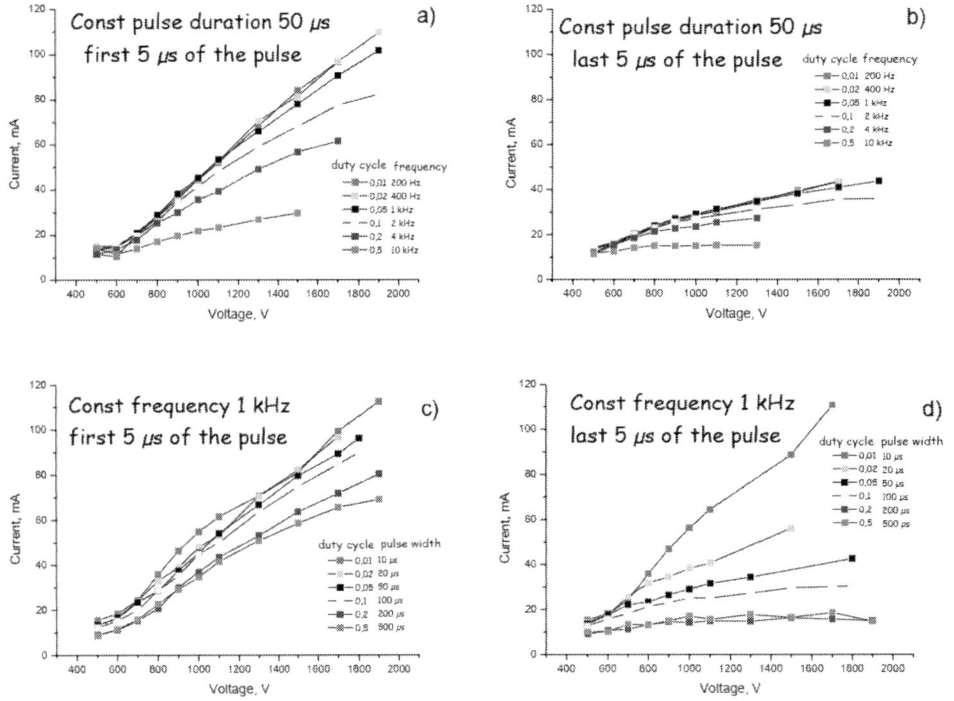

Figure 14. I-U characteristics for pulsed dc discharge at constant pulse duration (a, b: 50 µs) and constant frequency (c, d: 1 kHz) at 5 mbar.

It is apparent from the all plots in figure 14 that the averaged current value decreases with increasing duty cycle. This indicates that there can be a different temperature. But unlike the I-U plots for the continuous dc mode in the case of pulsed dc discharge most of the I-U characteristics are not curved. This means that the discharge gas temperature for a defined duty cycle remains nearly constant. For high pulse length, frequencies and duty cycles the current of pulsed dc discharges also starts to saturate at the end of the pulse (figure 14 b, d), because these conditions are similar to a continuous discharge.

The comparison of the I-U characteristics acquired at the beginning and at the end of the pulse shows clearly that at the beginning the current is higher than at the end. This agrees with the effect of gas heating, which however may not be the main reason for these different currents. In the case of pulses with short durations of 10 μs in figure 14 d) the times for current integration nearly overlap and no difference is possible.

I-U characteristics measured at constant pulse duration (figure 14 a, b) begin to change only at duty cycle 0.1 (2000 Hz). This means that cooling is efficient between the pulses, if there is more time than about 500 μs.

From figures 14 b) and d) follows that at 50 μs pulse length and 1 kHz frequency there is more influence of the pulse length on the discharge gas temperature than of the duty cycle at fixed pulse length. A similar effect was observed by D. Fliegel [31] for the breakdown voltage. His experiments showed that the breakdown voltage of pulsed discharge remains the same at different delays between two plasma pulses, but with increasing pulse duration the breakdown voltage decreases.

Measurements at varying pressure were also done and show the expected correlation between pressure and current. This correlation can be used for temperature estimation as shown in [26].

3.4. Pulsed radio frequency discharge

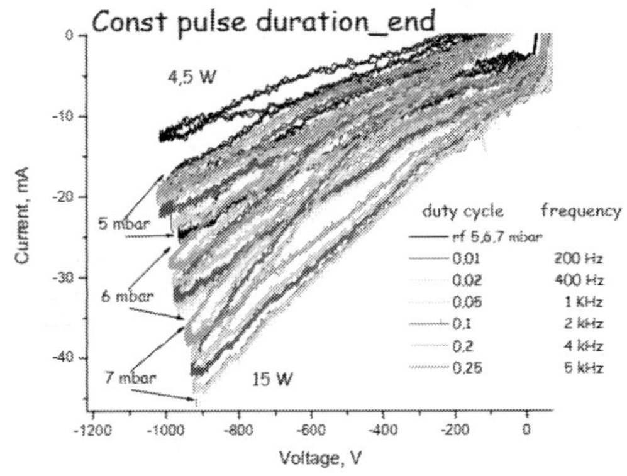

Figure 15. Ionic part of the I-U characteristics at different pressures and duty cycles (50 μs constant pulse length, cycle during last μs).

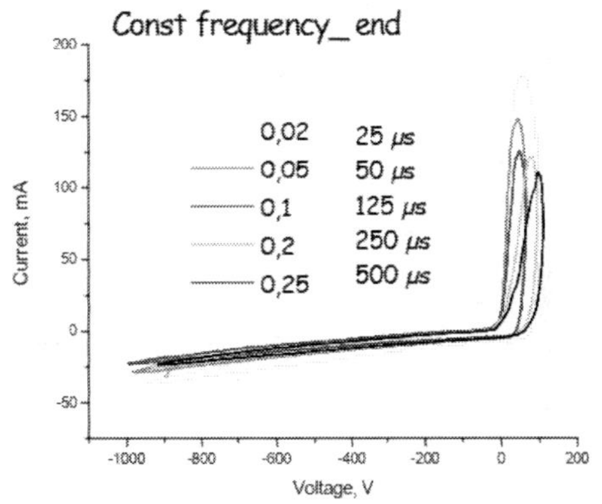

Figure 16. I-U characteristics for pulsed rf discharge at constant frequency (400 Hz) at 6 mbar (25 μs – 625 μs pulse length).

Equivalent measurements to the dc pulsed case were also performed for the pulsed rf discharge. However, 25 % was the highest duty cycle which could be applied, because the rf generator needs about 20 μs to become stable (dc bias development) and to switch off. More details about these results are presented in [26] and only some comparisons are done here.

All I-U characteristics were measured at constant rf generator anode voltage of 1100 V. Figure 15 shows the ionic part of the I-U characteristics at different pressures and duty cycles measured at the end of the pulse. Two I-U plots with the same slope can correspond to the low duty cycle and low pressure and to the high duty cycle and high pressure. For constant frequency and at 6 mbar the same effect is visible in figure 16. This phenomenon gives an approach to estimate the discharge temperature (see [26]). Figure 16 shows also changes in the electronic part. At high duty cycles the maximum current reduces and the hysteresis increases.

The difference between the beginning and end of the pulse is not as pronounced as in the dc case, because there are no high currents in the first microseconds at rf discharges. Similar to the dc case, the weakest effect occurs at the beginning of the pulse at constant frequency and the strongest effect is visible in this mode at the end of the pulse (see figure 16).

Additional experiments with varying anode voltage, constant pressure (6 mbar) and constant pulse length (50 μs) have shown that the voltage has nearly no influence on the slope of the ionic current, even if the power changes by a factor of 2.

4. Summary and Conclusion

Accurate measurements of voltage and current support the interpretation of analytical results and are necessary as input or fit parameters for modelling groups. From the comparison of different hardware for dc pulsers it becomes clear that several microseconds after pulse termination current may still flow, because of discharging capacities.

In continuous dc and rf mode, the transformation of I-U into P-U curves simplifies the determination of threshold voltages and saturation currents. From the comparison of those curves it was confirmed that in rf mode the effective voltage corresponds best to the voltage in dc mode.

In pulsed mode, the variation of pulse length and frequency influences the I-U curves and complicates the evaluation and presentation of the data. In general, a higher duty cycle reduces the current at the same voltage. The similarity of this effect with a reduction of pressure or gas density most probably is caused by gas heating.

In pulsed dc it was found that the time behaviour of voltage and current is independent of frequency and pulse length, if the duty cycle is kept constant. The current within the first 5 μs decreases during the pulse. This effect is less pronounced in rf mode, because the strong currents at the beginning are missing. The current decrease because of gas heating is similar and after about 500 μs off pulse the gas cools down. An increase of voltage does not change the slope of the current voltage characteristic to the same amount as a corresponding change of the duty cycle. Also the changes in the electronic part of the I-U characteristics need further attention. Therefore, continued cooperation with modelling groups is needed and planned.

Acknowledgments

The investigations about pulsed discharges were supported by the European Research Training Network Gladnet, contract nr. MRTN-CT-2006-035459.

References

[1] Wilken L, Hoffmann V, Wetzig K 2007 *Spectrochim. Acta Part B* **62** 1085.

[2] Hoffmann V, Praessler F, Wetzig K 1998 in: Vogt C, Wennrich R and Werner G (Editors), *Colloquium Analytische Atomspektroskopie* (Universität Leipzig) 29

[3] Engel A V 1955 *Ionized Gases* (Clarendon Press)

[4] Francis G 1956 The glow discharge at low pressure, *Handbuch der Physik* (Springer Verlag, Berlin)

[5] Ingold J H 1978 Glow discharges at DC and low frequencies in: Hirsch M N, Oskam H J (Editors), *Gaseous Electronics* (Academic Press) 19

[6] Payling R, Jones D, Bengtson A 1997 *Glow Discharge Optical Emission Spectrometry* (John Wiley & Sons, New York)

[7] Bengtson A 1994 *Spectrochim. Acta Part B* **49** 411

[8] Bogaerts A, Okhrimovskyy A, Gijbels R 2002 *J. Anal. At. Spectrom.* **17** 1076

[9] Hagelaar G J M, Pitchford L C 2002 *J. Anal. At. Spectrom.* **17** 1408

[10] Hoffmann V, Kasik M, Robinson P K, Venzago C 2005 *Anal Bioanal Chem* **381** 173

[11] Donkó Z 2001 *PHYSICAL REVIEW E* **64** 026401

[12] Belenguer Ph, Kasik M, Michellon C, Pitchford L C 2001 in *Proc 25th Int Conf on Phenomena in Ionized Gases,* 17.–22. July 2001, Nagoya, Japan, 81
[13] Donkó Z, Bánó G, Szalai L, Kutasi K, Rózsa K, Pinheiro M, Pinhão N 1999 *J. Phys. D: Appl. Phys.* **32** 2416
[14] Kasik M, Michellon C, Pitchford L C 2002 *J. Anal. At. Spectrom.* **17** 1398
[15] Bogaerts A, Gijbels R, Serikov V V 2000 *J. Appl. Phys.* **87** 8334
[16] Smith J 1962 *Brit. Journ. Appl. Physics* **13** 399
[17] Ferreira N P, Strauss A, Human H G C 1983 *Spectrochim. Acta Part B* **38** 899
[18] Li L, Millay J T, Turner J P, King F L 2003 *J. Am. Soc. Mass Spectrom.* **15** 87
[19] Wilken L, Hoffmann V, Wetzig K 2003 *J. Anal. At. Spectrom.* **18** 1133
[20] Bengtson A, Eklund A, Lundholm M, Saric A 1990 *J. Anal. At. Spectrom.* **5** 563
[21] Wilken L, Hoffmann V, Uhlemann H-J, Siegel H, Wetzig K 2003 *J. Anal. At. Spectrom.* **18** 646
[22] Bogaerts A, Wilken L, Hoffmann V, Gijbels R, Wetzig K 2001 *Spectrochim. Acta Part B* **56** 551
[23] Bogaerts A, Wilken L, Hoffmann V, Gijbels R, Wetzig K 2002 *Spectrochim. Acta Part B* **57** 109
[24] Bogaerts A, Gijbels R 2000 *J. Anal. At. Spectrom,* **15** 895
[25] Bogaerts A, Gijbels R, Jackson G P 2003 *J. Anal. At. Spectrom.* **18** 533
[26] Efimova V V, Voronov M V, Hoffmann V, Eckert J 2008 *Publ. Astron. Obs. Belgrade* **84** 369
[27] Paschen F 1889 *Ann Phys u Chem* **37** 69
[28] Dogan M, Laqua K, Massmann H 1971 *Spectrochim. Acta Part B* **26** 631
[29] Franz G 1990 *Kalte Plasmen* (Springer Verlag) 226
[30] Marshall K A, Casper T J, Brushwyler K R, Mitchell J C 2003 *J. Anal. At. Spectrom.* **18** 637
[31] Fliegel D 2007 *Thesis* (ETH Zürich)

Advances in the optical diagnostics of kinetic processes in atmospheric pressure dielectric barrier discharges

G Dilecce, P F Ambrico and S De Benedictis

Istituto di Metodologie Inorganiche e dei Plasmi-CNR, sede di Bari, Via Orabona, 4, 70126 Bari - ITALY

E-mail: giorgio.dilecce@ba.imip.cnr.it

Abstract. Dielectric barrier discharges at atmospheric pressure have recently received a growing interest because of their potentially advantageous use in a number of applicative fields, in both their main classes of 'volume' and 'surface' geometric arrangements. From a diagnostic point of view, the high pressure regime adds problems relevant to an enhanced role of collision quenching, vibrational and rotational relaxation, while maintaining many of the demands of a highly non-equilibrium system for a detailed characterization of its degrees of freedom. In addition, filamentary DBDs are intrinsically pulsed systems, in which the electron impact excitation is confined in small space regions and very short time intervals, such that a prominent afterglow phase, both spatial and temporal, is present in the discharge volume, and kinetic processes involving long-lived species, chemi-luminescent reactions, recombinations (aided by the large pressure) can competitively come into play in the excitation of electronic states. We have applied our ensemble of diagnostic methods, based on time resolved emission and laser spectroscopy, to the investigation of elementary kinetics in DBDs. Here we present our last two years results that include: Optical-Optical Double Resonance (OODR)-LIF measurements of $N_2(A^3\Sigma_u^+)$ density, in a volume DBD that is of relevance in the debate upon the mechanisms for establishing the glow regime; OODR-LIF and emission study on a Masuda type surface discharge; measurement of $N_2(C^3\Pi_u$, v) quenching and vibrational relaxation rate constants and its relevance to nitrogen Second Positive System emission diagnostics at atmospheric pressure; kinetics of $CN(B^2\Sigma^+$, v) formation and violet system emission in N_2-CH_4 discharges. The latter issue shows in addition a correlation between emissions and surface status (i.e. the presence of a deposit) that is a clear monitor of a gas-surface interplay.

1. Introduction

Dielectric Barrier Discharges (BDB) at atmospheric pressure (ATP) are discharges in which at least one electrode is covered by a dielectric layer, i.e. it is not in direct contact with the gas. The simple consequence of such an arrangement is that, after the breakdown, charge accumulation on the dielectric surface prevents the corona-to-spark transition and, finally, stops the discharge itself if the voltage does not increase continuously. If an AC voltage is supplied to the electrodes, the discharge starts again when the applied electric field is reversed. As a consequence a non-equilibrium regime with low gas temperatures is intrinsically maintained without the need for complex pulsed HV supplies. Strong non-equilibrium at ATP and simple design make then DBDs very attractive for a wide variety of applications. These range from well established industrial applications like ozone generation, CO_2 lasers, excimer lamps, to promising ones like polymer surface treatment, pollution control, hydrocarbons reforming, plasma assisted combustion, aerodynamic flow control, plasma display panels. It is not our aim in this short paper to report exhaustively on the physics and applications of DBDs: good reviews are largely available in the literature [1, 2, 3, 4, 5]. Rather, since we are interested in the diagnostic investigation of the microscopic kinetics occurring in DBDs, a field that is often neglected in the literature, we will remind and highlight those peculiarities that are of importance for both diagnostics and kinetics.

Various discharge geometries have been realized, that can be divided into two main classes: volume discharges, in which the two electrodes are faced to each other and the discharge burns in between, and surface discharges, in which the electrodes are on (or embedded in) the same dielectric slice and the discharge burns in a thin layer above the surface. In most cases the discharge is made of a group of microdischarges distributed in time and space, and the current looks like that shown in figure 1, in presence of a sinusoidal applied voltage. This is what is commonly referred to as the *filamentary* regime. DBDs' operating frequencies range from 50 Hz to 500 KHz, and operating voltages from few KV to few tens of KV, depending of course on the gas and on the electrodes geometry. Microdischarges properties depend solely on the gas composition, pressure and geometry, and this means that an increase of the discharge power results simply in an increase of the number of microdischarges. A simple geometrical scaling is then allowed, i.e., for example, doubling the electrodes area will result in a doubling of the current/power, provided the external circuit is able to supply sufficient power. We will see that, from the point of view of discharge kinetics, such an idea is over-simplified when dealing with long living species that depend on the history of the discharge, i.e. on how much time/how many cycles the gas mixture has been exposed to the discharge.

Typical parameters of the microdischarges are: duration 1- 20 ns, filament radius 50-100 μm, electron density 10^{14}-10^{15} cm^{-3}, electron temperature 1-10 eV. Filaments then carry large

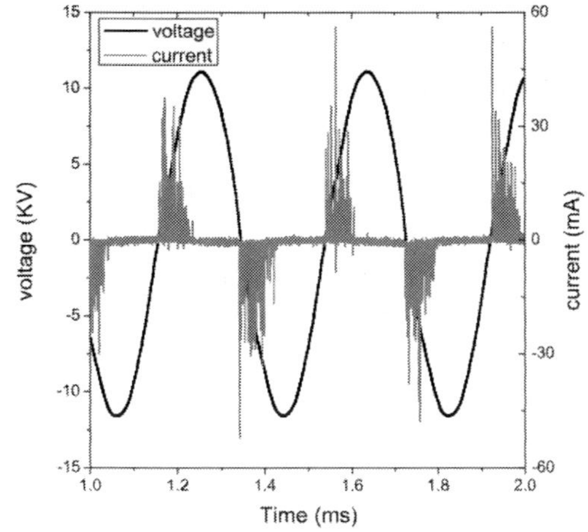

Figure 1. Voltage and current in a filamentary N_2 DBD, with 2 mm electrodes gap. The average current is several times lower than peak microdischarges current

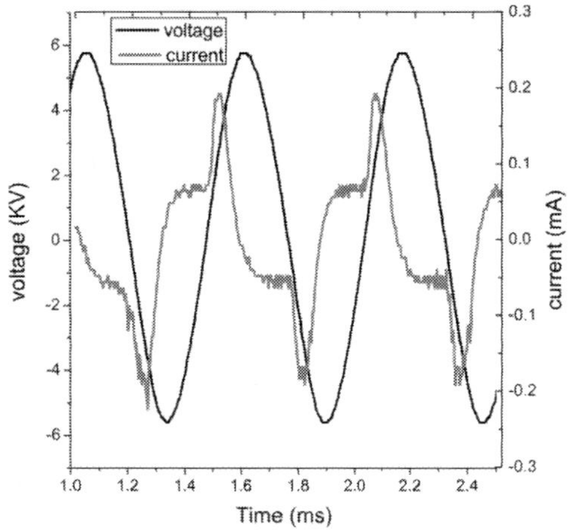

Figure 2. Voltage and current in a N_2 APTD, with 1 mm electrodes gap. The discharge current is superimposed to the electrodes displacement current.

electron energy and density, but they fill a very small fraction of the discharge volume and for a very short time. In other words, in the average volume two phases coexist: short intense and small discharges and large time and space afterglows. Even a time-resolved (gated) observation cannot discriminate between the two, and the measurement will be an average over the two phases. The only case in which a discrimination has been possible is that reported in [6], [7], in which a special geometric arrangement was able to produce a single filament discharge whose emissions were observed by cross-correlation spectroscopy.
Within a restricted range of parameters the DBD can exist in a diffuse mode, that can be an Atmospheric Pressure Glow Discharge (APGD) or an Atmospheric Pressure Townsend Discharge (APTD), the latter being found in nitrogen discharges (see [8] and references therein). The current is continuous, like in figure 2, and the discharge is diffuse. Due to the discharge spatial uniformity, this operating mode is very attractive for surface treatment applications. Its appearance seems to be linked to the survival of a sufficient charge density between two discharge cycles (voltage half-cycles), likely due to the presence of long-lived electronic excited (metastable) states. It is then a case in which microscopic kinetics can determine the macroscopic behavior of the discharge and can have a heavy influence on the applicability of DBDs to material processing. The APGD is still a mixture of temporal discharge and post-discharge regimes, but, contrary to the filamentary case, the two regimes can be discriminated by a gated measurement synchronized to the applied voltage.

2. Experimental approach

Our optical diagnostic approach to the study of discharge kinetics includes a set of time-resolved emission and laser spectroscopy tools:

- *Laser Induced Fluorescence (LIF)*, for the detection of non-emitting transient species, i.e. radicals (ground state) and electronic metastable states. LIF is essentially and absorption measurement, in which the state under observation (lower state) is pumped to an electronic upper state by absorption of laser photons through a resonant transition. The observable quantity is the fluorescence light emitted by the upper state after absorption. We have applied LIF in the following three configurations:
  - (i) LIF excitation spectra - laser wavelength scan, fixed full/partial fluorescence detection). Used for the determination of ro-vibrational distributions of the lower state;
  - (ii) LIF fluorescence spectra - spectrally resolved fluorescence detection at fixed laser wavelength. Used for the determination of collision phenomena in the upper state (electronic quenching and vibrational relaxation);
  - (iii) fixed laser wavelength and fluorescence detection, for analytical purposes, i.e. the determination of relative/absolute concentrations of transient species and their evolution in the discharge/post-discharge cycles;
- *Emission spectroscopy*, for the detection of emitting species, and in particular:
  - (i) high/medium resolution (up to few angstrom bandpass) for the determination of ro-vibrational distributions of the emitting state;
  - (ii) low resolution - single spectral feature (for example a full molecular band) detection, for the monitoring of the relative population of the emitting state and its evolution in the discharge/post-discharge cycles.

The recording of the time evolution of spectroscopically observable quantities is the basis for the application of a method, that we have largely employed in our past research in low-pressure discharges, for the identification of microscopic processes taking place in the discharge. In practice, if we have a reaction:

$$A + B \xrightarrow{k} C + M \tag{1}$$

whose reactants (A and B) and at least one product (C) are observable, the following relationships involving the temporal evolution of A, B and C holds:

$$[C](t) \propto [A](t)[B](t) \tag{2}$$

The verification/negation of such a correlation gives strong indication that reaction (1) is/is not the dominant mechanism for the production of the observable specie C. Note that the absolute concentration of A, B and C needs not to be known, but only their relative temporal evolution is required. A typical application example is the pooling reaction:

$$N_2(A^3\Sigma_u^+, v') + N_2(A^3\Sigma_u^+, v') \xrightarrow{k_2} N_2(C^3\Pi_u, v'') + N_2(X^1\Sigma_g) \tag{3}$$

The $N_2(C)$ state is responsible of the intense Second Positive System (SPS) emission. The nitrogen triplet metastable can be measured by LIF. If SPS evolves in time as the square of the metastable density (LIF signal), reaction (3) holds. More in detail, the time correlation method is used in this case to look for the time interval, typically the deep post-discharge, in which SPS emission is due to reaction (3). Another example is the energy transfer:

$$N_2(A^3\Sigma_u^+, v') + NO(X^2\Pi) \xrightarrow{k_1} NO(A^2\Sigma^+, v'') + N_2(X^1\Sigma_g) \tag{4}$$

whose occurrence can be verified again by comparing the time evolution of NO-γ system emission (from NO(A) state) to N_2(A) (LIF) and NO(X) (LIF) time evolutions.

Peculiarities of spectroscopic diagnostics at atmospheric pressure are tied to an increased influence of collision processes with respect to the low-pressure case (see [9] and references therein), that involve the ro-vibrational manifolds of the electronic states we are looking at. Three collision energy transfers have to be considered:

(i) *Rotational Energy Transfers* (RET). RET collisions at ATP are very fast, such that even the rotational distribution of fast radiative electronic states is a Boltzmann one with $T_{rot}= T_{gas}$. Also LIF processes can be heavily affected by RETs: in the typical duration of a pulsed tunable laser, about 10 ns, hundreds of RET collision occur, affecting the rotational redistribution, and then the saturation condition.

(ii) *Vibrational Energy Transfers* (VET or vibrational relaxation). VET collisions are usually slow in ground states, and normally do not affect ground state distributions of transient species (radicals). They can be fast in electronically excited states, depending on the availability of collision partners with energetically contiguous electronic states. We have encountered two different such cases. The $N_2(C^3\Pi_u$,v) in a N_2 discharge, that shows large vibrational relaxation rates, and the CN($B^2\Sigma^+$, v) in a N_2 + CH_4 discharge, for which VET rates are low.

(iii) *Electronic Quenching.* It is the process by which a collision removes the electron from the excited state manifold toward any other final product. Electronic quenching determines the fraction of the excited state population that can be observed through photon emission. Such a fraction is given by the quantum yield:

$$Y = \frac{A}{A + \sum_i [Q_i] k_i} \tag{5}$$

where A is the Einstein coefficient of the observed transition(s), $[Q_i]$ is the density of the i-specie quencher and k_i is the i-specie quenching rate coefficient. Quantum yields at ATP can be so low as to suppress (i.e. to make it difficult to detect) some emission systems. A classical case is that of nitrogen First Positive System (FPS), that is emitted by $N_2(B^3\Pi_g)$ state. Its quantum yield at ATP is about 10^{-4}, due to a quite low radiative rate (some 10^5 s^{-1}, depending on the band). On the contrary the quantum yield of $N_2(C^3\Pi_u)$ at ATP is about 10^{-2}, by virtue of a two orders of magnitude larger radiative rate. As a consequence, while at low pressure, where electronic quenching is not large (below 1 Torr) both SPS and FPS bands are readily observable, and the color of the discharge is orange, at ATP FPS emission is dramatically reduced, only SPS remains intense, and the color of the discharge turns blue-violet. An analogous case is that of CN emissions, that we have observed in our investigations on nitrogen - methane/benzene DBDs. Two intense systems are observed at low pressure, the violet and the red system, coming from $CN(B^2\Sigma^+)$ and $CN(A^2\Pi_i)$ states respectively. Also in this case, CN(B) state has a roughly two orders of magnitude larger radiative rate (about 1.5×10^7 s^{-1} [10]) than CN(A) state (some 10^5 s^{-1}, depending on the band [11]). At ATP practically only the violet system shows intense emissions.

3. OODR-LIF on $N_2(A^3\Sigma_u^+)$

The classical single-photon LIF scheme for the detection of N_2(A) involves the FPS N_2(B-A) transitions, both for excitation and detection [12]. As we have seen the quantum yield for fluorescence from the B-state is very unfavorable at ATP, and we already noticed in [13] that at pressures above 1 Torr the LIF signal became difficult to detect. We then proposed in [14] an Optical-Optical Double Resonance (OODR) LIF scheme in which the metastable is pumped to C-state by absorption of two photons resonant with FPS and SPS transitions:

$$N_2(A^3\Sigma_u^+, v''=0) + h\nu_{L1} \rightarrow N_2(B^3\Pi_g, v'=3) + h\nu_{L2} \rightarrow$$
$$\rightarrow N_2(C^3\Pi_u, v) \rightarrow N_2(B^3\Pi_g, v') + h\nu_F \qquad (6)$$

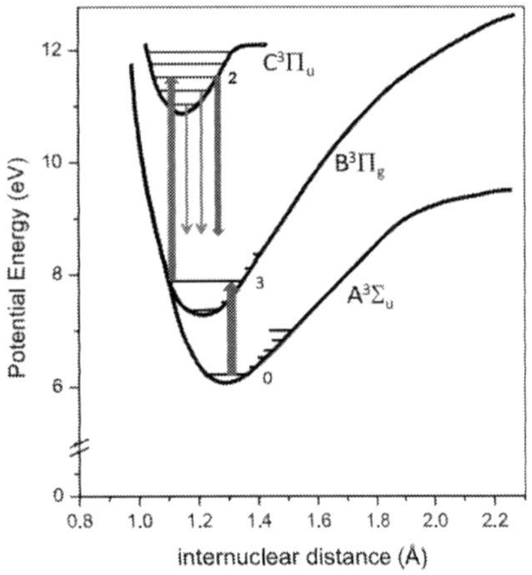

Figure 3. Potential energy diagram of the three nitrogen triplet states involved in the OODR-LIF scheme (6). The red and blue upward arrows indicate the photon absorption on FPS and SPS transitions. In the figure level v=2 is excited, that is the case of the scheme used for N_2(A, v"=0) density measurement. Downward arrows indicate the direct (thick blue) and collision fluorescence (thin blue) (see text).

Such a scheme takes advantage of the two orders of magnitudes larger quantum yield of C-state at ATP. The scheme is also depicted in figure 3. Following the photons absorption, fluorescence is observed from the same excited v-level (direct fluorescence) and from all the vibrational levels lower than v. The latter is the consequence of vibrational relaxation and we will call it collision fluorescence. Direct fluorescence has been used for both electronic quenching rate constants measurement and for N_2(A) OODR-LIF density measurements, while collision fluorescence has been used for vibrational relaxation rate constants measurements.

3.1. Collisions in the (C, v) manifold

The Second Positive System of N_2 is frequently used for optical diagnostics of gas discharges. In particular it can be used to infer the electron temperature T_e and the ground state $N_2(X)$ vibrational temperature, T_v, and it has often been used for modeling tests of N_2 discharge kinetics (see [15] and references therein). Discharge parameters are inferred from the (C, v) measured vibrational distributions in a condition in which C state is excited predominantly by electron impact (i.e. in the discharge), whose collision cross sections are well known. Relative (C, v) populations can then be calculated with T_e and T_v as parameters that can be varied until agreement is found with the experimental (C, v) distribution. At ATP the knowledge of the rate constants of all collision phenomena in the (C, v) manifold is of fundamental importance to this end. OODR pumping of all the v levels of C-state gave us the opportunity to measure all the electronic quenching and vibrational relaxation rate constants from the analysis of the direct and collision fluorescence pulses [16]. It is important to point out that this was the first complete measurement obtained by selective excitation of single v-levels. Previous measurements where limited to electronic quenching only, and obtained by non-selective excitation of the whole (C, v) manifold, while no reliable data where available for vibrational relaxation (see [16] and references therein). The results of [16] were based on Stern-Volmer linear plots in the pressure range 2-35 Torr, and the quenching of levels v$\neq$0 were measured after direct laser pumping of the levels. About the quenching of level v=0, following literature data that reported a non-linear pressure dependence of the quenching rate at low pressures (0.01 - 2 Torr), we investigated this issue by direct laser pumping of level v=0 [17]. We found that the quenching rate K(P) was not a linear function, but could be described by the empirical function:

$$K(P) = A_0 + k_\infty P - (k_0 - k_\infty)\, P/\left(1 + P/P_0\right) \tag{7}$$

with A_0=(2.7298±0.0116)×10^7 s^{-1} (the radiative rate); $k_0 = (2.22 \pm 0.56) \times 10^6$ $Torr^{-1}s^{-1}$; $k_\infty = (4.144 \pm 0.046) \times 10^5$ $Torr^{-1}s^{-1}$; $P_0 = (0.87 \pm 0.26)$ Torr. In the limits P$\rightarrow$0 and P$\rightarrow \infty$ (in practice P$\ll P_0$ and P$\gg P_0$) equation (7) becomes, respectively:

$$K(P) = A_0 + k_0 P \tag{8}$$

$$K(P) = A_0 + (k_0 - k_\infty)\, P_0 + k_\infty P \tag{9}$$

In practice, then, for pressures above few Torr, a linear dependence on P is empirically valid, with a rate constant equal to k_∞ and an 'effective' radiative rate given by $A_0 + (k_0 - k_\infty)\, P_0$. We did not extend this study to the other v$\geq$ 1, but, if a non-linear dependence analogous to the v=0 case occurs, the Stern-Volmer plots in the range 2-35 Torr reasonably represent the P$\rightarrow \infty$ situation, with the radiative rates equal to the 'effective' one defined above.

Table 1. Recommended rate constants for vibrational relaxation, k_{vw} and total collision quenching k_v^Q of $N_2(C^3\Pi_u$, v) by collision with N_2. Rate constants are in units $10^{-11}cm^3s^{-1}$. Empirical radiative rates A_v are also reported (in $10^7 s^{-1}$) (see text).

w/v	0	1	2	3	4
0	-	1.19±0.13	0.528±0.043	0.734±0.072	0.736±0.067
1	-	-	0.763±0.094	0.464±0.059	1.02±0.093
2	-	-	-	0.360±0.034	0.285±0.036
3	-	-	-	-	0.105±0.037
k_v^Q	1.24±0.06	3.14±0.21	4.28±0.21	6.34±0.27	9.86±0.46
A_v	2.91±0.015	2.67±0.02	2.72±0.03	2.62±0.03	2.36±0.07

We then report in table 1 the rate constants and radiative rates that, according to the previous discussion, can be empirically used for the calculation of the quenching rate by a linear law above a pressure of a few Torr.
Using the complete set of rate coefficients we were able in [15] to fit experimental (C, v) distributions measured in a DBD at ATP, with far greater precision than it was possible with old and incomplete data sets. We also showed that neglecting the vibrational relaxation led to a large overestimation of the electron temperature, and that the fit accuracy achieved by the new set allowed a quasi-independent determination of T_e and T_v. Note that the T_v that can be determined by SPS diagnostics is practically only the first level temperature, if the ground state N_2(X, v) excitation is not large, as it is the case at ATP.

3.2. $N_2(A^3\Sigma_u^+)$ density measurement

The pre-ionization source that causes the establishment of the APTD mode in a nitrogen DBD is believed to be secondary electron emission by N_2(A) metastable impact on the dielectric surface [8]. This idea is based on two facts: a) the strong structural similarity of the APTD - the momentary cathode acts as a source of secondary electrons; b) the oxygen argument - in presence of even small quantities of oxygen, which is known to be a strong N_2(A) quencher, the discharge turns to the filamentary mode. A measurement of N_2(A) density in a N_2, N_2 + small O_2 admixtures can then be of help in the understanding of this issue. In [18] we applied the OODR-LIF technique to this end. Absolute density calculations from LIF outcomes is normally a difficult task. In the case of OODF-LIF at ATP it becomes overwhelmingly difficult. Critical points for a modeling of the OODR process are:

(i) multi J-levels excitation. More than one ro-vibronic transition is involved in the laser absorption, so that an accurate knowledge of the lasers spectral profiles is required. A further complication is the non-negligible lorentzian pressure line broadening at ATP.
(ii) account for RET collisions in the B-state manifold. With the RET rate coefficient of 10^8 Torr s^{-1}, 760 RET collisions take place within the 10 ns laser pulse duration
(iii) RET collisions in the A-state manifold influence the saturation conditions, and should be taken into account as well.

We then applied a calibration method to a selected condition, and then measured the metastable density by the calibrated OODR-LIF in all the desired cases. The selected condition was the post-discharge of an APTD in pure N_2, in which we proved that reactions (3) and (4) hold simultaneously. It was then possible to apply the method proposed in [19], by which the metastable density is proportional to the ratio of SPS and NO-γ emission intensities and to the NO density. The proportionality constant contains various terms, among which the ratio k_1/k_2 of the rate coefficients of reactions (3) and (4). We introduced two major improvements with respect to [19]. First, NO density was measured by LIF instead of being presumed on the basis of known small NO additions to the gas feed. Second, we adopted an 'effective' k_2 value that takes into account the strong vibrational relaxation in the calculation of N_2(C) excitation by the pooling reaction.
By applying this strategy we found a metastable density of the order of 10^{13} cm^{-3} in the pure nitrogen APTD, with a population of N_2(A, v=1) at most 15% that of N_2(A, v=0). Addition of small oxygen concentrations to the discharge was observed to induce a transition to the filamentary regime, starting from 250 ppm of oxygen, that appears not to be caused by a marked decrease of the metastable density in the discharge, but rather by a considerable increase of its quenching rate. Such an increase, due to collision quenching by O_2 and O, strongly reduces the survival of the metastable between two discharge events in the voltage cycle, and reduces its ability to reach the surface as well. These observation are consistent with the idea that

the APTD regime is sustained by a space charge memory effect due to the nitrogen triplet metastable, as outlined before.

We also attempted to apply OODR-LIF measurements of the nitrogen triplet metastable in a Masuda-type surface discharge [20]. Surface discharges develop in a thin (order of 100 μm) layer above the surface. The laser diagnostic cannot then, for geometrical reasons, probe the discharge layer. We then measured the N_2(A) density at 1 mm from the discharge, in a so-called space afterglow. In addition, the calibration method described above could not be applied since it is based on the measurement of emissions that come instead from the discharge, i.e. from a volume different from that probed by the laser. We then applied the OODR-LIF calibrated in the volume discharge as above. Finally the metastable density in the afterglow was traced back to the discharge by estimations based on NO-γ emissions, that, also in this case, are due to reaction (4). We found a metastable density of the order of 10^{12} cm^{-3}, i.e. one order of magnitude lower than in the volume discharge. Such a lower density might be explained by larger diffusion losses due to the vicinity of the production zone (the discharge) to the surface.

4. CN($B^2\Sigma^+$) formation and emission in a N_2-CH_4 discharge

Dielectric Barrier Discharges are good candidates for applications of non-equilibrium plasmas to hydrogen production from hydrocarbons, liquefaction of gaseous fuels, CO_2 reforming (removal). Within this framework we approached fundamental studies in a N_2-C_6H_6 [21] and N_2-CH_4 DBDs [22], and finding also the formation of a liquid deposit in nitrogen(argon)-methane discharges [23]. The CN(B-X) violet system is, as we have seen, the only prominent emission in atmospheric pressure discharge in nitrogen plus small amounts of hydrocarbons. Its intensity is comparable to that of most SPS bands, and can be readily observed with a non-cooled not-intensified CCD. In our DBD and in both mixtures with benzene and methane we found that the violet system emission shows an unusually large vibrational excitation of CN(B, v) state, that is not normally found in low-pressure discharges. Furthermore, the violet system emission was still present and with even more pronounced vibrational excitation in pure nitrogen fed discharges operated with the dielectric surfaces covered by the brownish, dusty deposit produced after operation with N_2-CH_4 feed gas. The production of CN(B) then showed peculiarities typical of chemi-luminescent/recombination reactions involving the interplay between the gas-phase and the surface deposits, whose understanding requires a complex diagnostic approach. We addressed this study in a pulsed dielectric barrier discharge (a packet of sinusoidal applied voltage followed by a period of null voltage) by combining the results of the following, time resolved diagnostics [22]:

(i) LIF on CN(X), to explore the vibrational excitation of the formed cyano radical, to address the issue of vibrational relaxation in the CN(B, v) manifold and to follow the CN density time evolution in the discharge cycles;

(ii) OODR-LIF on N_2(A), to investigate a possible role of the nitrogen triplet metastable in the CN(B) excitation;

(iii) medium spectral resolution spectroscopy, by gated intensified CCD, to look at the vibrational excitation of CN(B, v) in the discharge and post-discharge at different gas feed compositions;

(iv) low resolution spectroscopy, to follow the time evolution of spectral features in the discharge cycles and correlate it to the time evolution of LIF outcomes.

Both CN(X, v) and CN(B, v) showed a highly supra-thermal vibrational excitation, that is largely more pronounced in the pure N_2 case than in the presence of CH_4 in the gas feed (see figure 4 for the violet system emission case and figure 5 for the CN(B, v) vibrational distributions deduced from the spectra). The vibrational relaxation in both manifolds was found to be weak.

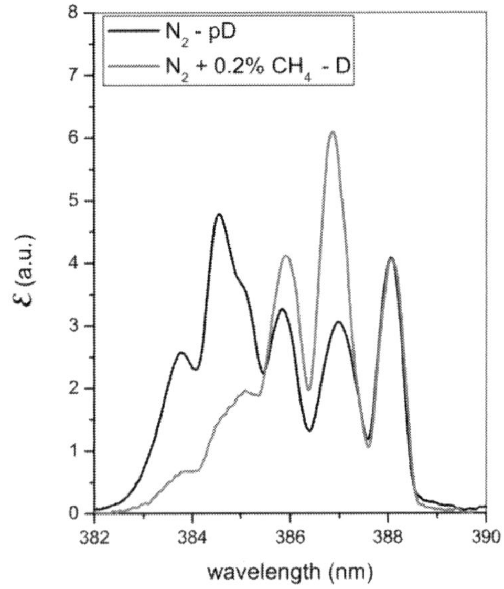

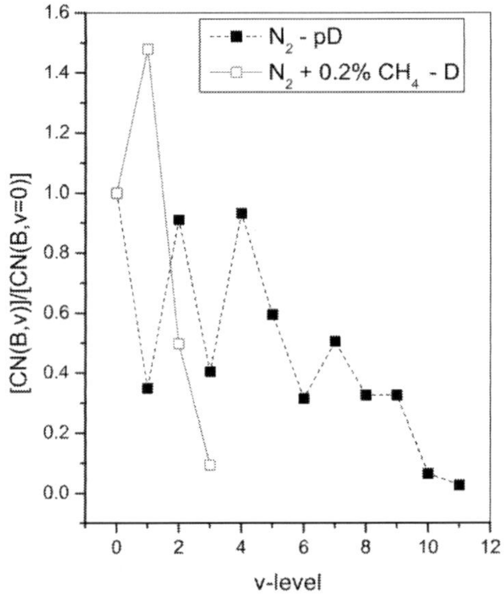

Figure 4. The Δv=0 sequence of CN violet system in pure N_2(+wall deposit) post-discharge (pD) and in N_2+0.2% CH_4 discharge (D).

Figure 5. Vibrational distributions deduced by fitting the spectra of figure 4 [22]. The distributions are normalized to the population of level v=0

The measured vibrational excitation can then be considered to be a 'quasi nascent' one, and is then related to the formation mechanisms. Finally a possible role of $N_2(A)$ in the CN(B) excitation could be safely excluded. The combination of all the results converged toward an interpretation involving the presence of two formation mechanisms, one of which is operative preferentially when the discharge is ON, the other being always operative in the voltage packet, and that can be isolated from the other mechanism in the post-discharge. We were able to assign quite safely the post-discharge mechanism to the recombination reaction:

$$C(^3P) + N(^4S) + M \rightarrow CN(B^2\Sigma^+) + M \qquad (10)$$

and the discharge mechanism likely to the reaction:

$$N + CH \rightarrow CN(B, A) + H \qquad (11)$$

C and CH come someway from the surface in the pure N_2 gas feed case, while they can be created, alternatively or in addition, in the gas phase when CH_4 is added to the gas mixture. Both mechanisms were found to reach a steady-state after a considerable number of cycles, this number being larger (i.e. slower kinetics) in the N_2 case than in presence of CH_4 in the gas mixture. The time required for steady-state clearly depends on how the species involved are created, being they extracted someway from the surface or produced in the gas phase. CH, for example, is produced in presence of methane in the discharge mainly by a stepwise process that forms CH_3, CH_2 and finally CH by successive electron impact events. Such an occurrence is a clear case in which the discharge kinetics can be influenced by the geometry of the discharge and by the gas flow (that determines the time/number-of-cycles the gas mixture is exposed to the active discharge), i.e. a case in which a simple geometrical scaling of the DBD is not correct.

References

[1] Kogelschatz U, Eliasson B and Egli W 1999 *Pure Appl. Chem.* **71** 1819–1828

[2] Napartovich A P 2001 *Plasmas Polym.* **6** 1 – 14

[3] Pietsch G J 2001 *Contrib. Plasma Phys.* **41** 620 – 628

[4] Wagner H E, Brandenburg R, Kozlov K V, Sonnenfeld A, Michel P and Behnke J F 2003 *Vacuum* **71** 417 – 436

[5] Fridman A, Chirokov A and Gutsol A 2005 *J. Phys. D: Appl. Phys.* **38** R1 – R24

[6] Wagner H E, Yurgelenas Y V and Brandenburg R 2005 *Plasma Phys. Controlled Fusion* **47** B641 – B654

[7] Kozlov K V, Brandenburg R, Wagner H E, Morozov A M and Michel P 2005 *J. Phys. D: Appl. Phys.* **38** 518 – 529

[8] Massines F, Segur P, Gherardi N, Khamphan C and Ricard A 2003 *Surf. Coat. Technol.* **174-175** 8–14

[9] De Benedictis S and Dilecce G 2002 *Pure Appl. Chem.* **74** 317

[10] Singh P D, Gomez Balboa A M and De Freitas Pacheco J A 1985 *Astrophys. and Space Sci.* **108** 153–160

[11] Halpern J B, Huang Y and Titarchuk T 1996 *Astrophys. and Space Sci.* **236** 11–17

[12] De Benedictis S and Dilecce G 1997 *J. Chem. Phys.* **107** 6219–6229

[13] De Benedictis S and Dilecce G 1993 *Chem. Phys.* **178** 547–560

[14] Dilecce G, Ambrico P F and De Benedictis S 2005 *Plasma Sources Sci. Technol.* **14** 561 – 565

[15] Dilecce G, Ambrico P and De Benedictis S 2007 *Plasma Sources Sci. Technol.* **16** S45–S51

[16] Dilecce G, Ambrico P F and De Benedictis S 2006 *Chem. Phys. Lett.* **431** 241 – 246

[17] Dilecce G, Ambrico P F and De Benedictis S 2007 *Chem. Phys. Lett.* **444** 39 – 43

[18] Dilecce G, Ambrico P F and De Benedictis S 2007 *Plasma Sources Sci. Technol.* **16** 511 – 522

[19] Simek M 2003 *Plasma Sources Sci. Technol.* **12** 421–431

[20] Ambrico P F, Simek M, Dilecce G and De Benedictis S 2008 *Plasma Chem. Plasma Process.* **28** 299–316

[21] Franceschi P, Guella G, Scarduelli G, Tosi P, Dilecce G and De Benedictis S 2007 *Plasma Process. Polym.* **4** 548 – 555

[22] Dilecce G, Ambrico P F, Scarduelli G, Tosi P and De Benedictis S 2008 *Plasma Sources Sci. Technol.* in press

[23] Scarduelli G, Guella G, Mancini I, Dilecce G, De Benedictis S and Tosi P 2008 *Plasma Process. Polym.* DOI: 10.1002/ppap.200800096

Smart Coating Technology by Gas Tunnel Type Plasma Spraying

Akira Kobayashi

Joining and Welding Research Institute, Osaka University
11-1 Mihogaoka, Ibaraki, Osaka 567-0047, Japan

Email:kobayasi@jwri.osaka-u.ac.jp

Abstract. Nano-science & technology is one of the most important scientific fields, and the material processing using the nano-technology is now advanced towards more precise and controllable smart stage. Regarding thermal processing, plasma system with high precise, has been expected for smart thermal processing. The gas tunnel type plasma system developed by the author exhibits high energy density and also high efficiency. Among the applications to the various thermal processing, one practical application is plasma spraying of ceramics such as Al_2O_3 and ZrO_2. The characteristics of these ceramic coatings were superior to the conventional ones. The ZrO_2 composite coating has the possibility of the development of high functionally graded TBC (thermal barrier coating). In this study, the performance such as the mechanical properties, thermal behavior and high temperature oxidation resistance of the alumina/zirconia functionally graded TBCs produced by gas tunnel type plasma spraying was investigated and discussed. The results showed that the alumina/zirconia composite system exhibited the improvement of mechanical properties and oxidation resistance. Now, one of the advanced plasma application, a smart coating technology, is expected to obtain the desired characteristics of ceramics with improved corrosion resistance, thermal resistance, and wear resistance.

1. Introduction

In order to apply Nano-science & technology to Material Science, the material processing should be developed towards more precise and controllable smart processing stage. Regarding an applicable heat source, the most important key should be the performance of the applied heat source. A plasma is one of the most superior heat sources, because of high temperature, high energy density, easy controllable, etc. Therefore more precise plasma system has been expected in order to establish a smart thermal processing.

The gas tunnel type plasma system developed by the author has high energy density and also high efficiency [1-3]. The outline of this plasma system and the applications to the various thermal processing are described briefly in the following chapters. One of typical applications is plasma spraying of ceramics such as Al_2O_3 and ZrO_2 [4]. The characteristics of these ceramic coatings by the gas tunnel type plasma spraying were superior to that by the conventional ones. Usually, the Vickers hardness of this sprayed coating became 20-30% higher than that of conventional plasma sprayed coating. And, the porosity was only half of the value of the conventional ones [5].

Plasma sprayed zirconia coatings are widely used as thermal barrier coatings (TBC) for high temperature protection of metallic components in the hot section of gas turbine such as burners, transition ducts, vanes and blades. It allows the high temperature operation and results to the increase in the efficiency of the engine and the durability of the critical gas turbine components. However, their use in diesel engine combustion chamber components has been quite rare, because of the durability problems in such conditions. The main problem is the TBC spallation at the interface between the coating and substrate due to the interface oxidation [6]. Although zirconia coatings have been used in many applications, the interface spallation problem is still waiting to be solved under the extreme conditions such as high temperature cycle and high corrosion environments. For that reason, there have been many investigations in developing high quality TBCs for diesel engines [7,8].

For TBC, the spalling of the coating is also very important problem as well as the coating quality. It is reported that the spallation rate can be relatively reduced by using suitable bond coat for the interface [9]. Nevertheless, it is difficult to find suitable bonding layers for all kind of substrate material. Alumina / zirconia composite coating was proposed as a potential candidate to improve the properties of thermal barrier coating system due to alumina's low melting point and high hardness. Also, extremely high porosity values (up to 25vol%) of TBC s have been obtained by functionally graded layer of alumina. TBC failure occurs easily at the interface between the metallic bondcoat and topcoat. During high temperature service an oxide scale consisting mainly of α-alumina forms along bond/topcoat interface.

The resistances for thermal shock and high-temperature corrosion are important properties of TBC. While the large porosity and the high melting point is advantage of ZrO_2 coating, the porosity has disadvantage for the adoption under the critical conditions such as high temperature and high corrosion environment. New type plasma spray methods are expected for using the excellent characteristics of ceramics such as corrosion resistance, thermal resistance, and wear resistance [10] by reducing the porosity and increasing the coating density.

Now, a high hardness ceramic coating could be obtained by means of the gas tunnel type plasma spraying, which were investigated in the previous study in detail [5,11,12,13]. The Vickers hardness of the zirconia (ZrO_2) coating was increased with decreasing spraying distance, and a higher Vickers hardness could be obtained at a shorter spraying distance. At L=30 mm, when P=33 kW, the Vickers hardness of ZrO_2 coating was about Hv=1200 [14], which corresponds to the hardness of sintered ZrO_2 (Hv=1,300). ZrO_2 coating formed has a high hardness layer at the surface side, which shows the graded functionality of hardness [15,16]. With the increase in the traverse number of plasma spraying, the hardness distribution was much smoother, corresponding to the result that the coating became denser. For TBC, the spalling of the coating is also very important problem as well as the coating quality.

In this study, the mechanical properties, thermal behavior and high temperature oxidation resistance, of high hardness ZrO_2 composite coating were investigated and the merit as TBC (thermal barrier coating) was clarified. The effect of alumina Al_2O_3 mixing ratio on the Vickers hardness of the ZrO_2 composite coating was also clarified in order to develop high functionally graded TBC. Moreover the adhesive characteristics of such high hardness zirconia-alumina (ZrO_2-Al_2O_3) composite coatings were investigated as well as its mechanical properties. Especially, the influence on the thickness of the zirconia composite coating was discussed. Moreover, the corrosion potential and deactivated corrosion current density are measured and analyzed corresponding to the microstructure of the zirconia composite coatings. The corrosion resistance Al_2O_3/ZrO_2 was evaluated as a function of Al_2O_3 percentage and coating thickness.

2. Gas Tunnel Type Plasma System

The schematic of gas tunnel type plasma torch developed by the author is shown in Figure 1. The working gas makes a strong vortex flow in the chamber, and forms low pressure gas tunnel along the torch center axis. This makes plasma production easier, and the strong vortex constricts and stabilizes the plasma jet. The feature of gas tunnel type plasma is shown in Table 1 as compared to the conventional ones. The gas tunnel type plasma system has high energy density and also high efficiency. [1,2,3]

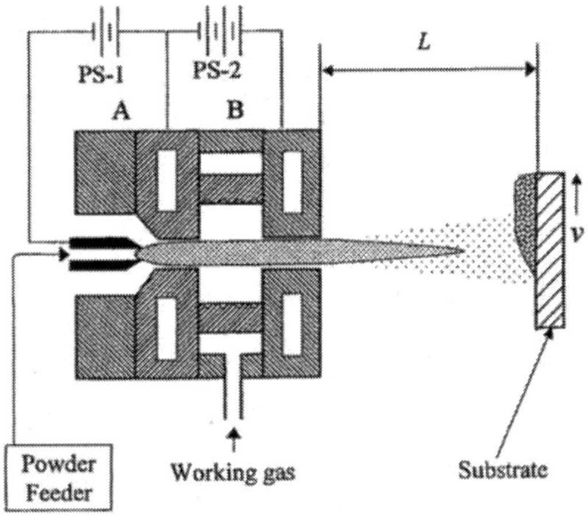

Figure 1. Schematic of the gas tunnel type plasma spraying torch.

Table 1. Comparison between gas tunnel type plasma jet and conventional ones.

	Gas tunnel type plasma jet	*conventional ones*
Temperature	15000 K	10000 K
Energy density	10^5 W/cm^2	10^4 W/cm^2
Heat efficiency	80%	50%

One example of application of the gas tunnel type plasma is the thermal spraying. Figure 1 shows the gas tunnel type plasma spraying torch. The experimental method to produce the high hardness ceramic coatings by means of the gas tunnel type plasma spraying have been described in the previous papers [4,5,11,12,13].

The spraying powder is fed inside plasma flame in axial direction from center electrode of plasma gun. So, the spraying powder flying around central axis was molten enough in the plasma, and the plasma spraying for high melting point ceramics is available.

This plasma system has many possibilities for the industrial applications to the various thermal processing, such as plasma spraying, surface modification. The typical applications are:

(1) Plasma spraying of ceramics (Al_2O_3 and ZrO_2 etc.)
(2) Surface modification of Ti materials (Nitridation)
(3) Other Applications such as nano-science, functional materials processing technology
(4) Application to environmental problems, such as CO_2 decomposition by plasma jet.

Moreover, the development of new type of smart plasma system is planned in order to apply to thermal processing of materials and so on.

3. Gas Tunnel Type Plasma Spraying

3.1. Characteristics of gas tunnel type plasma spraying

The gas tunnel type plasma spraying can make high quality ceramic coatings compared to conventional plasma spraying systems. The gas tunnel type plasma spraying torch [4, 5,11,12,13]is shown in Figure 1. The coating is formed on the substrate traversed at the spraying distance: L. In this case, the gas divertor nozzle diameter was d=20 mm. Table 2 shows the properties (hardness, porosity, etc.) of the Al_2O_3 coating by gas tunnel type plasma spraying [5,11]. The hardness was similar to sintered alumina: Hv=1,200 and high density, porosity was half of the value of the conventional ones. Even when the working gas is argon and low input of 20 kW, we can obtain enough high Vickers hardness of Hv=800.

Thus it can be easy to produce the high hardness ceramic coatings by means of the gas tunnel type plasma spraying.

Table 2. Comparison between gas tunnel type plasma spraying and conventional type for Al_2O_3 coating.
(Input = 45kW、Distance = 65-100mm)

	Gas tunnel type plasma spraying	*Conventional ones*
Vickers hardness	1200	800
Porosity	10%	20%

3.2. Experimental procedures

The gas tunnel type plasma spraying torch used was shown in **Figure 1**. The experimental method to produce the ceramic coatings by means of the gas tunnel type plasma spraying is as follows. After igniting plasma gun, the main vortex plasma jet is produced in the low pressure gas tunnel. The spraying powder is fed from center inlet of plasma gun. The coating was formed on the substrate traversed at the spraying distance of L.

Table 3. Experimental conditions.

Powder:	ZrO2 + Al2O3 Mixture
Traverse number: N	1~30
Power input, P (kW):	25~28
Working gas flow rate, Q (l/min): 180	
Powder feed gas, $Q\,feed$ (l/min):	10
Spraying distance, L (mm):	40
Traverse speed, v (cm/min):	25~1000
Powder feed rate: w (g/min):	20~35
Gas divertor nozzle dia., d (mm)	20

The experimental parameters for the plasma spraying are shown in **Table 3**. The power input to the plasma torch was about P=25 kW, and the power input to the pilot plasma torch, which was supplied by the power supply PS-1, was turned off after starting of the gas tunnel type plasma jet. The spraying distance was short distance of L= 40 mm

The working gas was Ar gas, and the flow rate for gas tunnel type plasma spraying torch was Q=180 l/min, and gas flow rate of carrier gas was 10 l/min. The powder feed rate of zirconia/alumina mixed powder was w=20~35g /min. The traverse speed of the substrate was changed the value from v=25 to 1000 cm/min. Also the traverse number was changed 1-30 times (: high speed traverse of v=1000cm/min, 30times).

The chemical composition and the particle size of Zirconia (ZrO_2) and/or alumina (Al_2O_3)powder used in this study was respectively shown in Table 4. This ZrO_2 powder was commercially prepared type of K-90 (PSZ of 8% Y_2O_3), and Al_2O_3 powder was the type of K-16T. The substrate was SUS304 stainless steel (3x50x50), which was sand-blasted prior to plasma spray.

Table 4. Chemical composition and size of zirconia and alumina powder used. (20~80% Al_2O_3 Mixture)

	Composition (wt%)					Size (μm)
ZrO2	ZrO_2	Y_2O_3	Al_2O_3	SiO_2	Fe_2O_3	
	90.78	8.15	0.38	0.20	0.11	10-44
Al2O3	Al_2O_3	Na_2O	SiO_2	Fe_2O_3		
	99.8	0.146	0.01	0.01		10-35

3.3. Analysis of Coating Properties

The microstructure of the cross section of zirconia composite coating was observed by an optical microscope in this research. The microscope is equipped with a CCD camera for image acquisition. Micrographs with two magnifications (200 X and 400 X) taken on polished cross sections are used for determining the total porosity and coating thickness by using image analysis software. The microstructure of the cross section of zirconia composite coating was observed by an optical microscope.

The Vickers hardness Hv_{50}, Hv_{100} of the sprayed coatings was measured at the non-pore region in those cross sections under the condition that the load weight was 50g, 100 g and its load time was 15s, 25 s. The Vickers hardness: Hv_{100} was calculated as a mean value of 10 point measurements. The distribution of the Vickers hardness in the cross section of the coating was measured at fixed distance from the coating surface in the thickness direction.

The adhesive strength between the ZrO_2 composite coating and the substrate was measured by using the tension tester original designed. The test piece for adhesive strength was 10mm square and the coating surface side and substrate side was respectively attached to each holder by polymer type glue. The load for the tester could be changed 0~200kg. The kgf/cm^2 was used as a unit for the adhesive strength of the composite coating. The adhesive strength of the ZrO_2 composite coatings was mainly measured in the case of different coating thickness.

The schematic of anodic polarization corrosion system is shown in Figure 2, which is a normal potentiostatic polarization corrosion tester, which is using a Hokuto Denko, HA303 power source. An Ag/AgCl reference electrode (SCE) was inserted in saturated KCl solution and was connected galvanically to the test cell by a self-made salt-bridge. A platinum wire used as the counter electrode was immersed in the reaction cell containing 500ml corrosion media of 0.5M HCl solution. HCl solution was chosen as corrosion media because Cl^- ions are assumed passing through the coating layer more easily than another commonly-used anodic oxidant SO_4.

The sample surfaces were firstly degreased by ultrasonic process in acetone for 5 minutes then were washed by distilled water before putting into test. The cleaned sample was held by a well-designed sample holder and immersed in the testing media for 15 min to stabilize its galvanical contact with the solution, then the sample potential was set to -0.5V and was swept to +0.5V at a rate of 10mV/s. All the tests were carried out at room temperature.

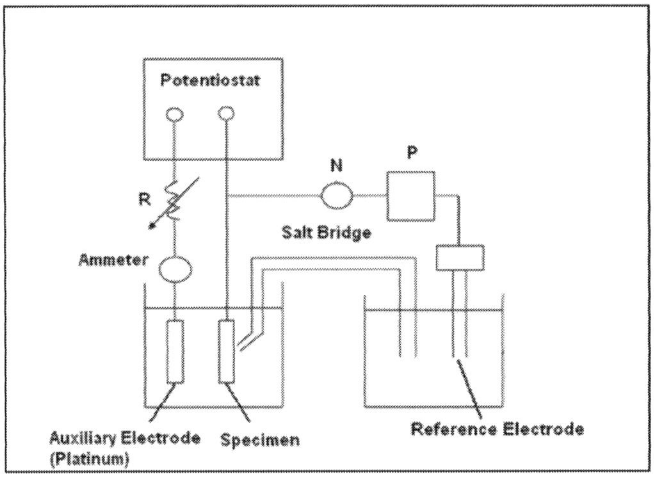

Figure 2. Anodic Polarization Corrosion Tester.

4. Results and Discussion

4.1. Microstructure and Vickers hardness of zirconia composite coating

Typical optical cross sectional micrographs for thermal barrier coatings are shown in Figure 3. Those are the zirconia composite coatings of 20% and 50%Al_2O_3 mixture, respectively. The coatings are porous and lamellar structure which is typical characteristic for as-sprayed coatings. The thickness was about 150μm.

The composition of the microstructure is represented by gray level variation. It consisted of 2 different layers, white and gray layers were deposited alternatively. The analysis by EPMA revealed that white was zirconia (ZrO_2) and gray was alumina (Al_2O_3). Pores appear to be dark, which permit them to be distinguished and quantified by image analysis.

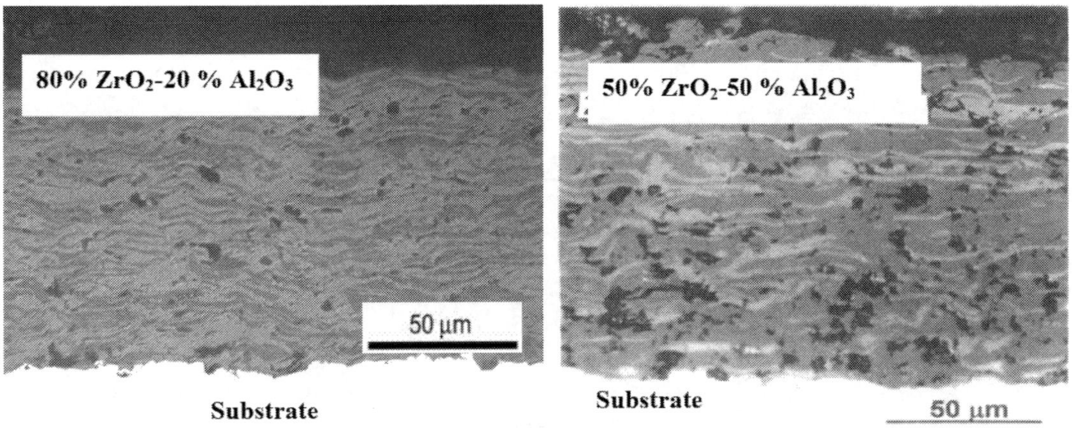

Figure 3. Micrographs of the cross-section of coating samples.

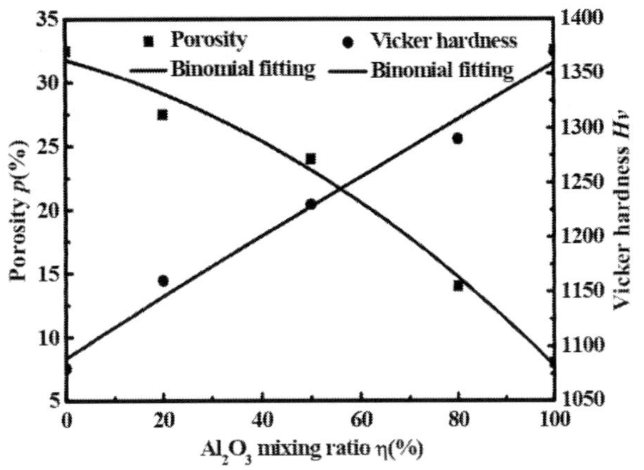

Figure 4. Dependence of Vickers hardness of zirconia composite coating on the alumina mixing rate. 2 times traverse at L=40mm when P=25kW.

4.2. Effect of alumina mixing ratio on the Vickers hardness of zirconia composite coating

Figure 4 shows the relationships between Vickers hardness and porosity of the ZrO_2 composite coatings and the alumina Al_2O_3 mixing ratio R (wt%), at the same spraying time. In this case, the coating thickness was approximately 200 μm at P=25 kW, L=40 mm, when the traverse number was two times.

The average Vickers hardness over the cross section of ZrO_2 composite coatings is increased with the increase in the Al_2O_3 mixing ratio. The increment of coating hardness corresponds to attendance of alumina particles with hardness higher than that of ZrO_2. The Vickers hardness of Al_2O_3 coating was Hv_{50}=1360.

The hardness distribution of the ZrO_2composite coating has remarkable graded functionality in the case of large Al_2O_3 mixing ratio. Because, the part near the substrate did not change so much, but the Vickers hardness near the coating surface became much higher. This leads to the development of a high functionally TBC.

The average porosity over the coating layer shows a decrease tendency with increasing alumina mixing ratio. In the meanwhile, the porosity profile (shown in Figure 6) over the coating cross-section gives an almost linearly graded distribution.

4.3. Graded Functionality of zirconia composite coating

The hardness distribution of the ZrO_2composite coating has remarkable graded functionality in the case of large Al_2O_3 mixing ratio. Because, the part near the substrate did not change so much, but the Vickers hardness near the coating surface became much higher.

Figure 5 shows the distribution of Vickers hardness: Hv_{50} of the zirconia/alumina composite coating shown in Figure 3 (coating thickness: about 150μm). The distribution of this composite coating has a highest value in the coating at the surface side: The maximum hardness was near to Hv_{50} = 1300 at the the coating surface of l=40 μm, and decreased linearly like towards the substrate side.

While, the porosity profile over the coating cross-section (Figure 6) gives an almost linearly graded distribution, increasing from the surface of the coatings towards the surface of the substrate. In as-sprayed condition the porosity variation ranges from 18.95% to 33.23% from the surface of the coatings to the surface of the substrate. Although lower porosity can increase the average hardness of the coatings, alumina prsent in the coatings is the origin of the improved hardness because higher mixing ratio of alumina results in lower porosity.

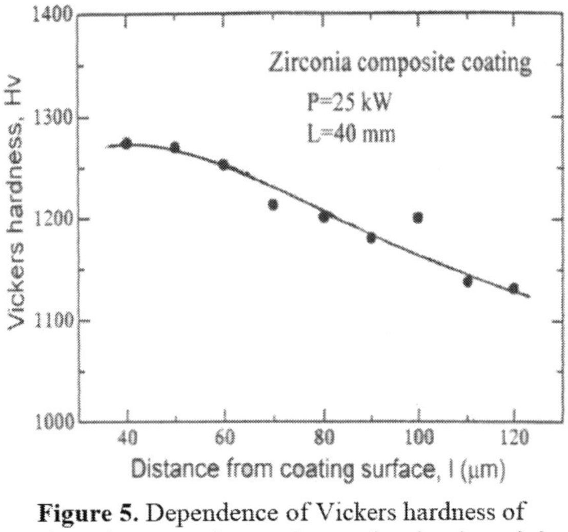

Figure 5. Dependence of Vickers hardness of zirconia composite coating on the alumina mixing rate. 2 times traverse at L=40mm when P=25kW.

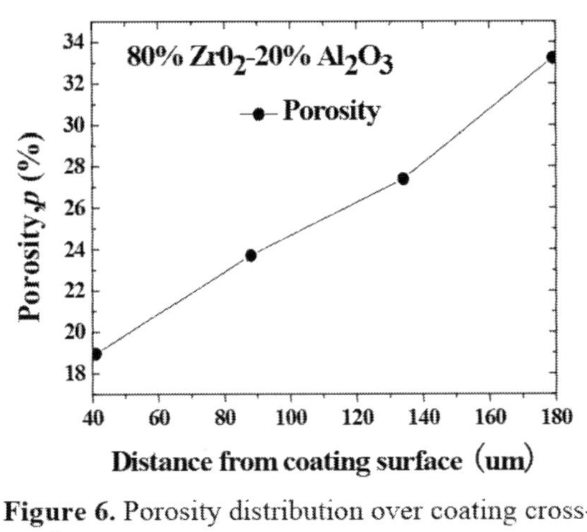

Figure 6. Porosity distribution over coating cross-section.

4.4. Influence of plasma thermal process on the coating

For an increase in the traverse number, the surface temperature of the coating during spraying became higher. Therefore it would be expected that coating density would be increased when the traverse number increases. The maximum Vickers hardness of ZrO_2 composite coating was almost the same when the coating thickness was the same. But the graded functionality became much better, and the distribution of Vickers hardness was much smoother as the traverse number was increased. This means that the structure at the surface of the coating was denser by the thermal process of the high energy plasma.

Regarding the microphotograph of ZrO_2/Al_2O_3 coating (Figure 7) produced by the gas tunnel spraying on the fixed substrate for 3s spraying time, the coating thickness was about 250 μm, and white and gray layers were deposited alternatively as the same as Figure 3.

The graded functionality of the structure is remarkable, and small pores are distributed disparately in the whole coating while large pores existed near the substrate. The surface side has fewer pores and dense, compared to the coating near the substrate. This was caused by the thermal process of the high energy plasma from the surface side of the coating.

In this case, the Vickers hardness was linearly decreased in the thickness direction towards the substrate side (Figure 8). The dense microstructure led to the suppression of the deviation of the hardness distribution.

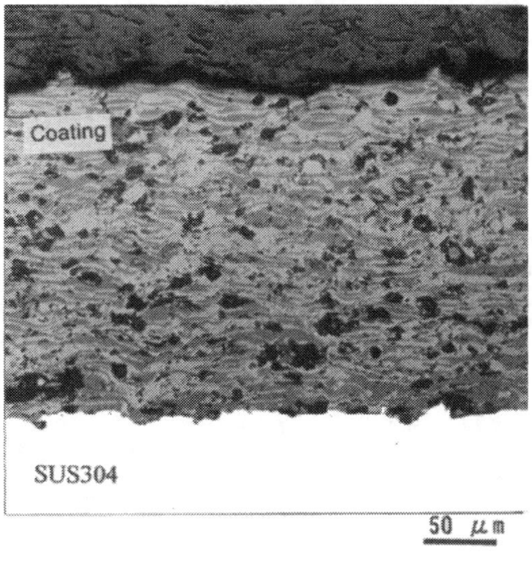

Figure 7. Microhotograph of cross section of zirconia composite coating. The traverse number was 30 time traverse. Sprayed at L=40mm when P=25 kW.

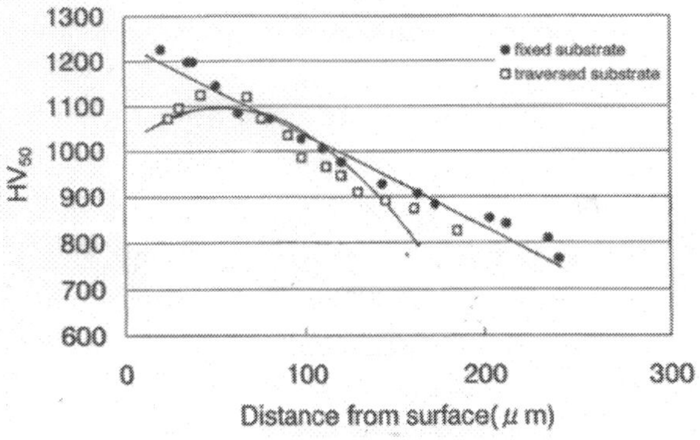

Figure 8. Distribution of Vickers hardness of zirconia composite coating sprayed by 30 times traverse at P=25kW, L=40 mm.

4.5. Adhesive strength of ZrO_2 composite coating

Figure 9 shows the dependence of the adhesive strength between coating and substrate on the coating thickness of ZrO_2 composite coating sprayed by 2-4 times traverse. The adhesive strength of the ZrO_2 composite coatings was decreased when the thickness was large. In the case of small coating thickness (100μm), the adhesive strength was large: more than 140 kgf/cm^2 for the coating thickness below 100μm. While, the value was $F = 100\sim120$ kgf/cm^2 when the thickness was more than 200μm.

Therefore the thick coating was much easier to break than thin coating, but the adherence was improved when the traverse number was large.

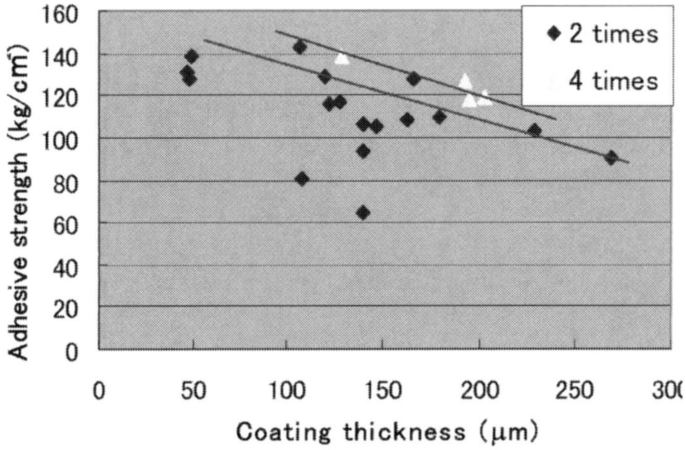

Figure 9. Dependence of the adhesive strength between coating and substrate on the coating thickness of zirconia composite coating sprayed at *L*=40mm, *P*=25kW. 2-4 times traverse.

4.6. Oxidation of ZrO_2 composite coating

After heat treatment at 1050 °C for 5 Hr, the ZrO_2 coating system showed spallation from the substrate. But the Al_2O_3 coating showed no spallation even failure occurred after exposure at the same heat treatment condition. The delamination failure is due to large thermal stresses developing within the coating and the phase transformation of Al_2O_3. Analytical model showed that plasma sprayed Al_2O_3 layer should be very thin since thicker layer would generate larger residual tensile stress.

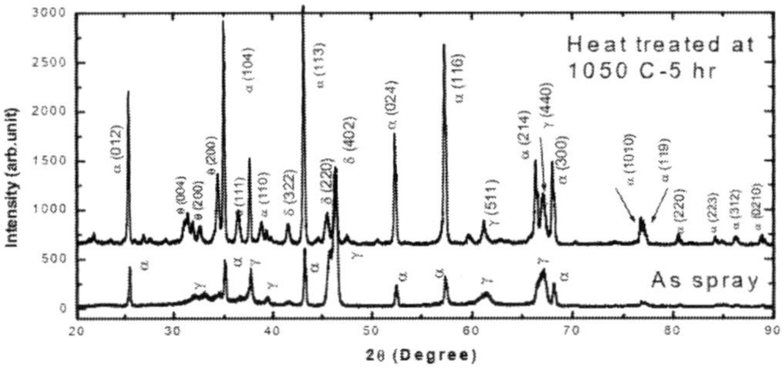

Figure 10. XRD patterns of as sprayed coating and coating after heat treatment.

To evaluate phase transformation of Al_2O_3 due to plasma spray process, free standing Al_2O_3 layer was oxidized at 1050 °C for 5 Hrs. X-ray diffraction was conducted to examine phases of the sprayed layers. A comparison of diffraction patterns of the as-sprayed and heat-treated Al_2O_3 is summarized in Figure 10. The α-Al_2O_3 that was formed during plasma spray undergoes a phase transformation to a more stable α-Al_2O_3 during heat treatment, although some fraction of γ phase was retained in the coating. Other phases, namely δ-, θ- Al_2O_3 were also identified. It is noted that the transformation of γ to α had never been direct such that δ and θ phases can be regarded as the intermediate phases, suggesting that the transformation of γ to (δ,θ) to α occurred in the annealed coating. The phase transformation of α to γ involves a volume change, resulting in additional residual stresses. Density values of coated samples believed that an additional volume change could be attributed to the porosity closing during heating.

4.7. Anodic corrosion polarization characteristics of ZrO_2 composite coating

Figure 11 presents the anodic corrosion polarization characteristics of the samples coated with different thickness of 80% ZrO_2+ 20%Al_2O_3 mixture coating. All the curves are obtained from their first polarization scan. From the curves, their corrosion potentials increase clearly with the coating thickness. However, their corrosion current shows a complicated tendency with the coating thickness, which is possibly due to the complex bonding states of the coatings to the substrates because the effective area of the substrate exposed to the corrosion media is responsible for the corrosion current.

Figure 12 displays the relationships of the corrosion potential of the tested samples to the alumina mixing ratios and thickness. As expected, the trends show that the corrosion potential goes up slightly with both the alumina content ratio and coating thickness. Theoretically, high corrosion potential means lower electrochemical activity and higher oxidation resistance. So, in conclusion, higher thickness and lower porosity sprayed coatings lead to increase of corrosion resistance because both higher thickness and lower porosity provide stronger diffusion resistance to prevent the anodic oxidants of the corrosion solution from accessing the interface of the coated samples.

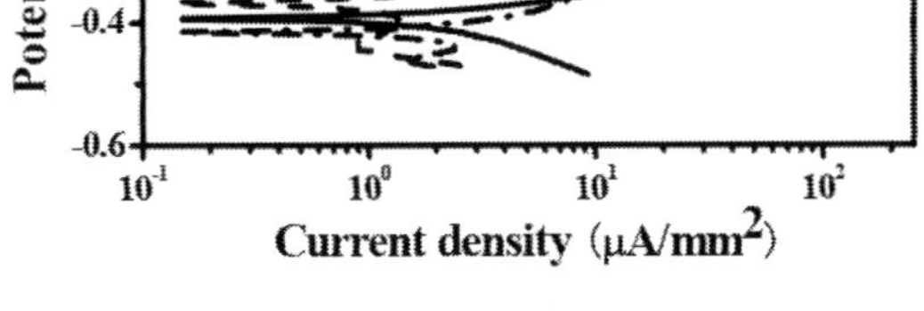

Figure 11. Corrosion curves of ZrO_2 composite coatings.

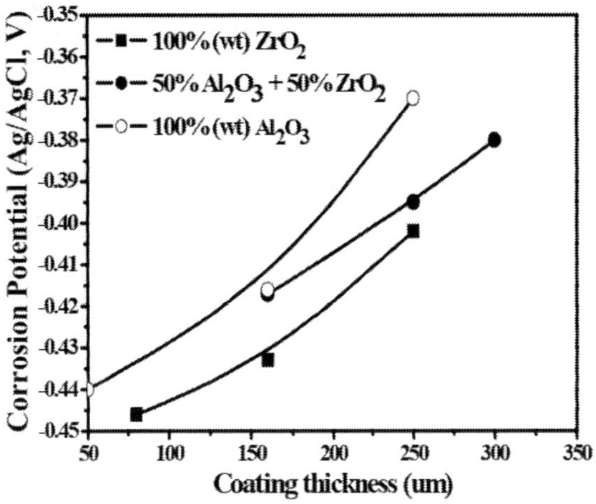

Figure 12. Corrosion potential versus thickness.

5. CONCLUSIONS

The following results were obtained during the application of the gas tunnel type plasma system to smart coating processing. The performance such as the mechanical properties, thermal behavior and high temperature oxidation resistance of the ZrO_2/Al_2O_3 composite coating produced by gas tunnel type plasma spraying was clarified.

(1) The gas tunnel type plasma system has high energy density and also high efficiency as compared to the conventional ones, and can be applied to the various thermal processing. One typical application is plasma spraying of ceramics such as Al_2O_3 and ZrO_2. And the characteristics of these ceramic coatings were superior to the conventional ones

(2) The ZrO_2/Al_2O_3 composite coating has graded functionality on the hardness and the porosity, and has a possibility of the development of high functionally graded TBC (thermal barrier coating). The effect of alumina mixing on the Vickers hardness of the ZrO_2 composite coating was also clarified in order to develop high functionally graded TBC.

(3) The ZrO_2 composite system exhibited the improvement of mechanical properties of thermal barrier coatings and oxidation resistance. The α-Al_2O_3 that was formed during plasma spraying undergoes a phase transformation to a more stable α-Al_2O_3 during heat treatment, although some fraction of γ phase was retained in the coating. The high temperature oxidation behavior of the functionally graded TBCs showed the effectiveness of Al_2O_3 layer functioning as an oxidation barrier.

(4) The higher alumina content and thicker coatings, the better the corrosion resistance, which is attributed to the diffusion resistance of the coating layers to corrosion reaction.

References

[1] Arata Y and Kobayashi A 1984 *J. High Temp. Soc.* **11-3** 124-131

[2] Arata Y and Kobayashi A 1986 *J. A. P.* **59-9** 3038-3044

[3] Arata Y and Kobayashi A 1986 *J. J. A. P.* **25-11** 1697-1701

[4] Arata Y, Kobayashi A, Habara Y and Jing S 1986 Gas Tunnel Type Plasma Spraying *Trans. of JWRI* **15-2** 227-231

[5] Arata Y, Kobayashi A and Habara Y 1987 *J. Applied Physics* **62** 4884-4889

[6] Vassen R, Kerkhoff G and Stoever D 2001 Development of a micromechanical life prediction model for plasma sprayed thermal barrier coatings *Mater. Sci. Eng.* A **303** 100-109

[7] Assanis D N 1989 *Journal of Materials Processing Technology* **4** 232

[8] Yonushonis T M 1997 *Journal of Thermal Spray Technology* **6(1)** 50

[9] Ramaswamy P, Seetharamu S, Varma K B R and Rao K J 1999 Thermal shock charateristics of plasma sprayed mullite coatings *J. Therm. Spray Technol.* **7(4)** 497-505

[10] Araya T 1988 *J. Weld. Soc. Jpn.***57-4** 216-222

[11] Arata Y, Kobayashi A and Habara Y 1987 Formation of Alumina Coatings by Gas Tunnel Type Plasma Spraying (in Japanese) *J. High Temp. Soc.* **13** 116-124

[12] Kobayashi A, Kurihara S, Habara Y and Arata Y 1990 *J. Weld. Soc. Jpn.* **8** 457-463

[13] Kobayashi A 1992 Property of an Alumina Coating Sprayed with a Gas Tunnel Plasma Spraying *Proc. of ITSC* 57-62

[14] Kobayashi A 1990 Formation of High Hardness Zirconia Coatings by Gas Tunnel Type Plasma Spraying *Surface and Coating Technology* **90** 197-202

[15] Kobayashi A and Kitamura T 1997 High Hardness Zirconia Coating by Means of Gas Tunnel Type Plasma Spraying(in Japanese) *J. of IAPS* **5** 62-68

[16] Kobayashi A and Kitamura T 2000 *VACUUM* **59-1** 194-202

The effects of traces of molecular gases (H_2, N_2 & O_2) in glow discharges in noble gases

E B M Steers[1,4], **P Šmíd**[1], **V Hoffmann**[2] **and Z Weiss**[3]

[1]London Metropolitan University, 166-220 Holloway Road, London, N7 8DB, UK
[2]Leibniz Institute for Solid State and Materials Research Dresden, P.O.Box 270116, D-01171 Dresden, Germany
[3]LECO Instrumente Plzeň spol. s r.o., Plaská 66, 323 25 Plzeň, Czech Republic

E-mail: e.steers@londonmet.ac.uk

Abstract: A brief introduction is given on the use of glow discharges in noble gases for spectrochemical analysis, with particular emphasis on Glow Discharge Optical Emission Spectroscopy (GD-OES). This explains the relevance of investigations on the effect of small quantities of molecular gases on the performance of discharges in noble gases, mainly in argon. The chief effects are on the electrical characteristics, the sputtering rate and the spectrum emitted. Earlier work is briefly reviewed, the options for spectral investigations discussed and the current state-of-the-art described for some samples (cathodes) of analytical importance (copper, iron and titanium) The main emphasis is on the addition of hydrogen to discharges in argon, with brief comments on other gas mixtures. A great deal of work remains to be done, both to produce empirical data for analytical use and to gain a fuller understanding of the discharge processes involved.

1. Introduction

Optical Emission Spectroscopy (OES) has been used for the analysis of chemical samples for very many years; initially, arc and spark sources were used in conjunction with prism or grating spectrographs with photographic plates. Later, low pressure glow discharge sources, such as demountable hollow cathode sources, were used in industrial laboratories, but relatively few systems were available commercially. Hollow cathode discharges can be used for metals by making the cathode of the sample material, or by introducing small pieces of the metal into, e.g., a graphite hollow cathode; solid powders can also be introduced directly into a graphite hollow cathode; drops of solution may also be introduced into a hollow cathode and the cathode heated to form a deposit on the walls. [1] The hollow cathode is thus versatile, but not very convenient if a solid metallic sample has to be machined to form a hollow cathode.

1.1 The Grimm source

In the late 1960's W.Grimm [2] introduced a new form of discharge source, shown diagrammatically in figure 1. In the original design, the diameter of the anode tube was 8 mm; typical operating conditions were 800 V, 80 mA, with an argon pressure of ~500 Pa. The source proved very convenient for routine analysis of bulk metallic samples; in all cases, calibration with samples of known

[4] To whom correspondence should be addressed

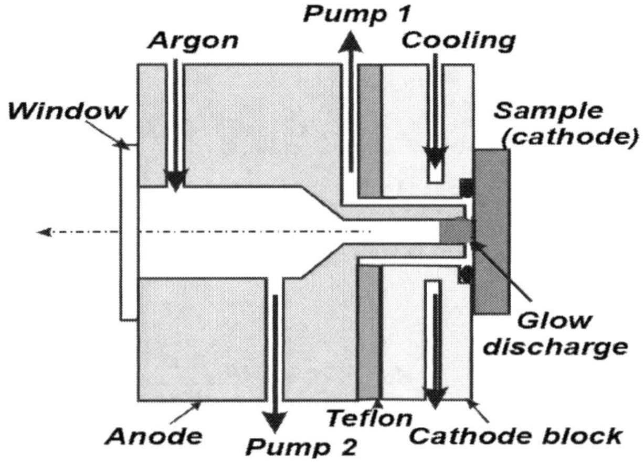

Figure 1. The Grimm source
A plane metallic sample acts as the cathode. Originally the anode tube had an inner diameter of 8 mm, but smaller diameters are now common.
The gaps between the outside of the anode tube and the cathode block, and between the end of the anode tube and the sample are each ~ 0.2 mm, and are evacuated by pump 1, so that when a discharge is initiated, it is restricted to an area of the sample equal to the cross-sectional area of the anode tube and no discharge occurs on the outside of the anode tube.

composition is required. The bombardment of the sample by ions and fast atoms causes sample atoms to be sputtered into the discharge and excited and ionised there. It was found that if appropriate discharge parameters are chosen, the sputtering is uniform across the sample surface and the crater formed has a flat base. Under these conditions the source can be used for the compositional analysis of the layers of e.g. a coated or layered material [3].

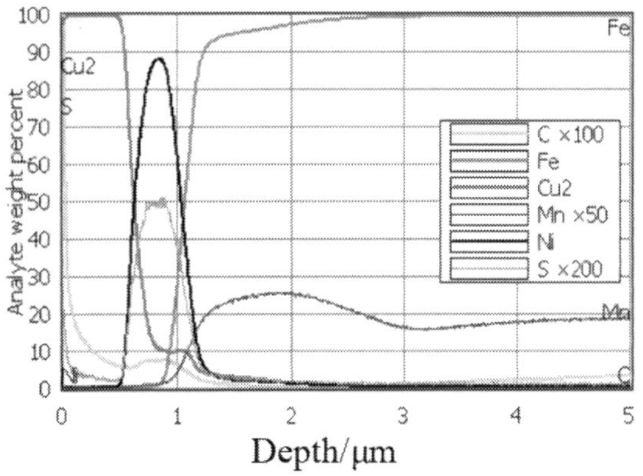

Figure 2. Typical compositional depth profile of a copper coated steel sample. The nickel layer ensures that there is good adhesion of the outer copper layer. (Recorded using LECO GDS 850A, 4mm anode tube, 750V, 20 mA)

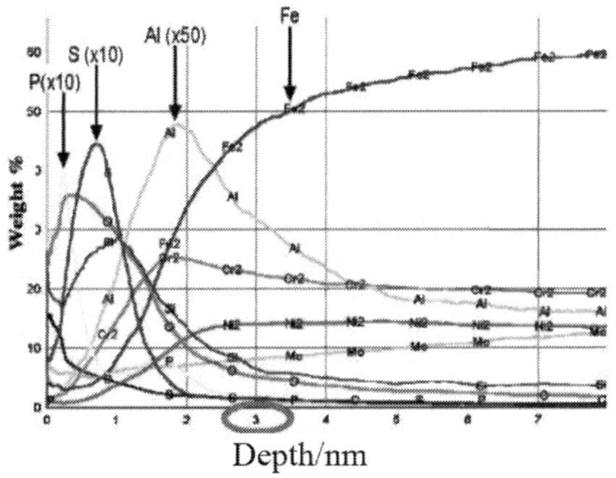

Figure 3. Illustration of the use of GD-OES to provide information of very thin layers on the surface of a stainless steel sheet. [4] The conversion of line intensity to composition by weight may not be accurate for such thin layers.

Instrumentation based on this source and consisting of source, vacuum and gas system and spectrometer is available from a number of commercial manufacturers; calibration algorithms are built into such instruments so that the sample composition, or the depth profile of a layered sample, is displayed directly. Figure 2 shows a routine example of such an analysis; the steel cover for an unused flue outlet has a copper coating with an intermediate layer of nickel about 1 μm thickness. Figure 3 shows that it is possible to record very much thinner layers if sufficient care is taken.

In current instrumentation, anode tubes with various diameters are available; the standard diameter is usually 4 mm, though 2.5 mm or even 1.0 mm is possible. The pressure is controlled by a servo system, so that the voltage and current may be set, and the pressure is automatically adjusted to the correct value within a few milliseconds. It should be noted that in many cases, particularly with older equipment, the recorded pressure may be a reproducible parameter, but is measured at some distance from the source and does not correspond to the actual pressure within the source. For a given voltage and current, the required pressure will depend on the sample material; nevertheless, it is found that such pressure variations have a smaller effect on recorded emission yields than current or voltage changes, so it is very common practice to work with fixed voltage and current – for a 4 mm diameter anode tube, 700 V and 20 mA are often described as "standard conditions".

The relationship between the measured intensity and the concentration of a particular element E within a sample matrix is given by the "Standard GD-OES equation" [5]:

$$I_{\lambda(E),M} = AR_{\lambda(E)}\, c_{E,M} q_M$$

where $I_{\lambda(E),M}$: line intensity, $c_{E,M}$ concentration of E in matrix M
$\lambda(E)$ wavelength for line of element E, q_M sputtering rate of matrix M
$R_{\lambda(E)}$ "emission yield", A instrumental constant

This simple equation must be modified in many cases to allow for interferences from overlapping lines of other elements, and at high concentrations for the effects of self-absorption [5]. The emission yield of a given line may be determined using samples of known composition and sputtering rate. Ideally $R_{\lambda(E)}$ should be independent of the matrix used – "matrix-independent emission yield" but in many cases this does not hold – the calibration curves of $C_{E,M}$ *vs* $I_{\lambda(E),M}$ depend on the matrix used. For example, differing calibration curves are obtained for the analysis of zinc in copper and aluminium based zinc alloys [6].

1.2 Later developments

Since the first introduction of the Grimm source, the potential of the technique has steadily increased [7]. In ~1990, rf powered sources were introduced permitting the analysis of non-conducting layers and solids. This technique has been successfully employed for the analysis of, e.g. coated and painted steel sheets for the automotive industry. The source is usually capacitively coupled to the power supply via a matching network; the sample acquires a negative dc bias, and so is bombarded by ions and fast atoms and sputtered material enters the discharge in the usual way. However, the use of the matching network means that the only electrical parameter that can be conveniently measured is the power delivered by the generator. The rf voltage reduces if non-conducting samples are used and the dc bias voltage cannot be measured and thus changes in the plasma impedance are not recorded.

The problem is more serious when relatively thick non-conducting samples are analysed, rather than thin non-conducting layers. On the other hand, rf power supplies means that the discharge is more readily initiated. Samples may be coated with a very thin non conducting layer; this does not present a problem with rf excitation, but with dc excitation, the starting potential must be sufficiently high to break down the surface layer; the potential then requires a short time to adjust to stable conditions, and in addition very thin surface layers can be damaged.

In more recent years, very thin layers have been analysed using GD-OES. Angeli et al. [8] have demonstrated that layers of nm thickness can be detected routinely nowadays (see also figure 3). The analysis of layers in a multilayer stack can also be carried out with the same resolution as obtained in more traditional methods of layer analysis, but in a much shorter time and therefore at lower cost, and to greater depths. More recently, Shimizu [9] has shown that it is even possible to identify the position of the atomic layers when a very thin layer of molecular material is deposited on a sample.

For the analysis of such thin layers, a very flat clean sample is obviously required. In addition, the spectra must be recorded from the initiation of the discharge, and any contamination on the sample or in the source may lead to the transient presence of molecular species, which may influence both the discharge conditions and the intensity of individual lines. Moreover, analyses are often required of layers containing compounds such as oxides and nitrides, where the sample itself contains an element forming a molecular gas. When the sample is sputtered, this gas enters into the discharge. The problem is particularly acute for the analysis of non-conducting solids such as glasses with a very high oxide content.

2. The effects of molecular gases - previous work in the field

Fischer, Naoumidis and Nickel [10] investigated the effect of nitrogen and oxygen on one or two lines of each of a number of elements (the lines available on a commercial GD-OES system) and found major intensity changes. The effect of hydrogen contamination on line intensities was first clearly demonstrated by Bengtson et al. in 1998 [11]. With a standard GD-OES system, a bulk sample of a chromium containing steel was analysed; two chromium lines were available. At the start of the discharge, these lines behaved in different ways until they came to a common value for the Cr content. Bengtson demonstrated that the presence of hydrogen increased the intensity of one Cr line, whilst decreasing that of the other. After the start of the discharge, the hydrogen arising from contamination is removed and the correct chromium content was obtained. Clearly the values obtained for a thin layer might be incorrect.

Hodoroaba [12-15] carried out a detailed study of the effect of hydrogen on the intensity of copper lines, using controlled additions of hydrogen in the range 0- 0.6 % H_2 v/v. A large number of Cu I, Cu II, Ar I and Ar II lines were selected using a monochromator, and the relation between line intensity and hydrogen concentration measured for each line. He operated using a fixed voltage (1000V), fixed pressure (~7.70 hPa) mode with a 2.5 mm diameter anode tube. The introduction of hydrogen caused an increase in plasma resistance, and hence a fall in the current. The sputtering rate also fell, but the sputtering rate per unit current for copper remained almost constant – other elements show major changes in sputtering rate. In some cases, the shape of the crater profile changed. Allowing for the changes in current and sputtering, Hodoroaba showed that the behaviour of each line – or rather lines from each upper energy level – investigated was different, but general trends could be identified:- the intensity of Cu I lines tended to increase, whilst that of Cu II lines fell dramatically. Ar I lines tended to be less intense, whilst Ar II lines increased in intensity. Hodoroaba [13] also compared the electrical parameters and line intensity for a titanium sample sputtered with various hydrogen concentrations in the plasma gas with those for a titanium hydride sample sputtered in pure argon, and was able to show that the latter case corresponded to using about 0.2% v/v H_2 in pure argon. The exact H_2 concentration will depend on the details of gas flow in the source, but he showed that his data corresponded to the appropriate range of hydrogen concentrations.

Hodoroaba [14] also made some mass spectrometry measurements, and showed that whilst the intensity of Ar II lines increased, i.e. the population of excited Ar ions increased, the total number of argon ions fell greatly (and the number of ArH^+ ions increased). On the other hand, whilst the population of excited Cu^+ ions fell greatly, the total population of Cu^+ ions increased. This work showed the importance of obtaining data by mass spectrometry as well as by OES. Using standard GD-OES instrumentation, Hodoroaba [13] also investigated the behaviour of the one or two lines available for each of a number of other elements, and obtained a wide range of intensity variations. In all cases, when hydrogen was introduced, the hydrogen continuum was observed; with standard GD-OES instrumentation, this means that signals may be observed at the wavelength corresponding to an element X, although this element is not present in the sample.

Subsequently, further investigations have been carried out by the Oviedo group on the effects of hydrogen, nitrogen and oxygen [16]; the majority of their OES work has involved studies on the one or two lines of each element available using commercial GD-OES systems. Wagatsuma has published a review on glow discharge plasmas [17] covering use of mixed noble gases as well as the presence of molecular gases in noble gas plasmas. Bogaerts [18] has undertaken modelling studies on analytical glow discharges containing hydrogen but the experimental conditions were those for the VG9000 cell for mass spectrometry (1000V, ~ 5 mA) but a paper dealing with typical Grimm-type conditions has now been submitted [19].

3. Work by present authors on the effects of molecular gases

All the authors of the current paper have been collaborating for some years in investigations on the effects of molecular gases in the plasma gas, with the majority of our work has been with hydrogen and nitrogen. This study is now taking place within the EC Marie Curie Analytical Glow Discharge

Research Training Network "GLADNET" [20], and we have started to investigate the effects of oxygen. To get a clearer understanding of the processes involved we have also used neon and in some gases krypton as the plasma gas, but most of results presented involve argon. All the results presented in this paper have been obtained with dc excitation, usually in the constant voltage, constant current mode, though the exact conditions vary. Preliminary investigations have shown that similar general trends are observed when rf excitation is used, but there may be differing effects on specific energy levels.

It had already been shown by Hodoroaba [13] that similar results are obtained when molecular gases are added to the plasma gas and when the element forming the molecular gas is sputtered from the sample. The former method has been used in most of our work. Normally commercially supplied argon containing 2% v/v of the molecular gas is mixed with pure argon using a system of mass flow controllers (MKS Instruments) In this way, accurate mixtures are obtained even at small levels of the molecular gas. When other noble gases (neon, krypton, xenon) have been the main plasma gas, it has been necessary to mix the noble gas with the molecular gas directly. All the gas piping is stainless steel; nevertheless, some traces of OH bands have been sometimes observed in the spectra; these have been eliminated by introducing a cooled trap after the mass flow controllers and immediately before the inlet valve of the Grimm source.

3.1 Electrical parameters

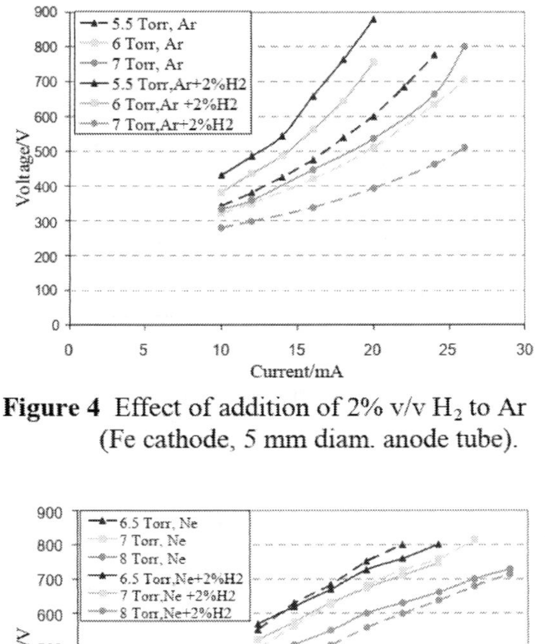

Figure 4 Effect of addition of 2% v/v H_2 to Ar (Fe cathode, 5 mm diam. anode tube).

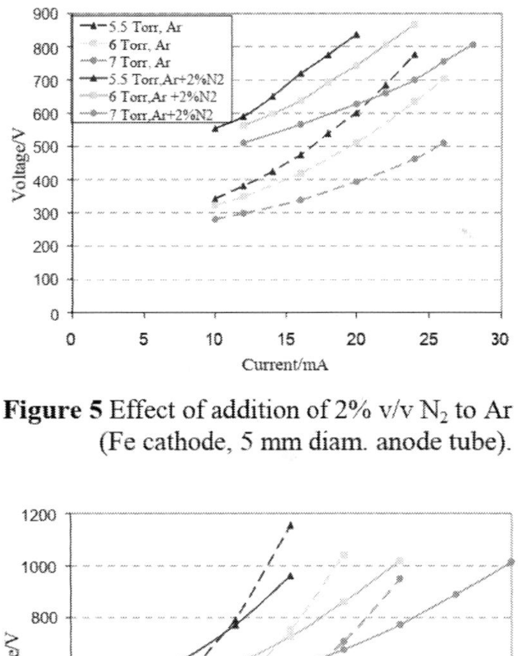

Figure 5 Effect of addition of 2% v/v N_2 to Ar (Fe cathode, 5 mm diam. anode tube).

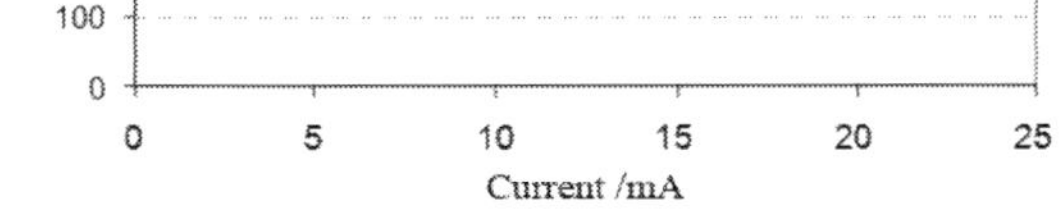

Figure 6. Effect of addition of 2% v/v H_2 to Ne (Fe cathode, 8 mm diam. anode tube).

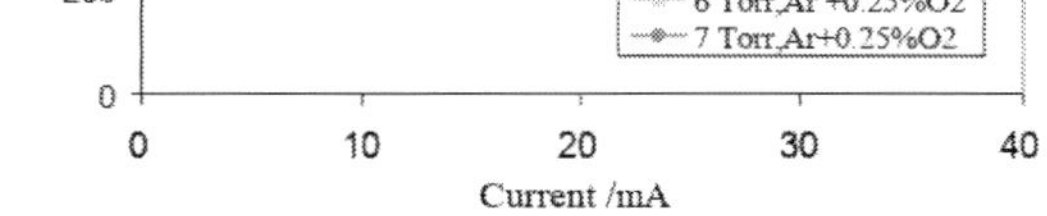

Figure 7. Effect of ***0.25*** % v/v O_2 added to Ar. (Fe cathode, 4 mm diam. anode tube).

In all cases with the typical discharge conditions used for spectrochemistry, the addition of hydrogen or nitrogen to argon causes an increase in the discharge resistance, so that under constant voltage, constant pressure conditions (as used by Hodoroaba) the current falls; under constant current, constant pressure conditions, the voltage rises whilst it is necessary to increase the total gas pressure to hold voltage and current constant, as in the majority of our work. The magnitude of the changes required depends to some extent on the cathode material. Changes in the V-i characteristics at various pressures caused by the addition of hydrogen and nitrogen to argon are shown in figures 4 and 5, respectively.

On the other hand, when hydrogen is added to neon, the discharge resistance changes are much smaller, and may be positive or negative (figure 6). The magnitude of the change depends on the discharge conditions. Similarly, when oxygen is added to argon, then, at a constant pressure, the sign of the resistance change may depend on the current (figure 7).

3.2 Sputtering rate

It has already been noted that the sputtering rate per unit current does not change much with hydrogen addition for copper; the situation is very different for iron and titanium; for a N_2 or H_2 content of about 1%v/v the sputtering rate falls to about 30% of the rate in pure argon for iron and about 15% for titanium [21]; these changes should be borne in mind when considering intensity changes. Correcting for sputtering rate changes by dividing the intensity by the sputtering rate, or using emission yields instead of measured intensities may give a more realistic picture of the intensity changes.

3.3 Spectral changes – interpreting the results

Two approaches may be used to investigate the spectral effects produced by a molecular gas.

(1) If a commercial GD-OES system with a polychromator is used, then only one or two lines can be investigated for each element; often a scanning monochromator can be used to select an additional line, but even so only a very limited number of lines can be investigated for each element. This may provide data for empirical corrections for analytical work, but as different lines of the same element may be affected in very different ways by a particular molecular gas, a full understanding of the effects cannot be obtaining in this way. It is possible to use a monochromator to make a series of sequential measurements, but this is a slow and time-consuming experimental procedure, and may well give rise to a misleading impression if an error is made in the gas mixtures used for a particular line.

(2) The preferable method is to record large sections of spectra simultaneously, using a range of gas mixtures. As photographic methods are no longer used in general practice (and in any case intensity calibration was tedious), very few suitable instruments were available until recently. The high resolution uv-vis Fourier transform (FT) spectrometers at Imperial College [22] are very suitable for this purpose, and have been used for studies on GD sources since ~1990. The best limit of resolution is ~0.03 cm^{-1}, so that at 200 nm (50,000 cm^{-1}), the chromatic resolving power (CRP) is ~ 1.6 x 10^6; the CRP is less than this value at higher wavelengths but in very many cases the instrumental contribution to the line profile is very small and line shapes may be studied. If such studies are not required, a lower resolution with better signal/noise ratio and shorter recording time can be used. Various photomultipliers are used for detection, so that typical observation ranges are 200-300 nm, 300-600 nm, etc. Recording times are of the order of a few minutes per spectral range. Using high vacuum conditions in the spectrometer, lines can be observed down to ~140 nm; the high wavelength limit is determined by the photomultiplier used, but we have made measurements up to ~ 900 nm.

More recently, some GD-OES systems have been produced with spectrometers with CCD arrays, and a number of grating and echelle instruments with CCD detectors are now available commercially. Such instruments are very suitable for such studies, although the limited resolution of some commercial systems makes line identification difficult, particular for complex samples.

With both FT and CCD spectra, the intensities of a large number of spectral lines can be recorded simultaneously in a pure plasma gas, and similar data recorded for a number of gas mixtures. It is then possible to study (a) the variation of intensity with molecular gas concentration for selected lines, and also (b) to plot the ratios of the intensity of a given line in a particular gas mixture to the intensity in pure noble gas against the energy of the upper level of the line. Such plots allow the identification of significant energies and excitation processes. The plots of intensity *vs* molecular gas concentration demonstrate that various upper levels show different responses to molecular gas concentration and also that the effects vary with different discharge parameters (voltage, current, pressure). The particular conditions (molecular gas concentration, discharge parameters) used for studies of type (b) will affect the clarity with which processes are detected.

The recording of complete spectra has also the great advantage that initially data from a limited number of lines can be evaluated, but data on other lines may subsequently be processsed when significant effects have been observed.

We have used this method for extensive investigations on the effect of hydrogen and nitrogen on systems with a range of metallic cathodes (copper, aluminium, iron, titanium, vanadium, silver, zinc, etc, and alloys of known composition). In this paper, we use work with iron and titanium cathodes to illustrate typical effects.

3.4 Effects on the ionised spectra of the sample

Early work by Hodoroaba [12] and others had shown that the addition of hydrogen to argon causes the suppression of Cu II lines excited by charge transfer. This effect also occurs for Fe II lines [23], as shown in figure 9, where lines in the Fe II spectrum with a total excitation energy (TEE) (ionisation energy + excitation energy of the ion) ~15.8 eV (Group A) are suppressed.

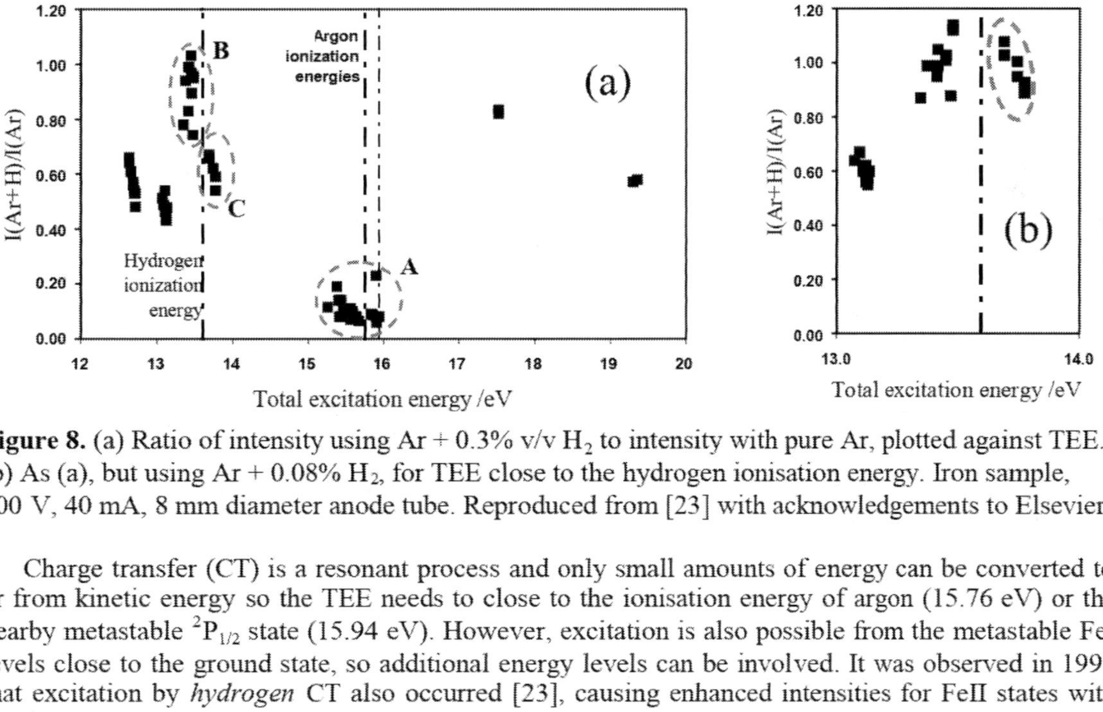

Figure 8. (a) Ratio of intensity using Ar + 0.3% v/v H_2 to intensity with pure Ar, plotted against TEE. (b) As (a), but using Ar + 0.08% H_2, for TEE close to the hydrogen ionisation energy. Iron sample, 700 V, 40 mA, 8 mm diameter anode tube. Reproduced from [23] with acknowledgements to Elsevier

Charge transfer (CT) is a resonant process and only small amounts of energy can be converted to or from kinetic energy so the TEE needs to close to the ionisation energy of argon (15.76 eV) or the nearby metastable $^2P_{1/2}$ state (15.94 eV). However, excitation is also possible from the metastable FeI levels close to the ground state, so additional energy levels can be involved. It was observed in 1998 that excitation by *hydrogen* CT also occurred [23], causing enhanced intensities for FeII states with TEE ~13.6 eV. Figure 8(a) shows the ratios of the intensity in an argon/hydrogen mixture to the intensity in pure argon, plotted against the energy of the upper state for a hydrogen concentration of 0.3% v/v, whilst the section of the plot figure 8(b) shows that the marked difference in the relative positions of groups B and C of levels for a hydrogen concentration of 0.08% v/v. This is wholly in agreement with the dependence shown in figure 10 of the intensity ratios for individual lines on the hydrogen concentration. Note that no allowance for changes in sputtering rate is made in figure 8, but it is included in figure 9. Hydrogen CT excitation has also been observed in Ti II and Cr II spectra.

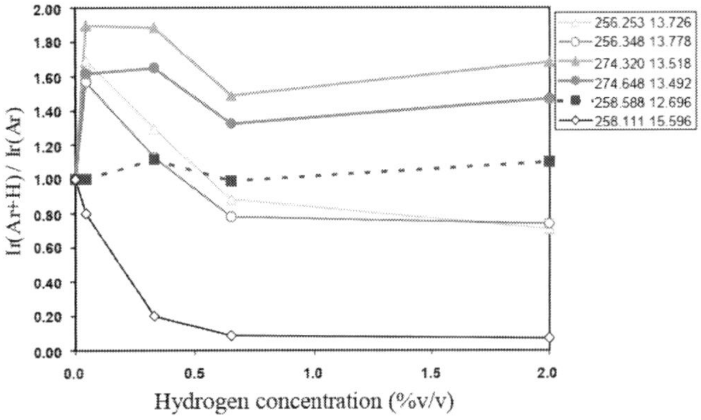

Figure 9. The ratio Ir(Ar+H) / Ir(Ar) plotted against hydrogen concentration for selected lines. Ir is the reduced intensity, i.e. the intensity divided by the sputtering rate. 700 V, 40 mA, 8 mm anode tube diameter.

Reproduced from [23] with acknowledgements to Elsevier

3.5 Effects on the atomic spectra of the sample.

In all cases, data has been obtained on the atomic spectra of the samples as well as on the ionised spectra, but until recently there was very little information on the excitation mechanisms involved. Added hydrogen increases the intensity of Fe I lines by a growing amount up to energy levels ~ 5 eV, with a marked peak or resonance effect at about 5.5 eV (figure 10). There is no similar effect when the added gas is nitrogen (figure 11). These data were obtained by FT spectrometry using two spectral regions "uv" (200-300 nm) and "visible" (300-600 nm); it is worth noting that if spectra from the "uv"

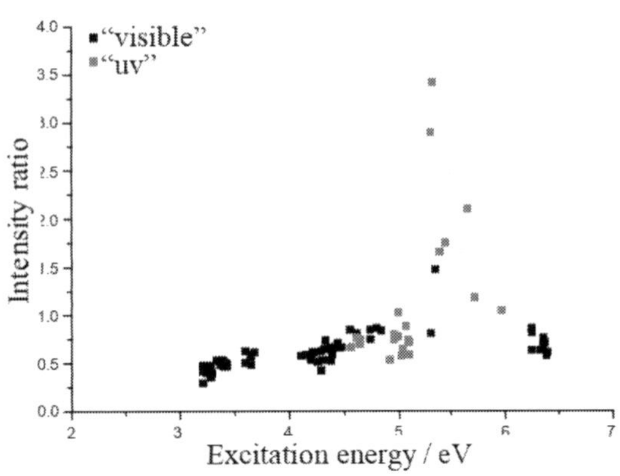

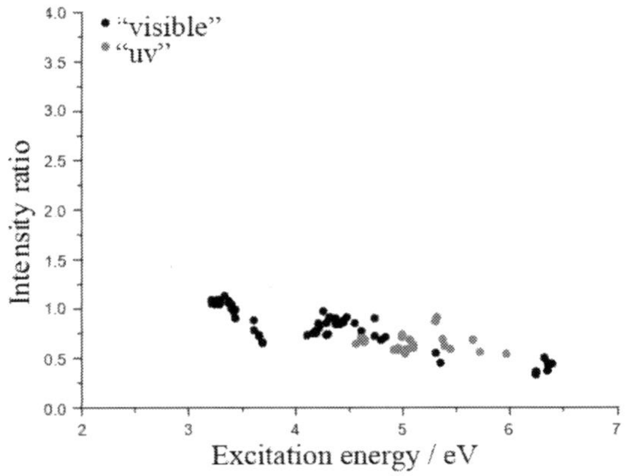

Figure 10. Ratio of intensities in Ar + 0.5% H_2 to intensity in pure argon, plotted against excitation energy of upper level. Fe cathode, 700 V, 10 mA, 4 mm diameter anode tube.

Figure 11. Ratio of intensities in Ar + 0.4% N_2 to intensity in pure argon, plotted against excitation energy of upper level. Fe cathode

Reproduced from [21] by permission of The Royal Society of Chemistry (RSC)

region had not been recorded, then the sharp resonance would have been less obvious. This emphasises the need to investigate the effect of the added gas on as many levels as possible – to do this it is usually necessary to cover a very wide wavelength region.

A similar, but more marked general trend is observed in the spectrum of Ti I (Fig 12), although the sharp resonant effect is not observed. For both Fe I and Ti I, the effects of the added hydrogen are greater at lower currents (and powers) (figure 13); this is a very important result analytically. For the measurement of thin layers a low sputtering rate is helpful; under these conditions, the time taken to remove any hydrogen surface contamination will be longer, and also this hydrogen will have a greater effect on the measured intensities.

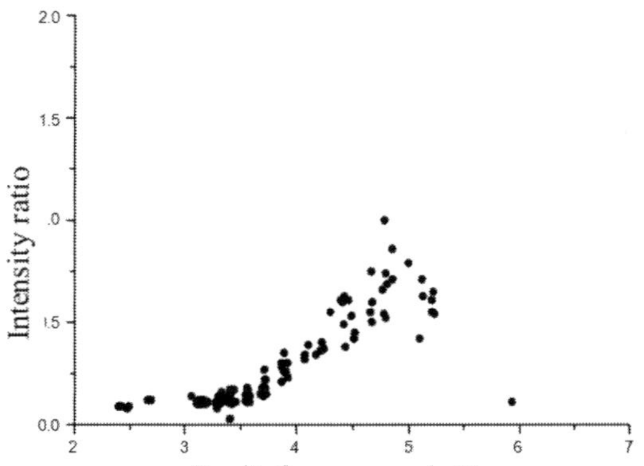

Figure 12. Ratio of intensities in Ar + 0.3 % H_2 to intensity in pure argon, plotted against excitation energy of upper level. Ti cathode, 700 V, 20 mA, 4mm anode tube diameter

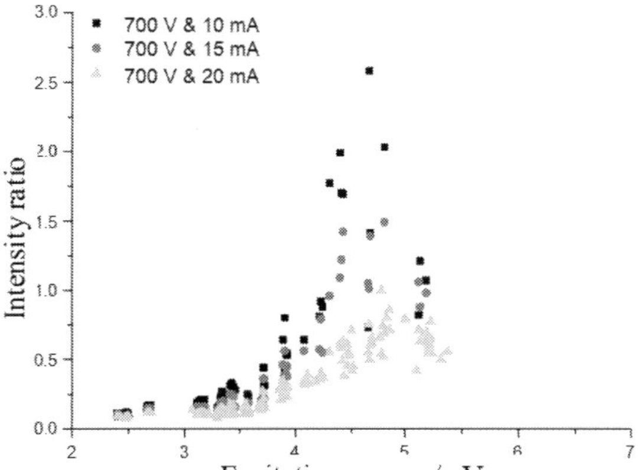

Figure 13. Ratio of intensities in Ar + 0.3% H_2 to intensity in pure argon, plotted against excitation energy of upper level. Ti cathode, for various currents, 700 V, 4mm anode

Reproduced from [21] by permission of The Royal Society of Chemistry (RSC)

We have suggested [21] that the increase for both Fe I and Ti I lines may be due to the process:

$$M_o + H_2^+ \rightarrow M^* + H + H^+$$

where M_o is the ground state sample atom and M* the excited sample atom. The total H_2^+ dissociation energy is ~18.6 eV; allowing for the hydrogen ionisation energy of 13.6 eV, the process could provide up to ~5 eV for the excitation of the sample atom M; an additional particle is available to satisfy conservation criteria for various amounts of surplus kinetic energy, so the process is feasible up to energies of about 5eV. We have as yet no suggestions on the origin of the resonant process observed with Fe I. Investigations on the behaviour of the atomic spectra of other elements are continuing.

3.6 Effects on the spectra of the plasma gases

Hodoroaba [12] reported the effect of hydrogen on about 30 lines in the ArI and ArII spectra and Šmíd [24] reported the effect of nitrogen on about 12 argon lines, but there has been no detailed study on the possible underlying discharge processes. In general, the effects produced by nitrogen are smaller; hydrogen increases the intensity of ArII lines whilst nitrogen causes a fall; on the other hand, for ArI lines, a fall occurs in all cases with hydrogen; for some lines nitrogen causes increased intensity. High resolution measurements show that hydrogen greatly reduced the self absorption of ArI transitions (e.g.811.531 nm) to the metastable argon levels, whilst nitrogen has a smaller, but still very significant effect.

Šmíd [24] measured the intensities of some N I lines in the vacuum uv region and showed that there is a good linearity between the intensities and the nitrogen content up to concentrations of about 0.2 %v/v but that the rate of increase then declines greatly; the lines studied are transitions to metastable levels but profile changes corresponding to reduction of self absorption could not be detected. The intensity of bands in the 2nd positive system of nitrogen also showed marked non-linearity with increasing hydrogen concentration. Hodoroaba [12] and Steers [23] both reported very non-linear behaviour for the intensity of the Hα and Hβ lines (656.2nm and 486.1 nm, respectively); these lines should not be subject to self absorption. It is clear that for both hydrogen and nitrogen, the degree of dissociation is highly dependent on the discharge conditions.

4. Conclusions

After a brief account of the use of glow discharges for analytical purposes, the complex spectral and other changes produced by the presence of molecular gases in the discharge gas are reviewed. The concentration ranges which have been studied are those which can occur in analytical applications. Some of the effects observed can be explained in terms of the excitation processes involved, but a large amount of further work using other sample materials and noble gases is required before a comprehensive picture can be obtained. At present, corrections can be made empirically in analytical work but a greater understanding of the processes involved will enable such corrections to be made on a more systematic basis. In addition, these results provide experimental data for comparison with modelling predictions.

Acknowledgements

Since February 2007, the investigations have been carried out within the European Union Analytical Glow Discharge Research Training Network "GLADNET", contract no. MRTN-CT-2006-035459. We thank Annemie Bogaerts (Antwerp University) for helpful discussions on discharge processes. P. Šmíd thanks the Deutsche Forschungsgemeinschaft for financial support whilst working at IFW Dresden.

References

[1] Caroli S (ed) 1985 *Improved Hollow Cathode Lamps for Atomic Spectroscopy* (Chichester, Ellis Horwood Limited)
Thornton K 1969 *Analyst* **94** 958-967
[2] Grimm W 1967 GermanPatent D1589389
Grimm W 1968 *Spectrochim. Acta, Part B* **23** 443-454
[3] Payling R, Jones D and Bengtson A (eds.) 1997 *Glow Discharge Optical Emission Spectrometry* (Chichester , John Wiley & Sons)
Marcus R K and Broekaert J A C (eds.) 2003 *Glow Discharge Plasmas in Analytical Spectroscopy* (Chichester , John Wiley & Sons)
[4] Marshall K, Brushwyler K, Maul C and Mitchell J 2003 paper presented at FACSS
[5] Weiss Z 2006 *Spectrochim. Acta Part B* **61** 121– 133
[6] Weiss Z 2006 *J. Anal. At. Spectrom.* 691-6
[7] Jakubowski N, Dorka R, Steers E and Tempez A 2007 *J. Anal. At. Spectrom.*, **22**, 722-735
[8] Angeli J, Bengtson A, Bogaerts A, Hoffmann V, Hodoroaba V-D and Steers E 2003 *J. Anal. At. Spectrom.* **18** 670-9
[9] Shimizu K, Payling R, Habazaki H, Skeldon P and Thompson G E 2004 *J. Anal. At. Spectrom.* **19** 692-5.
[10] Fischer W, Naoumidis A and Nickel H 1994 *J. Anal. At. Spectrom* **9** 375-380
[11] Bengtson A and Hanstrom S 2002 *ISIJ International* **42** supplement 1-5
[12] Hodoroaba V-D, Hoffmann V, Steers E and Wetzig K 2000 *J. Anal. At. Spectrom* **15** 951-8
[13] Hodoroaba V-D, Hoffmann V, Steers E and Wetzig K 2000 *J. Anal. At. Spectrom* **15** 1075-80
[14] Hodoroaba V-D, Hoffmann V, Steers E and Wetzig K 2000 *J. Anal. At. Spectrom* **16**, 43-49
[15] Hodoroaba V-D, Steers E, Hoffmann V, Unger W, Paatsch W and Wetzig K 2003 *J. Anal. At. Spectrom* **18** 521–6
[16] Martín A, Menéndez A, Pereiro R, Bordel N and Sanz-Medel A 2007 *Anal Bioanal Chem* **388** 1573-82
[17] Wagatsuma K 2001 *Spectrochim. Acta Part B* **56** 465-486
[18] Bogaerts A 2002 *J. Anal. At. Spectrom.* **17**, 768-779
Bogaerts A and Gijbels R 2002 *Spectrochim. Acta B*, **57**, 1071-99
[19] Bogaerts A, 2008 *Private Communication*
[20] www.gladnet.eu
Jakubowski N, Steers E and Tempez A 2007 *J. Anal. At. Spectrom.*, **22**, 715-721
[21] Šmíd P, Steers E, Weiss Z, Pickering J C and Hoffmann V 2008 *J. Anal. At. Spectrom.*, **23** 1223-33
[22] Pickering J C 2002 *Vibr. Spectrosc.* **29** 27–43
[23] Steers E, Šmíd P and Weiss Z 2005 *Spectrochim. Acta Part B* **61** 414-420
[24] Šmíd P, Steers E, Weiss Z and Vlček J 2003 *J. Anal. At. Spectrom.*, **18** 549-556

Supershort avalanche electron beams and x-ray in high-pressure nanosecond discharges

V. F. Tarasenko, E. H. Baksht, I. D. Kostyrya, M. I. Lomaev, D. V. Rybka

Institute of High Current Electronics SB RAS, 634055, Tomsk, 2/3, Akademichesky

VFT@loi.hcei.tsc.ru

Abstract. The properties of a supershort avalanche electron beam (SAEB) and X-ray radiation produced using a nanosecond volume discharge are examined. An electron beam of the runaway electrons with amplitude of ~ 50 A has been obtained in air atmospheric pressure. It is reported that SAEB is formed in the angle above 2π sr. Three groups of the runaway electrons are formed in a gas diode under atmospheric air pressure, when nanosecond voltage pulses with amplitude of hundreds of kilovolts are applied. The electron beam has been generated behind a 45 μm thick AlBe foil in SF_6 and Xe under the pressure of 2 atm, and in He under the pressure of about 12 atm. The paper gives the analysis of a generation mechanism of SAEB.

1. Introduction

For the first time an electron beam behind a foil was generated at a gas-filled gap breakdown at atmospheric pressure in 1974 in [1]. The recorded e-beam duration at a half-height was no more than 1.5 ns. Generation of electron beams was realized by using a nanosecond discharge in a non-uniform electric field. Nanosecond high-voltage pulses (>100 kV) were applied to a gap with a leading edge about 1 ns. In [1] it was supposed about the shorter duration of a beam current of runaway electrons, yet direct measurements of the beam current duration with subnanosecond resolution were not made. In a number of papers (see the review [2] and Refs in [2]) the authors tried to estimate the beam current duration by the duration of light pulses of scintillation plastic materials. Such measurements of the beam current pulse duration gave the value less than 1 ns.

Since 2003, an interest to investigation of e-beams generation in gas-filled diodes with high pressures has been rekindled (see the review [3-5] and papers [6-20], and Refs in [3-20]). In 2005, the advanced recording methods of electron beams and the use of digital oscilloscopes with wide bandwidth provided the measurements of the beam current duration with time resolution of ~ 100 ps. It was established that duration of a SAEB pulse at a half-height at atmospheric pressure of different gases is less than 100 ps (~96 ps in [7]). Such duration approached the limit resolution of the Tektronix TDS6604 oscilloscope used in [7]. The measurements performed in [8] gave the value of the beam current duration of runaway electrons at a half-height of 120-130 ps. It was supposed in [11] that the true value of the beam current pulse duration of runaway electrons should be ~10 ps. However, a question of duration and amplitude of a beam current, as well as of an electron energy distribution is still open. The authors of [1] reported that the number of electrons behind a foil, generated in atmospheric pressure air is ~10^9, yet they did not give any value of amplitude there. The following papers by that scientific group did not report about any higher value of the earlier reported number of electrons of ~10^9, but they specified that most registered electrons have anomalous energies, being greater than the gap voltage (see review [2] and Refs). Our estimation of the beam current amplitude of [1], judged by the number of electrons and duration of the beam current pulse at a half-height of ~100 ps gives the value of ~1.5 A. The our papers reported about the SAEB amplitude of tens-hundreds of amperes, and how oscilloscope type and recording methods used in experiments exert influence on amplitude and duration of a recorded signal. The beam current amplitude of 1.5 A with duration of 130 ps at a half-height was obtained in [8] by using the voltage pulses with an amplitude of 270 kV, and duration at a half-height of 250 ps in air under atmospheric pressure.

In order to obtain all above-said results meant for recording (estimation) the number of electrons or beam current amplitude and discharge current, the shunts made of TVO resistors or film resistors, collectors of various design, and a calorimetric method suggested in [21] to determine the nanosecond beam current amplitude were used. At that, the measurement results on peak current of runaway electrons reported by different authors, and by using different methods essentially differed. The electron energy distributions in air gas diodes at atmospheric pressure, obtained by different scientific teams differed as well [2, 7, 10, 20].

In this paper, the recent measurement results on duration and amplitude of e-beams generated at a nanosecond discharge in air at atmospheric pressure and others gases have been summarized.

2. The experimental set-ups and techniques

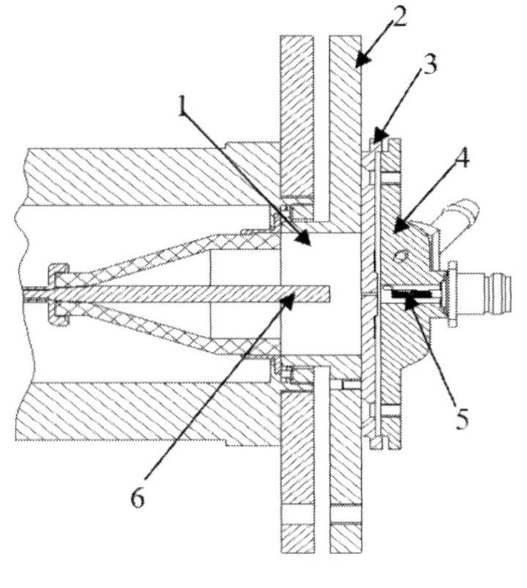

Figure 1. Schematic drawing of the gas-filled diode no.1 with collector no. 1: (1) gas-filled diode, (2) flange of the gas-filled diode, (3) collimator or a 250-μm-thick Cu foil with a 1-mm-diameter hole (a 50-μm-thick AlBe foil or a 10-μm thick Al foil is installed between 2 and 3), (4) collector's body, (5) receiving element of the collector (a 3-mm-diameter cylinder or a cone with a 5-mm-diameter base), and (6) cathode.

The experiments were carried out using a new subnanosecond pulser SAEB-150 (pulser no.1) and two nanosecond pulsers RADAN-220 [22] (pulsers no. 2 and no. 3) as well as three gas diodes. The pulser no. 1 generated a voltage pulse of ~130 kV at the voltage pulse FWHM <1 ns and voltage rise time of ~0.3 ns. The design of a gas diode connected to SAEB-150 and collector no.1 is shown in Fig. 1. The pulser no. 2 with the 20-Ohm impedance generated a voltage pulse in the discharge gap with the amplitude of ~230 kV and duration at FWHM of ~2 ns, at the voltage rise time in a transmission line of ~0.5 ns. The pulser no. 3 with a 20-Ohm impedance generated a voltage pulse in the discharge gap with the amplitude of ~270 kV and duration at FWHM of ~2 ns, at the voltage rise time of ~0.5 ns

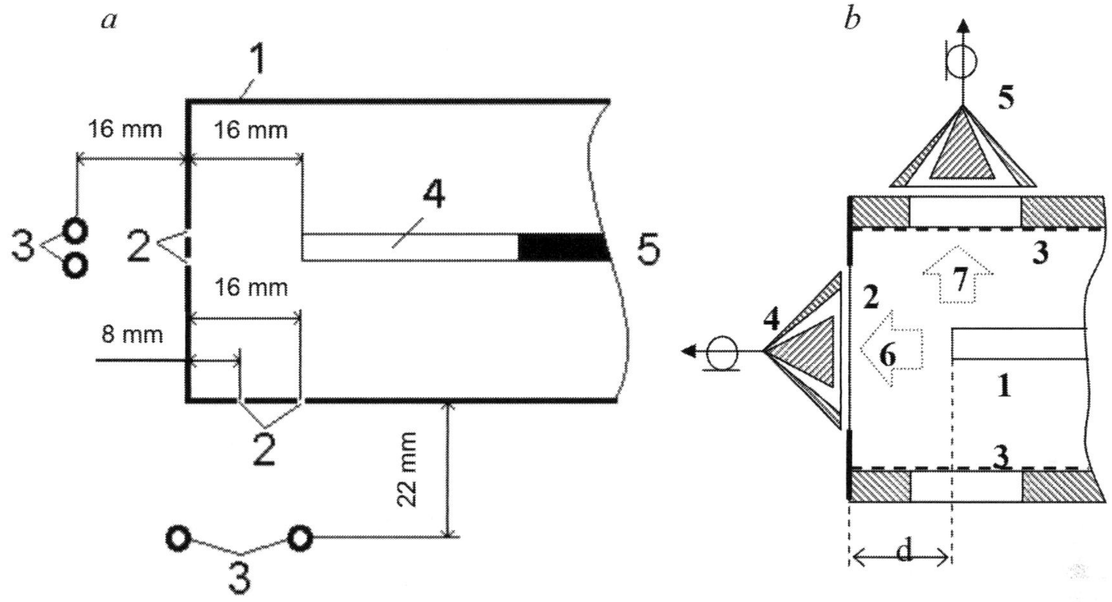

Figure 2. a - Schematic drawing of the gas-filled diode with the pinhole camera: 1 – diode's box, 2 – 350-μm-diameter holes; 3 – electrons beam's and x-ray radiation's autographs on a film, 4 – tube cathode, 5 – the place of a high-voltage nanosecond pulser connection.
b - Schematic drawing of the gas-filled diode and the registration system of a electron beam's spatial distribution: 1 – tube cathode; 2 – flat (foil, grid) anode; 3 – grid; 4, 5 – collectors to register an electron beam's current pulse along (4) and perpendicular (5) to longitudinal diode's axis; 6, 7 – electron beam's flows; d – cathode-anode distance.

(when connected to a gas diode with minimum inductance) or ~0.7 ns (when setting a gas diode with an additional chamber, Fig. 2).

The gas diodes demonstrated in Figs. 1 and 2. The design was similar to the one earlier described in [4-7]. The same way as in the most experimental studies devoted to formation of X-ray radiation and runaway electrons in the gas diodes, we used a flat anode and a small-size cathode in order to have an additional gain of electric field in the near-cathode region. The cathode was a stainless steel tube of ~6-mm diameter and 100-μm wall thickness, installed on a metallic rod of ~6 mm in diameter. The flat anode used for e-beam extraction was made of a 45-μm thick AlBe foil or Al-foil of 10-15 μm in thickness. The cathode – anode distance was varied from 6 to 30 mm. The gas diodes shown in Fig. 2 had additional chambers with windows or holes on the sidewalls. That provided recording of an electron beam and X-ray radiation transversely to the gas diode axis, although the inductance between the gas gap of the gas diode and the pulser increased. The gas diode shown in Fig. 2a had the holes (2), which served as an obscure objective. The gas diode shown in Fig. 2b had the grid-covered side windows and the second collector was placed behind them. Figure 1 shows a part of a gas diode and a design of the collector no.1 with the receiving part diameter of 3 mm and the time resolution being no worse than 50 ps [18]. The beam current recording was carried out both when the gap between the foil and receiving part of the collector was pumped out and the gap was filled in with the gas from the gas diode. At small gaps between the receiving part of the collector and foil any essential influence of air or more light gases at the atmospheric pressure on the beam current amplitude with the pulse length of ~100 ps was not observed. The diameter of the receiving part of the collector no.4 was 25 mm, fig.2b, which provided recording the beam current over the larger area of the foil. However, time resolution of that collector was worse, being of ~0.1 ns.

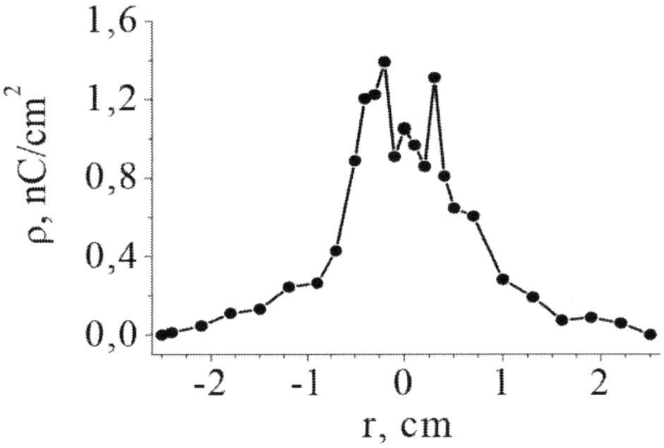

Figure 3. Charge dense distribution immediately behind a foil with the pulser no. 3 using.

To receive signals from a capacitive divider, collectors and a shunt, the oscilloscopes TDS-6604 with 6 GHz-band with 20 GS/s and DPO70604 with 6 GHz-band with 25 GS/s were used. Note, that in some regimes, the value of the current beam amplitude behind the foil could vary from pulse to pulse. Therefore, the maximum amplitudes of e-beam current are shown for all regimes. Amplitude instability of the gap voltage pulses and gas diode discharge current were low, not exceeding 10%. The discharge glowing was photographed by a digital camera. X-ray radiation was registering by Direct Reading Dosimeters (Arrow-Tech, Inc. Model 138). Such dosimeters made registration of X-ray radiation with quanta energies higher than 16 keV. With the diode geometry shown in Fig. 1, and the pulser RADAN-220 used, the dose was ~1.5 μR/pulse at a 2-mm distance from the anode (20-μm copper foil). Besides that, X-ray radiation and e-beam were registered by RF-2 film placed into a ~100-μm thick black envelope. The envelope was placed either just after the anode (by the edge or side surface of the gas diode covered with grid or (and) Al foil of 10-50 μs thickness, or at a definite distance from the anode made of copper foil of 250 μm thickness, Fig. 2a. In the second case (obscure chamber), the foil had the holes with the diameter of ~350 μm providing formation of images on RF-film. In the experiments with the film placed into a black envelope, images on the film appeared resulting from the e-beam and X-ray radiation influence. The experiments show maximal RF-film blackening due to the effect caused by the e-beam and X-ray radiation appearing at the e-beam deceleration by the envelope and foil. In the experiments with an open film (without the envelope), the discharge plasma image was formed on the film. The film was sensitive to UV and visible spectrum radiation. Time evolution of X-ray radiation pulses was registered using a semiconductor detector SPPD11–04 with the time resolution of ~1.5 ns. The detector makes registration of X-ray quanta with the energies ranging from 0.4 to 40 keV. Nevertheless, the measurements show that the detector had rather higher sensitivity to the beam electrons than to X-ray quanta. Therefore, an additional filter made of lavsan or polyethylene was set ahead of the detector.

3. Results of measurements of SAEB parameters

The main experimental results are presented in Figs.3-8 As it was determined previously [3-6], maximum parameters of SAEB are realized during formation of a volume discharge in a discharge gap. The present investigations confirmed this conclusion. Maximum beam currents behind the foil were recorded if the discharge contraction occurred after SAEB generation.

SAEB generation was investigated in most detail in the atmospheric pressure air. Maximum amplitude of the e-beam (number of electrons) was obtained with the pulser no.3 and the gas diode that had the minimum inductance of connection to the pulser. To determine a charge behind the foil, we used a

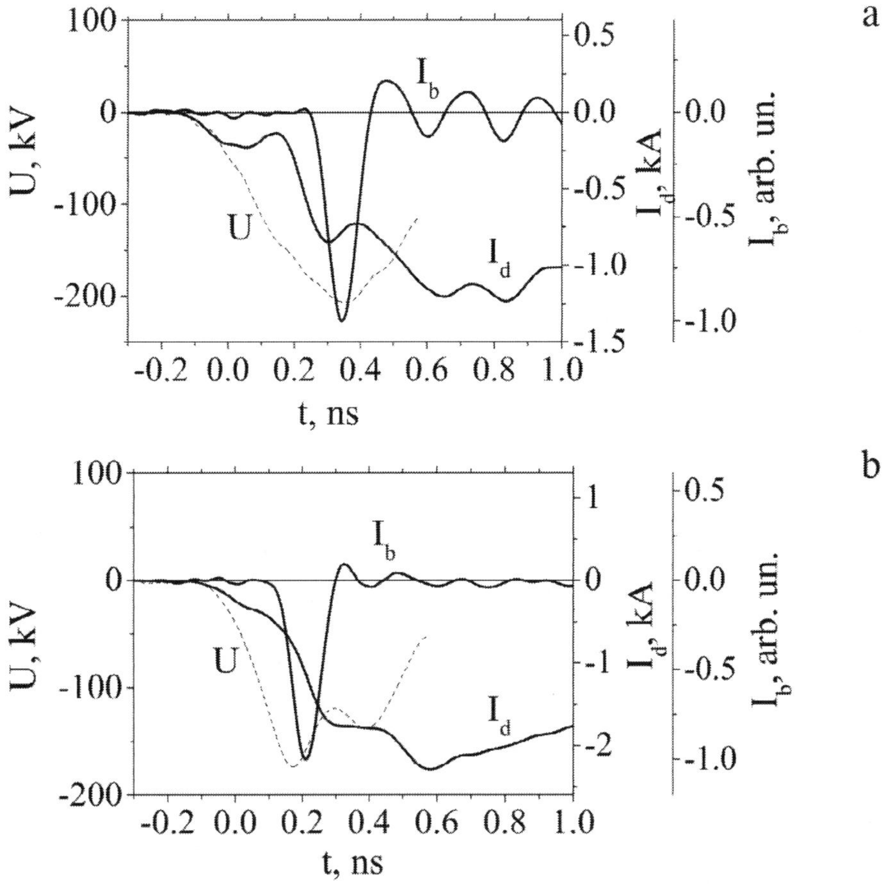

Figure 4. Waveforms of the voltage across the gap U, discharge current I_d and electron beam current I_b behind a foil at the pulser no. 1; spherical (a), and tubular (b) cathode using.

collector no.1 that had a 5-mm diameter receiving part and moved over the whole surface of the foil, Fig.3. Note, that the beam current behind the foil was recorded over the whole foil surface, even at the periphery, in the places near the sidewall of the gas diode case. The total charge value measured behind a 50-μm thick AlBe foil was of $\sim 3.5 \cdot 10^{-9}$ coulomb and behind a 10-μm thick Al foil it was of $\sim 4.4 \cdot 10^{-9}$ coulomb. This charge corresponds to the number of electrons behind a 10 μm thick Al foil that equal to $\sim 2.7 \cdot 10^{10}$. This value exceeds the number of the electrons obtained behind the 8- μm thick Al foil by more than an order of magnitude in [1]. Electrons with the energy of 10-40 keV being of $\sim 2 \cdot 10^{10}$ in number were not taken into account in the beam current amplitude. The SAEB amplitude (estimation by the maximum electron charge behind the foil) was of ~50 A at the beam extraction through a 10-μm thick Al foil. The beam current pulse length at a half-height was set to be equal to ~90 ps, i.e., corresponded to the duration measured in the given experiments. This amplitude of SAEB is confirmed by the measurements made using a collector no.2 that had a receiving part diameter equal to 20 mm. Increase of the collector receiving part diameter resulted in the proportional rise of the SAEB amplitude, however, the collector time resolution decreased (to 100 ps at the receiving part diameter of 20 mm). The obtained amplitude of SAEB exceeds the beam current of the runaway electrons behind a 15-μm thick Al foil by more than an order of amplitude if a graphite cathode is used in [8].

In the paper [11] it was supposed that the beam current duration of the runaway electrons equals to ~10 ps. At such duration of the beam current pulse and measurements of the electron charge behind

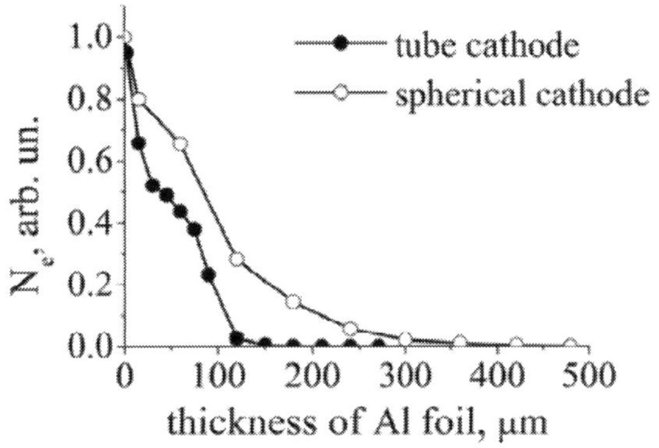

Figure. 5. Attenuation curves of the electron beam current corresponding to cathodes in the form of sphere and tube.

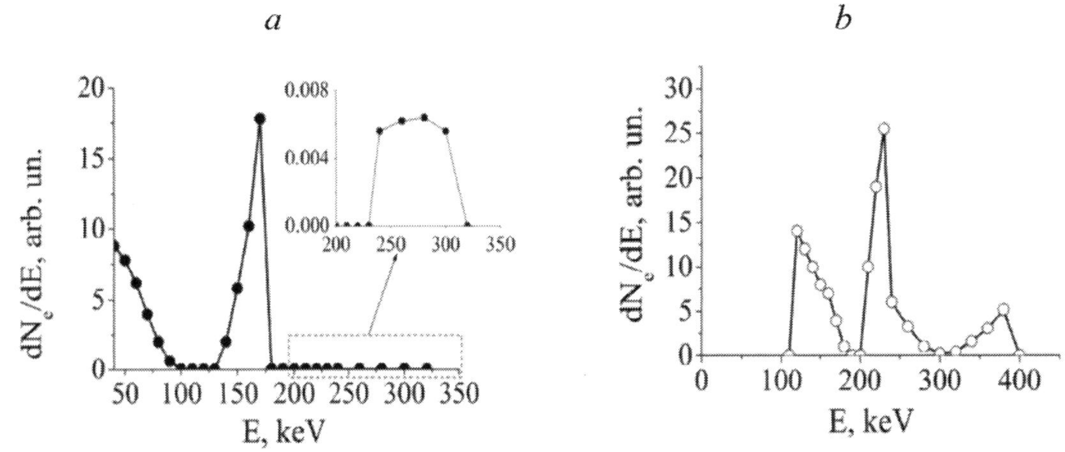

Figure. 6. Energy spectra of the runaway electrons corresponding to cathodes in the form of tube (a) and sphere (b).

the foil, the SAEB amplitude will increase in inverse proportion to its duration. E.g., when the pulse WFHM equals to 10 ps and the number of electrons is ~$2.7 \cdot 10^{10}$, the beam current amplitude will be more than 400 A. However, we consider that the real amplitude of SAEB is less and the beam current pulse duration is more than 10 ps. At present time, a design of a collector having a high resolution and a receiving part size enclosing the whole diameter of the foil in the gas diode is unknown. As we noted before, the beam current is recorded over the whole surface of the gas-filled diode foil window that was equal to ~50 mm in the experiments.

Fig. 4 presents the oscillograms of the voltage pulses at a gap, the discharge current and the beam current behind the foil with a pulser no. 1. To record the oscillograms (Fig.4) with the best time resolution was used. The oscillograms in Fig.4 are synchronized in time with each other. It is seen from the pictures that current through the gap increases practically at once after the voltage supply to the gap and equals to hundreds of amperes at the moment of SAEB generation commencement. The gap voltage for the pulser no. 1 at this moment of time corresponding to SAEB pulse commencement is approximately 200 kV. The beam current amplitude at a fixed duration of the voltage pulse edge grows with the voltage amplitude increase at the moment of SAEB generation. If the voltage pulse amplitude is fixed and the duration of its leading edge is changed, then at the pulse edge duration increase higher than 400 ps the SAEB amplitude usually decreased.

a

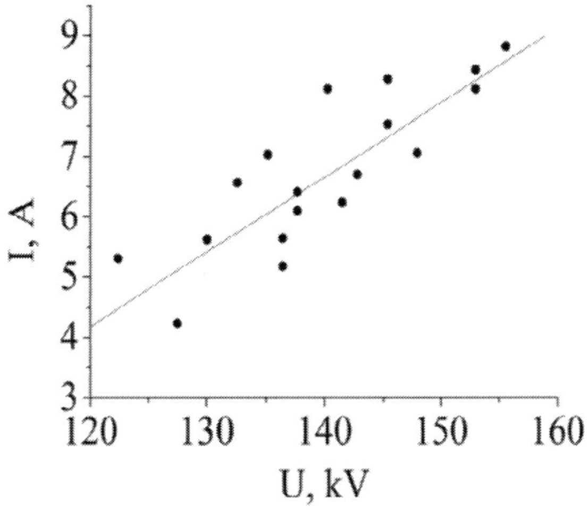

b

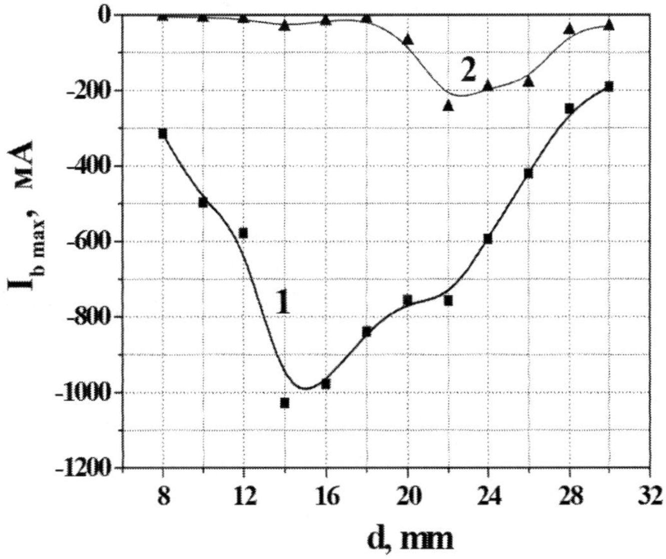

Figure 7. Dependence of the electron beam current amplitudes on the voltage amplitude (a); dependences of the electron beam current amplitudes (b) registered along (1) and perpendicular (2) to the longitudinal diode axis on the anode-cathode distance d.

Fig.5 presents the curves of the beam current amplitude attenuation when the aluminum filter thickness is increased and Fig. 6 presents the recovered distributions by energies of the beam current electrons obtained with the pulser no. 1. Two different cathodes were used in these experiments. Three groups of electrons were registered with both cathodes. Recovering electron spectra by the beam attenuation curves in the foils of different thickness, we used a refined technique described in [10]. Improvement of the electron spectra estimation technique owing to more precise definition of electron energy losses at their braking in the foil resulted in displacement of maxima at the distribution towards high energies. The shape of the spectrum has not changed essentially in this connection. The first group usually has the electron energy less than 50 keV. The second group of electrons corresponds to the generation of SAEB between the front of the dense plasma and anode. Maximum of electron energy at the electron distribution by energies in SAEB corresponds approximately to the maximum gap voltage. This implies that maximum at a voltage pulse on the gap and maximum of SAEB amplitude occur approximately at the same time. The third group of electrons under existing conditions had the energy essentially exceeding the one achieved by the electrons in vacuum at the maximum voltage on the gap. It was determined that the number of such electrons depends on the cathode design and the voltage pulse leading edge. When a cathode with a small curvature radius was used, e.g., in the form of a tube made of foil, their number was small, less than 10% of the number of electrons in the beam behind the foil. When the voltage pulse leading edge was increased up to 1.5 ns, anomalous energy electrons were not recorded. These measurements as well as the results obtained earlier [7, 10, 20], contradict to the results of [2] reporting about obtaining runaway electrons chiefly with "anomalous" energy in the atmospheric pressure air. We consider that in the paper [2] there was made a mistake in the gap voltage measurements and maximum at the electron distribution by energies corresponds to the maximum gap voltage. This is confirmed indirectly by the electron spectrum obtained in [2] at the air pressure of 0.2 Torr and the cathode design made in the form of a cone in this work.

Essential number of electrons with "anomalous" energy is generated when using a cathode with a large curvature radius. Such spectrum is shown in Fig.6b. Evidently, essential number of electrons with "anomalous" energy was obtained as well in the work [1] where a cathode with a large curvature radius was used.

When conducting the investigations, we discovered a new important feature of SAEB generation. It turned out that electron beam generation occurs not only in the anode direction but also transversely to the gas diode axis in the direction of its side walls (into the angle exceeding 2π sr). Measurements were made with a gas diode (Fig. 2(b)) for which the breakdown voltage was equal to ~250 kV at d = 12 mm. Time delay t_d of the gap breakdown depended on d as well and was equal to ~700 ps at d = 12 mm. Fig.7 presents the SAEB current amplitude recorded from the gas diode face end versus the value of the interelectrode gap between the cathode and plane anode. The amplitude of SAEB recorded through a lateral window versus the interelectrode gap value is presented here as well. Both dependences were recorded at a simultaneous registration of the beam current from the face end and from one side. Decrease of d resulted in decrease of SAEB current amplitude from one side; however, even at the gap of 8 mm the beam a side collector registered current. The beam current pulse shape from both collectors was similar. The FWHM of the pulses was of ~100 ps. Both signals came to the oscilloscope practically simultaneously. Pulse duration of the X-ray and the beam current recorded with a detector SPPD11–04 was equal to ~2 ns. Evidently, a real FWHL of X-ray pulse is essentially shorter but for its measurement it is necessary to have a detector with higher time resolution.

As it appears from the results of the experiments, SAEB is generated both in the main spacing at the gas diode axis and between the cathode (cathode holder) and the gas diode case. Generation of SAEB to the gas diode case occurred even at a small main gap (d = 8 mm). This gap was three times less than the distance from the cathode holder to the lateral wall (24 mm). A full solid angle to which SAEB was generated exceeded 2π sr in the given experiments. Correspondingly, SAEB amplitude at registration of run-away electrons in all directions essentially exceeds the beam current amplitude measured behind the foil installed transversely to a longitudinal axis of the gas diode. This difference will increase slightly at the lengthening of the cathode holder and the gas diode lateral sides. However, at the cathode holder lengthening the beam its length generates current at different time and this will result in the increase of the beam current pulse duration but its amplitude will have no changes.

The absorption curves plotted from the data concerning the beam current attenuation from the face end and one side of the diode turned out to be the same at the equal interelectrode gaps. This testifies to the similarity of the energetic spectrum of the electron beams directed both along and transverse to the diode axis.

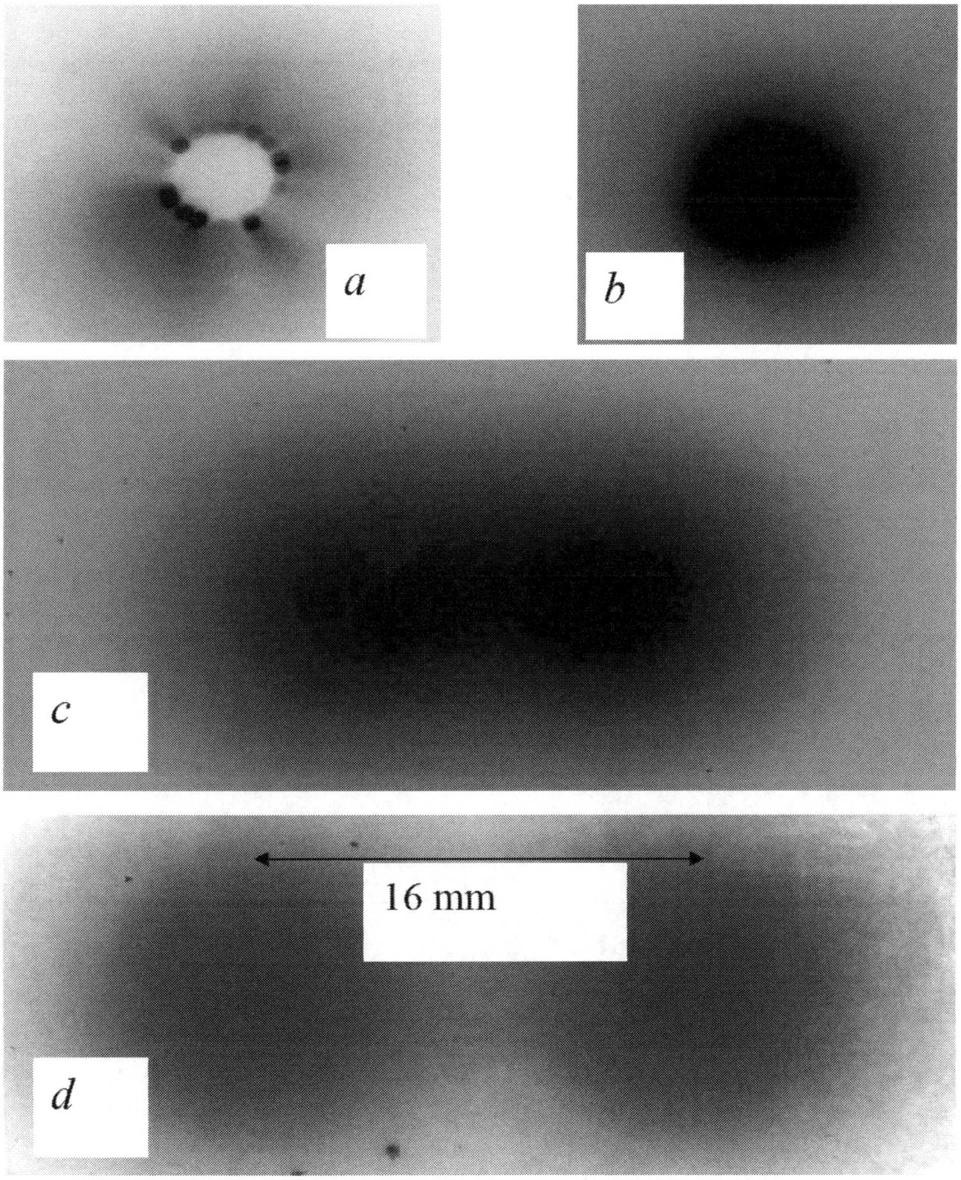

Figure. 8. Pinhole camera images obtained for conditions depicted on the figure 3. The images were taken with (b, c, d) and without (a) light-proof envelop at using one (a, b) and two (c, d) holes with a diameter of 350 μm in flat Cu anode with a thickness of 250 μm.

Testing of the results obtained at SAEB current measurement by a collector was made at a gas diode that was similar to the gas diode in Fig. 2(a). At the face end and at one side of this gas diode there were windows covered with a thin aluminum foil. Behind the windows there was placed a photographic film in a black paper envelope. This allowed recording the X-ray radiation and the beam current electrons from the face end and from the side of the gas diode by the film blackening. This result also confirms the electron beam generation into a solid angle exceeding 2π sr. Besides, with the gas diode no.2b there were taken the autographs of the electron beam registered by a photographic film as well as the photos of the discharge plasma glowing in the visible and UV range. Fig.8 presents the obtained images. At the discharge images in the rays of UV and visible spectrum regions, Fig. 8(a), one can see brighter cathode spots and diffuse jets at the cathode. The obtained pictures coincide with the photographs made by camera in the similar conditions. However, the image obtained at the simultaneous exposure of the electron beam and X-ray radiation to the film placed into an envelope differs essentially from the discharge photograph. Based on the analysis of the images in Fig.8, one can suppose that in these conditions an electron beam is generated from the region with the outer diameter of ~12 mm. This region has the shape close to the spherical one and occupies approximately a half of the discharge gap. Here, the electron beam generation occurs not only in the anode direction, Fig. 8(b) and Fig.8(c) but also transversely to the gas diode longitudinal axis in the direction of its lateral walls, Fig.8 (d) (to the solid angle exceeding 2π sr). It should be noted that the electron beam generation region is less than it is registered at the images possibly due to the electron scattering in the air.

Investigation of SAEB generation in other gases at the atmospheric pressure confirmed the well-known data [1-20], that its amplitude is maximal at the gas diode filling with light gases (helium and hydrogen). However, we managed to increase considerably the pressure of various gases at which SAEB was recorded behind the foil. Thus, in the given work with the pulser no.3 the electron beam behind the foil was obtained at the helium pressure of 12 atm.

At the gas diode filling with heavy gases, the runaway electron beam amplitude at the atmospheric pressure decreases in comparison with the air. We have managed to increase the pressure for these gases at which SAEB is generated. So, SAEB was obtained in SF6 and xenon at the pressure of 2 atm due to optimization of the gas diode with the pulser no. 3. Besides, increase of SAEB pulse FWHM was revealed for the mixtures with SF6. At the increased pressures (1.6-2 atm) the SAEB pulse duration increased up to ~150 ps. This result is important for determining a mechanism of the electron beam generation in a gas diode.

4. Interpretation of results

The obtained data can be explained in the following way. When voltage pulses with the amplitude of hundreds of kilovolts are supplied to the gap, then the high electric field strength is achieved due to the current gain at the micro- and macro heterogeneity. This results in the field emission from the cathode at the voltage pulse leading edge. Here, we will testify this conclusion by means of simple estimations. Amplification of the electric field β at its surface for a cathode in the form of a blade can be evaluated by the formula [23]: β~(r·d)-0.5, where β=E/U, E is the field strength near the cathode, V/cm; U is the applied voltage, V; d is the distance between the cathode and anode, cm; r is the radius of the cathode rounding, cm. In the used gas diode d is ~1 cm, r=50 μm, respectively, β≈14. Micropoints at the cathode provide the field strengthening 10-100 times more [24]. Thus, the electric field E near the cathode surface at the voltage pulse edge reaches the values of ≥5·107 V/cm [2, 11] that is sufficient for appearance of appreciable field-emission current at the voltage pulse edge. It should be taken into account as well that oxide films at a cathode surface appearing in the presence of air decrease the value of the voltage strength at which the field emission is observed.

Owing to the electric field amplification near the cathode, the emitted electrons can acquire the energy sufficient for their runaway (fast electrons in the near-cathode region of the amplified electric field). When moving away from the micropoints at the cathode and the cathode edge, the electric field strength is decreased and fast electrons lose their energy mainly on particle ionization in the near-cathode region. Simultaneously, secondary and field-emission electrons formed in the electric field that grows owing to the voltage increase at the gap, initiate the development of electron avalanches. Concentration of initial electrons is so great that the heads of avalanches are overlapped up to the streamer formation and relatively dense plasma of a diffusion discharge is formed. The edge of the plasma moves from the cathode to anode. Under these conditions, as it was shown previously [3-6], a diffusion discharge is formed in the gap at the dense plasma edge formed by the heads of avalanches moving to the cathode resulting in the excess negative charge.

When plasma reaches a definite size and, correspondingly, the gap between the plasma edge and anode decreases and the voltage in the spacing increases, a critical field E_{cr} is achieved between the plasma boundary and anode and an electron beam is generated. According to [3-5], the value of E_{cr} at which an essential part of electrons turns into the runaway mode can be determined from the expression $\alpha (E_{cr,p})d=1$, where α is the first Townsend coefficient, p is the gas pressure and d is the interelectrode gap. This criterion was obtained for the gaps with the uniform electric field at small (close to zero) initial energies of electrons. It should be noted that increase of the initial energy of electrons results in the critical field decrease and electrons turn into the runaway mode at smaller electric fields [25].

We suppose that under conditions of the given experiment the energy of a part of electrons at the dense plasma edge increases all the time due to their acceleration at the initial part of the voltage pulse edge and excess negative charge in the heads of avalanches and can reach the units of keV [2]. For these electrons, the electrical field in the gap for the conditions of Fig. 8 becomes sufficient for runaway. An excess negative charge of electrons at the dense plasma edge increases the electric field in the gap and the energy preliminary gathered by some part of electrons promotes decrease of the critical field value, as we noted before.

As it was determined in the paper presented, the source of electrons and X-ray radiation is a region that has the form close to the spherical one, Fig. 8. We believe that in this region the dense plasma is formed to the moment of SAEB generation with its front edge moving from the cathode to the anode and lateral walls of the gas diode. Owing to the fact that the dense plasma near the cathode has the form close to the spherical one, SAEB is generated for the electrons at the ionization wave front edge with the highest energy to the solid angle exceeding 2π sr. Further, the beam electrons continue accelerating at their motion to the anode (gas diode case). The number of runaway electrons can be increased at the expense of ionization cascade electrons. The energy of the electrons is increased both at the expense of the voltage increase at the gap and effect of polarization self-acceleration [26]. Thanks to the polarization self-acceleration, a small part of the beam electrons will have an essential energy. Note, that a part of energy of the runaway electrons is spent on the gas ionization in the gap providing the ionization wave motion [27]. Limitation of SAEB pulse duration is conditioned by the gap bridging with an ionization wave and electric field alignment in it. One can suppose from small lengths of SAEB that it reaches the anode practically simultaneously with the ionization wave front. Increase of SAEB current pulse duration at the increase of SF6 pressure can be explained by decrease of the rate of movement of the ionization wave in a heavy gas and, correspondingly, more late gap bridging by the ionization wave and electrical field alignment in it.

Mechanisms of electron beam generation in the gas diodes suggested in [2, 11] have no agreement with the above presented experimental results. However, we do not insist that the suggested mechanism of SAEB generation works absolutely in all experimental conditions. One can suppose that alteration of the experiment conditions can affect the physical processes in the gas diode. We plan to make a comparative analysis of different mechanisms of electron beam generation in the gas diodes in a separate work.

5. Conclusion

This paper reports the properties of supershort avalanche electron beam formed in different gases under atmospheric pressure and analyses the SAEB generation mechanism. It is shown that at a nanosecond discharge in the atmospheric pressure air and SAEB current recording through a small diameter area the pulse duration behind the foil is not larger than 80 ps. In order to record the pulse shape, it is necessary to use a small-size coaxial collector loaded to a high-frequency cable and to read a charge density distribution over the foil surface for determining the SAEB amplitude by means of this collector. The obtained electron distribution over the foil section should be compared with the

distribution obtained per pulse by means of a luminescent screen or a photographic film. An analysis of those data shows that at the beam current duration at a half-height of ~90 ps the beam current amplitude behind the 10-μm thickness Al-foil is ~50 A. It is demonstrated that SAEB is generated into an angle exceeding 2π sr in the air at the pressure of 1 atm. Decrease of the gap d between the cathode and anode results in the amplitude decrease of the beam current generated to the lateral walls of the gas diode. However, even at small gaps (d=8 mm) of the interelectrode intervals installed both along and transversely to the gas diode axis. The beam current is generated to the gas diode lateral surface placed at a certain distance from the cathode that exceeds several times the gap in the interelectrode intervals. Generation of the electron beam into the solid angle exceeding 2π sr is explained by the fact that at SAEB generation the dense plasma near the cathode has the form close to the spherical one. An electron beam was obtained behind AlBe foil of the thickness 45 μm at the pressure of SF6 and xenon in a gas diode up to 2 atm and at the pressure of helium up to 12 atm.
In conclusion we note that Tektronix Company created a unique oscilloscope DPO/DSA72004 (20 GHz-band with 50 GS/s). Researches made by means of this oscilloscope should allow obtaining new data concerning characteristic features of SAEB generation in gas diodes.

References

[1] Tarasova L, Khudyakova L, Loiko T, Tsukerman V 1974 *Tech. Phys.* **44** 564
[2] Babich L *High-Energy Phenomena in Electric Discharges in Dense Gases: Theory, Experiment, and Natural Phenomena*. - Arlington. VA. USA. Futurepast Inc., 2003. - 358p.
[3] Tarasenko V, Yakovlenko S 2004 *Phys. Usp.* Vol. **47** 887
[4] Tarasenko V, Yakovlenko S 2005 *Phys. Scripta* **72** 41
[5] Tarasenko V, Yakovlenko S 2005 *Plasma Devices and Operations* **13** 231
[6] Tarasenko V, Orlovskii V, Shunailov S 2003 *Russian Physics Journal* **46** 325
[7] Tarasenko V, Shpak V, Shunailov S, Kostyrya I 2005 *Laser and Particle Beams* **23** 545
[8] Mesyats G, Korovin S, Sharypov K, Shpak V, Shunailov S, Yalandin M 2006 *Phys. Tech. Lett.* **32** 35
[9] Krompholz H, Hatfield L, Neuber A, Kohl K, Chaporro J, Ryu H 2006 *IEEE Trans. of Plasma Science* **34** 927
[10] Tarasenko V, Kostyrya I, Petin V, and Shlyakhtun S 2006 *Technical Physics* **51** 1576
[11] Mesyats G 2007 *Phys. Tech. Lett.* **85** 119
[12] Kostyrya I, Tarasenko V, Tkachev A, and Yakovlenko S 2007 *Technical Physics Letters* **33** 309
[13] Bratchikov V, Gagarinov K, Kostyrya I, Tarasenko V, Tkachev A, and Yakovlenko S 2007 *Technical Physics* **52** 856
[14] Kostyrya I, Tarasenko V 2007 *Technical Physics Letters* **33** 424
[15] Baksht E, Lomaev M, Rybka D, and Tarasenko V 2007 *Technical Physics Letters* **33** 216
[16] Baksht E, Tarasenko V, Lomaev M, and Rybka D 2007 *Technical Physics Letters* **33** 373
[17] Erofeev M, Kostyrya I, and Tarasenko V 2007 *Technical Physics* **52** 1291
[18] Baksht E, Balzovskii E, Klimov A, Kurkan I, Lomaev M, Rybka D, and Tarasenko V 2007 *Instruments and Experimental Techniques* **50** 811
[19] Baksht E, Tarasenko V, Lomaev M, Rybka D, Tkachev A, and Yakovlenko S 2007 *Laser Physics* **17** 1124
[20] Tarasenko V, Rybka D, Baksht E, Kostyrya I, Lomaev M 2008 *Prib. Tech. Eksp.***2**. 62
[21] Shpak V 1980 *Prib. Tech. Eksp.* **3** 165
[22] Yalandin M, Shpak V 2001 *Instruments and Experimental Techniques* **44** 285
[23] Djuzhev N, Tishin J 2001 *Electronics: Science, Technology, Business* **1** 50
[24] Korolev Yu, Mesyats G *Field-Emission and Explosive Processes in Gas Discharges* Novosibirsk: Nauka, 1982
[25] Gurevich A, Zybin K 2001 *Physics-Uspekhi* **44** 1119
[26] Askar'yan G 1973 *Trudy FIAN* **66** 66
[27] Vasilyak L, Kostyuchenko S, Kudryavtsev N, Filigin I 1994 *Physics Uspekhi* **37** 247

Spectroscopic Study of Hydrogen Rotational, Vibrational and Translational Temperatures in a Hollow Cathode Glow Discharge

G Lj Majstorović

Military Academy, 11105 Belgrade, Generala Pavla Jurišića - Šturma 33, Serbia

E-mail: gordana.majstorovic@va.mod.gov.yu

Abstract. The optical emission spectroscopy is used for determination of hydrogen molecule rotational T_{rot} , vibrational T_{vib} and translational temperature T_{tr}. The experiment was carried out using copper or titanium hollow cathode discharges operated in pure hydrogen. The rotational temperature of excited electron energy levels is determined from Fulcher-α diagonal bands ($d^3\Pi_u \rightarrow a^3\Sigma_g^+$ electronic transition, P, Q and R branches), which were later used to derive rovibronic temperature of the hydrogen molecule ground state, which is assumed equal to T_{tr}. The vibrational temperature T_{vib} is determined for the same electronic transition ($d^3\Pi_u^- \rightarrow a^3\Sigma_g^+$, with $v' = 2{,}3\,;\Delta v\ = 0$). For all temperatures the radial temperature distribution is reported.

1. Introduction

Hydrogen hollow cathode glow discharges (HCGD) have been extensively used for study of fundamental discharge processes as well as for wide variety of applications [1]. For instance, recently, this type of discharge was utilized for producing hydrogen by reforming natural gas [2].

The translational or gas kinetic temperature in gas discharges is a parameter of utmost importance in the field of plasma chemistry while vibrationally excited neutral hydrogen molecules play significant role in the chemistry of weakly ionized hydrogen plasmas [3,4,5]. This is why the modeling of cold, reactive hydrogen plasma includes rotational and vibrational energy of the hydrogen molecule [5].

Several diagnostic techniques are developed to determine gas kinetic temperature T_g like coherent anti-stokes Raman scattering (CARS) [6, 7], laser-induced fluorescence (LIF) [8, 9] and optical emission spectroscopy (OES) [8, 10, 11, and 12]. Here we selected OES. This diagnostic technique provides information about atom, molecule and ion density in excited and ground state, as well as rotational, vibrational, and gas temperature including the excitation temperature of certain group of excited levels [13]. The technique is applied for measurements of the rotational T_{rot}, vibrational T_{vib} and determines translational temperatures in a hollow cathode glow discharge in hydrogen. The rotational temperature of excited electron energy levels is determined from the Boltzmann plot of intensities of rotational molecular hydrogen lines belonging to Fulcher-α diagonal bands ($d^3\Pi_u^- \rightarrow a^3\Sigma_g^+$ electronic transition; P, Q and R branches). Following procedure described elsewhere [14, 15] the temperature of ground state rovibronic levels is evaluated. Constrains of rotational temperature are discussed in detail. The vibrational temperature, after certain precautions, is also determined [16, 17, 18, 19], but from the ratio of relative intensities of the H_2 Fulcher-α diagonal bands ($d^3\Pi_u^- \rightarrow a^3\Sigma_g^+$ electronic transition; Q-branches with $v'=2,\ 3$). The applied method for vibrational temperature determination is based on the assumption that population of rovibrational levels in excited (n', v') state is in accordance with the Boltzmann's distribution.

The dependence of these temperatures and their radial distribution within HC was determined and discussed [20].

2. Experiment

The hollow cathode glow discharge (HCGD) source with two symmetrically positioned kovar anodes and copper (Cu) or titanium (Ti) cathode is used as a discharge source. The HC tube was 50 mm (Cu) or 100 mm (Ti) long with 6 mm internal diameter and 1 mm wall thickness. The construction details of HC discharge source are described elsewhere [21].

All HCGD experiments were carried out with hydrogen (purity of 99.999%). The continuous flow of H_2 was sustained between 50 and 450 cm^3/min (at room temperature and atmospheric pressure) in the pressure range 2-10 mbar by means of a needle valve and two-stage mechanical vacuum pump.

To operate the discharge in a DC mode, a current stabilized power supply (0–2 kV, 0–100mA) is used. An air-cooled variable 10 kΩ ballast resistor is placed in series with the discharge and power supply. For all measurements, the cathode was grounded.

During the discharge operation, the cathode was either air cooled with a fan (110 mm dia; AC 220V/13W), placed 150 mm from discharge tube, or gradually heated by changing cooling rate of the fan. The temperature of the outer wall of the HC tube is measured by a K-type thermocouple.

The radial distribution spectra recordings were performed in equidistant steps perpendicular to the discharge axis, with an estimated spatial resolution of 0.30 mm. For radial intensity measurements the discharge was run between HC and rear anode, located at 20 mm from the cathode. All spectroscopic measurements of the gas discharge in HC are performed end-on and, in this way, the Abel inversion procedure required to convert integral radial intensity measurement into radial intensity distribution, see e.g. [22] is avoided. The light from the discharge was focused with an achromat lens (focal length 75.8 mm) onto the entrance slit of spectrometer (2 m focal length with 651 g/mm refection grating; the reciprocal dispersion of 0.74 nm/mm in first diffraction order). Spectral line recordings were performed with an instrumental profile very close to Gaussian with the measured full half-width of 0.018 nm. Signals from CCD detector (29.1mm, 3648 pixels) are A/D converted, collected and processed by PC.

3. Basics of hydrogen molecular temperature measurement

In previous studies Fulcher-α band spectroscopy was used because of its strong visible emission intensity and few perturbations. Namely, the most of the excited electronic states of hydrogen are considerably perturbed and simple adiabatic models like Franck – Condon approximation and Hönl-London formulae are not valid for calculation of transition probabilities in this case. The choice of the Fulcher-α, rather then $G^1\Sigma_g^+ \rightarrow B^1\Sigma_u^+$ molecular system was encouraged by this fact. Mainly Q-branches of the Fulcher-α band system are considered to be free from perturbations. Due to Λ-type doubling, $d^3\Pi_u$ state degenerates into the $d^3\Pi_u^-$ and $d^3\Pi_u^+$ states. The $d^3\Pi_u^-$ can only have a Q-branch, whereas the $d^3\Pi_u^+$ has P and R branches, in the spontaneous rovibronic emissions to $a^3\Sigma_g^+$. Since the degenerated $d^3\Pi_u^+$ state interacts strongly with the $e^3\Sigma_u^+$ state, the P and R branches of Fulcher-α are perturbed and relative transition probabilities for these lines differ from Hönl-London factors [15]. Therefore, we used Q-branches from $d^3\Pi_u^-$ state. In order to prove that the observed electronic transition have resolved rotational structure with almost fully separated lines, Fortrat parabolae of all three branches, see figure 1a, versus line wave numbers (m is an integer, which numbers successive lines) are drawn in figure 1b.

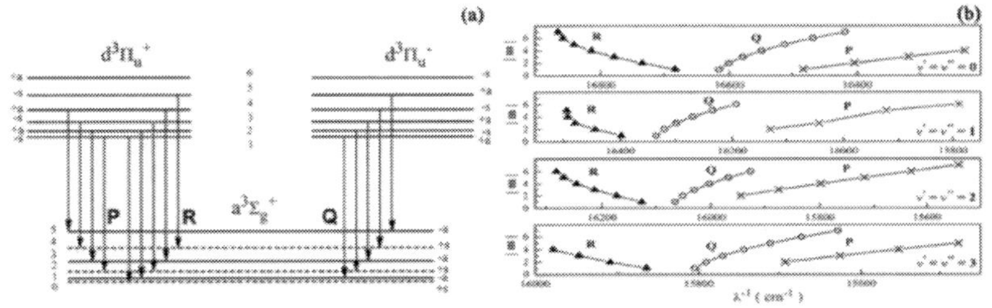

Figure 1: (a) Term diagram for the Fucher-α band transitions for $v' = v'' = 0$; and (b) Fortrat parabolae of the diagonal bands for $v' = 0,1,2,3$; where m is an integer, which counts successive lines in each branch.

3.1. Rotational temperature

Detailed description of the rotational temperature determination is given in [20] and, therefore, minimum details will be given here for completeness. If the populations of rotational levels in excited vibronic state (n', v') are close to the Boltzmann`s distribution, the logarithm intensity of a spectral line of rovibronic transition $(n', v', N' \rightarrow n'', v'', N'')$ is a linear function of the upper level energy. By ploting $ln\left(\frac{I^{n',v',N'}_{n'',v'',N''}}{\upsilon^4 g_{a.s.} H_{N'N''}}\right)$; ($\upsilon$ is a wave number, $g_{a.s.}$ - the statistical weight of the n', v', N' rovibronic level, caused by nuclear spin and the symmetry with respect to the permutation of the nuclei ('a' or 's'), $H_{N'N''}$ - Hönl-London (HL) factors) against term values for upper level a straight line is obtained, whose slope $\frac{hc}{kT_{rot}(n',v')}$ may be used for determination of the rotational temperature $T_{rot}(n',v')$ of excited state. In the case of the $\Pi \rightarrow \Sigma$ electronic transition $H_{N'N''}$ may be written as: (N'' -1) for P branch, ($2N''$ +1) for Q branch and (N'' +2) for R branch [23].

Within the framework of the model discussed in [15] the logarithm of the scaled rovibrational population density should be a linear function of the rotational energy in ground $X^1\Sigma_g^+, v = 0$ vibronic state:

$$\ln N^*_{n'v'N'} \equiv \ln \frac{N_{n'v'N'}}{g_{a.s.}(2N'+1)\tau_{n'v'N'}} = -\frac{hcE_{XON}}{kT_0(n',v')} + const. \qquad (1)$$

where is: $N^*_{n'v'N'}$ - scaled rovibronic population density, $\tau_{n'v'N'}$ - radiative lifetime of n', v', N' level and E_{XON} - rovibronic term value of ground vibronic state. Than $T_0(n',v')$ is the rotational temperature of the ground vibronic state determined from the rotational population density distribution of an excited (n',v') vibronic state. Further, according to [15], the temperature $T_0(n',v')$ can be considered as a valid estimation of the ground state rovibronic temperatutre i.e. H_2 translational temperature T_{tr}.

Here, it should be emphasized that the model predicts Boltzmann distribution in excited electronic-vibrational states. If one makes an additional assumption, that the rotational energy term $F_N \propto B_{nv}N(N+1)$ (according to the rigid rotator expression [14] then the following relation between rotational temperatures in the ground and electronically excited state may be obtained:

$$\frac{T_0(n',v')}{B_{X0}} = \frac{T_{rot}(n',v')}{B_{n'v'}} \qquad (2)$$

where $B_{n'v'}$ and B_{X0} are rotational constants of excited and ground vibronic states.

3.1.1. *Constrains of rotational temperature.* The rotational temperature $T_{rot}(n',v')$ of excited state may have a physical meaning, when the characteristic time of rotation-translation relaxation τ_{RT} for the n',v' vibronic state under the study is much shorter than the characteristic time of population and deactivation processes. In this case the rotational population distribution in the state n',v' is close to Boltzmann's distribution law with rotational temperature equal to translational temperature, i.e. gas temperature [14]. In an opposite extreme case, when the radiative lifetime is much smaller than τ_{RT}, rotational collisional mixing of neutral species during the lifetime of the state n',v' is not sufficiently efficient to ensure equilibrium of rotational distribution with the gas temperature. The degree of relation between the rotational population density distribution in n',v' state and the population of ground electronic state should be proven within the framework of excitation-deactivation balance equations, see e.g. [13].

3.2. Vibrational temperature

The procedure described in [14, 23] for vibrational temperature T_{vib} determination is used. If population density $N_{v'}$ of the vibrational levels is close to Boltzmann distribution law with $T \approx T_{vib}$, the intensity ratio of the two consecutive vibration bands of the same sequence is given as in [24]. If one introduces measured band intensities in two bands arising from different initial vibrational levels, it is possible to determine the vibrational temperature T_{vib} of an excited molecule:

$$\frac{1}{T_{vib}} = \frac{k}{hc[G(v')_2 - G(v')_1]}[ln(\frac{\nu_{v'v''1}}{\nu_{v'v''2}}) + ln(\frac{A_{v'v''1}}{A_{v'v''2}}) + ln\frac{I^{v'v''}_{em2}}{I^{v'v''}_{em1}}] \quad (3)$$

Where is: $G(v')$ - vibrational term value for upper vibrational level; $\nu_{v'v''}$ - corresponding transition frequency, which, to the first order approximation, can be the frequency of the band origin, $A_{v'v''}$ - the transition probability for spontaneous emission between vibrational states and $I_{em}^{v'v''}$ – band integral intensity. For transition probabilities and term values, see e.g. [25].

4. Results and discussion

The electronic transition $d^3\Pi_u \rightarrow a^3\Sigma_g^+$ is in the visible part of electromagnetic spectra (595-645nm) [26], see an example of recorded spectra in figure 2.

Relative intensities of spectral lines of P, Q and R branches of various bands of: $d^3\Pi_u^+ \rightarrow a^3\Sigma_g^+$; $d^3\Pi_u^- \rightarrow a^3\Sigma_g^+$ electronic transitions have been measured in above mentioned pressure range, see **Experimental**. Then, the relative population densities of rovibronic levels were calculated from (2) using data presented in Table 2 of [15]. Here, it should be mentioned that, in general case, the difference between semiempirical *N*-dependencies of the transition probabilities and those predicted by the Hönl-London formulae are significant and may even attain orders of magnitude. This may lead to systematic errors in rovibronic population densities calculated from (2). On other hand, there are certain cases when the difference between semiempirical *N*-dependencies of the transition probabilities and those predicted by the Hönl-London formulae vanish. Two examples of the population at the HC axis of the $d^3\Pi_u^+,0,N$ and $d^3\Pi_u^-,0,N$ levels evaluated from the intensities of R, P and Q branches of corresponding diagonal bands are presented in figures 3 and 4. In cases 3a and 4a, the Hönl-London factors were used for all branches as well as for Q branch in figures 3b and 4b. For this branch semiempirical data are identical with Hönl-London factors [15]. For P and R branches in figures 3b and 4b, semiempirical transition probabilities are used from Table 2 in [15]. Here, one should notice that in figures 3b and 4b, data points for all three branches evaluated with semiempirical transition probabilities represent a single data set for the Boltzmann plot. In the same figures solid lines correspond to best fits, while dashed lines determine upper and down limit slopes and they are used for estimation of the temperature measurement uncertinity.

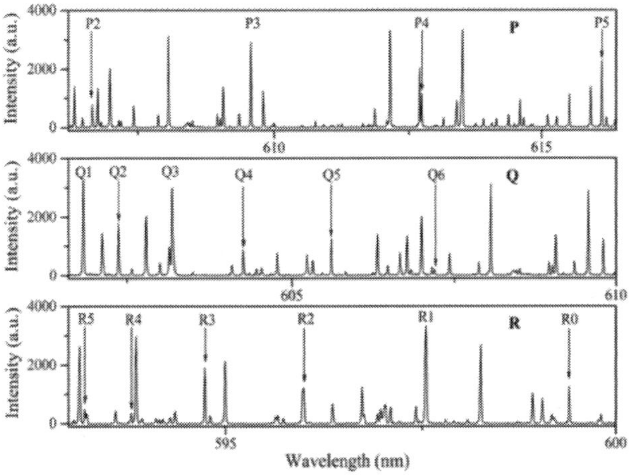

Figure 2: Emission spectra of rotational lines for Fulcher-α system diagonal band; R, Q, P branches with $\Delta v = 0,\ v' = 0$. Experimental conditions: titanium HC discharge in pure H_2 at p=2.5mbar; I=90 mA; U=400V.

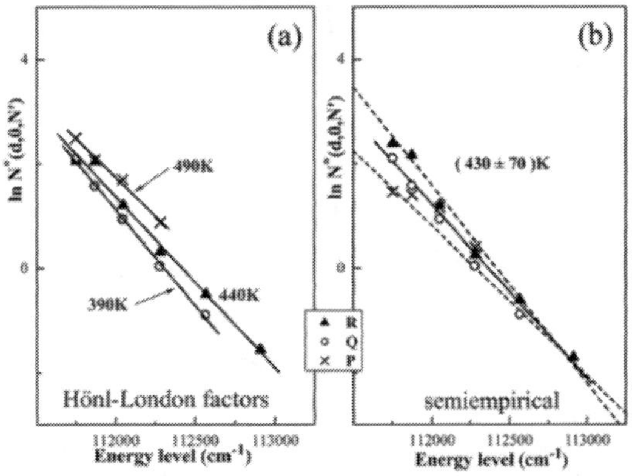

Figure 3: Semilogarithmic plot of population densities for $d^3\Pi_u,0,N'$ rovibronic levels of H_2 calculated from measured intensities with: (a) Hönl-London factors and (b) semiempirical transition probabilities [15]. Discharge conditions: copper HC discharge in H_2 at p=3mbar; I=45mA; U=455V.

One can notice from figure 3a that populations for the same rovibronic level evaluated from spectral line intensities of different branches using Hönl-London factors leads to the large spread of temperature values when Cu HC is used see figure 3a. With Ti HC, see figure 4a, the spread is well within an estimated experimental uncertainty of line intensity measurement.

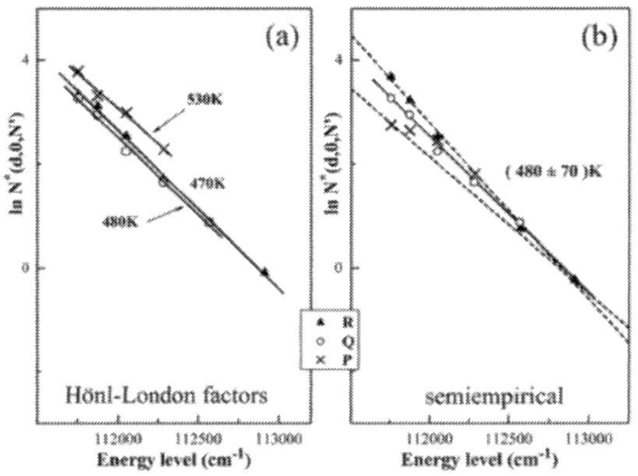

Figure 4: Semilogarithmic plot of population densities for $d^3\Pi_u, 0, N'$ rovibronic levels of H_2 calculated from measured intensities with: (a) Hönl-London factors and (b) semiempirical transition probabilities [15]. Discharge conditions: titanium HC discharge in H_2 at p=3mbar; I=90mA; U=400V.

In the case, when semiempirical data [15] are used for temperature evaluation, see figures 3b and 4b, large spread of temperature data occurs, but the average values are in agreement with those in figures 3a and 4a. It is interesting to note that, in contrast to Hönl-London factors approach, semiempirical results from the P branch in both experiments with Cu and Ti HC, see figure 3b and 4b, can't be correctly fitted with one straight line. This may be indication that semiempirical values for the first several lines of this transition should be reconsidered.

Here, it should be outlined that the Q-branch of the Fulcher-α band system ($d^3\Pi_u^- \rightarrow a^3\Sigma_g^+$ electronic transition) is considered as being most reliable for temperature determination. There are two reasons: (*i*) temperature derived from the Boltzmann plots set by using HL or semiempirical transition probabilities agree well within experimental uncertainty, see figures 3 and 4; and (*ii*) the spectral lines of the P and R branches are difficult to investigate experimentaly since they come from levels of the $d^3\Pi_u^+$ state, which is perturbed by the $e^3\Sigma_u^+$ state [26]

The rotational temperature of the ground vibronic state is evaluated from the results for the Q-branch ($v' = v'' = 0$) of the Fulcher-α band system ($d^3\Pi_u^- \rightarrow a^3\Sigma_g^+$ electronic transition) using the calculated term values this state. The results for both HCGD are presented in figure 6. The temperature at the axis of Cu and Ti HCGD are $T_0(d^-,0) = (800 \pm 100)$K and $T_0(d^-,0) = (980 \pm 100)$K, respectively. Although the current density is same for both hollow cathodes (j=4.8 mA/cm^2), different temperatures at the discharge axis are most likely caused by different power inputs, see **Experimental**.

Further, an attempt was made to investigate in which way the population of excited vibrational states of Q branch influence estimated values of T_0. Using the Hönl-London factors approach the rotational temperature of excited states is determined from the Q branch for different excited vibrational levels v'. The results for different radial positions along HC diameter are given in the lower part of figure 6. Here, it should be mentioned that radial temperature profiles, given in figure 6, are symmetrical within the experimental scatter and, therefore, the results are mirrored at the $r=0$.

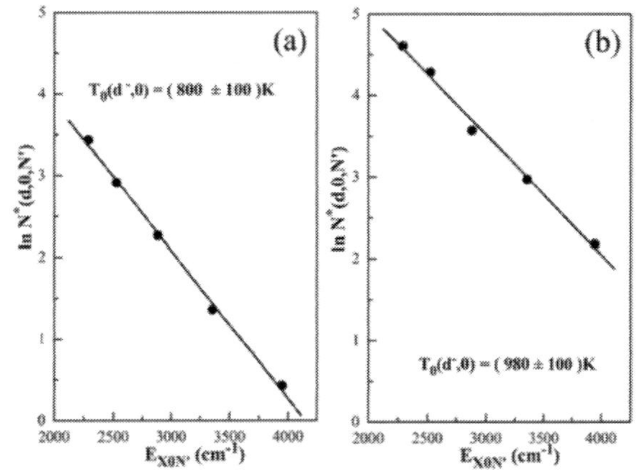

Figure 5: Semilogarithmic plot of rotational population densities of $d^3\Pi_u,0$ versus rotational energy of the molecular hydrogen ground state ($X^1\Sigma_g^+,0$): (a) copper HC and (b) titanium HC. Discharge conditions: same as in figures. 3 and 4.

Thus, only values of the rotational temperature derived from the population of Q branch vibrational state $v' = 0$ were recalculated to determine T_0 of the ground vibronic state $X^1\Sigma_g^+(v=0)$ at different radial position, see the upper part of figure 6. The results in figure 6 show that both temperatures, T_0 and T_{rot} (Q branch; $v' = 0$), change along HC diameter with a bending point at the radial position r=0.9 mm. Here, one should note that the radial temperature profiles are measured only up to r=1.8 mm, because intensities of first five lines in a Q-branch are too weak to allow accurate temperature measurement.

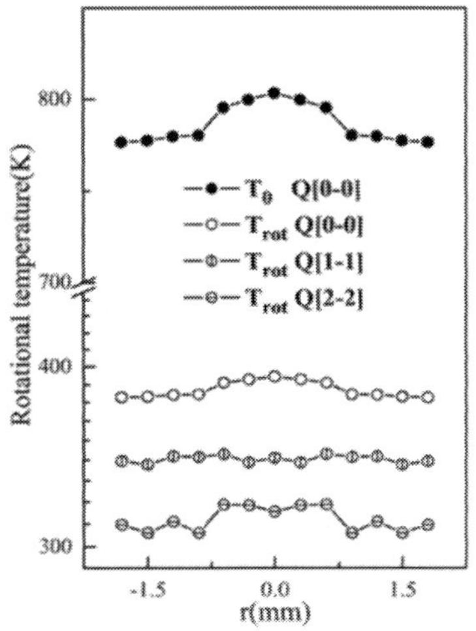

Figure 6: Radial distributions of the rotational temperature measured from Q-branch $v' = v'' = 0,1,2$ of the Fulcher-α band system and the rotational temperature T_0 recalculated from the population of Q branch $v' = 0$ for the ground vibronic state $X^1\Sigma_g^+(v=0)$. Experimental conditions: copper hollow cathode discharge in H_2 at *p=3 mbar; I=45 mA; U=454V.*

In accordance with above described model of low pressure hydrogen discharge [15], the translational gas temperature is expected to be twice as large as the rotational one of an excited state, see equation (2), where rotational constants of excited and ground vibronic states are related as $2B_{n'v'} = B_{X0}$. In our case, the temperature recalculated from equation (2) for the ground vibronic state $X^1\Sigma_g^+(v=0)$ is also two times larger than the rotational temperature of excited state $d^3\Pi_u^-$.

Although the population density distribution over vibrational levels in gas discharge plasmas has usually non-Boltzmann character, the Boltzmann plot technique is widely used for rough estimation of vibrational temperature T_{vib} or vibrational population distribution [14, 24]. The vibrational temperature is routinely determined from the measurement of the relative intensities of bands in a well-developed system.

The highest precision is achieved if the used bands are free from overlap. According to (3), apart from band intensities, the relative band transition must be available. For required relative H_2 transition probabilities and vibrational term data, see [25].

Radial intensity distributions for vibrational bands of the Fulcher-α system ($d^3\Pi_u^- \rightarrow a^3\Sigma_g^+$ electronic transition; Q-branch $v' = v'' = 2$ and $v' = v'' = 3$; spectral lines from Q1 to Q5) are shown in figure 7a. The maximum radial distance from the axis (r=2.1 mm) is determined by the intensities of the branch, which for higher radius become too weak to allow accurate vibrational temperature measurement. To determine T_{vib} of the H_2 molecule in excited state, integral intensities of these two bands are introduced in equation (3).

The distribution of T_{vib} in excited state al along HC diameter follows profile with shape similar to the one for rotational temperature, see figure 7, but with the bending point located now at the radial position r=1.2 mm, see figure 7b. The dependence of average vibrational temperature upon HC wall temperature shows that vibrational temperature does not change significantly with HC wall temperature range, in particular if one takes into account the change of HCGD voltage.

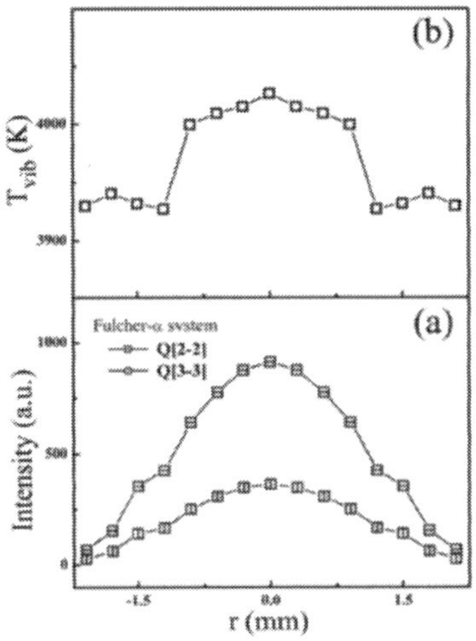

Figure 7: (a) Radial intensity distribution of the Fulcher-α system ($d^3\Pi_u^- \rightarrow a^3\Sigma_g^+$ electronic transition; Q-branch $v'=v''=2$ and $v'=v''=3$ and (b) Calculated vibrational H_2 temperature distribution. Experimental conditions: titanium HC discharge in pure H_2 at p=4 mbar; I=90 mA; U=400V.

Acknowledgment

This work within the Project 141032 "Low-temperature plasmas and gas discharges: radiative properties and interaction with surfaces" is supported by the Ministry of Science and Environmental Protection of the Republic of Serbia.

References

[1] Slevin P J and Harrison W W 1975 *Appl. Spectrosc. Rev.* **10(2)** 201

[2] Da Silva C F, Ishikawa T, Santos S, Alves Jr C and Martinelli A E 2006, *International J. of Hydrogen Energy* **31** 49

[3] Krasheninnikov S 2002, *Phys. Scr.* **T 96** 7

[4] Pigarov A Yu 2002, *Phys. Scr.* **T 96** 16

[5] Hollmann E M, Pigarov A Yu and Taylor K 2005 *J. Nucl. Mate.* **337-339** 451

[6] Péalat M, Taran J P E, Bacal M and Hillion F 1985, *J. Chem. Phys.* **82** 4943

[7] Biel W, Bröse M, David M, Kempkens H and Uhlenbusch J 1997 *Plasma Phys. Control. Fusion* **39** 661–681

[8] Chu H N, Den Hartog E A, Lefkow A R, Jacobs J, Anderson L W, Lagally M G and Lawler J E 1991 *Phys. Rev. A* **44** 3796

[9] Gicquel A, Hassouni K, Breton Y, Chenevier M and Cubertafon J C 1996 *Diam. and Rel. Mater.* **5** 366

[10] Gritsin S I, Kossyi I A, Malykh N I, Ral'chenko V G, Sergeichev K F, Silakov V P, Sychev I A, Tarasova N M and Chebotarev A V 1998 *J. Phys. D: Appl. Phys.* **31** 2942

[11] Gans T, Schulz-von der Gathen V and Döbele H F 2001 *Plasma Sources Sci. Technol.* **10** 17

[12] Goyette A N, Jameson W B, Anderson L W and Lawler J E 1996 *J. Phys. D: App. Phys.* **29** 1197

[13] Röpcke J, Kaning M and Lavrov B P 1998 *J. Phys. IV France* **8** 207

[14] Herzberg G 1950 *Molecular Spectra and Molecular Structure* vol **I** (New York: Van Nostrand-Reinhold)

[15] Astashkevich S A, Käning M, Käning E, Kokina N V, Lavrov B P, Ohl A and Röpcke J 1996 *J. Quant. Spectrosc. Radiat. Transfer* **56** 725

[16] Röpcke, J., Davies, P. B., Käning, M. and Lavrov, B. P.: *Low temperature plasma physics-Fundamental Aspects and Application* (Hippler R at al Wiley-VCH Berlin, NY Toronto etc.)

[17] Tomasini L, Rousseau A, Gousset G and Leprince P 1996 *J. Phys. D: Appl. Phys.* **29** 1006

[18] Laimer J, Huber F, Misslinger G and Störi H 1996 *Vacuum* **47** 183

[19] Xiao B, Kado S, Kajita S and Yamasaki D 2004 *Plasma Phys. Control. Fusion* **46** 653

[20] Majstorović G Lj, Šišović N M, Konjević N 2007 *Plasma Sources Sci. Technol.* **16** 750

[21] Šišović N M, Majstorović G Lj and Konjević N 2007 *Eur. Phys. J. D* **41** 143

[22] Abdel-Rahman M, Gans T, Schulz-von der Gathen V and Döbele H F 2005 *Plasma Sources Sci. Technol.* **14** 51

[23] Covacs I 1969 *Rotational Structure in the Spectra of Diatomic Molecules* (London: Adam Hilger LTD) pp. 131

[24] Philips J G 1961 *Optical Spectrometric Measurements of High Temperature* III (Dickeman P J, Chicago:Univ. Chicago Press) pp. 217

[25] Fantz U and Wünderlich D 2004 *Franck-Condons Factors, Transition Probabilities and Radiative Lifetimes for Hydrogen Molecules and Their Isotopomere INDC(NDS)-457*

[26] Crosswhite H M 1972 *The Hydrogen Molecule Wavelength Tables* (Gerhard Heinrich Dieke, New York: Wiley- Interscience)

[27] Lavrov B P, Ostrovsky V N and Ustimov V U 1981 *J. Phys. B* **14** 4701

Modeling of a dielectric barrier discharge used as a flowing chemical reactor

D. Petrović[1,2], T. Martens[1], J. van Dijk[3], W. J. M. Brok[3], and A. Bogaerts[1]

[1]Research group PLASMANT, Dep. of Chemistry, University of Antwerp, Belgium
[2]Institute of Physics, University of Belgrade, Serbia
[3]Department of Applied Physics, Eindhoven University of Technology, The Netherlands

E-mail: dragana.petrovic@ua.ac.be

Abstract. Our aim is to develop and optimize a model for a dielectric barrier discharge used as a chemical reactor for gas treatment. In order to determine the optimum operating conditions, we have studied the influence of the gas flow rate, reactor geometry and applied voltage parameters on the discharge characteristics.

For this purpose, a two-dimensional time-dependent fluid model has been applied to an atmospheric pressure DBD in helium with nitrogen impurities, in a cylindrical geometry. The numerical model is based on the continuity and flux equations for each type of particles treated, the electron energy equation and the Poisson equation. The gas flow is incorporated in the flux equations as a source term. The set of coupled partial differential equations is solved by the so-called modified strongly implicit method. The background gas flow is numerically treated separately, assuming in the model that there is no influence of the plasma on the flow. Indeed, the gas convection velocity is calculated using the commercial code Fluent and it is used as input into the 2D fluid model. The plasma characteristics have been studied in terms of gas flow rate, applied voltage amplitude and frequency, and geometrical effects. The electric currents as a function of time for a given applied potential have been obtained, as well as the number densities and fluxes of plasma species.

1. Introduction

Chemical treatment of gases and liquids, as well as surface treatment, are some of the industrial applications of atmospheric pressure dielectric barrier discharges (DBDs). They are also efficient generators of atoms, free radicals and chemically active species [1]. One of their main advantages is that due to the non-thermal character of DBDs, they operate at room temperature while the electrons are still highly energetic. This enables reactions that thermodynamically would not occur at such low gas temperatures and it can make DBDs efficient plasma-chemical reactors.

The average energy of the electrons generated in DBDs, as well as the discharge characteristics in general, strongly depend on the gap width, the applied frequency, the power input, the type of carrier gas and the type of dielectric used. Our aim is to develop a model for a dielectric barrier discharge used as a flowing chemical reactor for gas treatment. Due to small dimensions of DBD setups, plasma diagnostic studies are not easy and numerical modeling is very important for understanding the discharge behaviour. In the present paper, in order to determine the optimum operating conditions, we discuss some aspects of the influence of the gas flow rate, reactor geometry and applied voltage parameters on the discharge characteristics.

In Section 2 the numerical model and the basic equations are given. A short description of the setup under consideration and the operating conditions is in Section 3. In Section 4, results obtained in the numerical modeling are presented. Finally, conclusions are in Section 5.

2. Numerical model

For this purpose, a two-dimensional time-dependent fluid model for an atmospheric pressure DBD, as a part of the Plasimo code [2], has been applied. It was originally developed by Hagelaar for a DBD used in plasma display panels [3] and later transformed and extended by Brok and van Dijk, and integrated into the Plasimo code [4, 5].

2.1. Basic equations

The model is based on the continuity equations and equations of motion for each type of particles treated:

$$\frac{\partial n_\alpha}{\partial t} + \nabla \vec{\Gamma}_\alpha = S_\alpha, \quad S_\alpha = \sum_j c_{\alpha j} R_{\alpha j}, \quad R_{\alpha j} = k_j n_j n_\alpha, \tag{1}$$

and

$$\vec{\Gamma}_\alpha = \pm \mu_\alpha \vec{E} n_\alpha - D_\alpha \nabla n_\alpha + n_\alpha \vec{v}_b, \tag{2}$$

where the subscript α denotes the type of species. $\vec{v}_b$ is background flow velocity and all the other symbols retain their conventional meaning. Note that the mobility is taken as a positive value for the positive ions and a negative value for the electrons (and negative ions, if included in the model). For the neutral species, the first term in Eq. 2 is zero.

For the electrons an energy-balance equation is also solved:

$$\frac{\partial n_e \bar{\varepsilon}}{\partial t} + \nabla \vec{\Gamma}_{\bar{\varepsilon}} = S_{\bar{\varepsilon}}, \tag{3}$$

where $\bar{\varepsilon}$ is the mean electron energy and $\vec{\Gamma}_{\bar{\varepsilon}}$ is the electron energy flux density given by [6]:

$$\vec{\Gamma}_{\bar{\varepsilon}} = -\frac{5}{3} \mu_e \vec{E} n_e \bar{\varepsilon} - \frac{5}{3} n_e D_e \nabla \bar{\varepsilon}. \tag{4}$$

Note that the energy changes are usually a loss of energy (i.e. ionization and excitation) but a gain of energy can occur as well as a result of superelastic collisions. For the ions and neutral species no energy-balance equation needs to be included, because these species can be regarded as being at the same energy (temperature) as the background gas. The gas temperature is assumed to be 300 K, uniformly distributed over the discharge. Moreover, for these species, the local field approximation can be used, i.e. the mobility and diffusion coefficients of the heavy particles are taken to be functions of the local reduced electric field (local field divided by pressure). The values used are taken from Ref. [7].

In order to obtain a self consistent electric field distribution the Poisson equation:

$$\epsilon \nabla \vec{E} = \sum_\alpha q_\alpha n_\alpha, \tag{5}$$

is solved as well, not only within the plasma but also inside the dielectrics, where it is reduced to $\epsilon \nabla \vec{E} = 0$ because no species are present inside the dielectrics.

2.2. Gas flow

The gas flow is incorporated in the flux equations as a source term (the third term on the right side of Eq. 2) [8]. The fundamental assumption in the model is that the plasma does not influence the flow. This allows us a separate numerical treatment of the background gas flow and use the results as input for the plasma calculation.

On the other hand, the flow influences the plasma as a momentum source for the species. This momentum source enters in the drift-diffusion equations. This effect is particularly important for the neutral species, as there are no electric forces to modify their speed.

The gas convection velocity is calculated using the commercial computational fluid dynamics code Fluent [9], i.e., by solving the conservation equations for mass, momentum and energy. The solution was considered to converge when the residuals of the guiding equations have become smaller than 10^{-6}.

2.3. Modeling chemistry

The model is applied here to a DBD in helium with 100 ppm nitrogen impurity. The species included in the model are the background gases He and N_2, electrons, He^+, He_2^+, N_2^+ and N_4^+ ions, He_2 excimer, and He atoms excited to the metastable levels.

For most of the electron-induced processes the reaction rates are calculated on the basis of energy-dependent cross-sections, and with a separate program, called BOLSIG+, which is based on a Boltzman equation [10]. This program is applied for a wide range of different, fixed reduced electric fields. For each value of the reduced electric field the average electron energy, the mobility of the electrons, and the rates of the different electron-induced processes are tabulated. By use of these tables, the rates, mobilities, and diffusion coefficients (derived from the mobilities using the well-known Einstein relationship, assuming the bulk electrons are Maxwellian), will be applied corresponding to the electron energy calculated with the above electron-energy-balance equation. Details about the chemistry model and the reactions taken into account can be found in [11]-[13].

The effect of charge accumulation on the dielectric materials is also considered. The boundary conditions of the electrons are largely determined by secondary electron emission from the ions. Following the procedure of Hagelaar et al. [14] we use a constant secondary-electron emission coefficient of 0.05 for the helium ions and 0.001 for nitrogen ions and also consider that this ion-induced secondary-electron emission produces electrons with a mean initial energy of 4 eV. Reflection at the walls is neglected and electron desorption from the dielectrics is not considered, because it has been stated [15] that the latter process is very unlikely to happen.

The equations (1)- (5) are solved within the same time-step by the so-called modified strongly implicit method, developed by Schneider and Zedan [16], and using an extra stabilization method, as explained elsewhere [3], until convergence is reached within every time-step of the periodic cycle, and periodic behaviour over the discharge cycles is obtained.

3. Setup and conditions

The model was applied to a cylindrical DBD reactor consisting of two coaxial stainless steel electrodes with 2 mm thick alumina as a dielectric on the inner side of the outer electrode, as illustrated schematically in Fig. 1. The gap width is varied from 0.7 mm over 1.5 mm to 3 mm by changing the diameter of the inner tube. The length of the reactor is kept fixed at 9 cm and the outer diameter is 3 cm. A sine shaped high voltage is applied on the inner electrode whereas the other electrode was grounded. As it is already mentioned, we applied the model to a helium DBD with nitrogen impurity at atmospheric pressure and temperature 300 K.

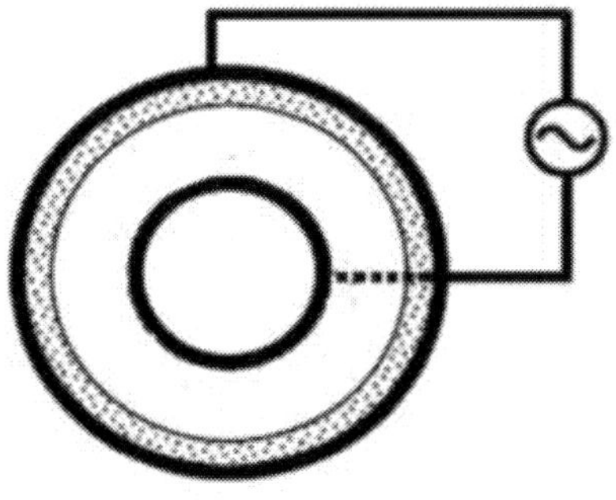

Figure 1. Schematic diagram of the setup.

4. Results

4.1. Discharge characteristics

The plasma characteristics have been studied in terms of applied voltage amplitude and frequency, and geometrical effects. The influence of the gap width on the electron number density is shown in Fig. 2. It can be seen that at 20 kHz and 20 kV applied voltage a decrease of the gap width leads to an increase of the electron number density in the system (see Fig. 2a), as has been experimentally observed as well [17]-[19]. The highly energetic electrons will lead to the generation of some highly active species including radicals that is very desirable for the purpose of a chemical reactor for gas conversion. However, applying different voltage characteristics (i.e., voltage amplitude and frequency) this regular behaviour can change, as can be seen in Fig. 2b. In this particular case, when the applied voltage frequency and amplitude are 75 kHz and 10 kV, respectively, the highest electron number density is for 1.5 mm gap distance.

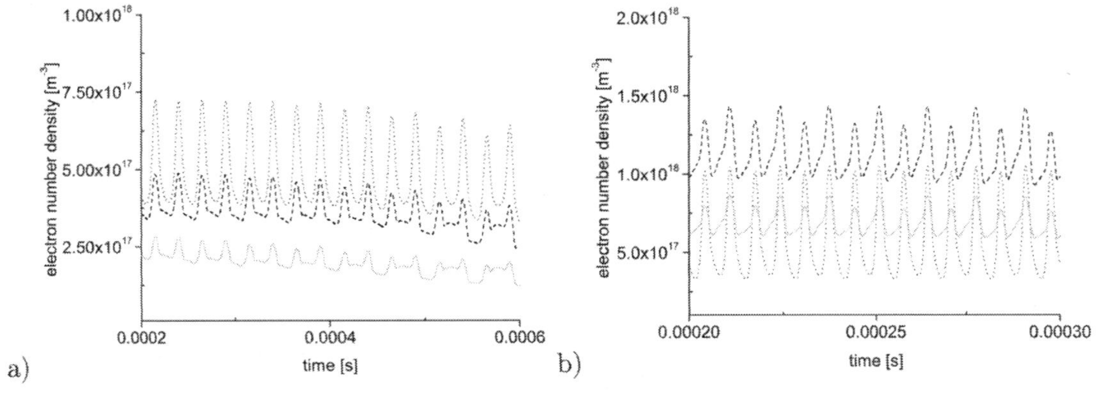

Figure 2. Calculated spatially averaged electron number density as a function of time, for the gap width 0.7 mm - red dotted line, 1.5 mm - black dashed line, and 3 mm - green solid line. The applied voltage frequency and amplitude are a) 20 kHz, 20 kV; b) 75 kHz, 10 kV.

The spatially averaged total electric currents as a function of time for different values of applied potential amplitude and frequency are shown in Fig. 3. A clear transition from one narrow peak, over multipeak to a broad profile can be seen. The total current consists of discharge and displacement currents and when the discharge component is dominant narrow peaks will appear (see Fig. 3a) that is usually reported in the literature [20]-[23]. Increase of the voltage amplitude might cause a multibreakdown in the system and produce a multipeak current profile (see Fig. 3b). On the other hand, increase of the applied frequency leads to an increase of the displacement current and broadening of the current peaks. In Fig. 3c a clear superposition of the displacement current (sine shaped profiles) and the discharge current (narrow peaks) is seen.

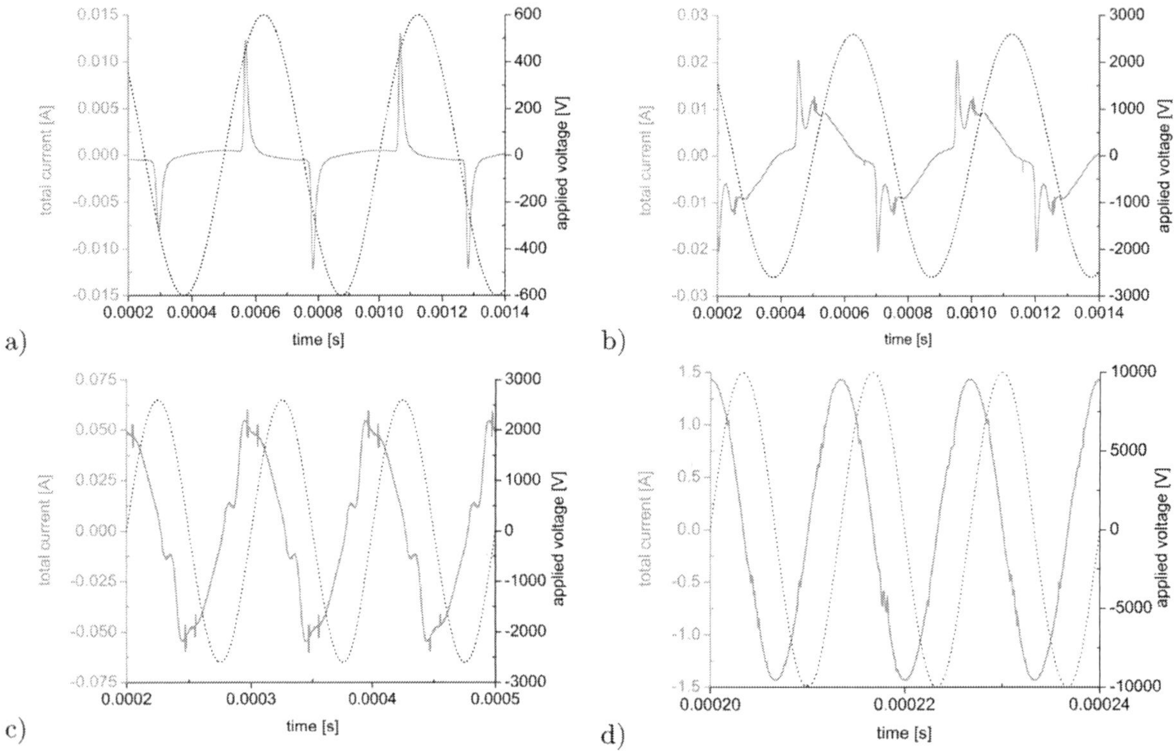

Figure 3. Calculated total currents (orange solid lines) as a function of time in an atmospheric DBD in helium with 100 ppm nitrogen impurity, for different applied voltage frequencies and amplitudes: a) 2 kHz, 0.6 kV; b) 2 kHz, 2.6 kV; c) 10 kHz, 2.6 kV; d) 75 kHz, 10 kV (black dotted lines). The gap width is 0.7 mm.

Further increase of the applied voltage frequency and amplitude will cause that the displacement current becomes dominant and the total current profile becomes very broad, with an almost sine shaped profile (see Fig. 3d).

4.2. Gas flow effect

The adjustment of flow rate allows the creation of uniform DBDs with optimized input power while a lower breakdown electric field is required [24]. Transition from a streamer discharge to a silent glow discharge due to the presence of a flow in the system is also reported in Ref. [25]. An increase of species fluxes (except for the electrons) due to convection in a Grimm-type glow discharge is shown in Ref. [26]. For gas conversion applications it has been observed that the increase of the gas flow rate reduces the conversion and decreases the yields. On the other hand, the effect of the flow rate on product selectivity depends on the particle type [27]-[29].

As it is already mentioned, the gas convection velocity was calculated using the Fluent code [9]. The feed gas is usually introduced into the reactor via a mass flow controller. Because of this, the mass flow at the reactor inlet was given as a boundary condition in the background gas flow velocity calculations. The flow was considered to be laminar and a steady state was assumed.

For a mass flow rate of 5 l/min, the calculated axial velocity profile at z=4.5 cm, as a function of radial position (i.e., in the gap between both concentric electrodes), is shown in Fig. 4. The flow velocity reaches a maximum (around 2.2 m/s) in the middle of the gap. At the walls, the flow velocity becomes zero due to the viscous drag. The radial component of the velocity is calculated to be negligible compared to the axial one and it is taken to be zero in further calculations. A two dimensional velocity profile in the total reactor (axial symmetry is assumed) and a zoomed in profile are shown in Fig. 5.

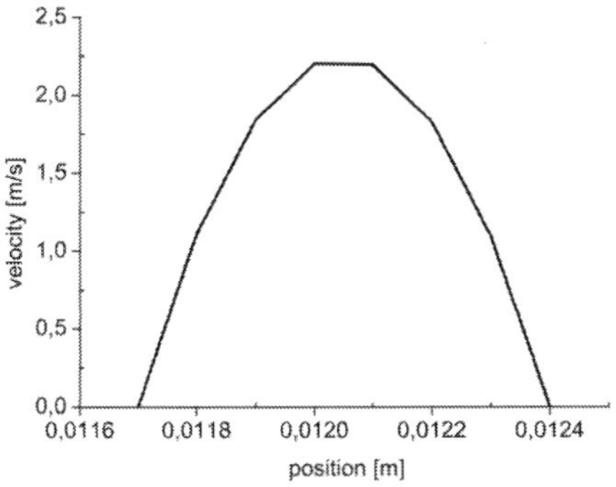

Figure 4. Axial velocity profile at z=4.5 cm, as a function of radial position, for a mass flow rate of 5 l/min.

The calculated background gas convection velocity is used as input in the 2D fluid model. The radial and axial flux of the He_2 species with gas flow in the system at the maximum gas velocity (r=0.0121 cm, z=4.5 cm) are shown in Fig. 6a and Fig. 6b, respectively. The radial flux of the He_2 species without gas flow at the same position is shown in Fig. 7a. As can be seen (compare Fig. 6a and Fig. 7a), there is no influence of the convection on the flux in the radial direction, which is expected since the radial component of the gas flow is negligible. On the other hand, without gas flow in the system, there was no flux of He_2 species in the axial direction. However,

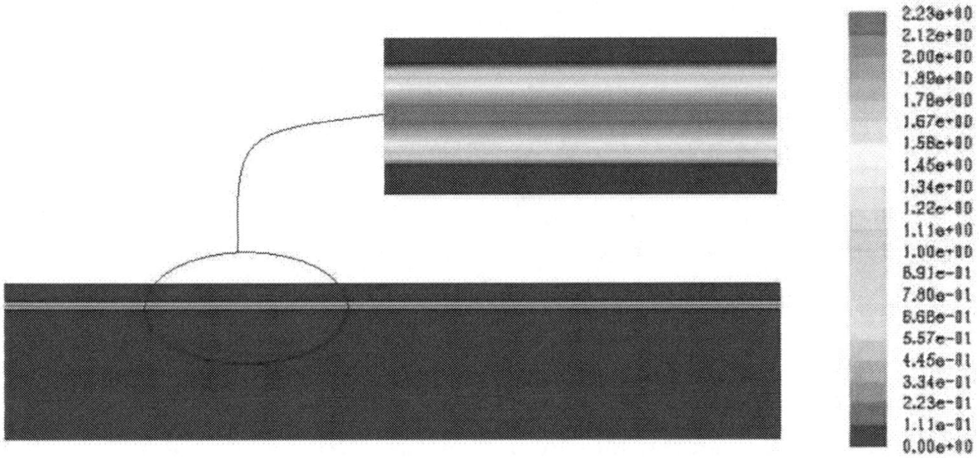

Figure 5. Contours of velocity magnitude (in [m/s]) in the total reactor and zoomed in view of the gap. The mass flow rate is 5 l/min. The gap width is 0.7 mm, the length is 9 cm, the diameters of outer and inner electrode are 30 mm and 24.6 mm, respectively, and the dielectric thickness is 2mm.

introduction of a flow of 5 l/min gives rise to an axial flux, which is one order of magnitude higher than the radial one (see Fig. 6b). A similar influence of the gas flow can be observed on

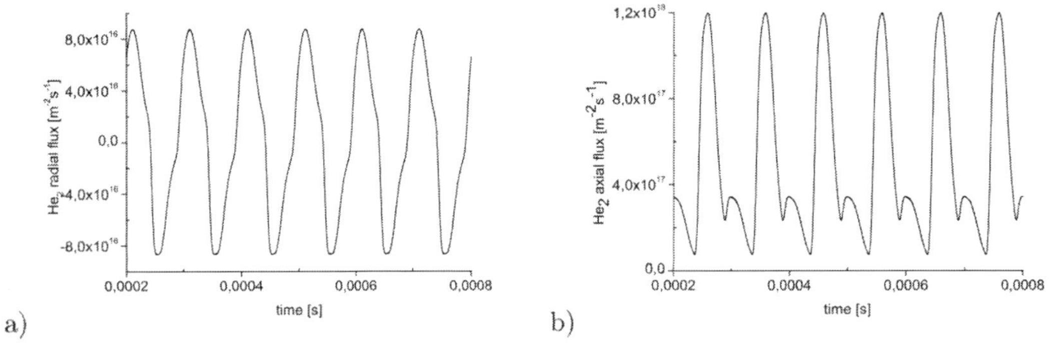

Figure 6. Calculated a) radial and b) axial fluxes of He_2 for a mass flow rate of 5 l/min. The applied voltage frequency and amplitude are 10 kHz and 2.6 kV. The gap width is 0.7 mm.

the ion fluxes, Fig. 8. While the radial flux remains unchanged, the axial flux appears in the system. However, opposite to the case of neutrals where the axial flux becomes dominant due to the flow, for the ions it remains two - three orders of magnitude lower than the radial flux. Indeed, as can be seen from these figures, the axial flux is of the order of 10^{16}, 10^{17} $m^{-2}s^{-1}$ (see Fig. 8) while the radial flux is of the order of 10^{19} $m^{-2}s^{-1}$ (see Fig. 7b).

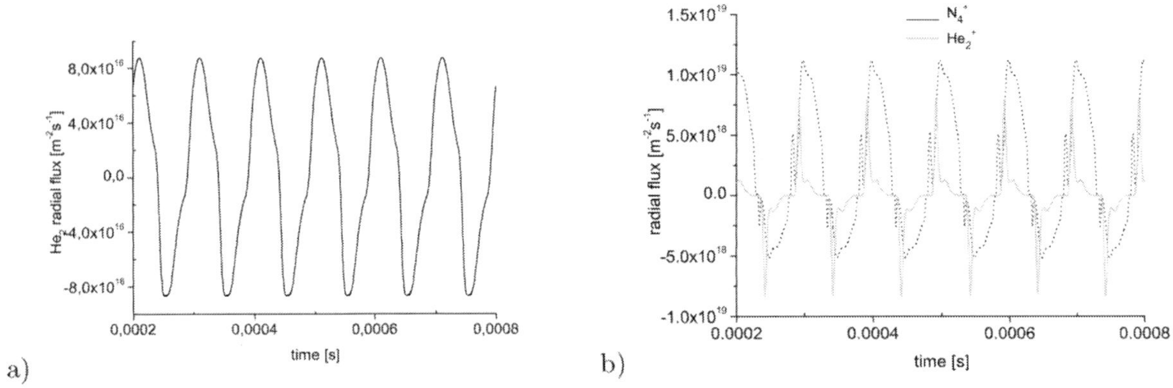

Figure 7. Calculated radial fluxes of a) He_2 and b) He_2^+ (green solid line) and N_4^+ (black dotted line) without gas flow. The applied voltage frequency and amplitude are 10 kHz and 2.6 kV. The gap width is 0.7 mm.

5. Conclusions

We have presented the influence of the gas flow rate, reactor geometry and applied voltage parameters on the discharge characteristics of a cylindrical DBD in helium with nitrogen impurity. This study shows that discharge voltage and frequency effects have a large impact on the electric current profiles, which can range from one narrow peak, over multipeak to a broad profile. A decrease of the gap width leads to an increase of the electron number density, while applying a higher voltage frequency can change this regular behaviour. A gas flow rate in the system causes appearance of the axial fluxes of neutral species, as well as of ions, while the radial fluxes remain unchanged.

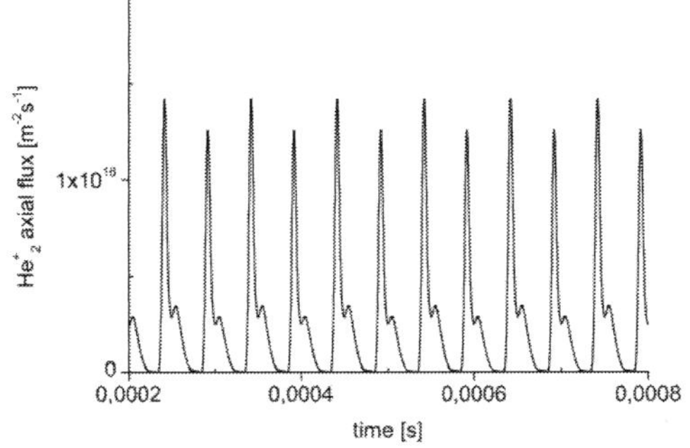

a)

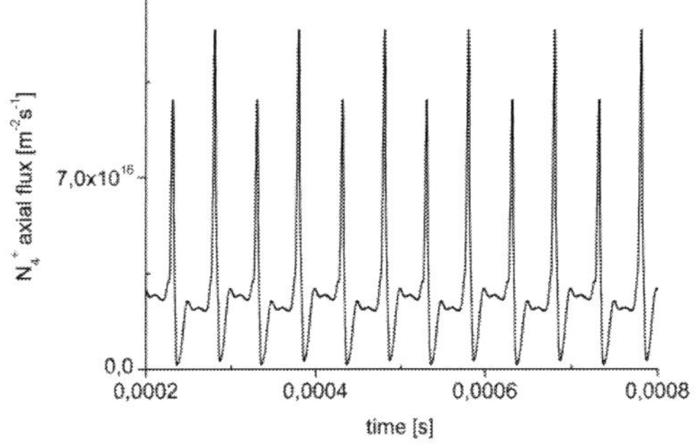

b)

Figure 8. Calculated axial fluxes of a) He_2^+ and b) N_4^+ for a mass flow rate of 5 l/min. The applied voltage frequency and amplitude are 10 kHz and 2.6 kV. The gap width is 0.7 mm.

References

[1] Kogelschatz U 2003 *Plasma Chemistry And Plasma Processing* **23**(1) 1

[2] http://plasimo.phys.tue.nl

[3] Hagelaar G 2000 *Modeling of Microdischarges for Display Technology*, PhD Dissertation, Eindhoven University of Technology

[4] Brok WJM, van Dijk J, Bowden MD, van der Mullen JJAM, Kroesen GMW 2003 *J Phys D: Appl Phys* **36** 19671979

[5] Brok WJM, Gendre MF, van der Mullen JJAM 2006 *J Phys D: Appl Phys* **40** 156162

[6] Raizer YP 1991 *Gas discharge physics* Springer, Berlin Heidelberg New York

[7] Ellis HW, Pai RY, McDaniel EW 1976 *At Data Nucl Data Tables* **17** 177210

[8] Broks BHP, Brok WJM, Remy J, van der Mullen JJAM, Benidar A, Biennier L, and Salama F 2005, *Phys. Rev. E* **71**(3), 036409.

[9] http://www.fluent.com

[10] Hagelaar GJM, Pitchford LC 2005 *Plasma Sources Sci Technol* **14** 722733

[11] Martens T, Bogaerts A, Brok W, van Dijk J 2007 *Anal. Bioanal. Chem.* **388** 1583

[12] Martens T, Bogaerts A, Brok WJM, van der Mullen JJAM 2007 *J. Anal. At. Spectrom.* **22** 1033

[13] Martens T, Bogaerts A, Brok W, van Dijk J 2008 *Appl. Phys. Lett.* **92** 041504

[14] Hagelaar GJM, Kroesen GMW, van Slooten U, Schreuders H (2000) *J Appl Phys* **88** 22522262

[15] Raizer YP, Shneider MN, Yatsenko NA 1995 *Radio-frequency capacitive discharges* CRC Press, Florida

[16] Schneider GE, Zedan M 1981 *Numer Heat Transfer* **4** 119

[17] Kogelschatz U, Eliasson B 1995 *Handbook of Electrostatic Processes*, Marcel Dekker, New York 581

[18] Kogelschatz U, Eliasson B 1997 *J. Phys. IV*, **7** 46

[19] Eliasson B, Kogelschatz U 1991 *IEEE Trans. Plasma Sci.* **19** 309

[20] Golubovskii YuB, Maiorov VA, Behnke J., and Behnke VG 2002 *J. Phys. D: Appl. Phys.* **36**

[21] Mangolini L, Anderson C, Heberlein J, and Kortshagen 2004 *J. Phys. D: Appl. Phys.* **36** 1021

[22] Massines F, Rabehi A, Decomps P, and Gardi RB. 1998 *J. Appl. Phys.* **83** 2950

[23] Wang Y-H and Wang D-Z. 2004 *Chin. Phys. Lett.* **21** 2234

[24] Nersisyan G, Graham WG 2004 *Plasma Sources Sci. Technol.* **13** 582

[25] Gherardi N, Massines F 2001 *IEEE Trans. Plasma Sci.* **29** 536

[26] Bogaerts A, Okhrimovskyy, R. Gijbels 2002 *J. Anal. At. Spectrom.* **17** 1076

[27] Pietruszka B, Anklam K, Heintze M 2004 *Appl. Catal. A: General* **261** 19

[28] Song HK, Choi JW, Yue SH, Lee H, Na BK 2004 *Catalysis Today* **89** 27

[29] Jian T, Li Y, Liu Ch, Xu G, Eliasson B, Xue B 2002 *Catalysis Today* **72** 229

Collisionless Kelvin-Helmholtz instability and vortex-induced reconnection in the external region of the Earth magnetotail

F. Pegoraro, M. Faganello and F. Califano
Dipartimento di Fisica Università di Pisa, Pisa, Italy
E-mail: pegoraro@df.unipi.it

Abstract. In a magnetized plasma streaming with a non uniform velocity, the Kelvin-Helmholtz instability plays a major role in mixing different plasma regions and in stretching the magnetic field lines leading to the formation of layers with a sheared magnetic field where magnetic field line reconnection can take place. A relevant example is provided by the formation of a mixing layer between the Earth's magnetosphere and the solar wind at low latitudes during northward periods. In the considered configuration, in the presence of a magnetic field nearly perpendicular to the plane defined by the velocity field and its inhomogeneity direction, velocity shear drives a Kelvin-Helmholtz instability which advects and distorts the magnetic field configuration. If the Alfvén velocity associated to the in-plane magnetic field is sufficiently weak with respect to the variation of the fluid velocity in the plasma, the Kelvin-Helmholtz instability generates fully rolled-up vortices which advect the magnetic field lines into a complex configuration, causing the formation of current layers along the inversion curves of the in-plane magnetic field component. Pairing of the vortices generated by the Kelvin-Helmholtz instability is a well know phenomenon in two-dimensional hydrodynamics. Here we investigate the development of magnetic reconnection during the vortex pairing process and show that completely different magnetic structures are produced depending on how fast the reconnection process develops on the time scale set by the pairing process.

1. Introduction

Magnetic field line reconnection is one of the most important and basic physics processes in a magnetized plasma [1], capable at the same time of restructuring the large scale topology of the magnetic field and of affecting the global energy balance of the system by transforming the magnetic energy related to the plasma current inhomogeneities into plasma kinetic and/or thermal energy. In fact magnetic reconnection is considered to be the driver of the most important energetic phenomena observed in laboratory plasmas, e.g., in thermonuclear fusion devices, and in space plasmas, e.g., in the solar atmosphere or in the Earth's magnetosphere.

The research effort in the understanding of the physics underlying magnetic reconnection in high temperature plasma regimes has progressed remarkably in recent years but magnetic reconnection still remains a formidable challenge concerning the problem of why and under what conditions the system enters a reconnection phase and how this process evolves nonlinearly. Recent studies have focussed in particular on the so-called "fast" magnetic field line reconnection since in low collisionality or collisionless plasma regimes this process is expected to occur on time scales that do not exceed by large factors the dynamical time scales of the plasma configuration as determined e.g., within the ideal magnetohydrodynamic description. When current layers are formed that have widths of the order of the ion skin depth $d_i \equiv c/\omega_{pi}$, with ω_{pi} the ion plasma frequency, ions inside these current layers decouple their motion from the evolution of the magnetic field. *Per se* this decoupling is not sufficient to allow for magnetic reconnection which requires electron decoupling. Electrons decouple from the magnetic field inside a thinner region that, depending on the plasma parameters, has a width determined either by a nonvanishing resistivity or by kinetic (phase space) effects, or, as is the case in the fluid electron model description adopted here, by finite electron inertia effects. The large spatial separation between these two decoupling regions allows magnetic reconnection to grow at a faster rate [2, 3, 4, 5, 6].

Theoretically, two different approaches are routinely used when investigating magnetic reconnection. Magnetic reconnection is described either as an initial value problem in a magnetic configuration with an inhomogeneous current distribution [7, 8] where a reconnection instability develops spontaneously (in a configuration that allows for modes with a positive delta prime in the jargon of one-dimensional reconnection studies) or a boundary value problem is considered where reconnection is driven in a "stable" magnetic configuration by an external driver, e.g., by injecting Alfvén waves [9] or by imposing a stationary plasma inflow that pushes regions of different magnetic polarity (regions where at least one on the magnetic field line components reverses its sign) one against the other [10, 11]. In both cases the system is "prepared" *ad hoc* in order to develop a reconnection process.

Observationally, the Earth environment is a laboratory of excellence for the study of the physics of magnetic reconnection because of the importance of this process in the dynamics and shaping of this region and because of the increasing wealth of data of ever improving quality on the electromagnetic field profiles and on the particle distribution functions obtained *in situ*. The understanding of the phenomena related to collisionless magnetic reconnection have made great progress due to the results obtained by space missions like Geotail and Cluster. Satellite measurements have given *in situ* evidence of reconnection in the magnetotail [12, 13, 14], in the magnetosheath [15] and even in the turbulent plasma in the Earth's bow shock [16]. The key role of whistler waves has been demonstrated in a "fast" reconnection event reported by the Cluster quartet [17], in agreement with theoretical predictions [18, 19]. Of special relevance is the first evidence of the presence in the magnetotail of coherent magnetic structures (defined as magnetic islands) with typical dimensions of the order of the ion inertial length and their link to the generation of energetic supra-thermal electrons [20]. Magnetic island type structures embedded in a current sheet in the magnetopause were reported in Refs. [21, 22] starting from Cluster data and using a magnetohydrostatic reconstruction technique. In addition, a denser plasma layer was observed on the Earth side of these structures. This denser layer is a signature of a solar wind plasma entry inside the magnetosphere.

In a magnetized plasma streaming with a nonuniform velocity, the Kelvin-Helmholtz instability plays a major role in mixing different plasma regions and in stretching the magnetic field lines leading to the formation of layers with a sheared magnetic field where magnetic field line reconnection can take place. If the Alfvén velocity associated to the in-plane[1] magnetic field is sufficiently weak with respect to the variation of the fluid velocity in the plasma, the Kelvin-Helmholtz instability generates fully rolled-up vortices which advect the magnetic field lines into a complex configuration, causing the formation of current layers along the inversion curves of the in-plane magnetic field component. Since the plasma dynamics is essentially driven by the vortex motion, the reconnection events that are produced in these layers are usually denoted as vortex-induced reconnection [23, 24, 25, 26, 27].

Satellite measurements have provided clear evidence of the presence of rolled-up vortices at the flank of the magnetopause [28]. In this region, the velocity shear between the magnetospheric

[1] In the plane defined by the velocity and by the plasma inhomogeneity directions

and the solar wind plasmas generates hydrodynamic Kelvin-Helmholtz vortices which tend to pair in their non linear regime [29]. This provides an efficient mechanism for the formation of a mixing layer leading to solar wind plasma entering the magnetosphere. Furthermore, the vortices drag in their flow the magnetic field component parallel to the solar wind direction. As a consequence, the magnetic field lines are increasingly stretched inside the vortices until they reconnect. Besides transforming plasma kinetic energy into accelerated particle energy, heating, magnetic reconnection also affects the transport properties of the plasma and vortex-induced reconnection [23, 30], has been considered to be an efficient mechanism of vortex disruption and to allow eventually solar wind plasma to enter the magnetosphere [31, 25, 26].

On the other hand, pairing of the vortices generated by the Kelvin-Helmholtz instability is a well know phenomenon in two-dimensional hydrodynamics [32, 33]. In fact the role of the magnetic field on vortex dynamics has been studied essentially in the limit of only one vortex, the largest one contained in the simulation box. It has been shown that, even if the magnetic field is weak and unable to prevent the formation of the Kelvin-Helmholtz vortex, nevertheless the vortex-induced reconnection process eventually leads to vortex disruption [25, 26, 34].
On the contrary, three recent papers have investigated the development of secondary instabilities during the vortex pairing process (a vortex-induced Rayleigh-Taylor instability in Ref. [35] and vortex-induced magnetic reconnection in Refs. [36, 37]). In particular it has been shown that completely different magnetic structures are produced depending on how fast these secondary instabilities develop on the time scale set by the pairing process.

2. Competing instabilities

Fluids and plasmas in nature do not usually manifest themselves in quiescent stationary or quasistationary states (what we generally call "equilibria") but exhibit a rich dynamical behaviour consisting of linear and nonlinear waves and instabilities, coherent nonlinear structures such as vortices and solitary perturbations, or fully developed turbulence.

In the following we will focus our attention not so much on the single instability mechanism and the single instability growth and saturation, but mainly on the competition between a restricted number of instabilities of the system, and on how this competition may become experimentally observable in specific plasma regimes of interest. Different instabilities can compete either because they are present from the start, as they can feed on the same energy source present from the start in the equilibrium configuration, or because they arise as secondary instabilities that feed on an energy source that was not present initially in the equilibrium configuration but that has been made available by the development of the primary instability.

The Kelvin-Helmholtz, the Rayleigh-Taylor and the magnetic reconnection instabilities are the main actors on the stage of the nonlinear dynamics of a magnetized plasma, as they can be driven directly or indirectly, i.e., in the form of secondary instabilities, by inhomogeneities in the plasma velocity, in the plasma density (or pressure) and in the plasma currents. The complex interaction between these instabilities is controlled by the time scales of the different processes at play, which involve both large spatial scales from which the initial drive of the instabilities originates, and small spatial scales where e.g., magnetic field line reconnection can occur. The timing between the two instabilities will determine the structure of the final configuration that the system can reach. The qualitative differences (not simply quantitative differences) between these possible final states are evident and can in principle be used as a diagnostic tool in order to determine experimentally how fast reconnection evolves in the different plasma regimes on the clock set by the evolution of the primary instability.

In this article we will investigate the interplay and the competition between a Kelvin-Helmholtz primary instability, with its nonlinear evolution characterized by the vortex pairing process, and the onset of secondary instabilities, such as magnetic reconnection in the case of a frozen magnetic field with field lines stretched by the differential plasma motion generated by the Kelvin-Helmholtz instability. We will also briefly consider the effect of a secondary Rayleigh Taylor instability driven by the centrifugal acceleration in vortices in a plasma with a strong density inhomogeneity.

3. Configurations with a velocity shear and a density gradient

First we report some recent results [35] on the interplay between the Kelvin-Helmholtz instability and a secondary Rayleigh-Taylor instability driven by the acceleration inside the rotating vortices in the presence of a plasma density gradient. Such a configuration is of interest for the case of the interaction of the solar wind with the Earth magnetosphere in regions where magnetic reconnection is not expected to occur.

Two-dimensional simulations [35] of the Kelvin-Helmholtz instability show that in an inhomogeneous compressible plasma with a density gradient (in a transverse magnetic field configuration) the vortex pairing process and the Rayleigh-Taylor secondary instability compete during the non-linear evolution of the vortices. Two different regimes have been shown to exist depending on the value of the density jump across the velocity shear layer. These regimes have different physical signatures that can be crucial for the interpretation of satellite data of the interaction between the solar wind and the magnetospheric plasma.

As mentioned in the Introduction, the Kelvin-Helmholtz instability has been shown [28] to play a crucial role in the interaction between the solar wind and the Earth's magnetosphere and to provide a mechanism by which the solar wind can enter the Earth's magnetosphere. In fact magnetic reconnection is believed to dominate the transport properties at the low latitude magnetopause when the field in the solar wind and the geomagnetic field are antiparallel (southward solar wind magnetic field). The reasoning goes that if magnetic reconnection were the only mixing mechanism in the magnetotail, the mixing between the solar wind and the magnetospheric plasma should not occur during northward magnetic field periods. Actually, an increase of the plasma content in the outer magnetosphere during northward magnetic field periods is not only observed but is even larger than during southward configurations [38, 39, 28]. For these reasons, the Kelvin-Helmholtz instability has been invoked as a possible mechanism in order to account for the increase of the plasma transport. In particular, the Kelvin-Helmholtz instability can grow along the flank magnetopause at low latitude, where a velocity shear exists and where the nearly perpendicular magnetic field does not inhibit the development of the instability [40, 41, 27]. The Kelvin-Helmholtz instability thus provides an efficient mechanism for the formation of a mixing layer and for the entry of solar plasma into the magnetosphere, explaining the efficient transport during northward solar wind periods. Several observations support this explanation and show that the physical quantities observed along the flank magnetopause at low latitude are compatible with a Kelvin-Helmholtz vortex [28, 42, 23]

The process of vortex pairing mentioned before is a basic two-dimensional hydrodynamic inverse cascade phenomenon [32, 33] and can be expected to be an efficient process in the nearly two-dimensional external region of the magnetopause at low latitude [29, 23] and is believed to be the major process causing the increase in the thickness of the mixing layer in the downstream region of the magnetotail. However, the density inhomogeneity in the layer between the solar wind and the magnetosphere strongly modifies the non-linear evolution of the Kelvin-Helmholtz instability as the Kelvin-Helmholtz vortices cause the onset of a secondary instability which quickly leads to the onset of turbulence [43, 44] in the system. In fact the centrifugal acceleration of the rotating Kelvin-Helmholtz vortex acts as an "effective" gravity force on the plasma. If the density variation is large enough, the Rayleigh-Taylor instability can grow effectively along the vortex arms. How quickly the vortex becomes turbulent is crucial since the turbulence caused by the onset of the Rayleigh-Taylor secondary instability may destroy the structure of the vortices before they coalesce and may thus be the major cause of the increase in the width of the layer with increasing velocity and density inhomogeneity.

Rolled-up vortices, generated by the fast growing mode of the Kelvin-Helmholtz spectrum and entering in the non-linear stage, could then evolve following an inverse cascade process, or by developing a secondary Rayleigh-Taylor instability. In order to separate the effect of the onset of a secondary vortex-induced reconnection process which will be discussed in the next section, here we consider the initial magnetic field to be perpendicular to the plane where the Kelvin-Helmholtz instability develops and to have no inversion points.

In Ref. [35] it has been shown numerically (see first part of the Appendix) that for a relatively moderate initial density jump corresponding in dimensionless units to $\Delta n = 0.5$, the evolution of the Kelvin-Helmholtz vortices is not disrupted by the onset of a secondary Rayleigh-Taylor instability. In Figs.1,2 the shaded isocontours of the plasma density in the (x, y) plane for an initial density jump $\Delta n = 0.5$ at $t = 370$ and at $t = 425$ show the formation and the evolution of these vortices. The two vortices start to interact following an inverse cascade and eventually merge generating a single vortex.

On the contrary in the case of a large initial density jump, $\Delta n = 0.8$, the evolution of the system is strongly affected by the development of secondary Rayleigh-Taylor instabilities inside the vortex arms. After the generation of the two Kelvin-Helmholtz vortices, Figs. 3,4, the Rayleigh-Taylor instability starts to develop inside the arms of the vortices, leading to the formation of a turbulent layer along the y-direction with typical transverse width of the order of the vortex size. The vortex pairing interaction is depressed by the onset of the secondary instability.

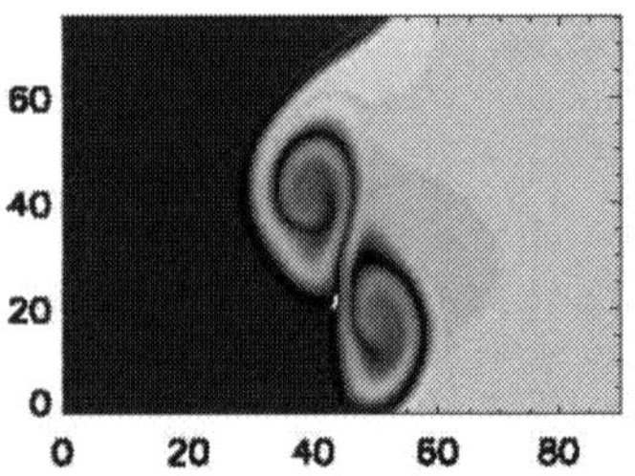

Figure 1. Plasma density isocontours for $\Delta n = 0.5$ at $t = 370$

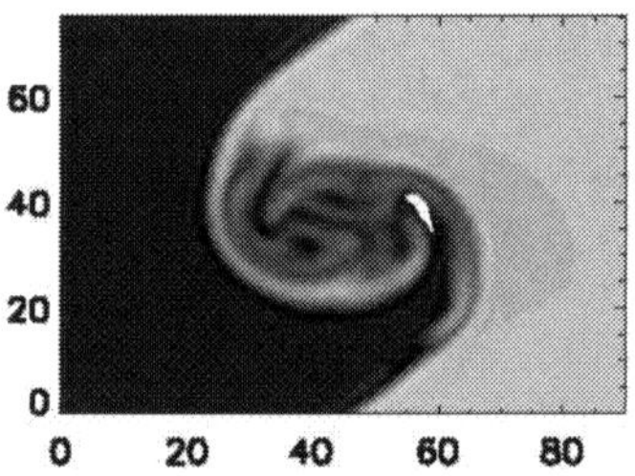

Figure 2. Plasma density isocontours for $\Delta n = 0.5$ at $t = 425$

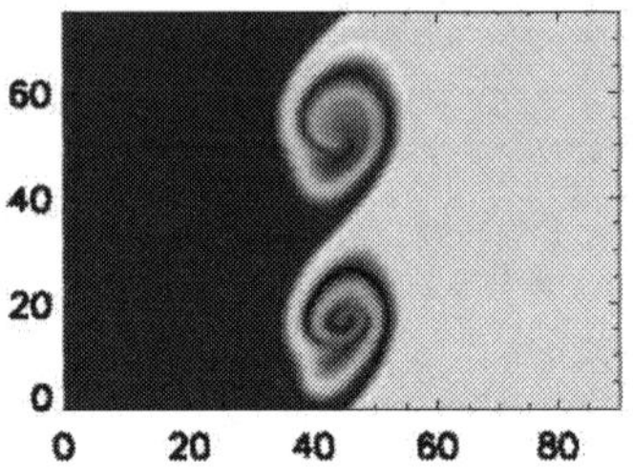

Figure 3. Plasma density isocontours for $\Delta n = 0.8$ at $t = 350$

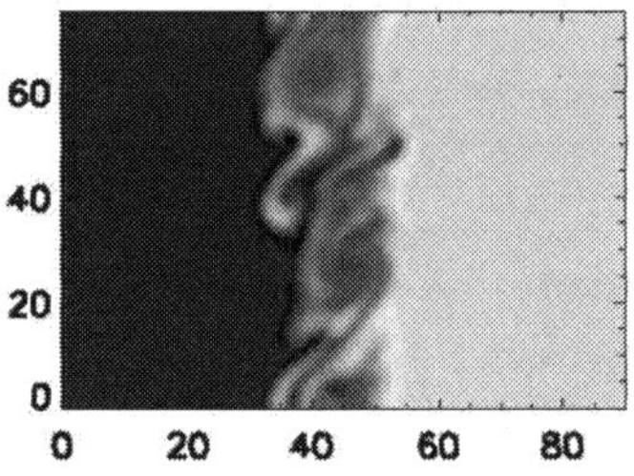

Figure 4. Plasma density isocontours for $\Delta n = 0.8$ at $t = 425$

In order to estimate the growth time of this secondary instability, we may consider each of the two vortices generated by the Kelvin-Helmholtz instability separately and assume such a structure to be stationary in the time interval $300 \le t \le 350$. During this period, the vortex propagates along the y-direction with a constant phase velocity and nearly constant amplitude.

We may model the vortex at $t = 300$ as an ' 'equilibrium" configuration and take the values n_1, u_1 and n_2, u_2 (as defined in the Appendix) inside two nearby vortex arms connected to the more and to the less dense parts of the plasma respectively, as the density and velocity values of two superposed inhomogeneous fluid plasmas in a slab geometry. In this model, the two plasma slabs are subjected to an effective gravity which corresponds to the centripetal acceleration arising from the arms curvature. Taking into account the profiles of the velocity shear and of the density of the vortex, we find that the estimate of the growth rate of the Rayleigh-Taylor instability is in agreement with that observed in the numerical simulations.

The competition between the vortex pairing process and the development of the Rayleigh-Taylor instability has important consequences from an observational point of view and can affect the transport properties of the system, see Ref- [35]. In the pairing case the density profile along the wind direction (along y) is characterized by well defined structures consisting of the well known step-like configuration that is directly related to the two vortex arms connected to the more and to the less dense parts of the plasma, and of a filament-like configuration that is related to the more complex central region of the vortex. This profile exhibits a well-defined periodicity, given by the wave length of the vortex, and is the signature of a rolled-up vortex. It corresponds either to the fast growing mode or to its sub-harmonics generated by the inverse cascade.

On the other hand, in the case where the secondary instability develops, the density profile along y exhibits a sequence of alternating high and low density filaments with no well defined wave length. This fact is related to the transition of the system to a turbulent state with the formation of a mixing layer via the development of smaller and smaller structures.

4. Configurations with a velocity shear and an in-plane magnetic field

In a magnetized plasma streaming with a nonuniform velocity, the Kelvin-Helmholtz instability plays a major role not only by mixing the plasma but also by stretching the magnetic field lines leading to the formation of layers with a sheared magnetic field where magnetic field line reconnection can take place. We refer again to the formation of a mixing layer between the Earth's magnetosphere and the solar wind at low latitudes during northward periods when the solar wind and the geomagnetic magnetic fields are parallel. As already discussed, velocity shear, in the presence of a magnetic field nearly perpendicular to the plane defined by the velocity field and its inhomogeneity directions, makes the Kelvin-Helmholtz instability grow along the flank magnetosphere.

If the Alfvén velocity associated to the in-plane magnetic field is sufficiently weak with respect to the fluid velocity jump, the Kelvin-Helmholtz instability forms fully rolled-up vortices which advect the magnetic field lines into a complex magnetic configuration. This advection causes the in-plane magnetic field component to develop inversion points where magnetic reconnection can take place. Since the plasma dynamics is essentially driven by the vortex motion, these reconnection events are usually denoted as vortex-induced reconnection. Although magnetic reconnection occurs only locally, this process changes the global topology of magnetic field, which is a necessary condition for plasma mixing, and thus the evolution of the vortices themselves.

Here we report the results of recent numerical investigations [36, 37] on the development of magnetic reconnection during the vortex pairing process and show that completely different magnetic structures are produced depending on how fast the reconnection process develops on the time scale set by the pairing process.

We consider a configuration (see second part of the Appendix) with a value of the plasma β parameter (defined as the ratio of the plasma pressure over the total magnetic field pressure) of order unity and show that in this regime the Hall term in Ohm's law, which arises from the decoupling of electrons and ions inside the current layers, allows magnetic reconnection to occur on time scales fast enough to compete with the pairing process. It is important to stress that in the numerical simulations in Refs. [36, 37] the conditions for magnetic reconnection are naturally provided, in an initially uniform in-plane magnetic field, by the motion of the Kelvin-Helmholtz vortices that grow and pair in the initially imposed shear velocity field.

We find that if the Hall term is removed from Ohm's law, the development of reconnection, and thus eventually of the Kelvin-Helmholtz vortices, is qualitatively, not only quantitatively, different. This result provides a clear cut example of the feedback between large and small scale physics, as the necessary conditions for reconnection to occur are produced by the motion of the large scale vortices, but the specific physical processes that make reconnection act faster or slower determine eventually the evolution of the entire system and the final magnetic field structure.

In Figs. 5-8 we show the magnetic field lines and a plasma passive tracer, advected by the velocity field, which is used in order to label the plasma domains initially on the right and on the left of the velocity null line.

In Fig. 5 we see the two vortices generated by the Kelvin-Helmholtz instability, corresponding to the fast growing mode wave number $m = 2$. As soon as the system enters the non-linear phase of the Kelvin-Helmholtz instability, the vortices start to pair following an inverse cascade process typical of two dimensional fluid systems. In Fig. 6 we see that the black and the gray domains are well separated by a ribbon (white in the figure) of nearly parallel, compressed magnetic lines. This ribbon is rolled-up by the rotation of the two vortices, and forms inside the folds between the vortices two current layers corresponding to two local magnetic inversion lines. We also see the formation of a first couple of X-points (one per vortex) in this region at $x_1 = 44$, $y_1 = 65$ and at $x_2 = 45$, $y_2 = 35$, respectively. This is the first reconnection event observed in our simulation. At $t = 437$ we show in Fig. 7 the formation of a second pair of X-points (one per vortex) in the same inversion region ($x_3 = 44$, $y_3 = 50$ and $x_4 = 45$, $y_4 = 47$). Magnetic reconnection develops at these X-points and forms magnetic islands with typical size $\sim d_i$, the maximum value compatible with the dimension of the current sheet. At the same time, the field line ribbon between the second pair of X-points shrinks and finally opens up. A new ribbon of field lines appears at $t = 460$, Fig. 8. This new ribbon no longer separates the gray and the black plasma regions. Indeed, during this process, significant portions of the gray plasma have been engulfed in the form of "blobs" into the black plasma region and viceversa. The inflow plasma velocity at the second pair of X-points is approximately 0.1 times the value of the local Alfvén velocity U_a in the x-y plane, in agreement with the values of the inflow velocity expected in the case of fast magnetic reconnection. Note that in the time interval given by a few growth times (a few times γ^{-1}) of the reconnection instability, the two vortices can only rotate by a few degrees so that the plasma displacement and the current rearrangement caused by the vortex rolling up are not sufficient to interfere with the development of the reconnection process.

In Ref. [37] it is shown that the ion decoupling region extends across the X-points over few d_i lengths. Inside this region the magnetic field is essentially frozen in the electron motion but the magnetohydrodynamic frozen-in law is not satisfied as the Hall term $\mathbf{J} \times \mathbf{B}$ becomes as large as the leading terms in the generalized Ohm's law. Inside a thinner region of width d_e the electrons are also decoupled from the magnetic field and magnetic reconnection can take place. Inside the thinner electron region the electrons are also decoupled from the magnetic field and magnetic reconnection can take place.

The role of the Hall term has also been demonstrated in Ref. [37] by running again the same simulation parameters simply omitting the Hall term in the generalized Ohm's law. Although the large scale motion of the vortices is still able to generate current sheets of comparable intensity and width, the process of magnetic reconnection is slower. In particular it does not succeed in forming the second pair of X-points quickly enough. The two vortices continue to roll-up while pairing and develop into a different magnetic pattern where magnetic islands are eventually generated all over the vortices leading to the disruption of the vortices. The rolling up and pairing of the vortices develops on a time scale comparable with the reconnection time

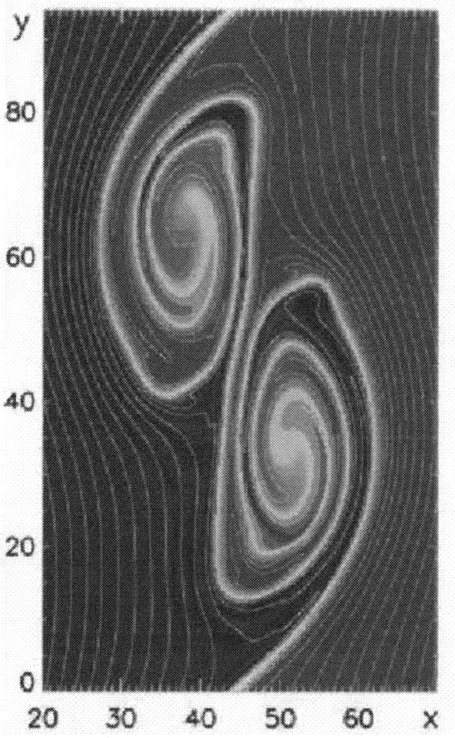

Figure 5. In-plane magnetic configuration at $t = 350$

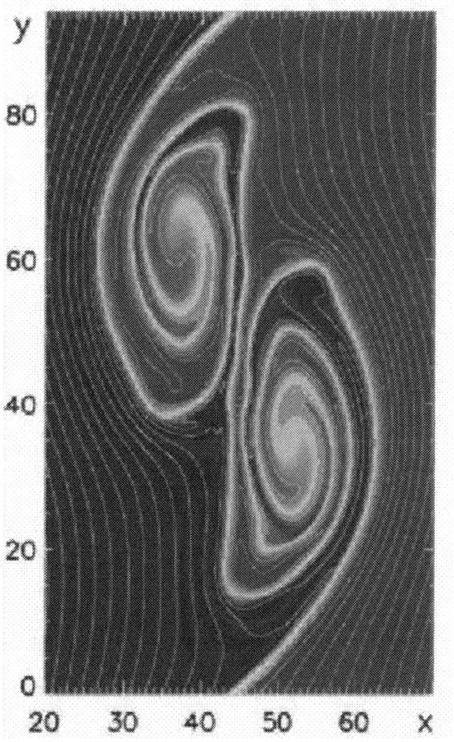

Figure 6. In-plane magnetic configuration at $t = 427$

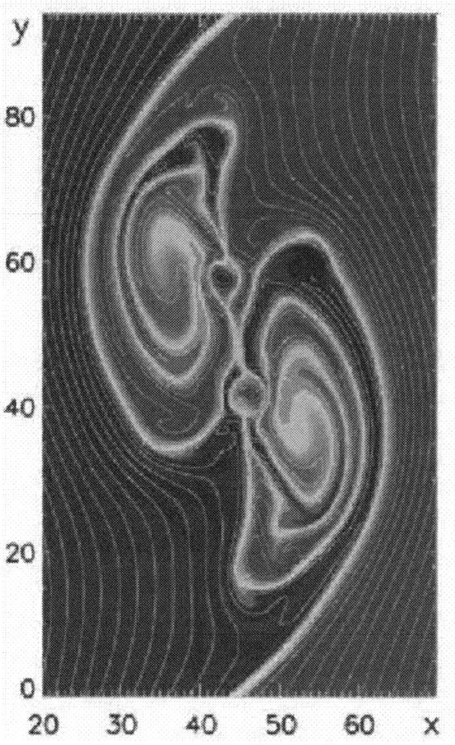

Figure 7. In-plane magnetic configuration at $t = 437$

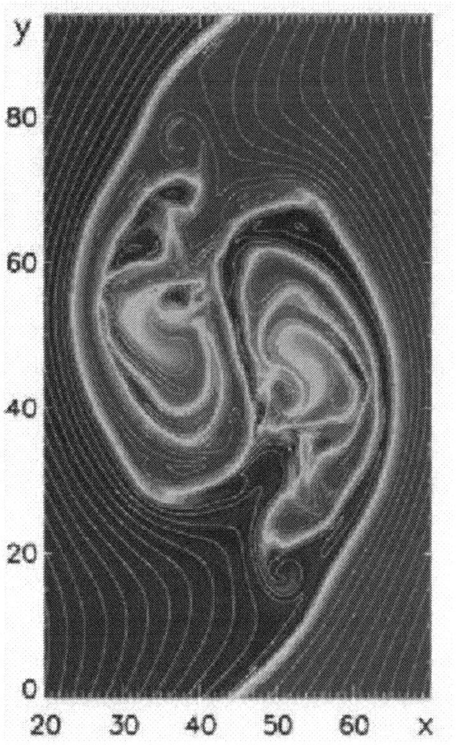

Figure 8. In-plane magnetic configuration at $t = 460$

and affects its evolution.
This competition between the development of the large scale magnetic configuration and the evolution and the reconnection instability determine the development of the entire system. If magnetic reconnection is not fast enough, the rolling up of the vortices destroys the favourable conditions for the reconnection instability to grow.

5. Conclusions

In this article we discussed how the competition between primary and secondary instabilities in a plasma can lead to qualitatively different physical states. These differences can be observed experimentally. Thus the outcome of this competition can give us important pieces of information on the time development of the instabilities as we can use one as a clock in order to measure the time development of the other.

We have considered a two dimensional plasma configuration with an initial large scale velocity shear and we have studied the interplay between the nonlinear development of the Kelvin-Helmholtz instability and the Rayleigh-Taylor instability driven by the centrifugal acceleration within the Kelvin-Helmholtz vortices in the presence of a density gradient. We have then studied the interplay between the nonlinear development of the Kelvin-Helmholtz instability and a magnetic reconnection instability driven by the current layers within the Kelvin-Helmholtz vortices in the presence of an initially uniform in-plane magnetic field. Actually in a real configuration both secondary instabilities should be considered as acting together.

In the case of the secondary magnetic reconnection instability we have considered a configuration with a value of the plasma β parameter of order unity and have shown that in this regime the Hall term in Ohm's law, which arises from the decoupling of electrons and ions inside the current layers, allows magnetic reconnection to occur on time scales fast enough to compete with the pairing process of the Kelvin-Helmholtz vortices.

In the literature, fast reconnection has been studied in a pre-imposed magnetic configuration consisting of a small region around a magnetic X-point, with a "forced" plasma inflow from the boundary towards the reconnection region. In the simulations presented in Refs. [36, 37] and reviewed here. the conditions or magnetic reconnection are naturally provided, in an initially uniform in-plane magnetic field, by the motion of the Kelvin-Helmholtz vortices that grow and pair in the initially imposed shear velocity field. We have found that if the Hall term is removed from Ohm's law, the development of reconnection, and thus eventually of the Kelvin-Helmholtz vortices, is qualitatively, not only quantitatively, different. This result provides a clear cut example of the feedback between large and small scale physics, as the necessary conditions for reconnection to occur are produced by the motion of the large scale vortices, but the specific physical processes that make reconnection act faster or slower determine eventually the evolution of the entire system and the final magnetic field structure.

Appendix

Simulation and plasma parameters for a configuration with a velocity shear and a density gradient

We consider a two dimensional description of the system, with the inhomogeneity direction along x, and the y-axis along the solar wind. This choice is justified since the evolution of Kelvin-Helmholtz instability is only weakly affected by slow equilibrium variation along the z-direction [41, 27]. We adopt a quasineutral plasma model described by the following set of single-fluid equations which we write in dimensionless conservative form as

$$\partial n/\partial t + \nabla \cdot (n\mathbf{U}) = 0, \tag{1}$$

with n the plasma density and $\mathbf{U}$ the fluid velocity.

$$\partial(n\mathbf{U})/\partial t + \nabla \cdot \left[n\mathbf{U}\mathbf{U} + P_T \bar{\bar{\mathbf{I}}} - \mathbf{B}\mathbf{B} \right] = 0, \tag{2}$$

with $P_T = P + B^2/2$, which is constant at $t = 0$, and the isothermal closure

$$\partial P/\partial t + \nabla \cdot (P\mathbf{U}) = 0. \tag{3}$$

The characteristic dimensional quantities are the mass density, the Alfvén velocity and the ion skin depth. The electric field is calculated by means of Ohm's law

$$\mathbf{E} = -\mathbf{U} \times \mathbf{B}, \tag{4}$$

where electron inertia and pressure terms are neglected. The equilibrium magnetic field at $t = 0$ is taken of the form $\mathbf{B}_{eq}(x) = B_{eq}(x)\mathbf{e}_z$. In this two dimensional transverse configuration $\mathbf{B}(x, y, t)$ follows a pure magnetohydrodynamic evolution also if the Hall term is included in Ohm's law. The magnetic field contributes only to determining the degree of compressibility of the plasma in the perpendicular plane which otherwise behaves hydrodynamically. For the low frequency range of interest the displacement current is not included.
The above equations are integrated by means of a numerical code. In this code, numerical stability is achieved by means of filters [45].
We consider an initial large-scale, sheared velocity field given by

$$\mathbf{U}_{eq} = \frac{U_0}{2} \tanh \frac{(x - L_x/2)}{L_u} \, \hat{\mathbf{y}} \tag{5}$$

and an equilibrium density of the form

$$n_{eq} = \frac{1}{2} \left[(2 - \Delta n) + \Delta n \tanh \frac{x - L_x/2}{L_n} \right] \tag{6}$$

We take $L_u = L_n = 3.0$ and the box-length in the x direction $L_x = 90$. We choose the box length $L_y = 24\pi$ in the periodic y direction, in order to have well separated linear growth rates for the modes $m = 1, 2, 3$, where $m = 2$ corresponds to the fast growing mode. The values of the two dimensionless parameters, the sound and Alfvén Mach numbers, are set as $M_s = U_0/C_s = 1.0$, $M_A = U_0/U_A = 1.0$, with $U_A = 1.0$ the value of the Alfvén velocity at $x = L_x$.

Simulation and plasma parameters for a configuration with a velocity shear and an in-plane magnetic field
We consider again a two dimensional description of the plasma configuration and adopt a two-fluid, quasineutral plasma model which is equivalent to a magnetohydrodynamic description with a generalized Ohm's law that includes the electron inertia and the Hall terms. In this model the electric field $\mathbf{E}$ is calculated by means of the following generalized Ohm's law [46] ($\mathbf{u_e}$ electron velocity, $\mathbf{u_i}$ ion velocity, $\mathbf{j} = ne(\mathbf{u_i} - \mathbf{u_e})$)

$$\left(1 - d_e^2 \nabla^2\right) \mathbf{E} = -\mathbf{u}_i \times \mathbf{B} + \frac{1}{n} \mathbf{j} \times \mathbf{B} - \frac{1}{n} \nabla P_e \tag{7}$$

where normalized quantities are used. We consider the same initial large-scale, sheared velocity field as given by Eq.(5) but. since we are primarily interested in the reconnection process, we consider a homogeneous density field in order to eliminate other secondary fluid instabilities.
The equilibrium magnetic field at $t = 0$ is homogeneous and given by

$$\mathbf{B}_{eq}(x, y) = B_{y,eq}\mathbf{e}_y + B_{z,eq}\mathbf{e}_z. \tag{8}$$

We take $L_u = 3.0$ and the box-length in the x direction $L_x = 90$. The box length in the periodic y-direction is $L_y = 30\pi$ in order to have well separated linear growth rates for the modes $m = 1, 2, 3$, where $m = 2$ corresponds to the fast growing mode of the Kelvin-Helmholtz instability. We take $d_e^2 = m_e/m_i = 1/64$.
The values of the dimensionless sound and Alfvén Mach numbers are set as $M_s = U_0/C_s = 1.0$, $M_{A,\perp} = U_0/U_{A,\perp} = 1.0$, $M_{A,\|} = U_0/U_{A,\|} = 20.0$, with $U_0 = 1.0$ and $U_{A,\perp}$, $U_{A,\|}$ the z and y component of the equilibrium Alfvén velocity, respectively. This choice allows the Kelvin-Helmholtz instability to develop into highly rolled-up vortices.

References

[1] Biskamp D 2000 *Magnetic Reconnection in Plasmas* Cambridge University Press, Cambridge.
[2] Shay M A *et al.* 1998 J. Geophys. Res **103** 9165.
[3] Uzdensky D A *et al.* 2006 Phys. of Plasmas, **13** 062305.
[4] Vekstein G *et al.* 2006 Phys. of Plasmas, **13** 122105.
[5] Bian N *et al.* 2007 Phys. of Plasmas, **14** 072107.
[6] Bian N *et al.* 2007 Phys. of Plasmas, **14** 120702.
[7] Furth H P *et al.* 1963 Phys. of Fluids **6** 459.
[8] White R B 1986 Rev. of Modern Phys **58** 183.
[9] Bulanov S V *et al.* 1990 Plasma Phys, and Contr. Fusion, **32** 377.
[10] Sweet P A 1958 *Electromagnetic Phenomena in Cosmic Physics* Cambridge University Press, Cambridge.
[11] Parker E 1963 Astrophys. J Suppl. **8** 177.
[12] Oieroset M *et al.* 2001 *Nature* **412** 414.
[13] Mozer F S *et al.* 2002 *Phys. Rev. Lett.* **89** 015002.
[14] Vaivads A *et al.* 2004 *Phys. Rev. Lett.* **93** 105001.
[15] Phan T D *et al.* 2007 *Phys. Rev. Lett.* **99** 255002.
[16] Retinó A *et al.* 2007 *Nature Physics* **3** 235.
[17] Deng X H *et al.* 2001 *Nature* **410** 557.
[18] Mandt M E *et al.* 1994 *Geophys. Res. Lett.* **21** 73.
[19] Shay M A *et al.* 1999 *Geophys. Res. Lett.* **26** 2163.
[20] Chen L J *et al.* 2007 *Nature Physics* **4** 19.
[21] Hasegawa H *et al.* 2005 *Ann. Geophys.* **23** 973.
[22] Stenberg G *et al.* 2007 *Ann. Geophys.* **25** 2439.
[23] Otto A *et al.* 2000 J. Geophys. Res. **105** 21175.
[24] Baty H *et al.* 2003 Phys. of Plasmas **10** 4661.
[25] Nakamura T K M *et al.* 2005 Geophys. Res. Lett. **32** 21102.
[26] Nakamura T K M *et al.* 2006 Adv. Space. Res. **37** 522.
[27] Hashimoto C *et al.* 2006 Adv. Space. Res **37** 527.
[28] Hasegawa H *et al.* 2004 *Nature* **430** 755.
[29] Miura A 1997 *Phys. Plasmas* **4** 2871.
[30] Liu Z X *et al.* 1988 *Geophys. Res. Lett.* **15** 752.
[31] Nykyri K *et al.* 2001 *Geophys. Res. Lett.* **28** 3565.
[32] Winant C D *et al.* 1974 J. Fluid Mech. **63** 237.
[33] Browand F K *et al.* 1976 J. Fluid Mech. **76** 127.
[34] Min K W *et al.* 1996 Geophys. Res. Lett. **23** 3667.
[35] Faganello M *et al.* 2008 Phys. Rev. Lett. **100** 015001.
[36] Faganello M *et al.* 2008 Phys. Rev. Lett. **101** 105001.
[37] Faganello M *et al.* 2008 to be published in Phys. Rev. Lett.
[38] Mitchell D G 1987 J. Geophys. Res. **92** 7394.
[39] Hasegawa H *et al.* 2004 Geophys. Res. Lett. **431** L06802.
[40] Miura A 1982 Phys. Rev. Lett. **16** 779.
[41] Brackbill J U *et al.* 2001Phys. Rev. Lett.**86** 2329.
[42] Fairfield D H *et al.* 2000 J. Geophys. Res. **105** 21159.
[43] Smyth W D 2003 J. Fluid Mech. **497** 67.
[44] Matsumoto Y *et al.* 2004 Geophys. Res. Lett. **31** 2807.
[45] Lele S K 1992 J.Comp. Phys. **103** 16.
[46] Valentini F *et al.* 2007 J.Comp. Phys. **225** 753.

Simulation science for fusion plasmas

S. Sudo[a], M. M. Škorić[a,1], T-H. Watanabe[a], Y. Todo[a], A. Ishizawa, H. Miura[a], R. Ishizaki[a], A. Ito, H. Ohtani[a], S. Usami, H. Nakamura, Atsushi Ito, S. Ishiguro[a], Y. Tomita[a], A. Takayama, M. Sato, T. Yamamoto, M. Den, H. Sakagami, R. Horiuchi[a], S. Okamura[a], and N. Nakajima[a]

National Institute for Fusion Science, National Institutes of Natural Sciences
[a]The Graduate University of Advanced Studies,
Toki-shi, 509-5292, Japan

E-mail: skoric.milos@nifs.ac.jp

Abstract. The world fusion effort has embarked into a new age with the construction of ITER in Cadarache, France, which will be the first magnetic confinement fusion plasma experiment dominated by the self-heating of fusion reactions. In order to operate and control burning plasmas and next generation demo fusion reactors, an advanced capability for comprehensive integrated computer simulations that are fully verified and validated against experimental data will be necessary. The ultimate goal is to predict reliably the behaviour of plasmas in toroidal magnetic confinement devices on all relevant scales, both in time and space. In addition to developing a sophisticated integrated simulation codes, directed advanced research in fusion physics, applied mathematics, computer science and software is envisaged. In this paper we review the basic strategy and main research efforts at the Department of Simulation Science of the National Institute for Fusion Science (NIFS)- which is the Inter University Institute and the coordinating Center of Excellence for academic fusion research in Japan. We overview a simulation research at NIFS, in particular relation to experiments in the Large Helical Device (LHD), the world's largest superconducting heliotron device, as a National Users' facility (see Motojima et al. [1]). Our main goal is understanding and systemizing the rich hierarchy of physical mechanisms in fusion plasmas, supported by exploring a basic science of complexity of plasma as a highly nonlinear, non-equilibrium, open system. The aim is to establish a simulation science as a new interdisciplinary field by fostering collaborative research in utilizing the large-scale supercomputer simulators. A concept of the hierarchy-renormalized simulation modelling will be invoked en route toward the LHD numerical test reactor.

1. Introduction

With the start of the construction of International Thermonuclear Experimental Reactor (ITER) in Cadarache, France, a multi-national collaborative, multibillion dollar effort, as the first in the next generation of burning fusion plasma experiments, a need for comprehensive integrated fusion plasma simulations is becoming crucial. Advances in simulation science enable us to critically evaluate and test our basic theoretical understanding through comparison with experiments. However, a future capability for integrated predictive whole machine modelling will bring about peak scientific

---

[1] Invited speaker, to whom any correspondence should be addressed.

objectives together with highest experimental productivity and the reduced operational risks. It is estimated that one ITER discharge will cost about 1 M$ with operational limits < 2000 pulses per year.

Here, we overview simulation research at the National Institute for Fusion Science (NIFS)- of the National Institutes of Natural Sciences (NINS), in particular relation to fusion plasma experiments in the Large Helical Device (LHD), as the National Users' facility. NIFS was established as an Inter-University Research Institute in 1989, and the coordinating Center of Excellence for academic fusion research in Japan, with the Mission to promote nuclear fusion science and research on its application, in particular, (1) To promote both experimental research based on the world largest superconducting device: Large Helical Device (LHD) and theoretical and simulation research. (2) To promote collaboration research with universities and institutes all over the country and also international collaborations. (3) To educate graduate students in the department of nuclear fusion science of the Graduate University for Advanced Studies, and also in the other universities. At NIFS, the current performance of the LHD experimental program toward a steady operation, reveals remarkable progress in terms of the proton temperature of 79 Million degree at the density of 20 Trillion/cc, peak density of 1100 Trillion/cc with the internal diffusion barrier (IDB), and the highest ever achieved, steady average beta of 5% and the largest total input energy of 1.6 GJ, in all magnetic confinement fusion devices (see, Motojima, [1]). More recently at NIFS, by merging the former Theory and Computer Simulation Center with the Computer and Network Center, new Department of Simulation Science (DSS) has become among the world largest research groups studying by theory and simulation science, properties of high-temperature fusion plasmas. Main goals of the simulation research are twofold: firstly, Understanding and Systemizing physical mechanisms in fusion plasmas and exploring Complexity science of a plasma, as a nonlinear, non-equilibrium, open system, as a basic research supporting fusion plasma studies. Secondly, the aim is to establish a Simulation Science as a new interdisciplinary field through collaborative research in utilizing the large-scale high-performance supercomputer simulators. Three simulation projects; namely, LHD and magnetic confinement simulation project, Laser fusion simulation project and Complex plasma simulation project have been launched. By developing predictive hierarchy-renormalized simulation model, each of above research goals is expected to be achieved in future, ultimately realizing the LHD Numerical Test Reactor (LHD-NTR) toward the concept of helical magnetic confinement fusion reactor, as priority for steady state operation revealed at NIFS, as a demonstration reactor after ITER. The *hierarchy-renormalized* simulation model is developed under active domestic and international collaboration; and it consists of the *hierarchy-integrated* simulation model and *hierarchy-extended* simulation model. The former is based on the transport model, which is suitable to describe the entire temporal behavior of experimentally observed macro physics quantities, while the latter model is focused at mutual interaction among neighboring hierarchies, as shown in Figure 1. The hierarchy-integrated simulation will be performed by renormalizing the results of the hierarchy-extended simulations, eventually leading toward the LHD-NTR. A basic strategy is schematically shown in Figure 2 [3].

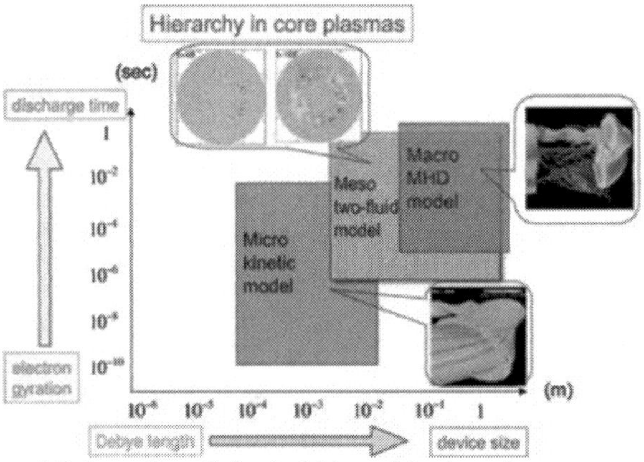

Figure 1 Schematics of physical hierarchies in a plasma

2. Multi-scale models and physical hierarchies

Multiscale modeling in magnetic fusion plasmas are among the most challenging tasks in the contemporary simulation science research. Multiphysics processes governing fusion plasmas span over a huge range of temporal and spatial scales resulting in intractability of direct brute force simulations, in any foreseen future. In the extreme cases, one finds the ratio of the transport time scale to the electron cyclotron time scale of $O(10^{14})$ and the device radius to the electron gyroradius of $O(10^{4})$. At the same time extreme anisotropy in the mean free path along and across the magnetic field can reach $O(10^{10})$, which together with high-dimensionality (3+3=6D) in real and velocity space, intrinsic nonlinearity and sensitive geometry present fundamental challenges to fusion simulations. Currently, scale separation is exploited to a large extent by solving the separate physical (hierarchy) models. Still, in reality, plasma phenomena are coupled since sharing same particle distributions and electromagnetic fields. Moreover, large number of critical cases is identified in fusion plasmas, where essentially coupling between multiscale and multiphysics problems must be performed (Figure 1). En route to developing a hierarchy-renormalized simulation model we present several examples of successful applications to gyrokinetic Vlasov simulations of turbulent transport, nonlinear simulations of energetic particle physics, two-fluid macro-instability simulation with inclusion of micro-turbulence effects, developing a high-resolution MHD code and pellet ablation. The complexity plasma simulations are illustrated with multi-hierarchy magnetic reconnection, blob transport, molecular dynamics and new numerical methodology for multi-scale plasmas. For analyzing huge simulation data, Virtual Reality (VR) system is developed for interactive visualization in 3D space and real time.

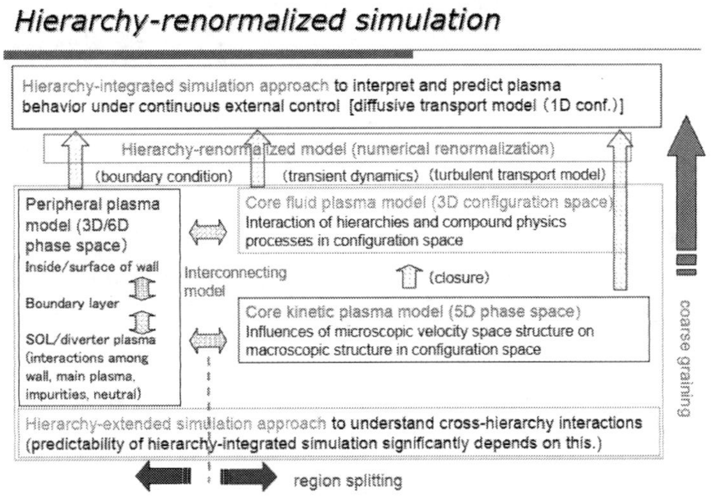

Figure 2

2.1. Gyro-kinetic simulation of the reduction of turbulent transport and zonal flow generation in Helical systems

In order to investigate effects of helical magnetic configurations on the ion temperature gradient (ITG) turbulence and zonal flows (ZF), we have performed gyrokinetic Vlasov (GKV) simulations [4,5] by utilizing the Earth Simulator at the Japan Agency for Marine-Earth Science and Technology (JAMSTEC) and the Plasma Simulator at NIFS. The GKV simulation code is extended so as to incorporate magnetic configuration models relevant to LHD experiments. In our recent work [6] for experimentally relevant conditions, we confirmed that an initially given zonal flow kept a higher level for a longer time in the inward-shifted configuration of the magnetic axis than that in the standard one. This is consistent with our analytical theory of zonal flows [7,8]. It should be emphasized that the inward-shifted case is optimized for reducing the neoclassical transport but with more unfavorable stability property than the standard configuration.

In the present study, the nonlinear GKV simulation implemented with the specified magnetic field parameters for the inward-shifted and standard plasma positions successfully confirms generation of large zonal flows enough to reduce the ion heat transport in the former case [9]. Contours of the electrostatic potential in the steady ITG turbulence are shown in Figure 3. For the inward-shifted configuration shown in Figure 3 (left), we clearly see radial structures of poloidal $\mathbf{E}\times\mathbf{B}$ zonal flows in the turbulent flow patterns on the poloidal cross section, while more isotropic $\mathbf{E}\times\mathbf{B}$ vortices are observed in the standard case [Figure 1 (right)]. The larger zonal-flow generation in the inward-shifted case agrees with the linear analysis of the zonal-flow response [6-8] which predicts a larger zonal-flow response to a given source in neoclassically optimized helical configurations such as the inward-shifted one. The obtained results agree with our theoretical prediction, and are consistent with experimental observation of better confinement in the inward-shifted LHD plasma [10].

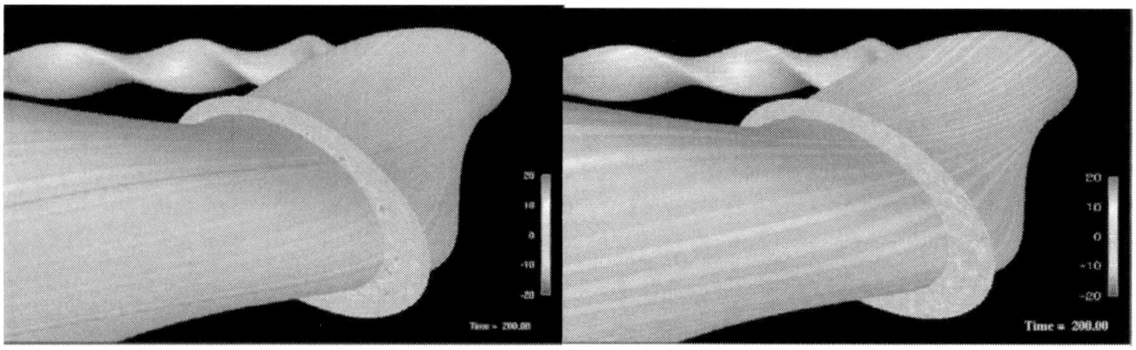

Figure 3 Contours of the electrostatic potential of ZF and ITG turbulence from GKV simulation for inward-shifted (left) and standard (right) LHD configurations. Normalization is chosen as $e\phi L_n/T_e\rho_i$.

2.2. Interaction between Alfvén eigenmodes and energetic particles

Alfvén eigenmodes are well known MHD type of oscillations in magnetically confined plasmas. However, in burning plasmas, fusion products in a form of alpha particles born in the 3.5 MeV energy range can resonate with Alfvén eigenmodes. This may further lead to (inverse) Landau growth and excitation of Alfvén eigenmodes and undesirable transport of energetic alphas. Many experiments using energetic ion beams generated during Neutral Beam Injection (NBI) or Ion Cyclotron Resonance Frequency (ICRF) heating of fusion plasmas have been performed in order to understand the interaction between Alfvén eigenmodes and energetic particles [11].

2.2.1. *Simulations of Alfvén eigenmodes with an extended Ohm's law for ITER-like plasma.*

The simulation code for MHD and energetic alpha particles, MEGA [11-12], has been extended with the drift model with the ion finite Larmor radius effect. In addition, the Ohm's law is extended with the effective resistivity where the electron damping is accounted by the Landau fluid model.

The extended MEGA code [13] is benchmarked with respect to the linear growth rate and the damping rate of the alpha-particle-driven n=4 toroidal Alfvén eigenmode (TAE) in the TFTR D-T plasma shot #103101. The initial energetic particle distribution is similar to that of a previous particle simulation reported in Ref. 14, where an isotropic distribution for the energetic alpha particles is used in the velocity space. The number of marker particles used is 5.2×10^5. The number of grid points is $101\times16\times101$ for the cylindrical coordinates (R, φ, z) where R is the major radius coordinate, φ is the toroidal angle coordinate, and z is the vertical coordinate. The simulation domain in the toroidal angle coordinate is $0<\varphi<\pi/2$ for this benchmark test of n=4 TAE. First, the linear growth rate was investigated with the effective resistivity turned off. The destabilized mode has a TAE spatial profile which consists of two major harmonics m/n=6,7/4 and frequency 215kHz. These results are consistent with the calculation with the NOVA-K code [5]. The linear growth rate obtained from this simulation, is 8.7×10^{-3} of the mode frequency. This linear growth rate is close to what is observed in the previous particle simulation [14] and calculated in the NOVA-K code [15]. Next, the damping rate was investigated with the alpha particle effects turned off and the effective resistivity turned on in the linear growth phase. The damping rate is $\gamma_d/\omega=4.8\times10^{-3}$, while we found $\gamma_d/\omega=1.6\times10^{-3}$ with the conventional MHD model. The increase in damping rate in the drift model arises from the effective resistivity. The damping rate with the effective resistivity is roughly half of the NOVA-K results [15].

We investigated an ITER-like plasma with the extended MEGA code [13]. The plasma profile investigated is based on the non-inductive scenario reported in ITER Technical Basis [16]. The major and minor radii are R=6.35m and a=1.85m. The ellipticity and the triangularity are 1.85 and 0.4, respectively. The toroidal magnetic field at the plasma center is B=5.18T. The number density is assumed to be uniform $6.7\times10^{19}m^{-3}$. The safety factor is $q_0=3.5$ at the plasma center, $q_{min}=2.2$ at r/a=0.7, and $q_{edge}=5.3$ at r/a=1.0. The thermal pressure profile is assumed to be uniform to focus on the instabilities driven by the alpha particles. The beta value of the thermal plasma is $\beta=2.5\%$. The electron temperature is 12.3keV, which is employed in the electron Landau damping model. The birth velocity of the alpha particle corresponds to $1.48v_A$; v_A is the Alfvén velocity at the plasma centre.

The spatial profile of the alpha particle beta value is calculated with the plasma profile of the non-inductive scenario taking account of the D-T reaction rate and the slowing-down of alpha particles. The alpha particle beta value at the plasma centre is $\beta_{\alpha0}=2\times10^{-2}$. The nonlinear evolution of toroidal Alfvén eigenmodes (TAE) in the ITER-like plasma was investigated [13]. The number of marker particles used is 2.1×10^6. The number of grid points is 101×100×101 for the cylindrical coordinates. The TAEs with toroidal mode numbers n=2-5 are destabilized. The TAEs peak around r/a~0.3. The toroidal electric field profile of n=3 TAE is shown in Figure 4. The saturation levels are $\delta B_r/B\sim10^{-3}$. The redistribution of alpha particles leads to the change in alpha particle beta profile by $\delta\beta_\alpha\sim10^{-3}$.

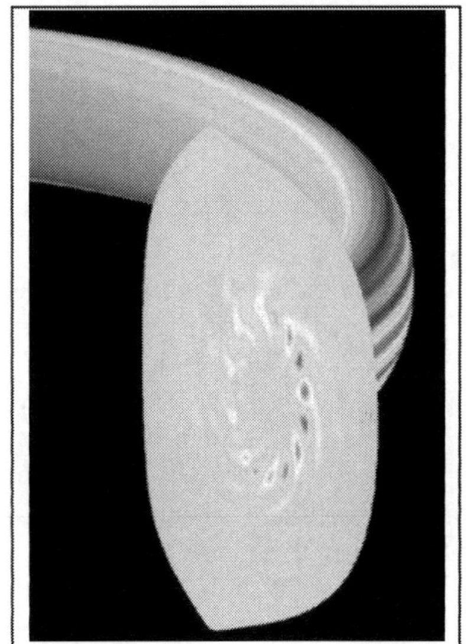

Figure 4. Toroidal electric field profile of n=3 TAE for ITER-like plasma.

2.3. Nonlinear evolution of MHD instability

We have been carrying out direct numerical simulations of the fully three-dimensional, compressible and nonlinear magnetohydrodynamic (MHD) simulations in the geometry of LHD. Purpose of the research is clarifying how hot plasma survives the interchange/ballooning instability in the LHD. For this purpose, the MHD In the Non-Orthogonal System (MINOS) code is developed. In the MINOS code, variables are described by the contravariant component on a helical-toroidal coordinate system. The 8th order compact finite difference code is adopted in the MINOS code to separate variables into directions parallel and perpendicular to magnetic field lines.

Direct numerical simulations by the use of the MINOS code are carried out for linearly unstable magnetic configuration, which is characterized by the position of the magnetic axis $R_{ax}=3.6$m and the beta-value of 3.7% at the peak position. Number of grid points range from (97, 97, 640) to (385, 385, 64) depending on the Reynolds number in the simulation. By the numerical simulations, we have found three mechanisms which can help the hot plasma to overcome the instability: (1) the local flattening of the pressure, (2) the kinetic energy release into the direction parallel to the magnetic field lines, and (3) the compressibility effect to reduce the linear growth rates. [16-20].

While these computational works have been carried out carefully, the influences of the small-scale motions on large-scale motions remain unknown. The difficulty comes from the characteristics of the interchange /ballooning instability. Because of these short-wavenumber instability, we need infinite number of grid points for ideal simulations. A simulation study with a large number of grid points (and thus a large Reynolds number) is now under progress [21].

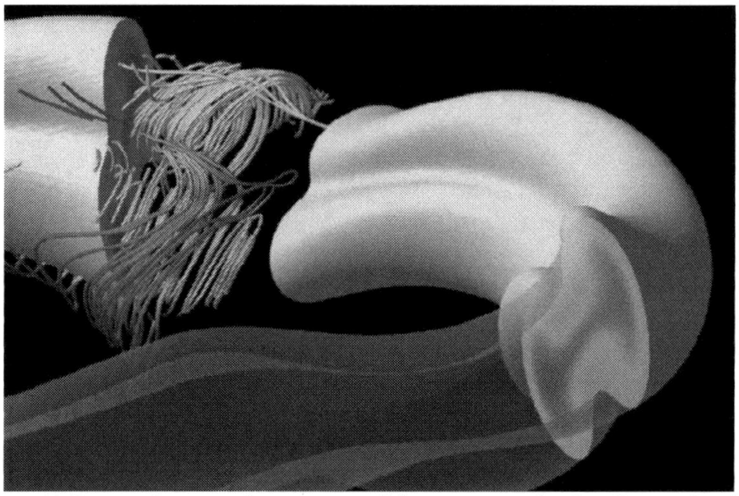

Figure 5. Iso-surfaces and contours of the pressure in the direct simulation of the MHD of LHD. Mushroom-like structures are formed on the contours of the pressure due to the growth of the m/n=2/1 unstable Fourier coefficient. Color lines represent stream lines.

2.4. Multi-scale interactions among micro-turbulence, macro-MHD instability, and zonal flows

Multi-scale interactions in magnetically confined plasmas were investigated analytically and numerically. The interaction between micro-turbulence and zonal flow is a typical multi-scale interaction because toroidal and poloidal mode numbers of micro-instabilities are much larger than one, while these numbers are zero for zonal flow. There are another multi-scale interaction between different instabilities which are independent in linear evolution when their free energies are different, for instance, there are interactions between micro-instability due to pressure gradient and macro-MHD instability that is caused by current density gradient and has low toroidal and poloidal mode numbers.

In our recent study we considered a situation in tokamak configuration that a current driven macro-MHD instability arises in a quasi-equilibrium including micro-turbulence and zonal flow. This is because a macro-MHD instability is destabilized when there arises one or more resonant q surfaces where the safety factor q is equal to a small rational number, while micro-instabilities can arise even if there is no integer q surfaces. Thus, micro-turbulence almost always exists before destabilization of macro-MHD instability. In order to examine this situation we carried out a three-dimensional numerical simulation of a reduced set of two-fluid equations and investigated mutual interaction between micro-turbulence and macro-MHD mode (double tearing mode) [21,22]. In the simulation micro-instabilities (kinetic ballooning mode) are destabilized at first, and then turbulent state is formed as shown in the left frame of Figure 6. We found that the excitation of macro-mode by the micro-turbulence is different from macro-MHD instability because the spatial profile of the former is significantly different from the spatial profile of double tearing mode. The energy transfer from micro-turbulence to macro-mode is similar to the energy transfer from micro-turbulence to zonal flow [22]. The macro-mode is a part of turbulence in the turbulent state. After a certain period of time, then, the induced macro-mode by the turbulence becomes eigenfunction of double tearing mode and grows up with magnetic reconnection as shown in the right frame of Figure 6, in the case that the equilibrium has free energy of macro-MHD instability such as a current density gradient [21]. When the free energy is small the double tearing mode saturates, and then we have a new quasi-equilibrium including not only the micro-turbulence and zonal flow but magnetic islands [21]. The magnetic reconnection of double tearing mode is caused by non-ideal effect due to turbulent mixing of magnetic flux at the resonant surfaces [23]. The effect of zonal flow produced by micro-turbulence on the stability of double tearing mode is not clear [24], while the zonal flow shear is suppressed by the appearance of double tearing mode as implied in Ref. [21].

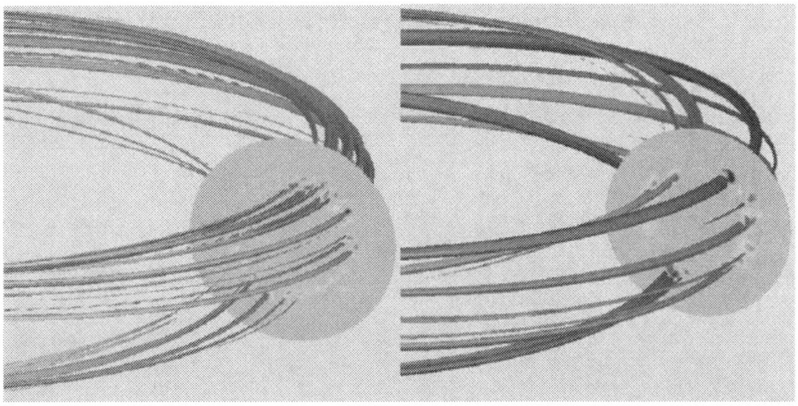

Figure 6. Three-dimensional color-contour of electrostatic potential. The quasi-steady state including electromagnetic turbulence is formed in the left frame. After the appearance of macro-MHD instability the spatial structure of turbulence is altered in the right frame.

2.5. Pellet ablation in a plasma

There are many important physics processes in connection with pellet fueling, not the least in the prediction of the pellet ablation rate and penetration distance in the plasma. Pellets ablating in hot plasma are surrounded by a cold, dense neutral ablation cloud which expands in all directions near the pellet, and then after the flow ionizes, it becomes constrained to follow the magnetic lines. To gain a fuller understanding of those atomic processes and multi-dimensional effects, a time-dependent code CAP has been developed which can rigorously treat the dynamical coupling between interior pellet motions and the exterior ablation flow [25]. The code uses an Eulerian approach based on the cubic-interpolated pseudoparticle (CIP) method [26].

In order to examine 2D effects resulting from anisotropic heating imposed by the magnetic field, the cylindrical axisymmetric coordinate system (r, z) is used. No effect of the magnetic field, except for setting the direction of the electron heat flux inducing the ablation, is included in the study. A deuterium pellet with an initial radius of 2 mm is used. Plasma electron temperature and density are 2 keV and 10^{14} cm^{-3}, respectively, which are assumed to be constant throughout the temporal evolution. The pellet surface pressure differential is of course due to the fact that the heat flux following magnetic field lines intercepting the pellet near the polar axis is entirely blocked by the pellet and thus fully absorbed, while the heat flux following field lines just grazing the pellet surface is partially transmitted across the equatorial midplane, so there is less absorption at the equatorial points. As a result, since the areal ablation rate is more enhanced at the poles than at the equatorial points, the ablation cloud becomes thicker in the polar direction than it is along the radial coordinate in the equatorial midplane. This leads to the ovalshaped density contours elongated in the polar direction as shown in Figure 7 at different times. The contour of the pellet density (which is uniform inside the pellet) has the opposite shape. Since the shear stress induced by the anisotropic pressure distribution can exceed the yield strength of the solid, the solid pellet becomes fluidized and deformed into an oval shape flattened in the polar direction and elongated in the radial direction. After a longer time, the pellet becomes flattened into a thin pancake while losing mass by ablation.

One of the key results of the study is that pellet deformation can shorten the lifetime of the pellet. The area of the pellet projected normal to the magnetic field lines increases at first, reaches a peak, and then falls to zero later on as shown in Figure 7; it reveals the essential dynamics of 2D pellet ablation. It is seen that the 2D ablation rate versus time curve (thick solid line) depends on the projected area because the total heating power available for driving ablation is simply proportional to the area. On the other hand, the dashed line plots the ablation rate for the 1D spherical symmetric isotropic heating. The ablation rate steadily decreases in accordance with the monotonically decreasing pellet size. The dashed line agrees with the thin solid line obtained by the transonic flow model [27]. In result, it is found that the effect of the pellet deformation can shorten the pellet lifetime.

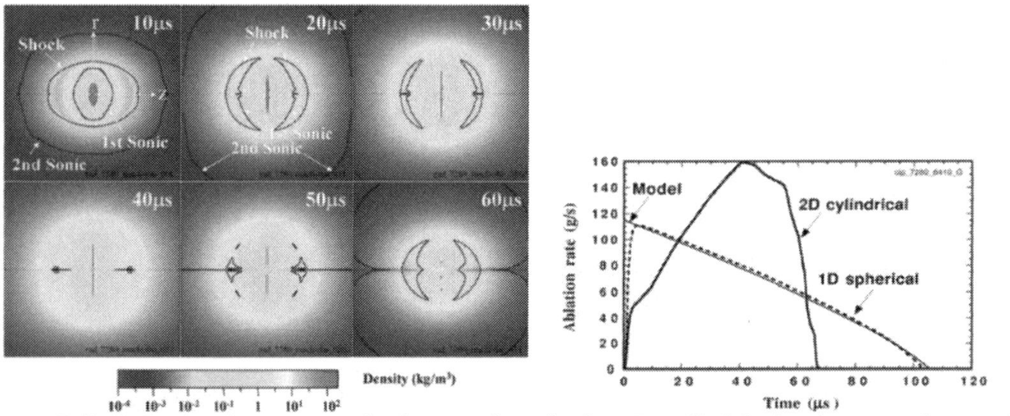

Figure 7.(left) Isodensity contours in the r-z plane in the 2D cylindrical axisymmetric coordinate system. Frames correspond to the time at 10, 20, 30, 40, 50 and 60 μs. Black lines indicate first sonic, shock and second sonic surfaces. (right) Time dependence of the pellet ablation rate for a constant heat flux over the pellet lifetime. The solid curve shows 2D cylindrical geometry and the dashed curve is 1D spherically symmetric case. The thin solid line is the ablation rate from the transonic flow model.

2.6. Multi-hierarchy simulation model for magnetic reconnection

Magnetic reconnection, which is a multi-scale phenomenon bridging between macroscopic and microscopic hierarchies, is widely believed to be one of important mechanisms controlling energetically active phenomena observed in high temperature, magnetized plasmas such as the solar flare, the magnetosphere, and fusion plasmas. For example, a microscopic process leading to the generation of electric resistivity such as wave-particle interaction and binary collisions is needed to excite magnetic reconnection. On the other hand, global plasma transport and global change of field topology take place as a result of magnetic reconnection, which, in turn, affect reconnection mechanism itself through complex nonlinear processes. Thus, its whole picture should be clarified by solving both microscopic physics and macroscopic physics consistently and simultaneously.

We are now developing multi-hierarchy simulation model for magnetic reconnection under MARIS (MAgnetic Reconnection Interlocked Simulation) project [28, 29]. The simulation model is based on domain decomposition method and consists of three parts (see left panel of Figure 8), i.e., MHD model to describe global dynamics of reconnection phenomena in the region far from reconnection point, electromagnetic PIC model (PASMO) to describe the microscopic processes in the vicinity of reconnection point, and interface model to describe the interaction between micro and macro hierarchies in the interface region. Both multi-time scale scheme [29] and shake-hand scheme [30] are also adopted for this interface model. Right panel of Figure 8 demonstrates the propagation of one-dimensional Alfven wave using developed simulation model where a uniform external magnetic field exists in y direction, mass ratio $m_i/m_e=100$, PIC domain is located in the center of the simulation box ($96<y<160$, spatial coordinate y is normalized by c/ω_{ce}), MHD domains in both sides of PIC domain ($0<y<64$, $192<y<256$), and the interface domains exist between MHD and PIC domains ($64<y<96$, $160<y<192$). It is clearly seen in Figure 8 that Alfven wave is propagating from MHD domain to PIC domain, and from PIC domain to MHD domain smoothly. Thus, it is concluded that the developed interlocked scheme works well.

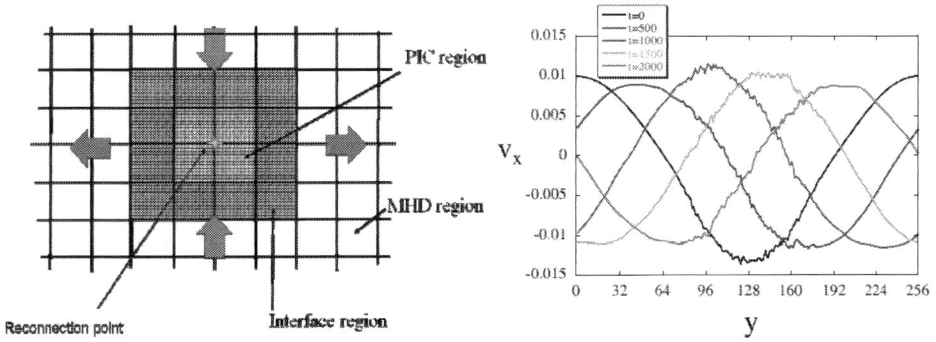

Figure 8 (left) Schematic diagram of multi-hierarchy model and (right) spatial profiles of fluid velocity v_x at $\omega_{ce}t = 0$, 500, 1000, 1500, and 2000. Wave propagates in the right direction

2.7. Three dimensional PIC simulation of blob dynamics

Recently, transport of edge plasma in magnetic confinement fusion devices has been recognized to be mostly convective and intermittent. As a theoretical explanation, blobs which are long-living macroscopic coherent structures have been referred. Some theoretical and numerical studies which are using two-dimensional modelling have been performed and their dynamics and stability have been revealed. However, these works have employed some assumptions to reduce the equations and the validation is not clear. We have planned three dimensional particle-in-cell (PIC) simulation studies to reveal detailed physical picture. The configuration of our 3D PIC simulation is following: an external magnetic field is pointing into the z-direction. Particle absorbing boundaries corresponding to diverter plates are placed at both ends of the z-axis. A particle absorbing plate corresponding to the first wall is also placed at one end of the x-axis, while a particle reflecting plane at the other one. Particle impinging onto the absorbing boundaries are removed from the system. In the y-direction, periodic boundary conditions are applied. Initially, a high density blob, like a piece of spaghetti, is placed along the magnetic field line in the center of the system. Figure 9 shows the temporal evolution of the ion density profile in the x-y plane at the center of the system. The blob moves across the magnetic field due to EXB drift, where electric field is self-consistently created by grad B drift. Its speed is about 0.1 c_s, where c_s is the ion acoustic speed [31]. More detailed study is under consideration.

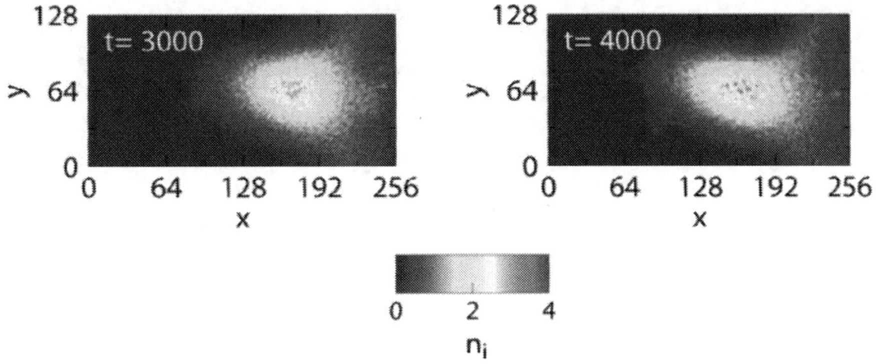

Figure 9. Time evolution of ion density profile in the x-y plane at $z=L_z/2$. The blob moves in the negative x direction due to EXB drift with speed about $0.1c_s$.

2.8. Molecular Dynamics Simulation of Hydrogen Atom Sputtering on a Graphite (0 0 0 1) Surface

Hydrogen atom sputtering on a graphite surface of divertor in the LHD was investigated using molecular dynamics (MD) simulation. Because graphite has a layer structure, two kinds of potential models were used. One is the modified Brenner reactive empirical bond order potential to represent a covalent bond when a distance of particles is shorter than 2.0 Å [32, 33]. The other potential model is the interlayer intermolecular potential developed by us to represent the layer structure with 'ABAB' stacking [34]. Time is developed by the second order symplectic integration. Because the farmer potential model is a complicated function, a time step is very small time 5 x 10^{-18} s. In this simulation, layered graphite is composed of eight graphene layers in which an interlayer distance is of 3.35 Å under the periodic boundary condition except for the direction perpendicular to graphene surfaces. Hydrogen atoms are injected onto a graphite (0 0 0 1) surface at a flux 2.5 x 10^{30} atom/m^2s. The MD simulation follows conditions of constant energy and volume, that is, temperature is not controlled.

The following dynamics of erosion was found by the MD simulation. The graphene layers were peeled off one by one from the surface when the bombardment of hydrogen atoms continues (Figure 10). However, under graphene layers maintained the layer structure of graphite. This phenomenon is called 'graphite peeling'. On the other hand, the graphite peeling did not occur in the MD simulations of graphite with vacancies and graphite without the interlayer intermolecular interaction. Both of these graphite change into an amorphous structure due to the bombardment of hydrogen atoms. This fact implies the following mechanisms. The repulsion of the interlayer intermolecular interaction defends against approach of the graphene layers and then the layer structure is maintained. However, a vacancy does not cause the interlayer intermolecular interaction and carbon atoms around the vacancy which have only two covalent bonds become more stable to create the third covalent bond with a carbon atom of the other layer than to keep the layer structure. In addition, it is also important that hydrocarbon molecule created from the graphite peeling had mainly a chain structure. The ends of the carbon chains were often terminated by the hydrogen atom, while the center of carbon chain does not attach hydrogen atoms.

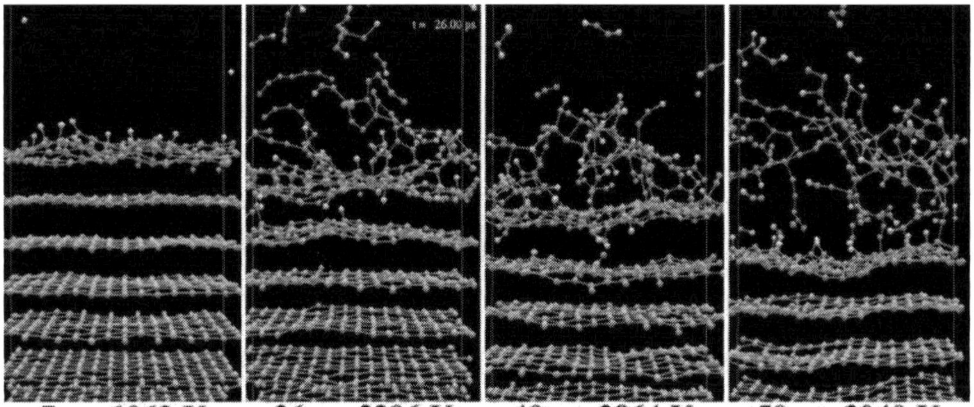

7 ps, 1062 K **26 ps, 2296 K** **40 ps, 2864 K** **50 ps, 3040 K**

Figure 10. Graphite peeling when incident energy is 5 eV. White and green spheres indicate hydrogen and carbon atoms. Green lines indicate covalent bonds.

2.9. Macro Projective Integration- EFREE- Method for Multiscale Plasmas

Multiscale problems are difficult to simulate due to interconnected physics of micro- and macro-scales which defies direct brute force numerical computations. A potential of a novel simulation framework in plasmas, Macro Projective Integration (called, Equation-free) method has been investigated [35]. Based on a micro-physics simulator (kinetic, i.e., particle-in-cell code-PIC), the macro-scale dynamics is determined by repeatedly extrapolating forward macro variables obtained from short bursts of micro simulations [36]. Different schemes for reconstruction and mapping between high-dimensional (micro) and low-dimensional (macro) phase space, using: moment representation, cumulative distributions, wavelets, etc., are analyzed [37-38]. Starting with proof-of-principle studies of simple one-dimensional nonlinear plasma paradigm, i.e. the ion sound waves, it is expected to be able to eventually move to more realistic multiscale fusion plasma models, as visualized in Figure 11.

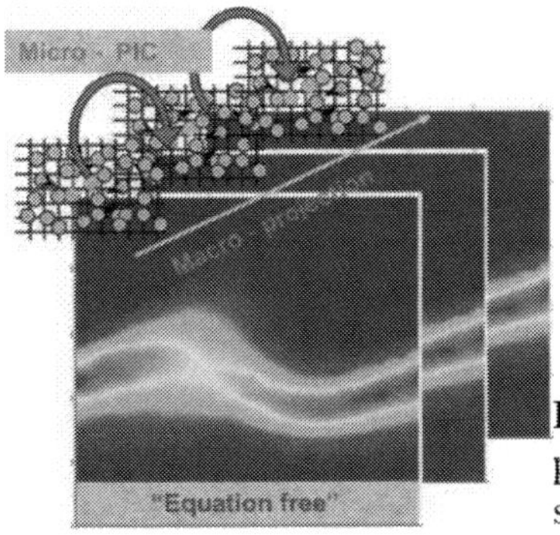

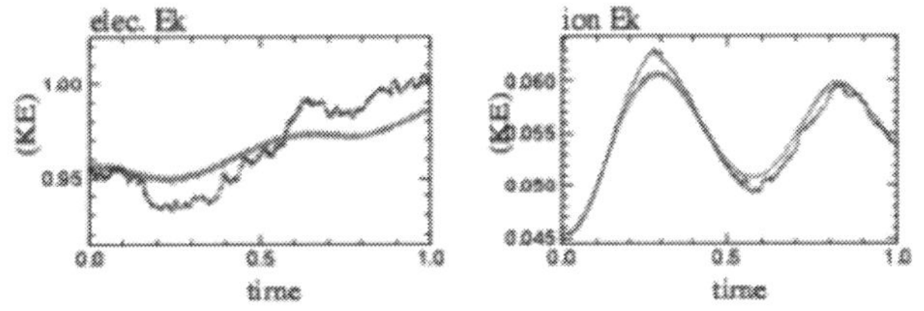

Figure 11. Concept of EFREE projective integration in a plasma (left), and electron and ion kinetic energy (right) in ion sound waves from a standard PIC (red) and EFREE method.

3. Conclusions

In conclusions, computer simulations are increasingly envisioned as the third methodology of science aiming to bridge a gap between the traditional theory and experiments by taking advantages in rapid progress in applied mathematics, computer science and high-performance computers. Moreover, the development of validated predictive simulation codes is understood to be crucial to ensuring the success of future burning plasma experiments, such as ITER. The future role of Petascale toward Exascale computing capability will be becoming of strategic importance for achieving above goals, as stressed in the comprehensive long-term proposal for a leadership Fusion Simulation Project (SciDAC) in US [39]. At NIFS, the present supercomputer is shortly to be replaced by the new Plasma Simulator system (Figure 12) in the competitive 100 teraflops range to be in the second stage upgraded to 315 TF. In particular, it may be stressed that in contrast to other leading supercomputer systems which typically share resources within the broad scientific community, NIFS Plasma Simulator, is likely the largest supercomputer system entirely committed to research on plasma and fusion science.

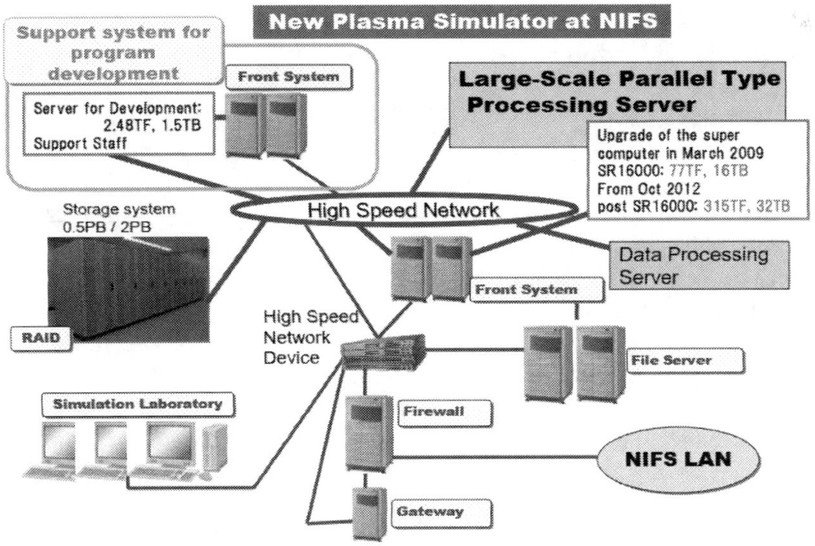

Figure 12. New Plasma Simulator supercomputer system at NIFS

Acknowledgements

This work was supported by the Grant-in-Aid for Science Research from the Ministry of Education, Science, Sports, Culture and Technology, Japan Society for the Promotion of Science, National Institute for Fusion Science, Graduate University of Advanced Studies, JAMSTEC and NIFS/NINS

Formation of International Network for Scientific Collaborations. One of us (M.M.Š) thanks H. Yamada for support and assistance in preparing the presentation, and G. Malović, L. Popović and Lj. Hadžievski under Ministry of Science of Serbia Project 141034, for generous effort in enabling the successful SPIG event and this contribution possible. Finally, we are much grateful to Professor Osamu Motojima for creative guidance and continuing encouragements.

References

[1] Motojima O *et al* 2003 *Nuclear Fusion* **43** 1674, and references therein
Motojima O 2006 in *Physics of Ionized Gases, AIP Conference Proceedings* **876** pp.349-360
[2] Sudo S 2007 Key lecture at the Commemorative Simulation Science Symposium, Toki
[3] Nakajima N, 2008 lecture at the 2nd Simulation Science Symposium, Tajimi
[4] Watanabe T-H and Sugama H 2006 *Nuclear Fusion* **46** 24
[5] Watanabe T-H, Sugama H and S. Ferrando-Margalet 2007 *Nuclear Fusion* **47** 1383
[6] Ferrando-Margalet S, Sugama H, and Watanabe T-H 2007 *Phys. Plasmas* **14** 122505
[7] Sugama H and Watanabe T-H 2005 *Phys. Rev. Lett* . **94** 115001
[8] Sugama H and Watanabe T-H 2006 *Phys. Plasmas* **13** 012501
[9] Watanabe T-H, Sugama H and Ferrando-Margalet S 2008 *Phys. Rev. Lett.* **100** 195002
[10] Yamada H *et al* *Nuclear Fusion* 2001 **41** 901
[11] Todo Y *et al* 2003 *Phys. Plasmas* **10** 2888
[12] Todo Y *et al.* 2005 *Phys. Plasmas* **12** 012503
[13] Todo Y *et al.* 2006 *J. Plasma Phys.* **72** 817
[14] Chen Y *et al.* 1999 *Phys. Plasmas* **6**, 226
[15] Fu G Y *et al.* 1998 *Phys. Plasmas* **5** 4284
[16] Miura H, Hayashi T and Sato T 2001 *Phys. Plasmas* **8** 4870
[17] Miura H, Nakajima N, Hayashi T and Okamoto M 2006 *J. Plasma Phys.* **72** (2006) 1095
[18] Miura H *et al* 2007 in *Theory of Fusion Plasmas, AIP Conference Proceedings* **871** pp.157-168
[19] Miura H, Nakajima N, Hayashi T and Okamoto M 2007 *Fusion Sci. & Technology* **51** 8
[20] Miura H *et al* to be presented at IAEA 22th Fusion Energy Conf. (Geneva, 2008) TH/P9-16
[21] Ishizawa A and Nakajima N 2007 *Phys. Plasmas* **14** 040702
[22] Ishizawa A and Nakajima N 2007 *Nuclear Fusion* **47** 1540
[23] Ishizawa A and Nakajima N 2008 to appear in *Theory of Fusion Plasmas, AIP Conference Proceedings of Joint Varenna-Lausanne International Workshop*
[24] Ishizawa A and Nakajima N 2008 *Phys. Plasmas* **15**, 084504
[25] Ishizaki R, Parks P B, Nakajima N and Okamoto M 2006 *Phys. Plasmas* **8** 4064
[26] Takewaki H, Nishiguchi A and Yabe T 1985 *J. Comput. Phys.* **61** 261.
[27] Parks P B and Turnbull R J 1978 *Phys. Fluids* **21** 1735.
[28] Horiuchi R, Ohtani H and Ishizawa A 2006 *J. Plasma Phys.* **72** 953-956
[29] Usami S, Ohtani H and Horiuchi R 2008 *Comm. Comp. Phys.* **4** 537-544
[30] Sugiyama T and Kusano K 2007 *J. Comput. Phys.,* **227** 1340
[31] Ishiguro S and Hasegawa H 2006 *J. Plasma Phys* **72** 1233
[32] Ito A, Nakamura H and Takayama A, to be published in *J. Phys. Soc. Jpn.*
[33] Brenner D W, O.A. Shenderova O A *et al* 2002 *J. Phys.: Condens. Matter* **14** 783.
[34] Ito A and Nakamura H 2008 *Commun. Comput. Phys* **4** 592
[35] Kevrekidis I G *et al* 2003 *Comm. Math. Sci.* **1** 718
[36] Shay M A *et al* 2007 *J. Comp. Phys.* **226** 571
[37] Maluckov A, Škorić M M and Ishiguro S 2008 *Commun. Comput. Phys.* **4** 556
[38] Škorić M M *et al* 2008 *Plasma Fusion Res.* **3** 021
[39] Tang W M 2008 *J. Phys: Conference Series* **125** 012047

Multilayer photoionization models of the nuclear and circumnuclear regions of type 2 AGN

P. Rafanelli, L. Vaona, S. Ciroi, F. Di Mille

Department of Astronomy, University of Padova

E-mail: piero.rafanelli@unipd.it

Abstract. We present the results of a new photoionization model, designed for simulating the spectra of the NLR of galaxies hosting an Active Galactic Nucleus (AGN). The validity of the model is verified reproducing the line ratios measured in the spectra of a sample of 1344 nearby ($z<0.1$) AGN of class 2, Seyfert 2 galaxies, extracted from the SDSS, and comparing them with the results obtained using different models proposed by other authors in literature. The reliability of the proposed model is discussed.

Introduction

We shortly introduce in a schematic form the general properties of Active Galaxies before approaching the specific problem of this review.

1.1. Morphology

Active Galaxies are morphologically characterized by a bright, semistellar Nucleus (Figure 1), the Luminosity of which is typically: $L_{AGN} \sim 10^{41} - 10^{47}$ erg $s^{-1} \sim 10^{8} - 10^{14}$ L_{SUN}, a value which is several order of magnitude larger than the typical luminosity of the nuclei of normal spiral galaxies: $L_{NUC} \sim 10^{6} - 10^{8}$ L_{SUN} and often even larger than the luminosity of the whole underlying galaxy: $L_{HOST} \sim 10^{9} - 10^{11}$ L_{SUN}. AGN with absolute magnitude fainter than -23 ($M_B > -23$) are called Seyfert galaxies and have usually redshift $z < 0.1$.

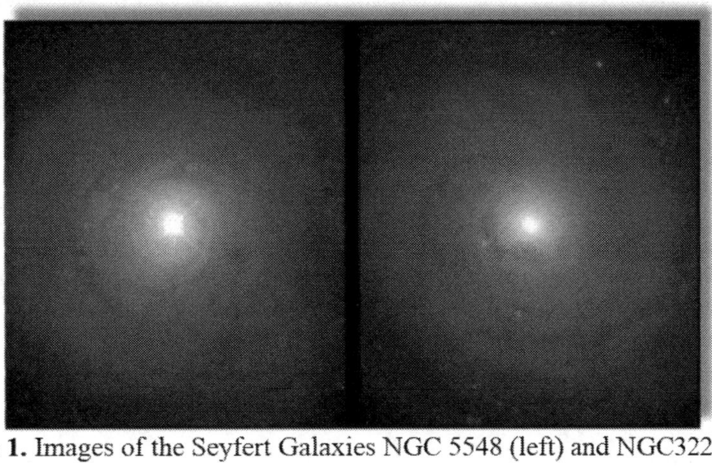

Figure 1. Images of the Seyfert Galaxies NGC 5548 (left) and NGC3227 (right)

1.2. Spectroscopic features

1.2.1 Emission line spectrum. The spectra of AGN are characterized by emission lines arising from forbidden transitions from metastable levels of heavy elements at different ionization degree and by recombination lines of HI, HeI, HeII, FeII and of heavy highly ionized elements (Figure 2).

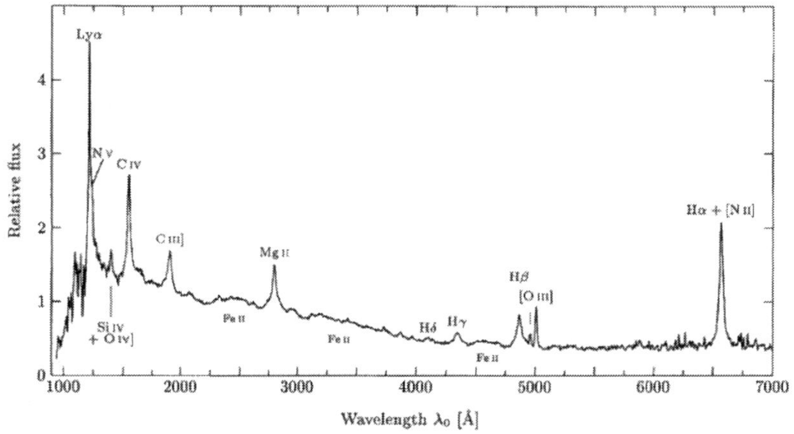

Figure 2. Typical composite spectrum of an AGN [1].

On the basis of the features of the emission line profiles there are at least two classes of AGN, namely Seyfert galaxies in the local Universe ($z<0.1$): Seyfert 1 and Seyfert 2 galaxies. Spectra of Seyfert 1 nuclei are characterized by a composite line profile of the permitted lines, characterized by a broad component (FWHM $\sim 10^4$ km s^{-1}) with superimposed a narrow one (FWHM $\leq 10^3$ km s^{-1}) and by narrow (FWHM $\leq 10^3$ km s^{-1}) forbidden lines. On the contrary Seyfert 2 galaxies are characterized by forbidden and permitted lines with a narrow profile (FWHM $\leq 10^3$ km s^{-1}), Figure 3. At first approximation the emission line spectrum of Seyfert 2 galaxies is very similar to the narrow line spectrum of Seyfert 1 galaxies. The absence of broad forbidden lines indicates that the typical density of the regions where the broad lines arise is larger than the critical density of the metastable levels from which the forbidden lines are emitted. These facts together with the typically different velocity fields derived from the line profiles of permitted and forbidden lines in the spectra of Seyfert 1 nuclei suggest that there are quite likely two different, kinematically separated, emitting regions around the nuclei of Seyfert 1 galaxies: a Broad Line Region (BLR) and a Narrow Line Region (NLR). In addition the similar features of Seyfert 2 spectra and narrow component Seyfert 1 spectra indicate that the two classes of active galaxies are quite likely different aspects of the same phenomenon. The typical parameters of BLR and NLR are reported in Table 1.

Table 1. Typical features of BLR and NLR in AGN.

Broad Line Region (BLR)	Narrow Line Region (NLR)
R~0.1pc	R~100pc
High velocity, FWHM ~ 10^4 km s^{-1}	Low velocity, FWHM ~ 10^3 km s^{-1}
High electron density: • No broad [O III] lines • Broad C III]1909 line **$N_e \sim 10^{9-10}$ cm^{-3}**	Low electron density: **$N_e \sim 10^{2-3}$ cm^{-3}**

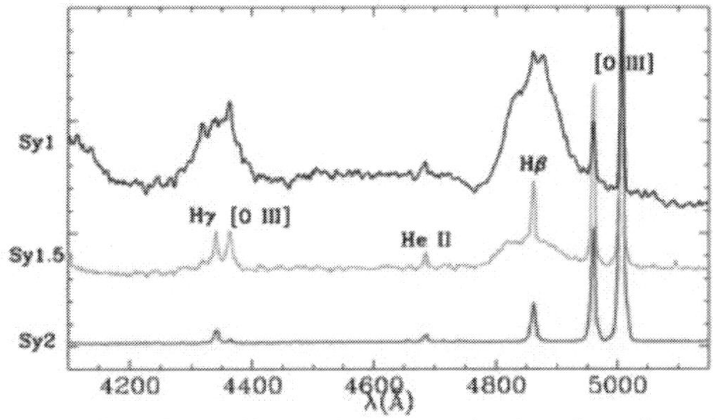

Figure 3. Spectra of Seyfert 1 (blue and green tracings) and Seyfert 2 (red) galaxies.

1.2.2 The continuum of AGN. The typical ultraviolet, optical and infrared continuum of AGN is reported in Figure 4. The continuum of Seyfert 1 galaxies is dominated in the optical by a featureless non thermal component. On the contrary the continuum of Seyfert 2 galaxies is dominated in the same range by the stellar component of the galaxy hosting the active nucleus. In a schematic and qualitative form we can summarize these features with the following statements:

- Type 1: continuum = power-law: $\mathbf{F}_\nu \propto \nu^\alpha$ ($\alpha \sim -1, -1.5$) + host galaxy
- Type 2: continuum = host galaxy

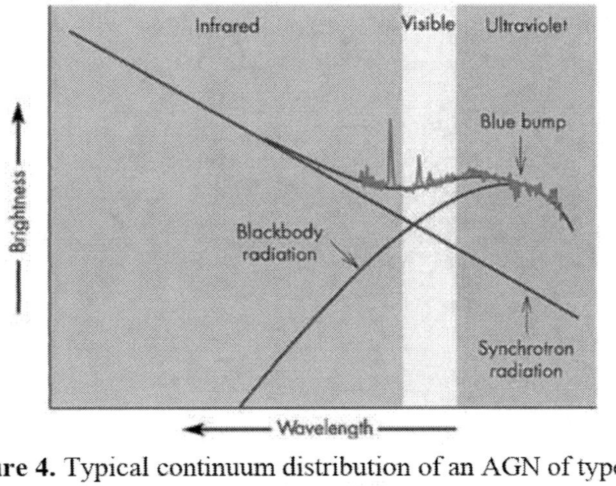

Figure 4. Typical continuum distribution of an AGN of type 1.

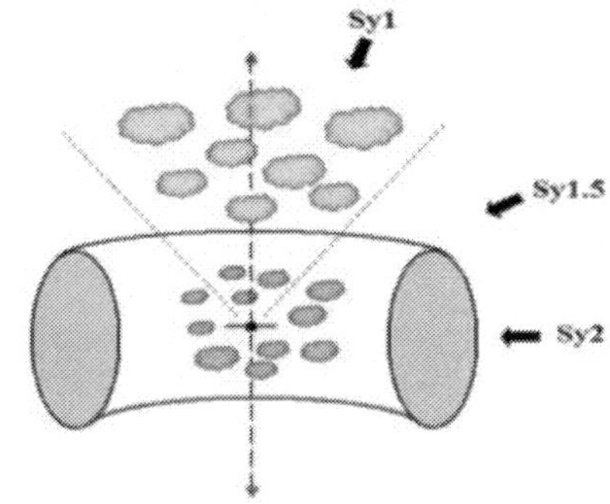

Figure 5. Unified model: blue and green clouds are the BLR and the NLR respectively. Supermassive Black Hole is at the center of the toroidal ring.

1.3. The standard unified model

The mostly accepted model for explaining the observed properties of Seyfert galaxies is sketched in Fig. 5. Seyfert 1 and Seyfert 2 have essentially the same nuclear structure: a dusty torus surrounding a Super Massive Black Hole, SMBH, ($M_{SMBH} \sim 10^7$–10^9 M_{SUN}), is accreting mass (Accretion Rate: 0.2 - 2.0 M_{SUN} yr $^{-1}$) which is in part transformed in radiating energy at a rate of 10^{45} - 10^{46} erg s $^{-1}$. This radiation illuminates the BLR located within the torus and the most extended NLR outside of it. In this way the different spectroscopic features of Seyfert 1 and Seyfert 2 nuclei depend only by an orientation effect of the line of sight, that is unable to see the BLR when its direction is skew to the axis of the torus, whereas can reach directly the BLR and the central emitting engine when oriented close to the axis of torus itself. The radiation emitted by the SMBH is forced by the dusty torus to shine within a conus which illuminates the NLR and the surrounding gaseous regions of the hosting galaxy.

The study of these regions, photoionized by the radiation emitted from the nucleus, is crucial for understanding the physics of the nuclear engine and the origin of the activity. A model describing correctly their features and tested over a large homogeneous sample of spectroscopically classified Seyfert 2 galaxies is therefore necessary.

This is the aim of our work, which is essentially devoted to identify a statistically significant sample of spectroscopically selected objects of the Seyfert 2 class and to investigate the physical features of the gas of their NLR up to a distance of 3 kpc (z=0.1, H= 75 km s^{-1} Mpc^{-1}) from the nucleus.

2. Selection of the objects

1344 Seyfert 2 galaxies have been extracted from the SDSS-DR6 (6860 square degree) in order to simulate and to reproduce their spectra using a photoionizaton code. The first selection criterium was to impose the presence of the following emission lines in the spectra of all galaxies: Hβ, [OIII] λ5007, Hα, [NII] λ6584, [SII] λλ6717-6731

In order to define a complete sample of emission line galaxies no constraint was imposed to the redshift and at the end we extracted a set of **216166** galaxies. The z versus Petrosian r-magnitude, corrected for Galactic extinction is shown in Figure 6. A cut at magnitude 17.7 reduced to **201425** the number of galaxies. The condition that [NII] λ6584 and Hα are separated was a good filter for selecting spectra with a narrow Hα line profile.

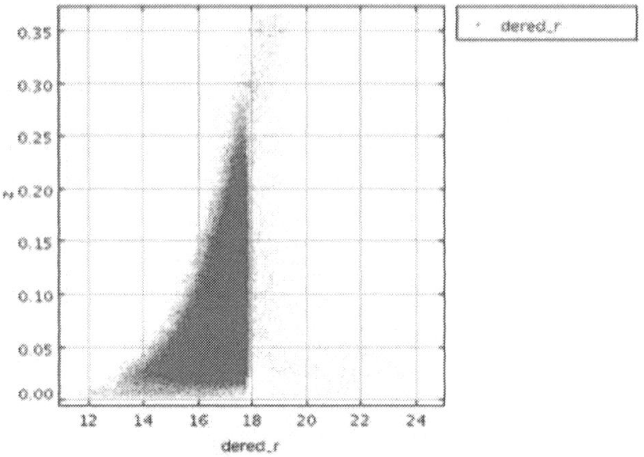

Figure 6. Redshift z as a function of the magnitude r of the extracted objects.

2.1. Classification criteria

2.2. The classification of the emission line galaxies has been performed using the Baldwin, Phillips, Terlevich (BPT) diagrams [2], based on the four optical line ratios [OIII]/Hβ, [NII]/Hα, [SII]/Hα,

[OI]/Hα and the borderline between starburst and AGN evaluated by Kewley et al. [3]. This line divides the plane of the diagram in two regions: the upper-right region is occupied by AGN, the lower-left region by HII- and Sturburst-like objects.

The diagnostic line ratios, like log[NII]6548-83/Hα-log[OIII]5007/Hβ etc., have been derived measuring the intensity of the emission lines after subtraction of the stellar component from the spectra of the selected galaxies. This procedure allows to avoid the influence of the underlying absorption component in the determination of the intensity of the recombination lines. The effects of such subtraction are shown in Figure 7, obtained analyzing the data published by Kewley et al. [3]. As an example it is shown that the point representing the line ratios of an object moves from top to bottom and from right to left after subtraction of the galactic stellar component.

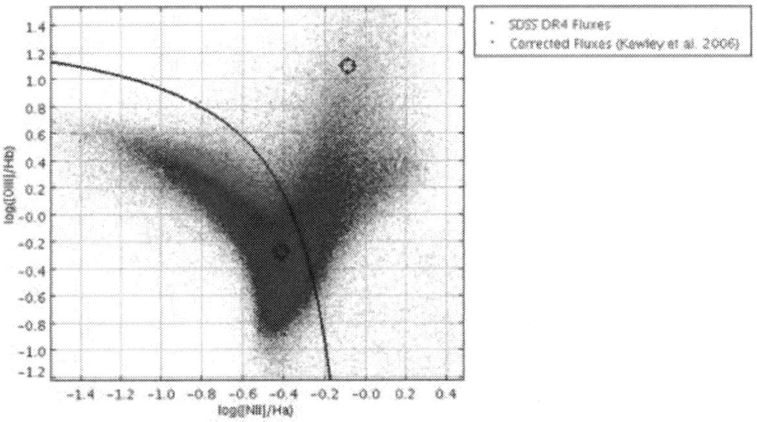

Figure 7. BPT diagnostic diagram. Red and blue points represent objects before and after subtraction of the stellar continuum from their spectra

It is clear that the absence of any subtraction of the stellar continuum tends to move in the BPT diagrams the line ratios measured in starburst objects in the AGN region. A pre-classification of the whole sample before any correction from the influence of the underlying galaxy permits to isolate most of the starburst objects, leaving the identification of the remaining galaxies to a more accurate classification after the subtraction of the stellar component. The Figure 8 and Figure 9 show the results of this pre-classification obtained using the line intensities measured automatically by the SDSS.

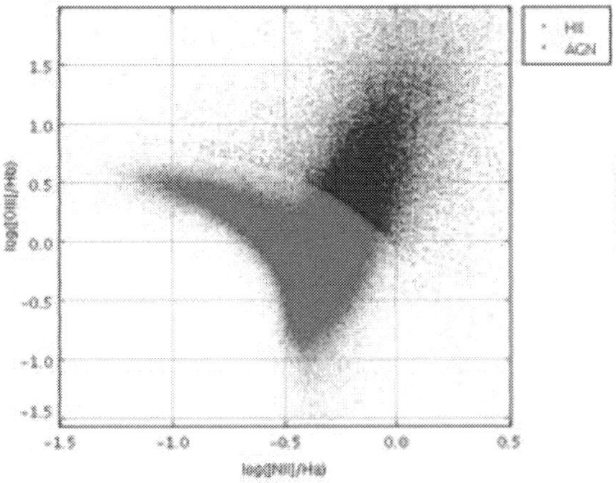

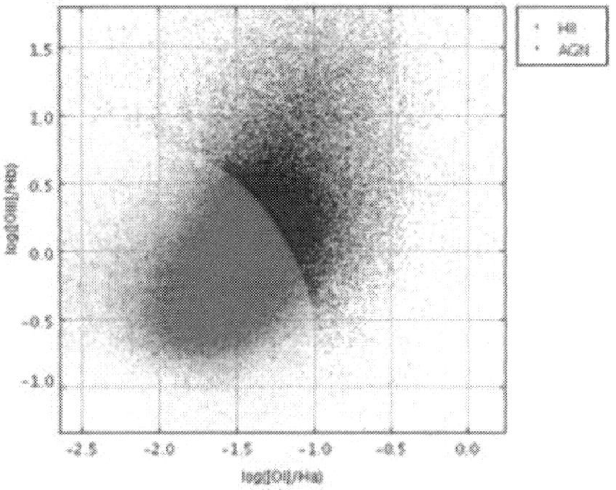

Figure 8. – 9. BPT diagnostic diagrams applied to the line ratios measured automatically in the spectra of 201425 preselected galaxies.

The objects classified as **suspected AGN** are now **31724.** They are: **Seyfert galaxies,** some Starburst galaxies, many LINERS (a subclass of low ionization AGN, that we want to avoid)

In order to reduce the number of LINERS we make the following considerations:

1) If we take the L([OIII]λ5007) luminosities of LINERS and Seyfert galaxies, measured in the ~ 85000 emission line galaxies classified by Kewley et al. [3], using the diagnostic diagrams after subtraction of the stellar component, we find a distribution which shows that galaxies with L [OIII] > $0.5 \cdot 10^{40}$ erg s^{-1} mainly belong to the Seyfert class, as suggested by Figure 10.
2) Since the [OIII] line flux is independent from the subtraction or not of the underlying stellar component, it is a good parameter for reducing the presence of LINERS among AGN.

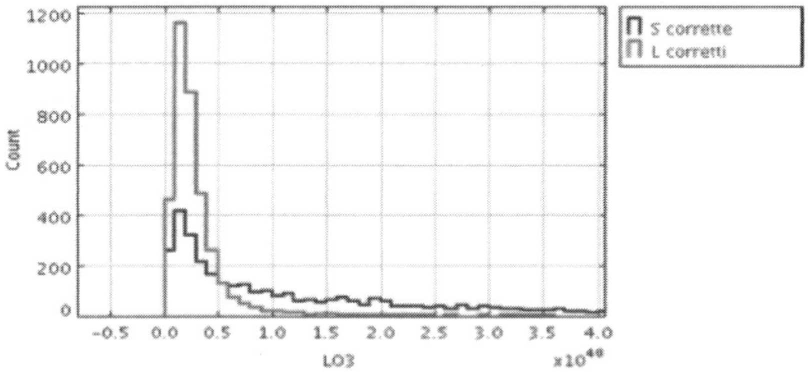

Figure 10. Number of Liners and Seyfert as a function of their [OIII] λ5007 luminosity measured in the Kewley sample [3].

The extraction of the objects with L [OIII] > $0.5 \cdot 10^{40}$ erg s^{-1} lowers the number of objects of our sample to 14372 galaxies, which are good Seyfert 2 (S2) candidates. We impose now the condition that $z \leq 0.1$ in order to limit our sample to nearby objects, for which the morphology can be studied in detail [4] and we further put as constraint that the error in the determination of F([NII] λ6584) be smaller than 5%. We have now 2458 objects remaining and completeness is guaranteed up to m ~ 16.0. All these spectra have been now corrected by stellar component using the spectral synthesis code STARLIGHT [5] and the line fluxes have been measured manually. The line ratios have been then plotted in the BPT diagrams, Figure 11, and we have identified 1344 Seyfert 2.

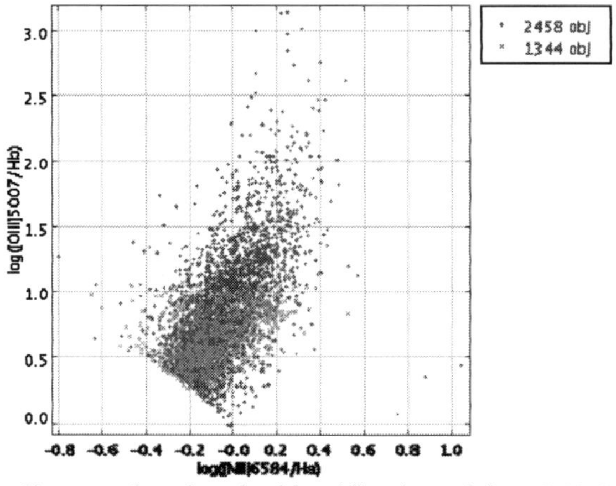

Figure 11. BPT diagnostic diagram showing the identification of the 1344 Seyfert 2 galaxies of our sample.

3. Analysis of the Sample

3.1. Internal extinction

Internal extinction of each galaxy has been evaluated using the measured Balmer decrement **Hα/Hβ** and assuming that its theoretical value is **2.95** [6]. The so derived values of **Av** have the distribution shown in Figure 12, peaked around **~ 1.3** and are linearly correlated with the stellar absorption values evaluated with STARLIGHT, Figure 12. All emission line fluxes have been then corrected by extinction.

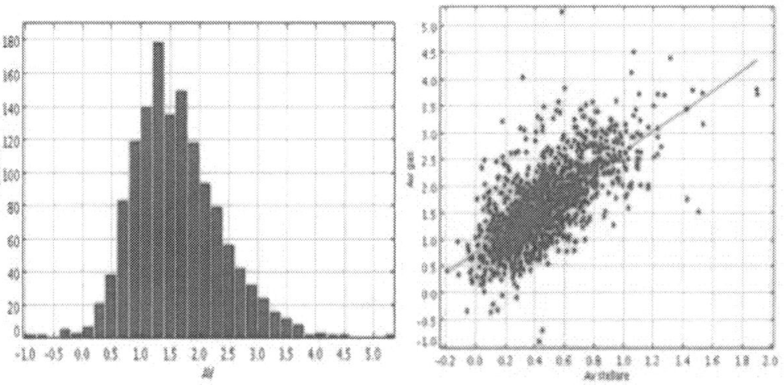

Figure 12. Distribution of the visual extinction A_V evaluated from the measured Balmer decrement (left) and comparison with the same parameter derived by STARLIGHT (right).

3.2. Electron Temperature and Density. Ionization mechanism

The electronic densities and temperatures have been measured using different line ratios in order to look for different regions with different physical properties.

The used line ratios are:

Ne → [SII] 6717/6731, [ArIV] 4710/4740

Te → [OIII] (4959+5007)/4363, [SII] (6717+6731)/4074, [OII] 3727/7325

The Full Width Half Maximum (FWHM) of the different emission lines has been measured in each one of the **1344** spectra. The electronic temperatures derived using the [OIII], [OII] and [SII] emission line ratios are all comparable and all are lower than 20000 K:

T_e ([OIII]) ~ 14900 ± 3900 K, T_e ([OII]) ~ 11400 ± 2800 K, T_e ([SII]) ~ 12800 ± 6000 K

An equilibrium temperature < 20000 K is consistent with **photoionization** as main ionization mechanism of the gas. This is confirmed by the BPT diagnostic diagrams where we have plotted the line ratios of Supernova Remnants, namely regions ionized by shocks and the line ratios of photoionized regions, Figure 13. It is evident that our AGN belong to the second class of objects.

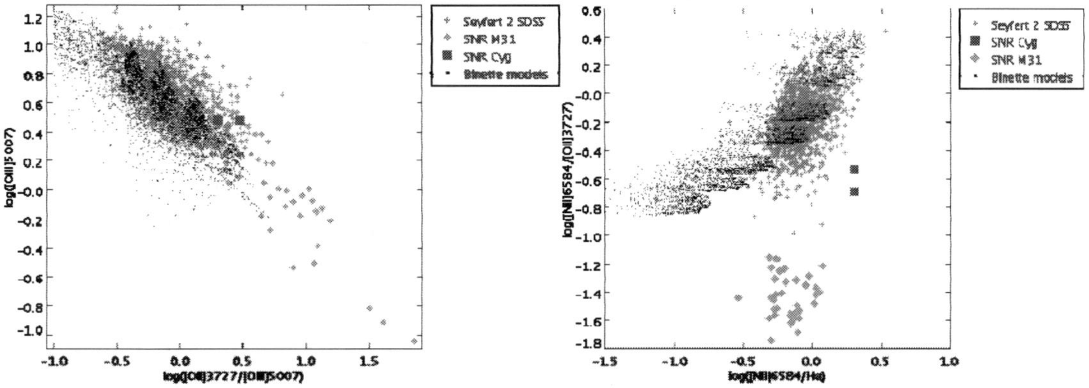

Figure 13. BPT diagnostic diagrams showing regions ionized by shock (SNR) and by radiation [7].

The electronic density distribution measured using the [SII] 6717/6731 line ratio is shown in Figure 14. The peak value is at log N_e ([SII]) ~ 2.5 cm^{-3}. Density measured in objects where [ArIV] is also detected shows the presence of two different density regions peaked at: log N_e ([SII]) ~ 2.5 cm^{-3} and log N_e ([ArIV]) ~ 3.7 cm^{-3}.

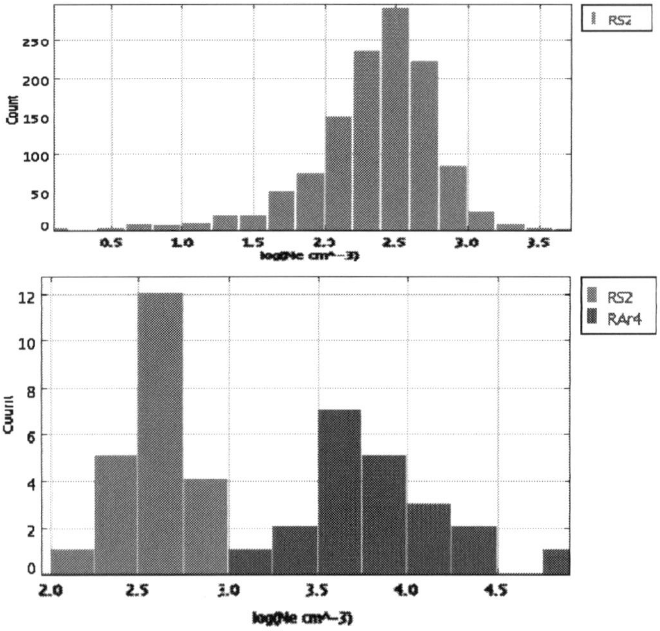

Figure 14. Density distribution in our sample. N_e ([SII]) in the upper histogram, N_e ([SII]) and N_e ([ArIV]) in the lower histogram.

A plot of the electron densities log N_e ([SII]) and log N_e ([ArIV]) as a function of the extinction for the objects in which both lines are present, Figure 15, shows that the electron density remains almost constant by increasing Av. This suggests that Av is a good measure of the geometrical thickness of the gas we are investigating in the NLR. The peak value of the electron density at log N_e ([SII]) ~ 2.5 cm^{-3} is found again plotting log N_e ([SII]) as a function of **Av,** derived from the Balmer decrement Figure 16. Here we find again that **1.3** is the most frequent value of **Av,** an indication on the most frequent size of the NLR. The electron temperature does not show any correlation with **Av**, Figure 17.

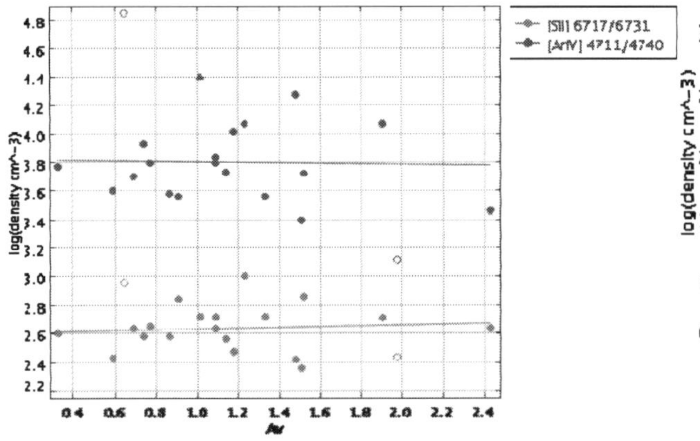

Figure 15. Log N_e ([SII]) and log N_e ([ArIV]) as a function of the extinction Av.

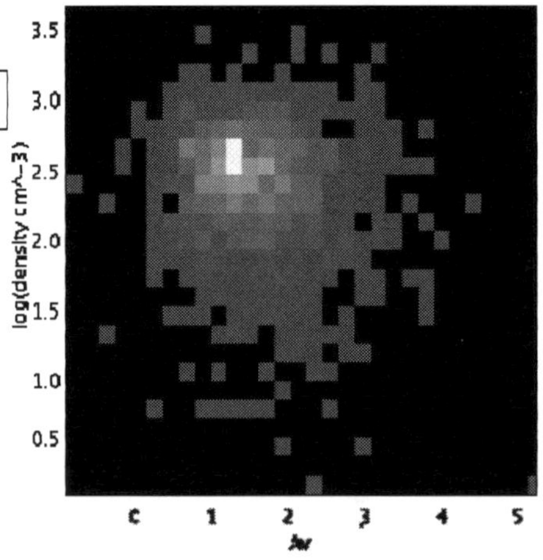

Figure 16. Log N_e ([SII]) as a function of Av.

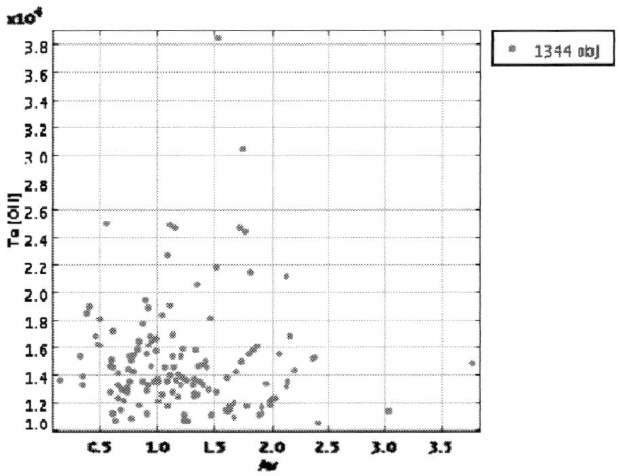

Figure 17. T_e[OIII] as a function of Av.

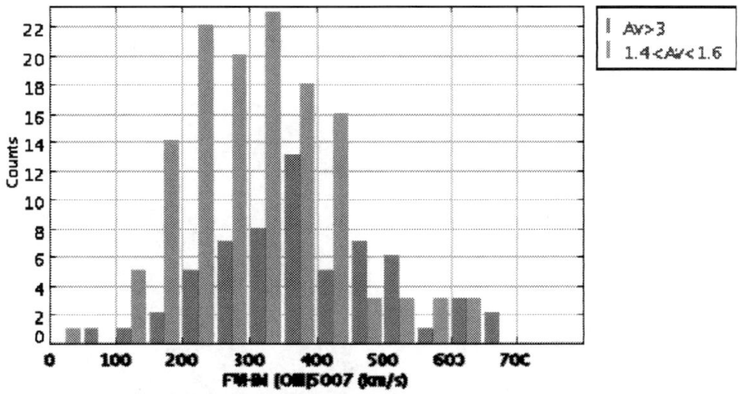

Figure 18. FWHM of [OIII] λ5007 for different Av ranges.

It is remarkable to note that the FWHM of the [OIII] λ5007 nebular line is larger for objects with Av > 3.0 than for objects with 1.4< Av <1.6, Figure 18, namely for objects with a smaller geometrical depth. Assuming that the velocity dispersion is due to gravitational effects, namely to the orbiting velocity of the gas around the nucleus, this suggests the possibility that thin NLRs are more far away from the nucleus than thick NLRs are.

3.3. Luminosity of the emission lines

The measured luminosities of some selected forbidden and recombination lines are plotted in Figure 19 - 20, as a function of Av. They show a straightforward linear correlation between line luminosity and extinction, namely geometrical thickness, which suggests in the case of recombination lines the absence of self absorption or opacity in the lines itself.

4. Photoionization models

4.1. A little bit of history

The simulation of the NLR spectra has been developed in the recent past by few authors and for very small samples of objects. As an example we report schematically the models proposed by two authors in Table 2. Figure 21 represents the geometrical structure of the Matter bound + Ionization bound slabs, proposed by Binnette et al.[7] and Whittle et al. [8]. Radiation reaches a high density Ionization Bound (IB) cloud after having interacted with a thin Matter Bound (MB) cloud.

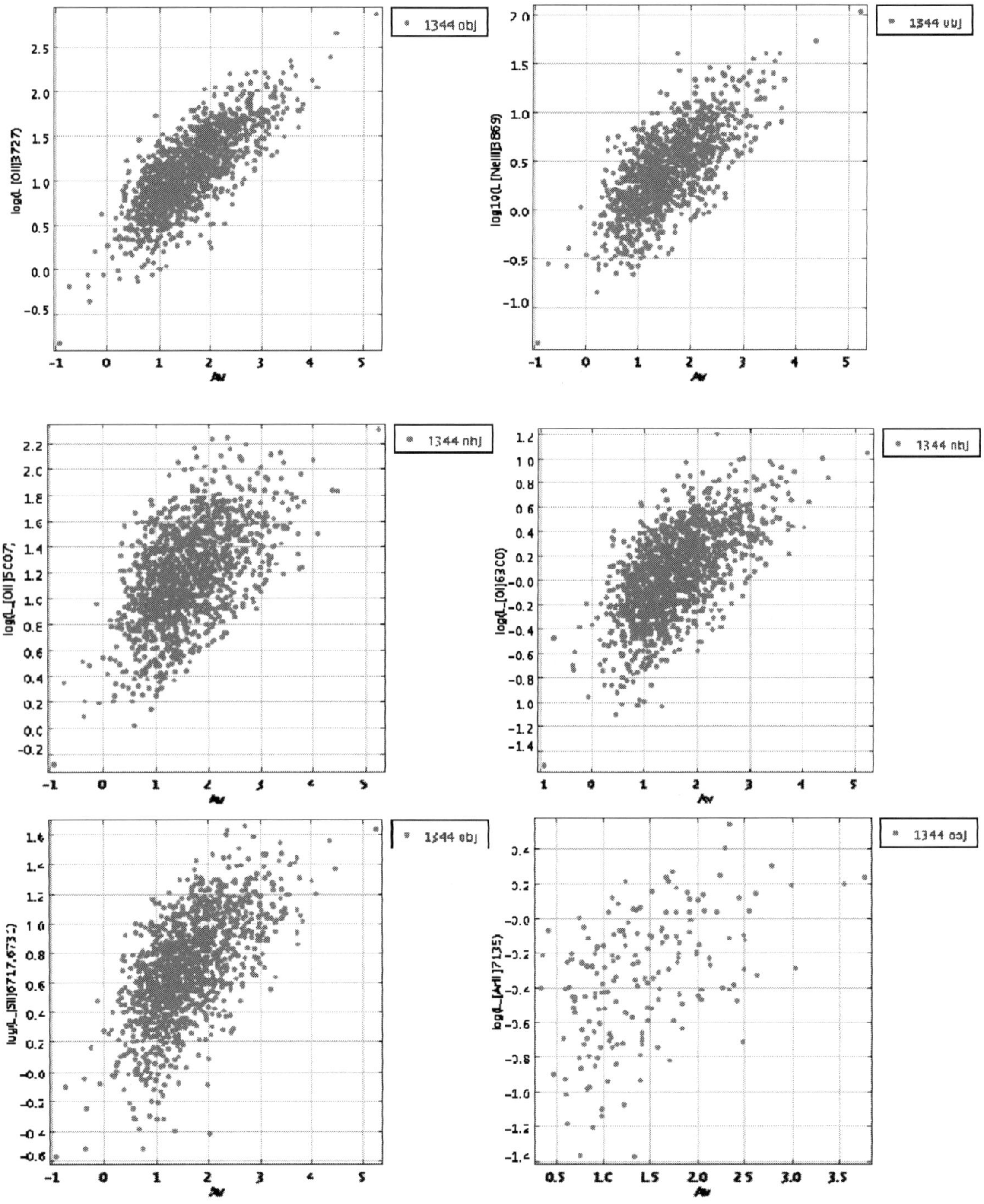

Figure 19. Luminosity versus Av for a selected number of forbidden lines.

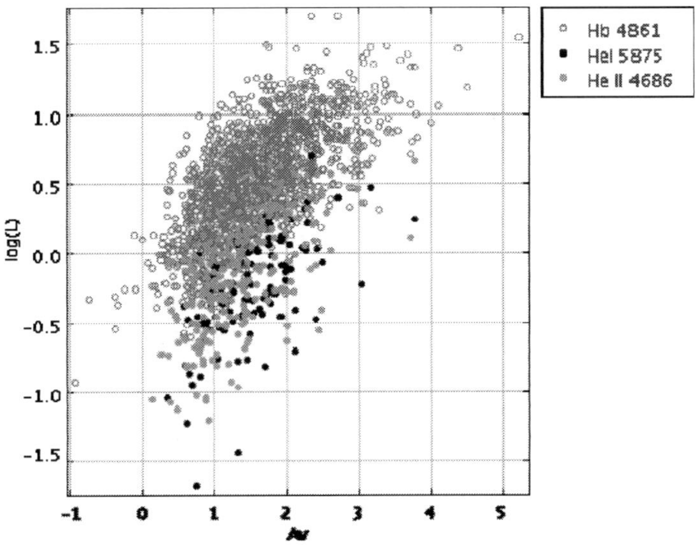

Figure 20. Luminosity versus Av for three recombination lines of different ionization potential.

Table 2.

Model	Author
• Matter bound + Ionization bound slabs → 5 objects	→Binette et al. (1996)
• Single gaseous cloud photoionization models → Mkn78	
Single gaseous & dusty cloud photoionization models → Mkn78	
Matter bound + Ionization bound models → Mkn78	
Single cloud shock models → Mkn78	→Whittle et al. (2005)

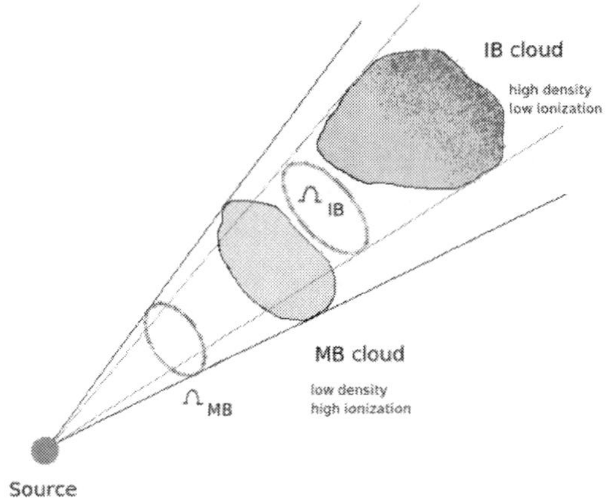

Figure 21. Matter bound + Ionization bound slabs as proposed by Binette (2006).

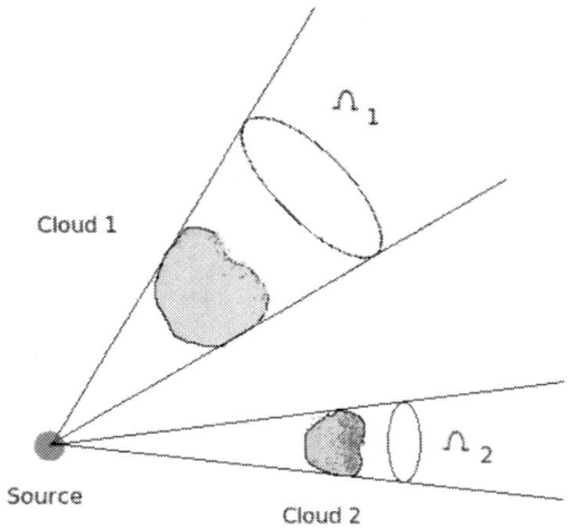

Figure 22. Two-cloud model proposed in this paper.

4.2. Two-cloud models

We propose in our approach a new model in which two different clouds photoionized by the central source are considered, Figure 22. For each object we look for the better agreement between the combination of a grid of input parameters and the output of the photoionization code CLOUDY, version 06.02.09c [9] applied to two clouds.

For each model we constrain the spectral index α of the central source and the chemical composition Z of the clouds to be the same. The line fluxes that we try to reproduce are quoted in Table 3.

Table 3. Lines to reproduce: Ion - λ (Å).

[O II] 3727 DOUBLET	[NE III] 3869	[S II] 4070 DOUBLET	Hγ 4340	[O III 4363	HE II 4686
[AR IV] 4711	[AR IV] 4740	Hβ 4861	[O III] 4959	[O III] 5007	[N I] 5199
[FE VII] 5720	HE I 5876	[FE VII] 6087	[O I] 6300	[N II] 6548	Hα 6563
[N II] 6583	[S II] 6716	[S II] 6731	[AR III] 7136	[O II] 7325 DOUBLET	

In order to define a grid of models, we have combined for each single cloud the following set of values of the 5 input parameters of the photoionization code CLOUDY, which should reproduce the observed line fluxes of the emission lines listed in Table 3:

Z = 0.5; 0.75; 1.0; 1.5; 2.0; 3.0; 4.0
α = -1.0; -1.30; -1.6; -1.9
log(U) = -1.6; -2; -2.4; -2.8; -3.2; -3.6
log n_e [cm^{-3}] = 1.0, 1.5, 2.0, 2.5, 3.0, 4.0
log n_c (H+) = 19.0 →22.4 step 0.4

The number of models to be considered is **10584**. The combination of the previous input parameters, assuming the presence of two different clouds, gives a number **of ~ 10** $^{+7}$ models, which correspond to five possible values of the ratio of the solid angles under which the central source subtends the clouds.

This number of models is reduced **to ~ 5 · 10** $^{+6}$ if we consider the line ratios which are in the AGN region of the BPT diagrams.

For comparison purposes we have calculated 15000 simulations of spectra using the Binette MB+IB model and the input parameters reported in Table 4. Only 5500 of them fit reasonably the diagnostic diagrams

Table 4. Grid of input parameters used for the simulations with the MB+IB model.

Cloud	Z	α	log(U)	log(n_e cm^{-3})	log n_c (H^+)
MB	0.5; 0.75; 1.0; 1.5; 2.0; 3.0; 4.0	-1.0; -1.3; -1.6; -1.9	-1.2; -1.4; -1.8; -2.0; -2.2; -2.4	1.7	19.0→22.4 step 0.2
IB	"..."	Transmitted ionizing spectrum	Calculated by CLOUDY	3.4	To reach T_e < 4000K

4.3. Comparison of the single and double cloud models with the MB+IB models

The reduced χ^2**-Test** applied to the set of spectra (1344) obtained isolating for each object the model, which best fits its spectrum, shows that the **two-cloud** models better fit the observed data than the **one-cloud** and **mb+ib** models, Figure 23. A confidence of the fits of the order of 90% is verified for a reduced χ^2 value $\chi^2 \leq$ **1.5.** The number of objects, which are characterized by a $\chi^2 \leq$ **1.5** is

868/1344 ➔ two-cloud 498/1344 ➔ one-cloud 503/1344 ➔ mb+ib

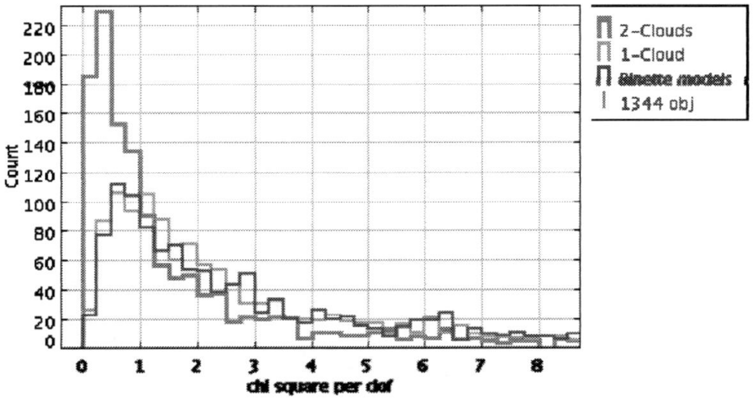

Figure 23. χ^2**-Test** for the three families of models.

The goodness of the results is shown qualitatively by the BPT diagrams, Figure 24, where we have plotted observed and simulated line ratios.

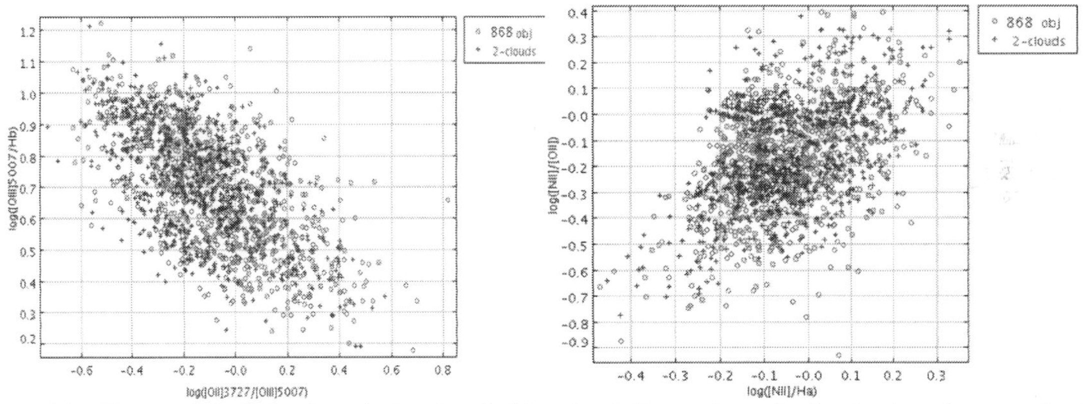

Figure 24. Observed (red dots) and simulated (blue dots) line ratios, obtained using the two-cloud models.

5. Conclusions

The work we have presented in this review is still in progress. We can then give only some preliminary results.

First of all the identification through a spectroscopic survey of 1344 Seyfert 2 galaxies has allowed to identify an homogeneous sample of objects well suited for statistical studies (completeness up to m~16.0) on the morphology (z<0.1) of the galaxies hosting the AGN and for the investigation of the physical properties of the NLR, the primary purpose of this work. The line profiles, the dependence of their width (FWHM) from Av, the finding that Av is quite likely a good indicator of the geometrical size of the NLR along the line of sight, the linear correlation between Av, luminosity of the forbidden lines and luminosity of the recombination lines are all crucial sources of information for revealing the geometrical and dynamical structure of the NLR and then for investigating on its origin.

Table 5. Results of the simulations, a and b indicate the two clouds.

	One cloud models		Binette's models		Double cloud models	
	mean	σ	*mean*	σ	*mean*	σ
Z	2.6	0.9	2.3	0.7	2.4	0.8
α	-1.7	0.2	-1.6	0.3	-1.6	0.3
log(U_a)	-3.0	0.2	-1.5	0.3	-2.7	0.5
log(Ne_a)	2.3	0.9	1.7	""	2.9	0.8
log(Nc_a)	20.9	0.6	20.7	0.6	21.0	0.6
log(U_b)			-3.4	0.3	-3.3	0.3
log(Ne_b)			3.4	""	1.8	0.6
log(Nc_b)			∞	∞	20.0	0.7

Not less important are the attempts to reproduce the observed spectra of the NLR of our sample of galaxies using different combinations of clouds and photoionization codes. The results of such simulations, when reliable, allow to restrict the range of values of the physical parameters, which govern the features of the NLR. Within the same class of objects we can for example identify subclasses, characterized by different physical parameters. This fine tuning of the physical parameters is then a tool for isolating phenomena which otherwise would hang over.

An example is shown in Table 5 where the combinations of the average values which best fit the observations are given for each model considered in this work. The mean values of the parameters and their dispersion give the possibility to isolate objects which deviate from the mean model and these will be object of our further investigation.

However looking at the mean values of our two-cloud model it is evident that the metallicity of the clouds is very high $Z \sim 2.4$, the two clouds have different electron density log Ne(a) ~ 2.9, log Ne(b) ~ 1.8, but more or less the same column density, namely different geometrical depth as already argued in section 3.2.

References

[1] Meusinger, H., Brunzendorf, J. 2001A&A, 374, 878
[2] Baldwin, A., Phillips, M.M., Terlevich, R. 1981, PASP, 93, 5
[3] Kewley, L.J., Groves, B.,Kauffmann, G., Heckman, T. 2006, 372, 961
[4] Rafanelli, P., Violato, M., Baruffolo, A. 1995, AJ, 109, 1546
[5] Cid Fernandes, R., Mateus, A., Sodré, L., Stasińska, G., Gomes, J. M. 2005, MNRAS, 358, 363
[6] Osterbrock, D.E., Ferland, G.J. 2006, "Astrophysics of Gaseous Nebulae and Active Galactic Nuclei", University Science Books
[7] Binette, L., Wilson, A.S., Storchi-Bergman, T. 1996, A&A, 312, 365
[8] Whittle, M., Rosario, D.J., Silverman, J.D., Nelson, C.H., Wilson, A.S. 2005, ApJ, 129, 104
[9] Ferland, G.J., Korista, K. T. Verner, D. A. Ferguson, J. W., Kingdon, J. B., Verner, E. M. 1998, PASP, 110, 761-778

Spectroscopy methods applied to the research in plasmas at atmospheric pressure

M D Calzada
Grupo de Espectrsocopía de Plasmas, Edifio A. Einstein (C-2)
Campus de Rabanales. Universidad de Córdoba, 14071 Córdoba (Spain)

E-mail: fa1cazal@uco.es

Abstract. Plasmas operated at atmospheric pressure are being the object of an increased attention due to their potential and current use in various fields of science and technology (excitation source for elemental analysis, purification of noble gases, hydrogen production and sterilization of medical instruments). In order to ensure that the industrial and technological applications of plasmas are being carried out with a maximum of efficiency it is necessary to know and control the processes that take place in the plasmas during their applications. These processes depend on the densities of different plasma species such as electron density and both atoms and molecules at excited levels together with the energy available in the discharge. This energy is fundamentally in form of the kinetic energy of electrons, that is, electron temperature and heavy particles like atoms and ions, called gas temperature. Passive spectroscopy techniques can be used to measure the densities and temperatures in plasma. Classical and new methods can be used for this purpose. An example of the use of the spectroscopy to know the processes that take place in plasmas corresponds to the study of the capacity of a surface wave discharge to dissociate the alcohols molecules for obtaining hydrogen.

1. Introduction

Plasmas operated at atmospheric pressure are being the object of an icreased attention due to their potential and current use in various applications such as excitation source for elemental analysis [1], purification of noble gases [2] and more recently hydrogen production [3] and sterilization of medical instruments [4].

In order to ensure that the technological applications of plasmas are carried out with a maximum of efficiency it is necessary to know the ability of the discharge to induce the desired microscopic processes. This ability depends on the densities of different plasma species such as electron density and both atoms and molecules at excited levels together with the energy available in the discharge. This energy is fundamentally in form of the kinetic energy of electrons (electron temperature) and heavy particles like atoms and ions (gas temperature).

Passive spectroscopy techniques can be used to measure the temperatures and densities in plasma. These techniques are based on the analysis of the radiation emitted by the own plasma. So, the use of passive spectroscopy does not perturb the internal kinetics of the discharge when the observation and control of plasma application are being carried out. Therefore, an important part of the effort in experimental and theoretical research in plasmas is devoted for devising, developing and proving spectroscopy methods for the purpose mentioned above. New methods have been recently developed in order to measure the electron density of plasmas at atmospheric pressure which can be considered as a complement to the classical methods.

Moreover, these techniques also allow us to know and control the processes that take place in the plasmas during their applications in specific fields. In this way, an example it is the study of the ability of a surface wave discharge (special type of microwave discharge) to dissociate the alcohols molecules for obtaining molecular hydrogen. This hydrogen could be used to generate electricity by fuel cells.

2. Spectroscopic methods applied at atmospheric pressure: temperatures and densities

Methods of passive spectroscopy are used for measuring temperatures and densities in plasmas generated at atmospheric pressure because the wide range and intensity of the spectral lines emitted by the atoms and ions into the discharge. These methods are based on the profile analysis of these spectral lines. So, starting from intensities and broadenings of them, we can obtain information about the values of the plasma parameters (temperatures and densities).

The shape and width of a spectral line emitted by the plasma are a consequence of the processes that happen in the discharge. Moreover, the optical device used in the laboratory to register the radiation introduces additional broadening on its profile. The line width is considered as the full width at half maximum (FWHM) and it is represented by $\Delta\lambda$ in the case where the profile of the line is given as a function of the wavelength.

At atmospheric pressure the spectral line profile can be approached to a Voigt function (Voigt profile). This function is the result of the deconvolution of a Gaussian function with a Lorentzian function, their broadenings being $\Delta\lambda_G$ and $\Delta\lambda_L$, respectively. The Gaussian part of the profile is due to the *Doppler* ($\Delta\lambda_D$) and the *instrumental* broadenings ($\Delta\lambda_I$); the Lorentzian part to the *Stark* ($\Delta\lambda_S$) and the *van der Waals* ($\Delta\lambda_W$) ones. To unfold the Voigt profile into its Lorentzian and Gaussian components it should be used adequate software. The other broadenings that contribute to the total profile width can be considered negligible for different reasons: a) *natural broadening* because it is very small compared to the rest of the broadenings, for instance, it is approximately equal to 10^{-4} nm for argon atomic lines, b) *resonant broadening* because the transitions used in the different methods that are going to describe do not involve the ground state and c) *broadening from self-absorption* when a proper choice of spectral lines is made.

2.1 Gas temperature

This temperature (T_g) is a measurement of the energy acquired by the heavy particles (atoms and ions) of the discharge, fundamentally, from the collisions with the plasma electrons.

It is habitual to determine the gas temperature from the ro-vibrational spectra emitted by molecular species like N_2^+, C_2, CN and OH, which are present in the plasma at trace level and assuming that these molecules are in equilibrium with the plasma atoms. The OH radical is the most common for this purpose; then, the rotational temperature obtained from the slope of the plot $lg(I\lambda/A)$ as a function of the transition of the upper state energy can be assumed to be equal to the gas temperature. In this expression I is the maximum value of the line intensity and A the corresponding transition probability. For surface wave plasma, the corresponding axial distribution of the gas obtained from the OH radical is almost constant along the plasma column with values approximately equal to 1400-1500 K, 1600 K and 2200 K for Ar [5], Ne [6] and He [7], respectively.

However, in some cases the ro-vibrational spectra of the molecular species are too weak. Then, another possibility is to determine T_g directly from the Doppler or the van der Waals broadenings of the spectral lines emitted by the plasma because each one of these broadenings is a function of the gas temperature. When the instrumental broadening dominates the Gaussian component of the spectral line, the Doppler broadening can not be used to obtain the gas temperature value; it is the case for an argon plasma in which $\Delta\lambda_D$ gets values at 1400-1500 K of T_g between 0.002 nm for 522.1 nm and 0.003 nm for 737.2 nm lines whereas $\Delta\lambda_I$ is of the 10^{-2} nm order [8]; the van der Waals broadening is then considered the most appropriate tool for this purpose [8-9].

As mentioned above, the Lorentzian part of a spectral profile is the sum of the Stark and the van der Waals broadening. In argon plasma created by a surface wave at atmospheric pressure, $\Delta\lambda_L$ of the spectral profiles coming from lower energy levels can be considered equal to $\Delta\lambda_W$ [9] and the gas temperature derived from this broadening has a similar trend to that the one inferred from the OH radical [9].

A recently published method proposes the measurement the gas temperature from lines whose Stark broadening is comparable to their van der Waals one [10]. This method is based on the consideration of a linear dependence on electron density for the Stark broadening of atomic lines. This linear dependence can be assumed when the asymmetry of the spectral line is eliminated, which corresponds to the ion contribution to the Stark broadening of the line. In an argon surface-wave plasma column, the axial profile of gas temperature can be considered constant and thus, the van der Waals too [5,9]. Taking into account these considerations, the origin ordinate of the straight line corresponding to the variation of the Lorentzian width of an Ar atomic line with the electron density is equal to the Lorentzian width for zero electron density value. This ordinate can be considered equal to the van der Waals broadening and used to calculate the plasma gas temperature. In Table 1 the gas temperature obtained by using three Ar atomic lines are shown, which are in agreement with those obtained from the ro-vibrational spectra of OH radical [5].

Table 1. Values of gas temperature from the origin ordinate of the straight lines for three Ar atomic lines

Lines	522.1 nm	549.6 nm	603.2 nm
T_{gas} (K)	1260 ± 130	1100 ± 60	1020 ± 50

2.2 Electron temperature

Electron temperature (T_e) is one of the plasma parameter more difficult to measure. It coreesponds to the energy of the plasma electrons and being use in the ionization and excitation processes that take place in the discharge.

For plasmas in partial Local Thermodynamic Equilibrium (pLTE), the distribution of population of the atomic system top levels obeys the Saha distribution which also implies a Boltzmann balance at equilibrium.Therefore, the excitation temperature (T_{exc}) derived from the Boltzmann-plot is considered equal to the electron temperature [5,11]. This procedure utilises the assumed linear dependence of $lg(I\lambda/gA)$ with the upper level of the considered transitions. Here, I is the intensity of each spectral line, λ the wavelength, g the statistical weight of the upper level and A the probability of spontaneous emission. From the slope ($-$ $0.625/T_{exc}$) of the straight line fitted through these points, T_{exc} is obtained, then, it is considered equal to electron temperature.

However, in plasmas that are far from pLTE, the Boltzmann-plot method can not be used and an alternative is the line-continuum ratio method that utilizes the intensity (I) corresponding to an emitted spectral line and the adjacent continuum (ε_c) to this line, I/ε_c being a function of T_e [12].

2.3 Electron density

The electron density (n_e) is one of the most important plasma parameters because the electrons control de processes of excitation and ionization that take place in the discharge directly or through processes in various stages in which the electrons also play a fundamental role in the intermediate stages.

The Stark broadening of the spectral lines is the most popular tool to determine plasma electron density. In this way, a deconvolution process is necessary in order to extract the profile Lorentzian part from the total width of the spectral line. In plasmas with electron density value higher that 10^{15} cm^{-3} the contribution of the van der Waals broadening to the Lorentzian width of the spectral profile can be considered negligible; in this case the Lorentzian width is approximately equal to the Stark broadening. However, for plasmas with electron density lower than 10^{15} cm^{-3} it is necessary to calculate the contribution of the van der Waals broadening to the Lorentzian width and eliminate this broadening from that one [10].

2.3.1 Stark broadening of the Balmer lines: H_α, H_β and H_γ

The hydrogen Balmer lines such as H_α, H_β and H_γ are classically used to measure the electron density, since the Stark broadening of these lines is the strongest and very sensitive to electron density variations.

The expression that connects this broadening with electron density is derived from the Kepple-Griem theory [13-14] without to take into account the influence of ion dynamics on the spectral profiles of the Balmer lines. So, the theoretical profiles for these lines have a different structure respect to the ones of the experimental profiles and, consequently, the electron density values are different depending of the utilized Balmer lines. In order to resolve this discrepancy, computational models have been or are being developed taking into account the ion dynamics effect on Balmer line profiles. Among them, we find the Gigosos-Cardeñoso model [15] which provides a series of tables that give the relation between the Stark broadening of the line profile and the electron density. The use of these tables permits to obtain the same electron density value independent of the utilized line [16].

2.3.2. Stark broadening of the atomic lines

When hydrogen Balmer lines are not intense enough, the use of the Stark broadening of non-hydrogenic spectral line emitted by the discharge itself is another possibility to measure the electron density [17]. Studies carried out in a surface wave discharge at atmospheric pressure have shown that the line whose lorentzian broadening present a variation with the electron density can be used for this purpose. For an argon plasma, these lines belong to the upper levels of the atomic system [9]. The values stemmed from the Ar lines are approximately 1.5 or 1.8 times higher than the obtained from the H_β line. The observed differences could be due to the uncertainties of the parameters that appears in the equation given by [14] and for each one of the atomic lines.

2.4 Metastable atom populations

It is habitual to use absorption methods to measure the population of the metastable atoms for plasmas. These methods are based on the interaction between the radiation emitted by an external radiation source (for instance a lamp) and the absorbing medium.

However, an alternative is the self-absorption method; the discharge being radiation source and absorbing medium, at the same time. In the application of this method, the population of a specific level is determined from the ratio of the two partially self-absorbed lines, which both end at the same level whose population wants to be measured. One of the two lines is much less self-absorbed than the other one and is used as reference line. The ratio of the intensities is a function of the absorption coefficient of the line more self-absorbed; this coefficient depends of the population of the metastable levels [18-19].

Self-absorption method was utilized to measure the metastable populations for an argon surface wave discharge generated at atmospheric pressure obtaining values in agreement to those appearing in the literature by using absorption methods [19].

3. Other spectroscopic methods for measuring electron density and gas temperature

In addition to the methods described in the above section, other methods have recently been developed and applied in order to measure such parameters as electron density and gas temperature for plasmas at atmospheric pressure, and these are also based on the analysis of spectral lines.

3.1 H_α *and* H_β *calibration method*

For plasmas generated under certain experimental conditions, only H_α line is detected and its use requires the influence of ion dynamics on the profile of the spectral line to be taken into account. In order to avoid the use of the H_α line, a method called the H_α-H_β *calibration method* can be used [20].

In this method, the representation of the Stark broadenings of the H_α and H_β lines at the introduction of a small amount of hydrogen into the discharge gives a calibration curve. In this way, once the Stark broadening of H_α line is measured, the corresponding one for H_β can be found from this curve and then the electron density can be calculated. This method was applied to an argon plasma torch open to the air for which H_β was not detectable, offering a simplified way to estimate electron density when the H_β line is not detected [20].

3.2 Intersection method

Yubero *et al.* [21] have studied the possibility for measuring n_e and T_e simultaneously from Stark broadenings of the Balmer lines H_α, H_β and H_γ lines. The method is based on the dependence of Stark broadening of the hydrogen Balmer lines on n_e and T_e. In this way, for a fixed value of Stark broadening for each one of the three Balmer lines, a plot for n_e dependence on T_e can be depicted. Taking into account the error bars for the Stark broadenings three zones can be depicted and their crossing gives a region where the electron density and temperature values are common for the three lines.

This method was applied to an argon plasma column generated by a surface wave [21], obtaining values for electron density in agreement with those obtained for this discharge in other works [5, 9, 16] but of 30-40% higher than in the case of T_e [22]. Thus, this method could be useful for electron density measurement but not for electron temperature.

3.3 Gas temperature from plasma atomic lines

Under operative condition of atmospheric pressure, the Lorentzian width of a spectral line can be written as the sum of the Stark and the van der Waals broadenings. In an argon surface wave plasma column, the axial profile of gas temperature can be considered constant and thus, the van der Waals broadening can be too [5, 9]. Moreover, Stark broadening has a negligible dependence on electron temperature for this plasma type. Taking into account these considerations, a linear dependence between Lorentzian width and electron density is established. The origin ordinate of the straight line corresponds to the Lorentzian width for zero electron density and it can be considered approximately equal to the van der Waals broadening and used to calculate the plasma gas temperature. The 522.1, 549.6 nm and 603.2 nm spectral lines of ArI were used for this purpose observing a significant discrepancy in the case of 522.1 nm [10]. Analyzing the shape of this line, a significant asymmetry was observed. If the contribution of the asymmetry is eliminated from the total width of the line, a new representation is made for Lorentzian width versus electron density and a new value of the origin ordinate is obtained. In this case, the gas temperature derived from this origin ordinate was in agreement with those obtained from the other considered atomic lines and the gas temperature measured from the ro-vibrational spectra of OH radical [5].

4. Control of processes by emission spectroscopy of: molecular hydrogen production
In this section, an example of the use of the emission spectroscopy to control the processes during the decomposition of alcohol molecules by the plasma is shown.

The discharge was created inside a quartz tube open at one end (1.5 and 4 mm inner and outer diameters, respectively) and plasma flame being the configuration chosen due to its ruggedness withstanding the introduction of alcohol [3].

Alcohols (methanol, ethanol, propanol and butanol) were introduced into the plasma by the method called *bubbling* at room temperature in order not to involve additional energy in the process. The total Ar gas flow was divided into two parts: one of them was used as carried gas bubbling into the recipient containing the alcohol to drag its molecules, giving place to a mixture of gas and alcohol. This mixture was joined to the rest of argon flow and introduced into the quartz tube, in which the plasma was generated.

The plasma capability to break down the bonds of alcohol molecules was analyzed through the radiation emitted by the discharge. This radiation was collected by an optical fiber and coupled to the entrance slit of a Jobin-Yvon Horiba monochromator, type Czerny-Turner, with a focal distance of 1 m and holographic grating of 2400 lines/mm. A computer-controlled CCD camera was used to register the spectra.

Figures (1a) and (1b) show the characteristic spectra emitted by an Ar plasma without introducing alcohol and when it is introduced into it. For the pure Ar plasma, atomic lines of argon, hydrogen lines and molecular bands of OH and NH radicals were observed; the hydrogen lines and molecular bands are due to the impurities present in the plasma gas. However, when alcohols are introduced new species are detected in the discharge, such as CH, CN and C_2 together with an increase in the intensity of the H_α line. Besides, a black solid substance appeared deposited on the discharge tube. All these radicals and increase of hydrogen line come from the dissociation of alcohol molecules. This result shows the good capability of the surface wave plasma to break down the alcohol molecules introduced into it [3].

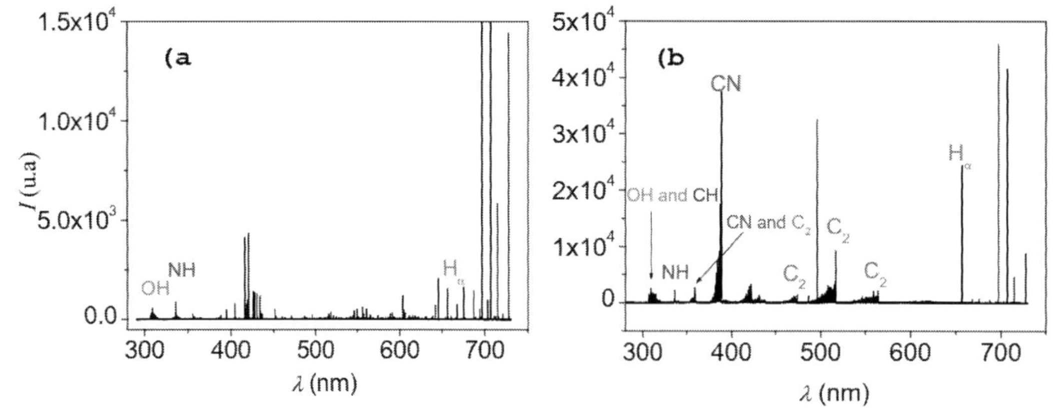

Figure 1(a)-1(b). Spectrum emitted by the plasma (a) without alcohol and (b) when alcohols have been introduced into the plasma. From Ref. [3]

The kinetics of the decomposition of alcohols can be studied from the observation of the hydrogen lines and the molecular bands that appear in the spectra. Studying the emission of the hydrogen line (H_α), (Figure 2) one shows the variation of the H_α line when different alcohols are introduced into the plasma. One observes that the highest intensity for this line corresponds to the methanol. It is due to H/C ratio and vapour pressure are also highest for a given temperature. As the experiments were carried out under the same temperature and pressure conditions for the four alcohols, the amount of the inlet methanol was the highest one.

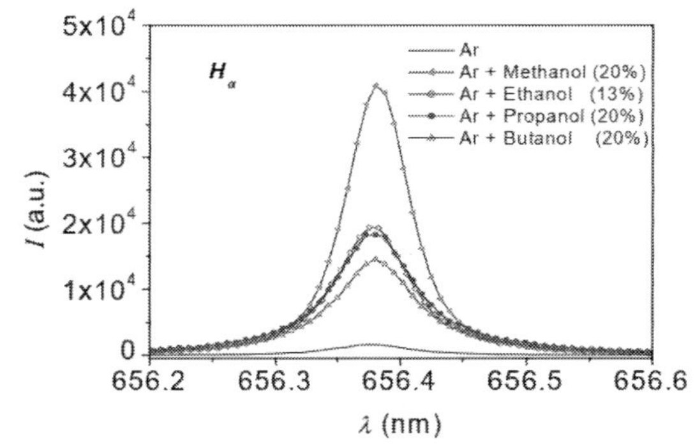

Figure 2. The H_α line emitted by a plasma of pure argon and when the four alcohols were introduced into it. From Ref. [3]

For comparison, a more detailed study of molecular bands appearing in the spectra corresponding to cases of methanol and ethanol can help to understand the kinetics in the plasma during the decomposition process of alcohols. In Figure (3a), the intensities of C_2 bands are significant for ethanol, which explains the great amount of black deposit in the discharge tube when this alcohol is introduced. Also, the result for CN bands is in agreement with the higher number of carbon atoms in the ethanol molecules (Figure (3b)). However, in case of OH radical, the result appearing in Figure 4 shows a greater decomposition of the methanol molecules which is according to the higher intensity of H_α line shown in Figure 2.The CH, CN and C_2 bands in the spectra indicates the existence of dissociation reactions of the alcohol molecules and, at the same time, recombination reactions which leads to the formation of these new species.

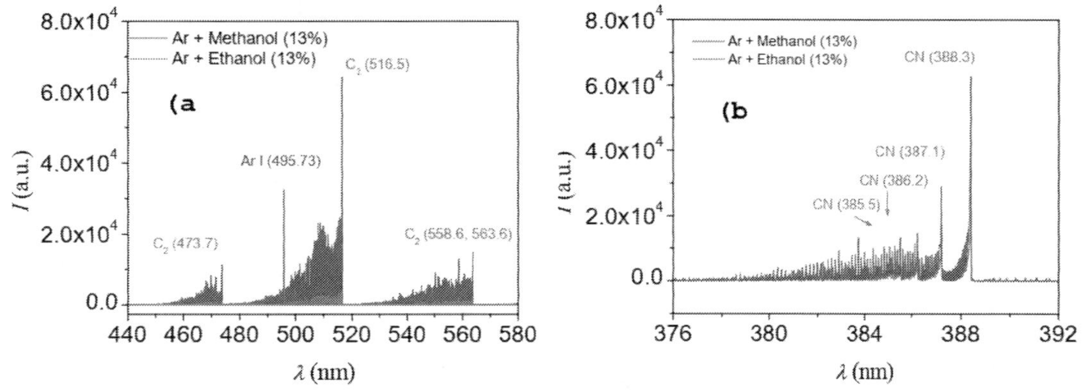

Figure 3(a)-3(b). (a) The C_2 band and (b) the CN band emitted by the plasma when methanol and ethanol were introduced into it. From Rfe. [3]

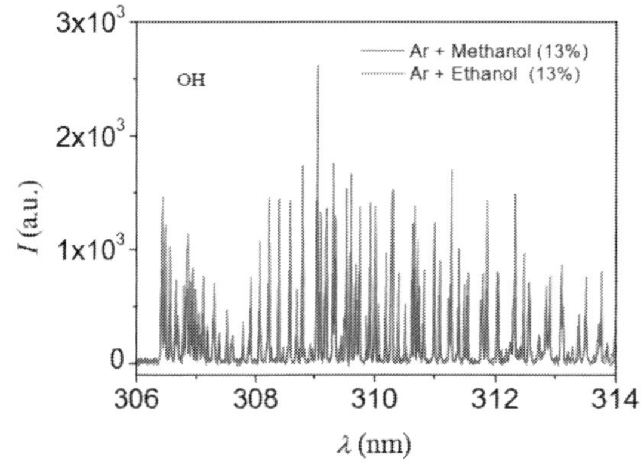

Figure 4. The OH band emitted by the plasma when methanol and ethanol were introduced into it.

From an environmental point of view, the detection or not of CO bands from the alcohol decomposition is very important. In our case, CO bands were not detected and it seems to say us that the use of this discharge type at high pressure could be considered a non-contaminating source of molecular hydrogen [3].

5. Conclusions

Classical and new methods to measure the plasma parameters (densities and temperatures) have been described and results from the application of them in surface wave discharges at atmospheric pressure. Besides, the decomposition of alcohol molecules by this discharge type has been shown as an example on the use of emission spectroscopy to control the processes that take place in the plasma.

ACKNOWLEDGEMENTS

This work has been partially supported by the Vice-Rector Office of Scientific Policy of the Cordoba University, the projects No. FTN2005-02595 and ENE2005-00314(FTN) from the Ministry of Education and Science (Spain) and the European Community (FEDER funds).

[1] Pohl P, Jimenez Zapata I, Amberger M A, Bings N H and Broekaert J A C 2008 *Spectrochim. Acta B* **63** 415

[2] Kabouzi Y, Moisan M, Rostaing J C, Trassy C, Guérin D, Kèoroak D and Zakrzewski Z 2003 *J. Appl. Phys.* **93** 9483

[3] Jimenez M, Yubero C and Calzada M D 2008 Study on the reforming of alcohols in a surface wave discharge (SWD) at atmospheric pressure 2008 *J. Phys. D: Appl. Phys.* **41** 175201

[4] Boudam M K, Moisan M, Saoudi B, Popovici C, Gherardi N and Massines F 2006 *J. Phys. D: Appl. Phys.* **39** 3494

[5] Calzada M D, Moisan M, Gamero A and Sola A 1996 *J. Appl. Phys.* **80** 46

[6] Sainz A 2006 Doctoral Thesis Universidad de Córdoba (Spain)

[7] Kabouzi Y, Calzada M D, Moisan M, Tran K C and Trassy C 2002 *J Appl. Phys.* **91** 1008

[8] Calzada M D 2005 *Mem. S. A. It.***7** 198

[9] Christova M, Castaños-Martinez E, Calzada M D, Kabouzi Y, Luque J M and Moisan M 2004 *Appl. Spectros*. **58** 1032
[10] Yubero C, Dimitrijević M S, García M C and Calzada M D 2007 *Spectrochim. Acta B* **62** 169
[11] Sainz A, Margot J, García M C and Calzada M D 2005 *J. Appl. Phys.* **97** 113305
[12] Sola A, Calzada M D and Gamero A 1995 *J.Phys.D: Appl. Phys.* **28** 1099
[13] Kepple P and Griem H R 1968 *Phys. Rev.* **173** 317
[14] Griem H R 1974 *Spectral Line Broadening by Plasmas* (New York: Academic)
[15] Gigosos M A and Cardeñoso V 1996 *J.Phys.B: At.Mol.Opt.Phys.* **29** 4795
[16] Luque J M, Calzada M D and Sáez M 2003 *J.Phys.B:At.Mol.Opt.Phy.* **36** 1573
[17] Konjević N 1999 *Phys. Reports* **316** 339
[18] Jolly J and Touzeau M 1975 *J. Quant. Spectrosc. Radiat. Transfer.* **15** 863
[19] Santiago I, García M C and Calzada M D 2005 *Appl. Spectrosc.* **59** 1457
[20] Yubero C, García M C and Calzada M D 2006 *Spectrochim. Acta B* **61** 540
[21] Yubero C, Calzada M D and García M C 2005 *J. Phys.Soc.Jpn* **74** 2249

Kinematical parameters in the coronal and post-coronal regions of the Oe stars

A Antoniou[1], E Danezis[1], E Lyratzi[1,2], L Č Popović[3], and M S Dimitrijević[3]

[1]University of Athens, Faculty of Physics, Department of Astrophysics, Astronomy and Mechanics, Panepistimioupoli, Zographou 157 84, Athnes – Greece
[2]Eugenides Foundation, 387 Sygrou Av., 17564, Athens, Greece
[3]Astronomical Observatory, Volgina 7, 11160 Belgrade, Serbia

e-mail: ananton@phys.uoa.gr, edanezis@phys.uoa.gr, elyratzi@phys.uoa.gr, lpopovic@aob.bg.ac.yu, mdimitrijevic@aob.bg.ac.yu

Abstract. In this progress report we present the main results of our research. Using a new model we studied the kinematical parameters such as the random velocities of the ions, which create the spectral lines of C IV, N IV and N V in the spectra of 20 Oe stars, as well as the rotational and radial velocities of the regions, where the above ions are created. We calculated the values of the above parameters and we present the relations between them as well as their variation as function of the spectral subtype. We also present the random velocities of the ions for each one of the C IV, N IV and N V regions as a function of the photospheric rotational velocities. Finally, we propose an explanation for the large widths that we observe in the studied spectral lines, as these widths can not be explained as large rotational or random velocities.

1. Introduction

The spectra of Hot Emission Stars (Oe and Be stars) contain peculiar line profiles. In order to explain this peculiarity, we propose and use the Discrete Absorptions Components (DACs) [1] and Satellite Absorptions Components (SACs) [2] theory.

DACs or SACs arise from spherical density regions surrounding the star or lying far away from it, that having spherical (or apparent spherical) symmetry around the star or its own center [3].

In the case of DACs or SACs phenomenon we need to calculate the line function of the complex line profile. Recently, our group proposed a model in order to explain the complex structure of the density regions of hot emission stars and some Active Galactic Nuclei (AGNs), where the spectral lines which have SACs or DACs are created [2,4].

As we know, in order to find the mechanism responsible for the structure of DACs or SACs density regions we need to calculate the values of a group of parameters, such as the rotational, the random and the radial velocities, the Full Width at Half Maximum (FWHM), the optical depth, the absorbed or emitted energy, the column density, and the Gaussian standard deviation, as well as the relation among them. Directly from the model we can calculate the apparent radial velocities (V_{rad}) of the absorbing or emitting density regions, the Gaussian typical deviation (σ) of the random thermal motions, the apparent rotational velocities (V_{rot}) of the absorbing or emitting density region and the optical depth (ξ) in the center of the spectral line. From the above parameters we can calculate the percentage contribution (G%) of the random velocities to the broadening of the spectral line, the Full Width at Half Maximum (FWHM), the random velocities of the ions (V_{rand}) that produce the spectral lines, the absorbed or emitted energy (Ea, Ee) and the column density (CD).

We point out that with the proposed model we can study and reproduce specific spectral lines. This means that we can study specific density regions in the plasma surrounding the studied object.

In this paper we present some kinematical parameters of C IV, N IV and N V regions in the spectra of 20 Oe stars. We calculate the values of the rotational, random and radial velocities, the relation among them, as well as the variation of the radial velocities as a function of the spectral subtype. Finally, we give a possible explanation of the measured high values of the rotational velocities.

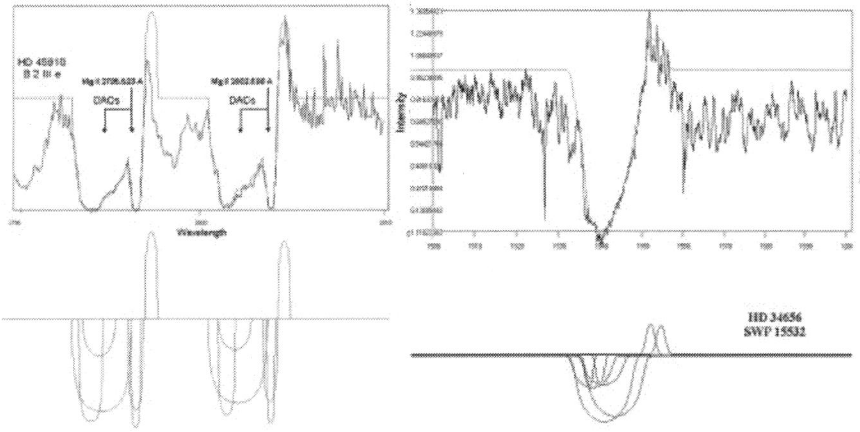

Figure 1. DACs in the spectrum of a Be star (left) and SACs in the spectrum of an Oe star (right). Below the fit one can see the decomposition of the observed profile to its DACs and SACs.

2. Observational Data

For this analysis we used the high resolution spectra (0.1 to 0.3 Å) taken with International Ultraviolet Explorer (IUE) found at the VILSPA database (http://archive.stsci.edu/cgi-bin/iue).

In Figure 1 we present the UV resonance lines of Mg II in the spectrum of HD 45910 and the UV resonance lines of C IV in the spectrum of HD 34656 which have DACs or SACs, respectively. In Table 1 we present the studied stars. As we know it is not possible to find stars between O0 and O3 spectral subtypes.

Table 1 The Studied Stars

The studied stars	**Spectral subtype**	**The studied stars**	**Spectral subtype**
HD24534	O9.5 III	HD57061	O9.0I
HD24912	O7.5 III ((f))	HD60848	O8.0Vpe
HD34656	O7 II (f)	HD91824	O7V((f))
HD36486	O9.5 II	HD93521	O9.5II
HD37022	O6 Vp	HD112244	O8.5Iab
HD47129	O7.5 III	HD149757	O9V(e)
HD47839	O7 III	HD164794	O4V((f))
HD48099	O6.5 V	HD203064	O8V
HD49798	O6p	HD209975	O9.5I
HD57060	O8.5If	HD210839	O6.0I

3. Studying the rotational velocities of the density layers of matter

Using the mentioned model we are able to calculate the values of the apparent rotational velocities of the independent regions which produce the main and the satellite components of the studied spectral lines.

In Figure 2 we present the rotational velocities' mean values of the ions of the C IV (λλ 1548.155, 1550.774 Å) and N V (λλ 1238.821, 1242.804 Å) resonance lines. In the C IV regions we detected two levels of values. The first level has values between 800 and 1800 kms^{-1} and the second level has values between 50 and 200 kms^{-1}. We detected the same phenomenon in the N V region. The first level has values between 1200 and 1800 kms^{-1} and the second level has values between 200 and 400 kms^{-1}.

An important phenomenon that we can detect in the UV spectra of some hot emission stars [5 - 10] are the very broad absorption or emission lines that we can not explain as rotational or random velocities of the density layers that construct these lines.

In order to explain this very large width we propose that around a central density region which produces the main absorption lines (and which may have the form of spiral streams and have accepted values of rotational and random velocities), we can detect micro-turbulent movements, which produce narrow absorption components with different shifts (see Figure 3). These narrow lines create a sequence of lines, on the left and on the right side of the main components. The density of these lines and their widths, which are added, give us the sense of line broadening (SACs phenomenon, see [4, 11]). As a result, what we measure as very broad absorption line, is the composition of the narrow absorption lines that are created by micro-turbulent effects. If this hypothesis is correct, the calculated width gives only the maximum value of the radial velocities of these very narrow components. Their appearance depends on the inclination of the rotational axis of the stellar disk.

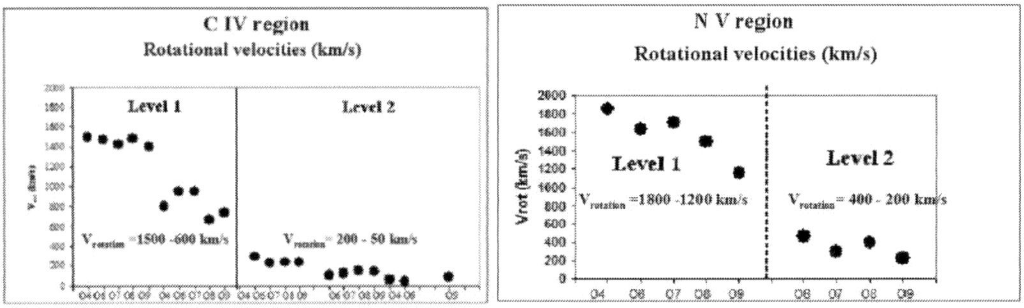

Figure 2. The rotational velocities' mean values of the ions of the C IV (λλ 1548.155, 1550.774 Å) (left) and N V (λλ 1238.821, 1242.804 Å) (right) resonance lines.

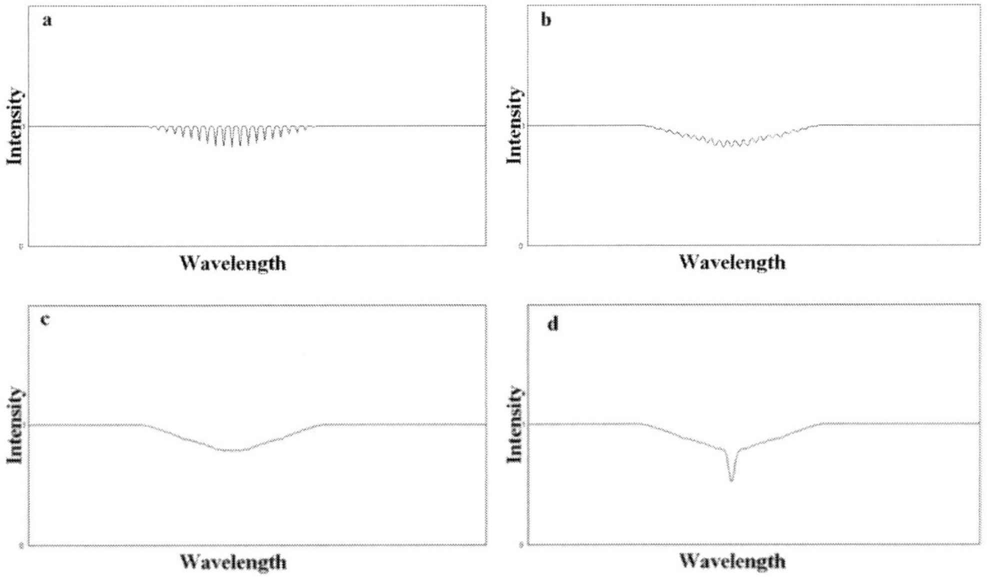

Figure 3. In (a) to (c) one can see how a sequence of lines could produce an apparent very broad absorption spectral line as an effect of SACs phenomenon. This means that when the width of each of the narrow lines is increasing (from a to c), the final observed feature looks like a single very broad absorption spectral line. In (d) one can see a combination of the apparent very broad absorption spectral line with a classical absorption line.

4. Studying the random velocities of the density layers of matter

Using our model we calculated the random velocities of the ions in the layers that produce the C IV, N IV and N V satellite components in the spectra of 20 Oe stars, which have different photospheric rotational velocities. In Figures 4, 5 and 6 we present the random velocities (V_{rand}) of the C IV, N IV and N V ions as a function of the apparent photospherical rotational velocities (V_{phot}), respectively.

In the C IV region the obtained values are about 160 kms^{-1} for the first component, 115 kms^{-1} for the second, 80 kms^{-1} for the third and 90 kms^{-1} for the fourth component. In the N IV region the obtained values are about 267 kms^{-1} for the first component and 133 kms^{-1} for the second one. Finally, in the N V region the measured values are about 190 kms^{-1} for the first component, 160 kms^{-1} for the second and 108 kms^{-1} for the third one. In each region we detected similar average values of the random velocities of the ions for each component for all the studied stars. This happens because the ionization potential of each studied ion, for all the studied stars, is the same, so the respective random velocities should be the same. As the values of the random velocities do not depend on the inclination of the rotational axis, we expect similar average values of the random velocities for each component, for all the studied stars.

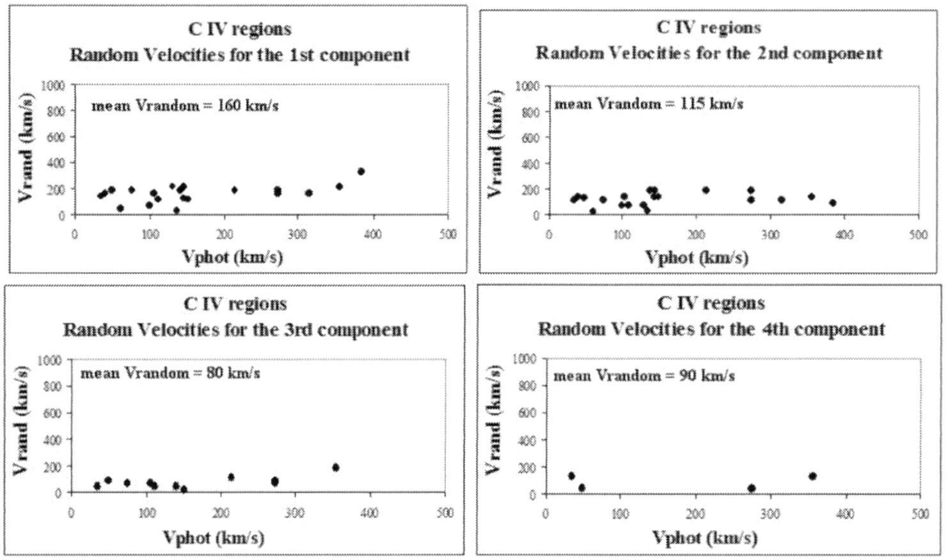

Figure 4. Random velocities (V_{rand}) of the C IV ions as a function of the apparent photospheric rotational velocities (V_{phot}).

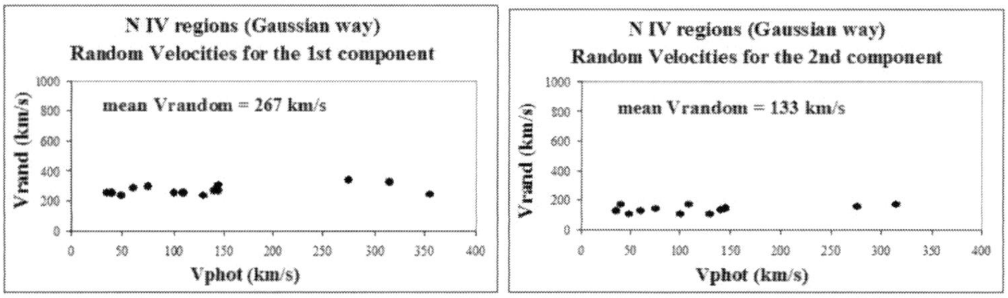

Figure 5. The same as in Figure 4 but for the N IV ions.

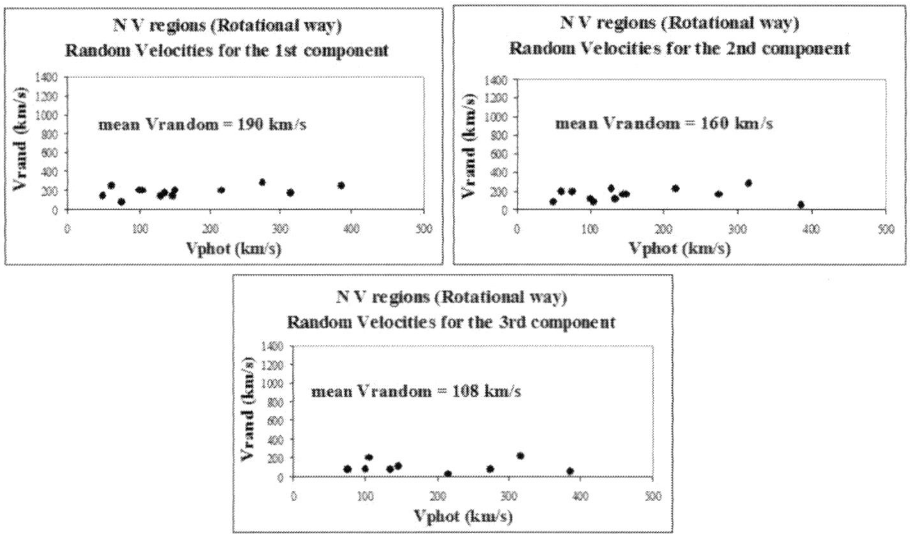

Figure 6. The same as in Figure 4 but for the N V ions.

5. Studying the relations among the kinematical parameters

In Figures 7 and 8 we present the relations among the studied kinematical parameters (V_{rand}, V_{rad}, V_{rot}), for C IV and N V density regions of the studied 20 Oe stars.

In Figure 7 we see the random (left) and the rotational velocities (right) as a function of the radial velocities in the C IV regions of the studied 20 Oe stars and in Figure 8 we present the corresponding results for the N V regions. In most cases we detect that the random and the rotational velocities increase when the absolute radial velocities increase.

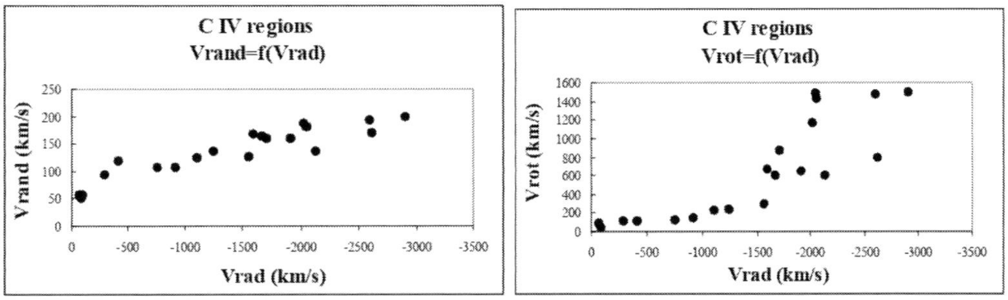

Figure 7: Random velocities (V_{rand}) (left) of the ions and rotational velocities (V_{rot}) (right) in the C IV regions, as a function of the radial velocities (V_{rad}).

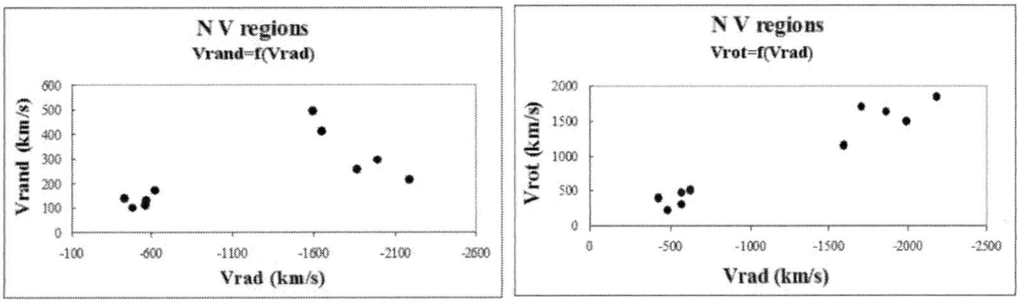

Figure 8. The same as in Figure 7 but for the N V regions.

6. Studying the radial velocities as a function of the spectral subtypes

Using the mentioned model we are able to calculate the values of the apparent radial velocities, of the independent regions which produce the main and the satellite components of the studied spectral lines as a function of spectral subtype.

In Figure 9 we present the radial velocities' mean values of the ions for the C IV (λλ 1548.155, 1550.774 Å) (left) and N V resonance lines (λλ 1238.821, 1242.804 Å) (right) as a function of the spectral subtype. In the C IV region we detected two levels of radial velocities. The first level has values between -3000 and -1500 kms^{-1} and the second one between -500 and -20 kms^{-1}. In the N V region we also detected two levels of values. The first level has values between -2300 and -1500 kms^{-1} and the second one between -500 and -100 kms^{-1}. There are two mechanisms which may create the radial velocities. One of these mechanisms creates high radial velocities and the second one low [2]. From many observations of Be stars, it has become clear over the last decade(s) that the slow part of the wind is equatorially condensed in a disk-like structure. Fore the hottest Be stars this disk co-exists with a fast stellar wind (most likely at higher latitudes). In this fast wind DACs are seen at high velocities (more than 1000km/s towards the observer). The fast winds are radiation driven from relatively slow speeds near the star, to very high speeds far away from the star.

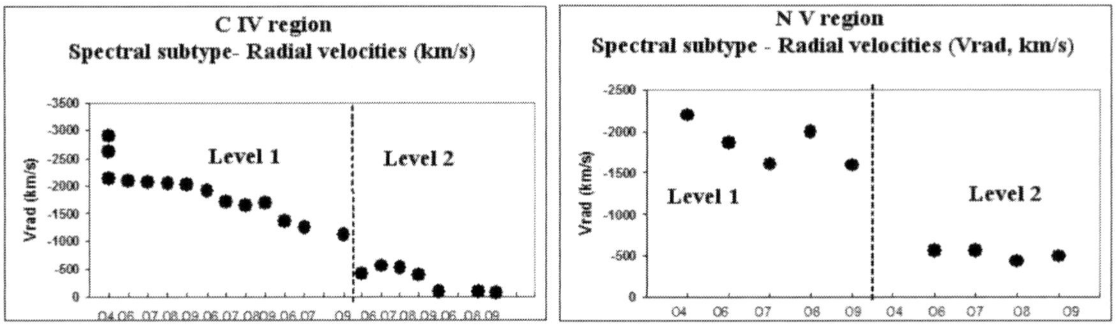

Figure 9. Radial velocities (V_{rad}) in the C IV region (left) and in the N V region (right) as a function of the spectral subtype. We detect two levels of radial velocities in both regions. The first level has high values (between -3000 and -1500 kms^{-1} in the C IV region and between -2300 and -1500 kms^{-1} in the N V region) and the second level has low values (between -500 and -20 kms^{-1} in the C IV region and between -500 and -100 kms^{-1} in the N V region).

7. Conclusions

The results deriving from this study are the following:

1. Using the proposed model, we can calculate the values of some parameters, such as the rotational, the random and the radial velocities, the FWHM, the optical depth, the absorbed or emitted energy, the column density and the Gaussian typical deviation, as well as the relation among them. This means that now we can try to understand the mechanism that is responsible for the DACs or SACs phenomenon.
2. The acceptance of SACs and DACs phenomena as the reason of the spectral line complex structure lead us to accept smaller values of the FWHM, the rotational and random velocities and the column densities and different values for the radial velocities and the optical depths because now the idea is that the complex line profile does not present a single spectral line, but a group of satellite components (DACs or SACs). This idea leads us to a different mechanism for the construction of the density regions that produce the DACs or SACs regions.
3. Finally, we propose an explanation for the large widths that we observe in the studied spectral lines, as these widths can not be explained as large rotational or random velocities. Around a central density region that produces the main absorption lines (that may have the form of spiral streams and which have accepted values of rotational and random velocities), we can detect micro-turbulent movements, which produce narrow absorption components with different shifts. These narrow lines create a sequence of lines, on the left and on the right of the main components. The density of these lines and their widths, which are superposed, give us the sense of line broadening [12]. As a result, what we measure as very broad absorption line, is the composition of the narrow absorption lines that are created by micro-turbulent effects. If this hypothesis is correct, the calculated width gives only the maximum value of the radial velocities of these very narrow components. Their appearance depends on the inclination of the rotational axis of the stellar disk.

Acknowledgments

This research project is progressing at the University of Athens, Department of Astrophysics, Astronomy and Mechanics, under the financial support of the Special Account for Research Grants, which we thank very much. This work also was supported by Ministry of Science Technological Development of Serbia and, through the projects "Influence of collisional processes on astrophysical plasma line shapes" and "Astrophysical spectroscopy of extragalactic objects".

References

[1] Bates B and Halliwell D R 1986 *Mon. Not. R. Astr. Soc.* **223** 673

[2] Danezis E, Nikolaidis D, Lyratzi E, Antoniou A, Popović L Č and Dimitrijević M S 2005 *Mem. Soc. It. Suppl.* **7** 107

[3] Lyratzi E, Danezis E, Popović L Č, Dimitrijević M S, Nikolaidis D and Antoniou A 2007 *Publ. Astron. Soc. Japan* **59** 357

[4] Danezis E, Nikolaidis D, Lyratzi E, Popović L Č, Dimitrijević M S, Antoniou A and Theodosiou E 2007 *Publ. Astron. Soc. Japan* **59** 827

[5] Andrillat Y and Fehrenbach Ch 1982 *Astron. Astrophys.* **48** 93

[6] Andrillat Y 1983 *Astron. Astrophys.* **53** 319

[7] Marlborough J M 1969 *Astrophys. J.* **156** 135

[8] Poeckert R, Marlborough J M 1979 *Astrophys. J.* **233** 259

[9] Doazan V 1970 *Astron. Astrophys.* **8** 148

[10] Antoniou A, Danezis E, Lyratzi E, Popović L Č, Dimitrijević M S 2008 *Publ. Astron. Soc. Japan* (submitted)

[11] Lyratzi E and Danezis E 2004 *AIP Conf. Proc.***740** 458

[12] Danezis E, Lyratzi E, Popović L Č, Antoniou A and Dimitrijević M S 2008 *Publ. Astron. Obs. Belgrade* **84** 463

Processes with neutral hydrogen and deuterium molecules relevant to edge plasma in tokamaks

Iztok Čadež, Sabina Markelj, Zdravko Rupnik and Primož Pelicon

Jožef Stefan Institute, Jamova cesta 39, 1000 Ljubljana, Slovenia

E-mail: Iztok.Cadez@ijs.si

Abstract. Neutral hydrogen molecule is an important constituent of edge plasma in tokamak reactors. These molecules are continuously produced by ion neutralisation and recombination on plasma walls and are subsequently reemitted in the plasma. We initiated an experimental program to study processes in which vibrationally excited hydrogen molecules are involved. Two specific experimental techniques have been developed, one for vibrational spectroscopy of hydrogen molecules (H_2 and D_2) and another one for H and D depth profiling in materials. New information has been acquired on hydrogen recombination on tungsten and also on dissociative electron attachment in hydrogen, both relevant to edge plasma. Experimental set-ups are shortly presented as well as some of the results on above processes.

1. Introduction

Detailed understanding and characterization of plasma-wall interaction and edge plasma in present tokamaks and future fusion reactors is becoming more and more important due to the ITER project. Different atomic and molecular collision processes decisively determine various processes of primary importance for stable and reliable operation of fusion reactor. Such processes are for instance, the physical and chemical sputtering of the wall material, fuel retention in the wall and its recycling, plasma detachment in tokamak divertor, photon emission from the edge plasma, mitigation of transient phenomena etc. An extensive overview of different aspects of plasma-wall interaction is given in [1].

Neutral hydrogen atoms and molecules are present in the edge plasma. They are continuously generated by ion recombination on the wall of the fusion reactor and on other plasma facing components and then reemitted to the plasma. They can be created also by volume recombination of ions and electrons in the relatively cold edge plasma. The interaction of excited neutral molecules with the walls and their importance for the edge plasma is still not well understood since there is not many experimental studies of relevant processes. Optical emission spectroscopy of edge plasma, especially from limiter and divertor regions, brought evidence of abundant presence of molecules and their involvement in volume processes. Direct correlation of the observed phenomena to the processes with neutrals is mainly possible by numerical simulations and very important activity exists in this field. As experimental results accumulate and numerical methods become more sophisticated also more processes have to be included in the models. For adequate data bases completion more experimental and theoretical works are needed.

We have constructed a set-up for vibrational spectroscopy of hydrogen molecules (H_2 and D_2) that is based on the properties of the dissociative electron attachment in hydrogen in order to facilitate dedicated experimental studies of processes with hydrogen molecules. For the same purpose we also developed a technique for in-situ measurement of H and D concentration in materials. This is done by ion beam analytical method ERDA (Elastic Recoil Detection Analysis), utilizing 4.2 MeV probing beam of $^{7}Li^{2+}$ ions. A short description of experimental techniques and results on production of vibrationally excited hydrogen molecules on tungsten and copper and on the edge plasma relevant process of dissociative electron attachment in hydrogen are presented.

2. Molecules in the edge plasma of tokamaks

Hydrogen molecules are the primary fusion fuel that is injected into the plasma chamber before or during discharge ignition and also during different rump-up plasma scenarios. Plasma fuelling is partially accomplished also by hydrogen desorption from the plasma wall during initial pulse rise. The details of atomic and molecular processes in this, initial stage of plasma ignition have negligible importance for the subsequent characteristics of the plasma behaviour so that they are never treated by modelling in detail. In a very short time core plasma becomes fully ionized and besides ions of the basic fusion constituents, D^{+}, T^{+} and He^{2+} only some multiply charged ions of impurities might be present. Any neutral particle that happens to enter core plasma (atoms used for plasma heating and current drive or for diagnostic purposes, fuel pellets or stray neutrals from edge plasma) is promptly ionized.

Contrary to the initial evolution of plasma discharge in tokamak, the detailed atomic and molecular processes are very important for plasma behaviour closer to the plasma wall. This is due to the lower plasma temperature in this region and also to more complex composition of the plasma. The most characteristic case of this situation are the regions where carbon is used as a plasma facing material and very reach plasma chemistry occurs in the near wall region. Chemical erosion and subsequent carbon re-deposition are strongly dependent on the different reactions among hydrogen and different hydrocarbon molecules.

Plasma behaviour is particularly sensitive to a large set of atomic and molecular processes in tokamak divertors. By means of magnetic field configuration plasma is there intentionally directed to the specially designed target plates and very high power load is produced. While the normal power load to the plasma wall in the main reactor is bellow and close to 1 MW/m^2, in divertor this load often reaches over 30 MW/m^2. Such high power has to be spread over larger surface in order to achieve sufficient life time of the wall components and prevent failure.

Plasma wall material has to fulfill very strict requirements. First, it must be able to sustain high power load varying from relatively moderate one on the main chamber wall to the extreme one on the divertor targets. On the other hand material must not be a source of impurity ions that would contaminate the core plasma. Next, very important criterion is that the material has low hydrogen uptake in order to keep total amount of retained tritium in fusion reactor below legal safety level as long as possible. Due to the above criteria the following choice of wall material for ITER has been made: the main wall of the plasma reactor will be made of beryllium, majority of the divertor will made of tungsten castellated structures and only the divertor target plates exposed to the highest power flux will be made of graphite (CFC – carbon fibre composite). On this way the best compromise for all requirements is achieved although numerous problems remain to be clarified – migration of materials, the role of the mixed compounds and layers to the lifetime of components, dust production etc.

Neutral molecules are present during discharge also in the surrounding region, outside the edge plasma but still inside the reactor vacuum vessel. While strong magnetic field holds core plasma strictly out of contact with materials and its configuration allows gradual decrease of plasma temperature to the cold plasma regime where it is in contact with limiting materials (PFC – plasma-facing components) there are still regions where quasi neutral gas cushion exists between this edge plasma and vacuum vessel wall. Such hidden regions are for instance gaps between different first wall construction elements and castellation in divertor target plates, hidden space bellow the divertor dome, pumping and gas injection ducts, diagnostic ports and similar. Here, the characteristics of atoms and molecules play role in additional energy transfer between edge plasma and vacuum vessel, and also in the fuelling and pumping conditions.

As stated above, the recombination of hydrogen ions on the plasma wall is the source of neutral molecules that are subsequently reemitted into the plasma. Neutral particles that are emitted to the plasma do not feel electric and magnetic fields and can penetrate in plasma depending on their ionization length. This is dependent on plasma density and temperature and also on the nature and internal excitation of the particle. Highly vibrationally excited molecule will most probably be dissociated to negative ion and atom and these fragments will subsequently be ionized by ion or electron collision. Contrary to this, a ground state hydrogen molecule will most probably be ionized by electron impact directly to positive ion.

Different individual processes involving neutral hydrogen molecules are important for edge plasma. These processes influence plasma behaviour on a complex way as they occur simultaneously and their relative importance depends on plasma and electron density, particle concentrations and temperatures and E and B field configuration. For properly taking into account individual cross section (CS) and reaction rate complex collisional-radiative models have to be employed. Many of recent developments of modeling codes (e.g. [2]) are based on the work of Sawada and Fujimoto [3]. For such codes to be efficient a detailed data bases are needed that contain reliable data on numerous individual process. Many such data do exist (e.g. see [4] for hydrogen data) but many are also missing especially for heavier hydrogen isotopologues (see Janev in [1]). Only very few data on reaction CSs and rate coefficients exist for vibrationally excited H_2 [5]. For heavier isotopologues the situation is even less favorable and much effort is needed to fill the gaps in the needed data sets.

3. Experimental techniques

In order to perform dedicated laboratory studies of processes with vibrationally excited hydrogen molecules we have developed and tested specific experimental techniques described in this section.

3.1. Vibrational spectroscopy of hydrogen molecules

Experimental method for vibrational spectroscopy that we employ in our studies is based on properties of the lowest energy dissociative electron attachment (DEA) in hydrogen ("4 eV process") and it is described in detail elsewhere [6, 7]. It is based on a very strong rise of the cross section for DEA, $e+AB \rightarrow AB^- \rightarrow A^-+B$ with ro-vibrational excitation of hydrogen molecule in its ground electronic state. A and B stand for any of hydrogen isotopes, in our case H or D, as DEA has the same properties for all hydrogen isotopologues. Excited ro-vibrational states of the ground electronic state of hydrogen can not be detected by simple emission spectroscopy as radiative transition probabilities are extremely small.

There are three main DEA processes in the low incident electron energy range. The first process has threshold at 3.72 eV ("4eV process") and proceeds through attractive potential of the lowest, $X^2\Sigma_u^+$ resonant H_2^- state. The second process, between 5 and 13 eV, proceeds through the repulsive $^2\Sigma_g^+$ state and leads to higher kinetic energy fragments. Finally, the third process with the threshold at 13.92 eV ("14eV process") proceeds through attractive resonant state of the $^2\Sigma_g^+$ symmetry that dissociates to H^- (D^-) and electronically excited (n=2) H (D) atom.

We have built a new set-up with magnetically collimated electron beam and specially designed extraction system [8]. Schematically it is shown in figure 1. Negative hydrogen ions, H^- or D^- are created by DEA in the region where electron beam intercepts sample gas target. These ions are extracted from interaction region by combined action of a rather broad electrostatic field and homogeneous guiding magnetic field and a ribbon shaped ion beam is formed. Ions are then mass filtered by properly positioned narrow selection slits and detected by channel electron multiplier. Details of extraction electrode system, model simulations and experimental results obtained with new system are described in detail in [8].

In order to measure the vibrational distribution in the target gas one records a dependence of ion yield on electron beam energy in the energy region between 0.1 and 5 eV. Vibrational distribution is obtained from such experimental spectrum by deconvolution using recent theoretical cross sections for DEA in hydrogen [9]. This deconvolution is also described in detail in [8].

Particular convenience of the described vibrational spectrometer is that it also allows detection of atomic hydrogen (H or D) [8]. This is done by measuring the yield of H^+ (or D^+) for electron energy just above the ionization threshold for $e + H(D) \rightarrow H^+(D^+) + 2e$. For switching from negative ion to positive ion mode one needs to change the polarity of extraction field, to reverse the direction of magnetic field and to change voltages on the channel electron multiplier.

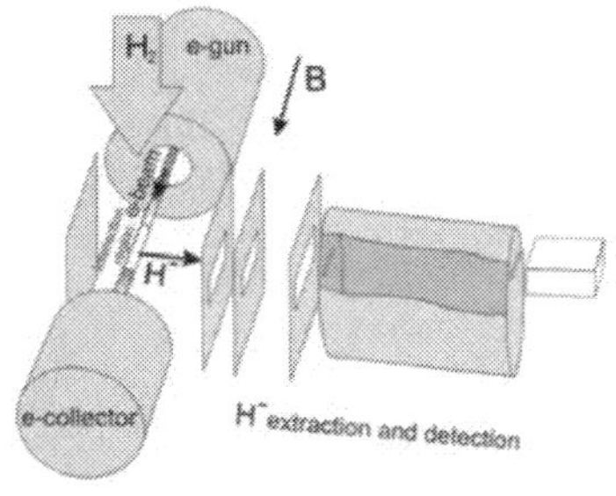

Figure 1. Schematic representation of the vibrational spectrometer. In the present experiment we can reliably deduce populations up to v=8 in H_2 and up to v=12 in D_2.

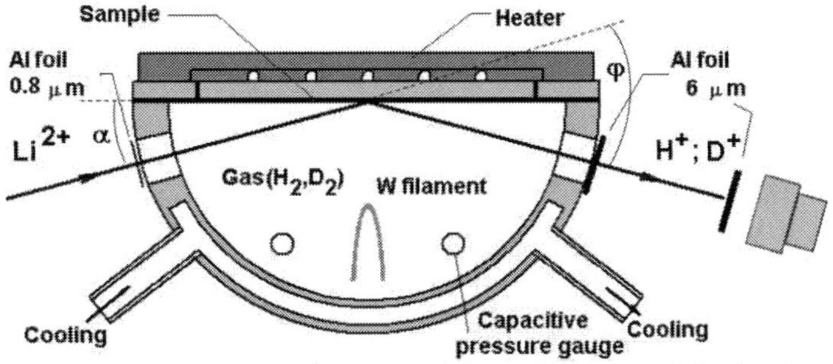

Figure 2. Schematic representation of HEC used for in situ measurements of H and D concentration. Thin Al foils are used on the entrance and exit windows of HEC in order to prevent hydrogen from leaking to the vacuum chamber. Typical hydrogen pressure in the cell was in the μbar range. Al foil on the exit window together with a second one placed in front of the detector serve to absorb scattered incident Li ions and prevent them to reach the detector. The thickness of the absorber foil is determined so to efficiently stop Li ions but to allow H and D to pass.

3.2. Determination of H and D concentration in materials

In order to perform a more detailed characterization of the surfaces where studied reactions with hydrogen atoms and molecules occur and to facilitate the interpretation of experimental results we have developed a method for in situ measurement of hydrogen concentration depth profile. For this purpose we use Elastic Recoil Detection Analysis (ERDA), a well established analytical method with fast ions. For this we use collimated beam of 4.2 MeV $^7Li^{2+}$ ions that impinges on sample surface at small angle, 15°, and H and D ions that are knocked out from the sample are detected in forward direction, again at 15° with respect to the sample surface [10]. The depth variation of the absolute concentration of H and D in the sample is obtained by analysing energy spectrum of the ejected ions. This method is set at the ERDA/RBS experimental station on 2MV tandem accelerator at Jožef Stefan Institute.

In order to be able to follow the evolution of hydrogen concentration profiles simultaneously with exposing surface to the controlled hydrogen atmosphere a special hydrogen exposure cell (HEC) was constructed [11]. The following experimental parameters can be varied: hydrogen atmosphere (pressure and composition - H_2 or D_2 can be introduced or a mixture of two), temperatures of the sample and of the wall and most important a partial dissociation of hydrogen gas can be achieved by thermal dissociation of molecules on a resistively heated tungsten filament. The hydrogen exposure cell used for the in situ measurements of hydrogen is shown in figure 2.

3.3. Sources of particles

For our present laboratory experiments we need sources of particles that are diagnosed by the above described diagnostic techniques. For this we are mostly using sources of vibrationally excited hydrogen molecules that operate on a traditional way: hydrogen flows through a gas cell where hot tungsten filament is present so that partially dissociated neutral atmosphere is created. Free atoms then recombine on cooled wall of the cell and vibrationally excited molecules are created. Neutral hydrogen gas that exits from such a cell contains ground state molecules, excited molecules and atoms. Detailed composition of the effusing gas depends on the cell geometry, dissociation filament temperature, temperature and composition of the recombining wall.

Test source (TS) used for the present studies of vibrational distribution of molecules created by atom recombination on metals and its position on the vibrational spectrometer (DTVE-B) is shown in figure 3. A disk of the studied material (25 mm dia.) is placed close (5 mm) and parallel to the exit orifice (6 mm dia.) so that recombination at the sample disk determines to a large extent the vibrational composition of the effusing gas. On this way, the vibrational relaxation due to wall and volume (lower density in front of the orifice) collisions is minimized. Sample disk is mounted on the cold finger cooled by a water flow and dissociation filament is placed around this cold finger behind the sample.

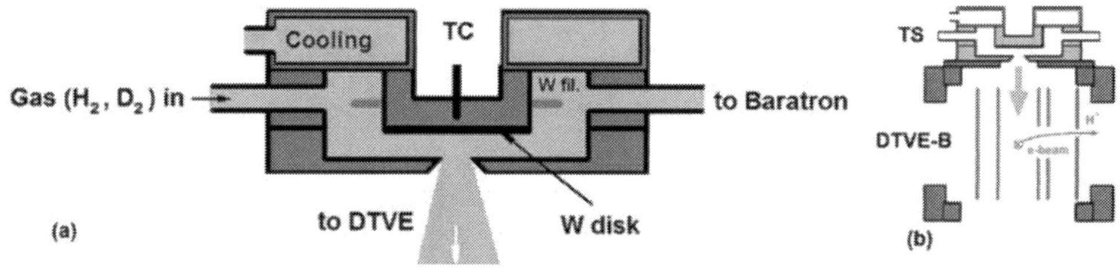

Figure 3. Test source used for the measurements of vibrational distribution of molecules produced by atom recombination (a) and detection geometry of test source (TS) as mounted on the body of the vibrational spectrometer (DTVE-B).

Another source of hot hydrogen molecules working on the same principle as TS but having different geometry was made for studies of influence of hot molecules injected into plasma on plasma emission. The same source was also used for the study of a possible role of vibrationally excited hydrogen molecules in chemical erosion of carbon films [12].

Yet another device that we are starting to use is a commercial source of atomic hydrogen HABS manufactured by MBE GmbH.

4. Results

4.1. Recombination of hydrogen atoms at metal surface

Production of hydrogen molecules by atom recombination at high-Z metal surfaces, tungsten in particular, has relevance to fusion development research because the largest part of divertor of ITER tokamak will be made of tungsten. Experiment with all-tungsten wall in Asdex Upgrade tokamak at Garching, also promoted the interest for all various reactions on tungsten. Reactions on some other high-Z materials such as Ta and Mo and also on Cu are also important to be studied due to other uses in reactors. Metal surfaces where reactions occur can be under very different conditions – from very high temperatures above melting point and under high plasma load and particle fluxes in the 10^{25} ions/m^2 range to rather low temperatures and shadowed from the direct plasma impact. Therefore detailed understanding of reactions on tungsten is very important and promptly needed.

4.1.1. Atom recombination at metal surfaces - hot atoms

Vibrationally excited molecules are produced by a recombination of hydrogen atoms on a surface that is exposed to the flow of hydrogen atoms. When the surface temperature is rather low (few °C) and atoms are impinging on the surface the mechanism of recombination is Eley-Rideal like. An incoming atom recombines with another atom adsorbed on the surface. There is another limiting recombination mechanism, Langimour-Hinshelwood, where two atoms adsorbed on the surface recombine and desorb as a molecule. The chemisorption energy (E_{ch}) of hydrogen atom adsorbed on a metallic surface varies from 0.2 eV to 1 eV. Therefore when two chemisorbed atoms recombine they need to obtain the lacking energy from the surface and if the surface temperature is too low, the atoms cannot recombine. In the case of the Eley-Rideal mechanism the free atom brings 2.24 eV of potential energy (one half of the dissociation energy of hydrogen molecule) and molecule desorbes with a maximum available energy of 2.24 eV – E_{ch}. Latest studies of interaction of H atoms with clean D covered metal surfaces have shown that the desorption rates of molecules are not Eley-Ridel like [13-15]. Beside that homonuclear products (D_2) were detected, which cannot be explained with the mechanism described above. This finding was explained by assuming a hot-atom mechanism [16-18]. A hot-atom concept is based on a fact that when an atom approaches the surface it gains considerable amount of energy (about 2.5 eV), due to the attractive potential between the atom and the metal. This energetic atom can then rather easily move along the surface, slowly giving its energy to the adsorbates and the substrate. This hot-atom can then either recombine with an adsorbed atom from the surface or stick to the surface by exchanging its energy with an adsorbed atom, creating another hot-atom. With the developed kinetic models [17, 18], the experimental desorption rates of HD and D_2 were nicely reproduced. Still the developed models can explain only the formation of molecules with maximum energy of 2.24 eV-E_{ch}, allowing the population of vibrational levels up to v=4. Population of higher levels can then only be explained through recombination of two hot-atoms which are in the limiting case almost free with total available energy of about 2x2.24 eV. This high exceeding energy can then be transferred to the translational and internal degrees of freedom of the desorbing molecule formed by recombination. This could explain the population of vibrational states higher than v=4. This process was neglected in the models due to the low-atom beam flux in treated experiments and expected rapid adsorbate relaxation [18]. New experiments are needed where recombination could be studied under higher atom flux and establish conditions where the recombination between two hot-atoms is possible.

4.1.2. Results

In order to study vibrational distribution of molecules created by surface recombination on metals we performed special dedicated experiments with the gas cell TS described above. Such geometry is used in order to enhance detection efficiency. Inconvenience of this method is that the vibrational distribution that is measured is not only a result of initial atom surface recombination but also of subsequent volume and wall collisions of created molecules. Therefore nascent vibrational distribution is necessarily modified to some extent depending on the geometry and the pressure in the cell. Comparing results with different sample materials allows one to estimate the contribution of secondary effects to the direct contribution from recombination on the sample.

A set of experimental measurements were performed recently using the TS with a 0.3 mm thick and 25 mm dia. disc sample made of tungsten (Goodfellow, hot rolled) [19]. Vibrational distributions for H_2 (squares) and D_2 (circles) molecules are shown in figure 4. Results are shown for two different kind of dissociation filaments, 0.2 mm dia. thoriated tungsten wire (labeled (1) in the figure) and 0.3 pure tungsten filament (labeled (2)). Two lines with different pre-exponential factors are also drawn in the figure, one corresponding to vibrational temperature T_v=3400 K and another to T_v=3700 K. As seen from the figure the lower temperature fits well vibrational distribution for D_2 while the higher one fits well the distribution for H_2.

Experimental distribution of H_2 obtained on another, older experiment is also shown in figure 4. It was obtained from some previous measurements performed at the experiment in Paris described in details in [20]. Main difference between the two cases is the shape and conditions in the $H_2(v)$ cell. In the old experiment tungsten was evaporated in-situ to the inner wall of the cell so that clean thin polycrystalline layer is deposited. In the present case we use commercial hot rolled tungsten conventionally polished and cleaned. Also, the pressure was higher in the old experiment (0.04 mbar) with respect to the present one (0.001 mbar). Besides these differences the vibrational distributions are similar but the older one is better described by lower vibrational temperature, T_v=3000 K.

Vibrational distribution as obtained by recombination of hydrogen atoms on copper is shown in figure 5. Recombination disc for this experiment was made of commercial OFHC copper

(Goodfellow). In this case much lower vibrational temperature was obtained and as a consequence of this a very small production rate of D^- was measured. This effect clearly illustrates very big importance of material property on behavior of different hydrogen isotopologues in edge plasma.

Our present measurements in main points confirm previous findings for studied materials performed with H_2 [20] but extend the study to more fusion relevant D_2. For both presented cases apparently lower vibrational temperature is obtained for D_2 than for H_2. It is important to note that measured vibrational distributions are often not reproducible and that they also evolve with time. All this indicates that detailed conditions on the surface are very important for the process of atom recombination and therefore for the resulting vibrational distribution of reemitted molecules.

In order to improve our understanding of recombination process we have undertaken complementary studies with the same quality of tungsten by in-situ ERDA analysis of hydrogen concentration depth profile using HEC [11], [19]. Conditions in HEC are comparable and very similar to the surface exposure conditions in TS. The only important difference between HEC and $H_2(v)$ cells used for production of hot molecules is that the first one has quasi stationary gas atmosphere while for the later ones there is always an hydrogen flow through the cell. Other characteristics, apart from the geometry are the same – dissociation filament, partially dissociated hydrogen atmosphere, atom recombination on the sample surface.

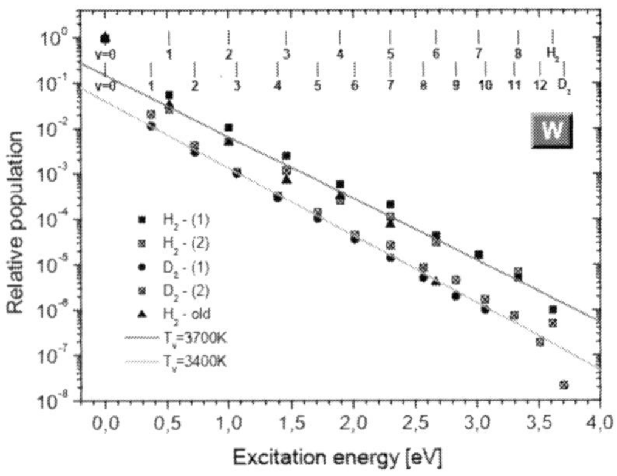

Figure 4. Relative vibrational populations of H_2 and D_2 molecules formed by atom recombination on tungsten as a function of excitation energy. Positions of individual vibrational state for the two molecules are indicated in the figure.

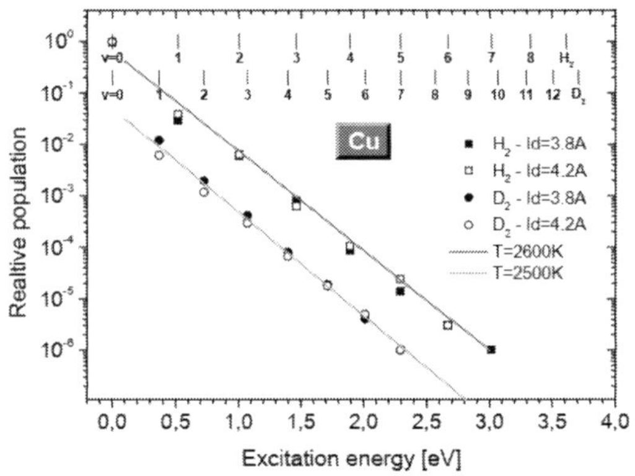

Figure 5. Same as in figure 4 but for recombination of H and D atoms on copper.

Rather high initial H-concentration on the surface was measured (about $30x10^{15}$ H-atoms/cm^2) which is at least three times higher than could be expected even for rough surface. As the sample was only conventionally cleaned it is apparent that impurities remained on the surface. When D_2 is introduced in HEC and dissociation filament is turned on an isotope exchange reaction occur. Due to the incertitude in absolute values of D concentration [11] we can not for the moment compare the absolute size of the steps in H (decrease) and D (increase) concentration time dependence. After some time D concentration becomes stationary but H concentration still remains rather high indicating that only a part of hydrogen on the surface was actively taking part in the exchange reaction. Apparently there are two kinds of hydrogen atoms on the surface, one is strongly bound in impurity compound and another one is more mobile and easily exchangeable with incoming D atoms. Slow decrease of H concentration after initial rapid exchange decrease is observed and this is presumably an indication of the cleaning action of D atoms of the impurity layer.

Above study shows how detailed information is needed in order to well understand and interpret observed vibrational distributions of molecules created at the surface by atom recombination. Detailed further study of these phenomena is under way by means of another experimental arrangement using in situ exposure of tungsten surface to the H (D) atom beam.

4.2. Dissociative electron attachment in hydrogen

Besides being an important atomic collision process for the edge plasma the DEA in hydrogen is a benchmark case for theoretical calculations of resonant processes in electron molecule scattering (e.g. [9]). However, surprisingly no absolute measurements were done since pioneering measurements of Rapp et al. [21] and Schulz and Asundi [22] in sixties. This is also valid for the study of spectacular dependence of cross sections for DEA in hydrogen on target ro-vibrational excitation that was performed only by Alan and Wong [23] in 1978. All modelling of plasmas where DEA is taken into account are performed by using theoretically calculated CSs. All experimental absolute CS values for the low energy, 4eV process are obtained by normalization to the absolute values of Rapp et al. [21].

Our present experimental set up is not intended for measurements of absolute CS and we rely on theoretical CSs from [9] when extracting vibrational distributions from H^- and D^- yields. However, detailed measurements with cold gas target allowed us to establish with high statistical weight the ratios between 14 eV and 4 eV processes in both H_2 and D_2. All measurements of H^- and D^- yields as a function of electron energy show an important signal background that starts at about 11 eV and whose relative intensity varies for different experimental arrangements used by different groups. We have shown by changing ion detection conditions that this background is not due to the negative ions from DEA but to some other particles reaching ion detector and inducing secondary electrons that are subsequently multiplied and detected as counts. These secondary particles are most probably metastable molecules produced by electron impact excitation of triplet states $a^3\Sigma_g^+$ and $c^3\Pi_u$ of hydrogen. This is indicated by observed steep threshold rise of the background that is characteristic for excitation of triplet states and also by the energy position of background threshold being close to the threshold of the triplet excitation at 11.8 eV. At higher electron energy UV photons produced by electron impact excitation of the lowest singlet states in hydrogen can contribute to the background as well. Presence of this strong background, that is very strong in [21], is systematically neglected in literature and therefore we proposed revised set of values for the 14 eV DEA in H_2, HD and D_2 by subtracting the contribution of the background [24]. Background was determined by polynomial fit of the wings of published CS where 14 eV DEA can not contribute to the signal. A preliminary set of values for the 4 eV process was also presented [24] as obtained from the measurements of the ratio of H^- and D^- signal from two processes in H_2 and D_2 respectively. During recent work on vibrational spectrometer we have determined this ratio with higher statistical weight and can state that ratio of DEA CS for 14 eV process to the CS for the 4 eV process is 11.5 ± 1.4 and 1040 ± 120 for H_2 and D_2 respectively. The original 14 eV CS from [21], the corrected 14 eV CS and new experimental values for the 4 eV process obtained from CS ratios are presented in Table 1.

Table 1. Cross sections for dissociative electron attachment in hydrogen for 14 eV and 4 eV process for cold molecules

	DEA CS for 14 eV process Rapp et al. [21] $[10^{-24}\ m^2]$	DEA CS for 14 eV process [21] corrected for background $[10^{-24}\ m^2]$	DEA CS for 4 eV process – present result $[10^{-26}\ m^2]$
H_2	2.1	1.6	14
HD	1.5	1.0	-
D_2	1.0	0.45	0.043

Strong isotope effect observed in DEA in hydrogen was first reported by Schulz and Asundi [22] and it is an important characteristic of this process. Calculated cross sections [9] reproduce observed strong decrease of CS with molecular mass and extrapolate this dependence also to heavier isotopologues. Isotope effect is illustrated by spectra of H^-/H_2 and D^-/D_2 shown in figure 6. For both cases background contribution to the signal is observed and it is relatively more important in the case of D_2 as CS for DEA is smaller while the CS for electronic excitation producing the background is approximately the same.

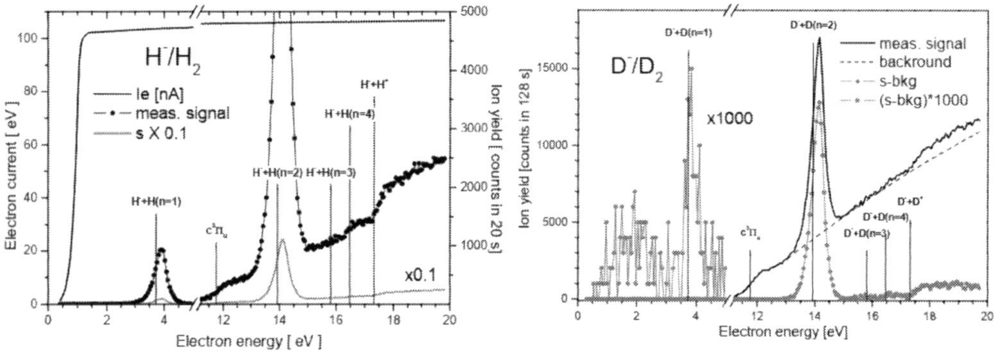

Figure 6. Experimental spectra for DEA to cold molecules for H_2 and D_2. Positions of different thresholds are also indicated in the figures.

Even though the CS for the 4 eV DEA process is negligible for heavy hydrogen isotopologues in the ground state the increase of CS with internal vibrational excitation is much faster for them so that for all hydrogen isotopomeres CSs become very high, in the range of $10^{-15}\ cm^2$ when excitation energy is above 2 eV. This property makes DEA in hydrogen very relevant for the edge plasma modelling if molecules are vibrationally excited.

5. Conclusions

New experimental techniques that have been recently developed in our laboratory are presented in this report. A short overlook of recent studies of vibrational distribution of molecules created by hydrogen atom recombination at metal surfaces is presented as well as measurements of H and D concentration on the surface of W. New evaluation of DEA CS is presented and new experimental values for 4eV process in H_2 and D_2 are given. Further work on these processes as well as on some other processes of relevance to fusion edge plasma are in progress.

Acknowledgements

This work was supported by contracts with the Agency for Research of Republic of Slovenia (ARRS) and Association EURATOM-MHEST. Authors gratefully acknowledge the support by CNRS, France and Université Pierre et Marie Curie, Paris, France by giving us on disposal the entire previous experiment and accompanying material.

References

[1] Clark R E H and Reiter D H (Eds.) 2005 *Nuclear Fusion Research - Understanding Plasma-Surface Interaction* 2005 (Springer Series in Chemical Physics, vol.78) (Berlin: Springer)

[2] Reiter D H EIRENE code ((http://www.eirene.de/), Greenland P T 2001 CRMOL - Collisional-radiative models for molecular hydrogen in plasmas *Forschungszentrum Jülich Report* Jül-3858

[3] Sawada K and Fujimoto T 1995 *J. Appl. Phys.* **78** 2913-24

[4] Janev R K, Reiter D and Samm U 2003 Collision Processes in Low-Temperature Hydrogen Plasmas, *Forschungszentrum Jülich Report* Jül-4105

[5] Celiberto R, Capitelli M and Larucchuita A 2002 *Phys.Scripta* **T96** 32-44, Krstić P S 2002 *Phys. Rev.* A **66** 042717, Krstić P S and Janev R K 2003 *Phys. Rev.* A **67** 022708

[6] Popović D, Čadež I, Landau M, Pichou F, Schermann C, and Hall R I 1990 *Meas. Sci. Technol.* **1** 1041-6

[7] Čadež I, Hall R I, Landau M, Pichou F, Winter M and Schermann C 2004 *Acta. Chim. Slov.* **51** 11-21

[8] Markelj S, Čadež I, Rupnik Z 2008 *Int. J. Mass Spectrom.* **275** 64-74

[9] Horáček J, Čižek M, Houfek K, Kolorenč P and Domcke W 2004 *Phys. Rev.* A **70** 052712 and private communication

[10] Pelicon P, Razpet A, Markelj S, Čadež I and Budnar M 2005 *Nucl. Instr. and Meth.* B **227** 591-6

[11] Markelj S, Čadež I, Pelicon P, Rupnik Z 2007 *Nucl. Instr. and Meth.* B **259** 989-96

[12] Markelj S, Čadež I, Schwarz-Selinger T 2008 18th Int. Conf. on Plasma Surface Interaction, Toledo, Spain May 26-30 *Abstr.* P1-06

[13] Kammler Th, Lee J, Küppers J 1997 *J. Chem. Phys.* **106** 7362-71

[14] Wehner S, Küppers J 1998 *J. Chem. Phys.* **108** 3353-9

[15] Kammler Th and Küppers J 1998 *J. Chem. Phys.* **111** 8115-22

[16] Kammler Th, Wehner S, Küppers J 1998 *J. Chem. Phys.* **109** 4071-7

[17] Kammler Th, Kolovos-Vellianitis D, Küppers J 2000 *Surface Science* **460** 91-100

[18] Jackson B, Sha X, Guvenc Z B 2002 *J. Chem. Phys.* **116** 2599-608

[19] Čadež I, Markelj S, Pelicon P, Rupnik Z 2008 18th Int. Conf. on Plasma Surface Interaction, Toledo, Spain May 26-30 *Abstr.* P3-68, submitted for publication in *J. Nucl. Mater.*

[20] Schermann C, Pichou F, Landau M, Čadež I and Hall R I 1994 *J. Chem. Phys.* **101** 8152-8

[21] Rapp D Sharp T E and Briglia D D 1965 *Phys. Rev. Lett.* **14** 533-5

[22] Schulz G J and Asundi R K 1967 *Phys. Rev.* **158** 25-9

[23] Allan M and Wong S F 1978 *Phys. Rev. Lett.* **41** 1791-4

[24] Čadež I, Rupnik Z and Markelj S 2005 *XXVIIth Int. Conf. on Physics of Ionized Gases*, Eindhoven, the Netherlands, 18-22 July, *Proc.* P02-236

Variability of GRB light curve: Shock wave model modification

Saša Simić [1], Luka Č. Popović [2]

[1] Faculty of Science, Department of physics, Radoja Domanovića 12, 34000 Kragujevac, Serbia
[2] Astronomical Observatory, Volgina 7, 11000 Belgrade, Serbia

E-mail: ssimic@kg.ac.yu

Abstract. The main topic of our investigation is a mysterious phenomenon of gamma ray bursts. If someone analyses the observation of this events, one can see that the most interesting data are among the gamma phase, during the first couple of seconds. In this stage central engine generate a high intensity and high energy radiation, observed in the form of pulses in the light curve. If we could understood the main physical processes in this phase, we could put some constraints for the model which try to explain the core of this phenomena, deeply hidden from observation by high optical thickness of surrounding material.

Observation of the GRB light curve can be easily done with the help of modern satellites, so that the data of this kind are vastly dispersed. We try to analyse them and to explain in details physical process which create the light curve pulses. Our research is based on the broadly accepted scenario of gamma phase in GRBs, which predicts that GRB core generate highly relativistic mater in some amount of time. This mater form shock waves of different velocities, due to highly differential motions. The shock waves can interact with each others, and in that moment radiation significantly increases, creating the observed pulses in the GRB light curve. We develop a phenomenological model based on the model developed by Huang and coauthors, to explain evolution of a shock wave expanding from some distance. In it, we implemented the ability to simulate collision between incoming shock and density barrier. We also propose that the density barrier is created by the material ejected from the core and spread around by the shocks. As a result of simulation we can get synthesized pulses, and by comparing them with the observational data we can acquire values of basic parameters used in our model.

1. Introduction

Investigation of the Gamma Ray Bursts (hereafter GRB) phenomena in the last decade has been pointed to high energy astrophysics involving relativistic jets and shock waves. Crucial research problem left unclear is functioning and nature of the GRB core, particularly mechanism for generation of relativistically moving material. We have try in our investigation to put some additional constraints on the GRB engine in the form of values of basic parameters of created shock waves. We have analyzed the shapes of pulses observed in the light curve of this event and simulate them numerically. Then in the process of comparisons between observer and synthesized pulses, we extract expected basic parameters of involved shock waves.

The main scenario presently accepted for this phenomena says that in some kind of explosion probably similar to supernova [1], (collapsar models) huge amount of material is ejected in the surrounding space. There is also another explanation of cause for this ejection, mostly based on the collision scenario, for example between two rotating neutron stars or black holes [2],

[3]. This highly accelerated material spread around center of explosion either isotropically (high total energy) or collimated into jets of matter (lower total energy). Differential motion of this relativistic mater create shock waves in its front. They present highly compressed and fast moving energetic, parts of ejecta which are created by the GRB core in some number. They have different velocities and by that ability for mutual collision and merging. In the moment of interaction the amount of radiation is highly increased creating the observed pulses in the light curve (internal shock wave model - [4]). Internal engine of GRBs is active a while, after it's shut down, marking the ending of the first phase of GRB event, known as a gamma phase. The duration of the first phase is usually around a couple of tens of seconds. The ejected material in form of shock waves continues to move away from the center of explosion merging in to one more massive but slower shock wave, activating the second phase of GRB called the afterglow (external shock wave model - [4].

This one is characterized with the lower energies and smooth continuous light curves, lasting for even years with transition from the X-ray band toward the lower energy radio bands.

Our goal is to examine the shock waves in the first phase of the event, simulating the expansion and collision of them with a suitable mathematical model. Additional comparison of theoretical and observational light curve, can give us values of the basic parameters used. In the next chapter §2 we will present model developed by us [5], [6] and give some fundamental constraints and assumptions. In the chapter §3 we will discuss process of extract and fitting of the observational data, as well as present and discuss obtained parameters.

2. Employed model and assumptions

Let us first describe the physical scenario of these processes. As in [7], we will consider an ultra relativistic flow of matter as an array of well defined collimated shells with random energy and mass, the so called '*small shocks*', because of its lower mass compared with the mass of the afterglow shock wave. They are created by differential motion of ejected material, which can produce shocks of different energies and Lorentz factors. As a result, collision can occur between two expanding waves of different velocities. This can produce low intensity pulses in the light curve.

In the process of interaction the resulting shock wave is created by merging of initial two, so the mass of newly created shock wave is its sum. On the contrary, the Lorentz factor is highly decreased, which can produce effect of accumulation of material at the certain distance, thus creating a density perturbation. The newly created mass ejection rush into such density perturbation merging with it and moving it further from the center of explosion producing the pulses of higher intensity and duration. Depending on the intensity and width of a such created density barrier we can simulate the wider-tinner or higher-lower pulses.

Also, considering the energy density of shock wave and barrier, we can get more symmetrical shape light curve pulses in the case of energetic shock wave and low energy barrier, and high asymmetrical pulses in an opposite situation. This is the main mechanism we use to explain high temporal variability of the observed γ-ray light curve. Concerning the scenario mentioned above, we see that the selection of the shock parameters has random nature in a given interval of values.

The central engine continuously creates new small mass shocks that cause the flow of shock waves with different initial parameters. With time, they are able to accelerate particles of the Inter-Stellar Medium (ISM) surrounding the GRB and to accumulate in another, massive shock wave which continues to spread with lower velocity. Then, the second phase of the GRB event starts, with the creation of the afterglow.

For proper description of evolution of relativistically moving matter we have use a phenomenological system of first order differential equations which evolve three important variables, Lorentz factor, mass and distance of shock wave [6]. First two of them are rewrited from the paper of [9], and third is derived by us, to incorporate sharp jump of basic variables in the moment of collision. Then we have:

$$\frac{dR}{dt} = c\sqrt{\Gamma^2 - 1}\Big[\Gamma + \sqrt{\Gamma^2 - 1}\Big], \tag{1}$$

$$\frac{d\Gamma}{dm_s} = -\frac{\Gamma^2 - 1}{M_{ej} + 2(1-\varepsilon)\Gamma m_s + \varepsilon m_s}, \tag{2}$$

$$\frac{dm_s}{dt} = 2\pi n m_p (1 - \cos\theta)\frac{R^2}{\Gamma^3}\left(3\Gamma\frac{dR}{dt} - 2R\frac{d\Gamma}{dt}\right), \tag{3}$$

where the parameter ε takes values from 0 for the adiabatic expansion, to 1 which describes a fully radiative case, and M_{ej} is the mass of a primary ejected material. Eqs. (1) - (3) are derived for an observer reference frame, and they have to be solved simultaneously, together with the density equation. Initial values of parameters and variables are highly dependent on the properties of the shocks.

For a specific density of surrounding media we use equation developed by [8], which we modified to include a density distribution to describe the barrier of accumulated ejected material. The barrier is assumed to have Gaussian profile of density, which is completely satisfied in the first order approximation. The final form of equation is:

$$n = n_0\left(\frac{R_0}{R}\right)^s (4\Gamma + 3)\left(1 + a\cdot\exp\left[-\left(\frac{R - R_c}{b}\right)^2\right]\right) \tag{4}$$

where R_c is the distance of the shock encounter, width at half maximum (b) represents the width of the slower shock, and the intensity of the Gaussian (a) represents the slower shock number density.

In principle, one can expect that the primary shock can collide with an ISM which has certain density distribution. In such highly relativistic physical system the relative motion of charged particles of ISM can generate intensive magnetic field in the reference frame of moving fluid. The geometry of magnetic field is considering to be mostly turbulent. We calculated it in a similar way as in [9], by assuming that the energy of the magnetic field is a certain fraction, ξ_b, of the total energy of the shock wave. The parameter ξ_B determines this magnetic energy part, and in the first approximation we will consider it constant during the interaction. In the comoving reference frame the expression for the magnetic field is taken as:

$$B' = \sqrt{8\pi\xi_b n_0 \Gamma m_p c^2 (4\Gamma + 3)\left(\frac{R_0}{R}\right)^s \left(1 + a\cdot e^{-\left(\frac{R-R_c}{b}\right)^2}\right)}, \tag{5}$$

The emission mechanism of shock waves is mainly based on synchrotron radiation, but for higher energy bands additional flux can be gained by the Inverse Compton (IC) radiation [4]. In the first approximation we will neglect flux gained by the IC effect and preserve our calculation on the synchrotron source. To calculate the intensity of the radiation by particles in the shock wave we will use the formulae given by [11], then the total emitted flux can be calculated as e.g. in [10]. Also, we should note that the shock waves contain relativistic electrons and baryons which contribute to the synchronistic radiation, but taking into account the difference in velocities of these constituents, one can neglect the contribution of baryons to the total emitted flux. Then in the comoving reference frame the total flux is:

$$P'_\nu = A\cdot\int_0^{\theta_m}\tan\theta d\theta\int_{\gamma_{emin}}^{\gamma_{emax}}\gamma_e^{-(p+1)}F(\nu'/\nu'_c)d\gamma_e \tag{6}$$

where A is:

$$A = \frac{\sqrt{3}e^3 B'}{m_e c^2} \frac{m_s}{m_p} \frac{C_1}{\ln(\cos(\theta_m))} \tag{7}$$

and $F(x) = \int_x^\infty K_{5/3}(x)dx$ where $K_{5/3}$ is the Bessel function of the second order. Here, ν_c' is the critical frequency of the radiation expressed by $\nu_c' = 3\gamma_e^2 eB'/4\pi m_e c$.

In equation 6 we implemented the effects of curved surface of an emitting shock wave. Also, in the case of small shocks the cooling time is longer then dynamical time of expansion of shell, so the effect of electron cooling can be neglected.

With presented model we are able to simulate the collision of shock waves in the first phase of GRB, which produce the pulses in the light curve (see [6]). In the figures below we present evolution of the main magnitudes describing the shock wave in the case where there is a collision with slow density barrier.

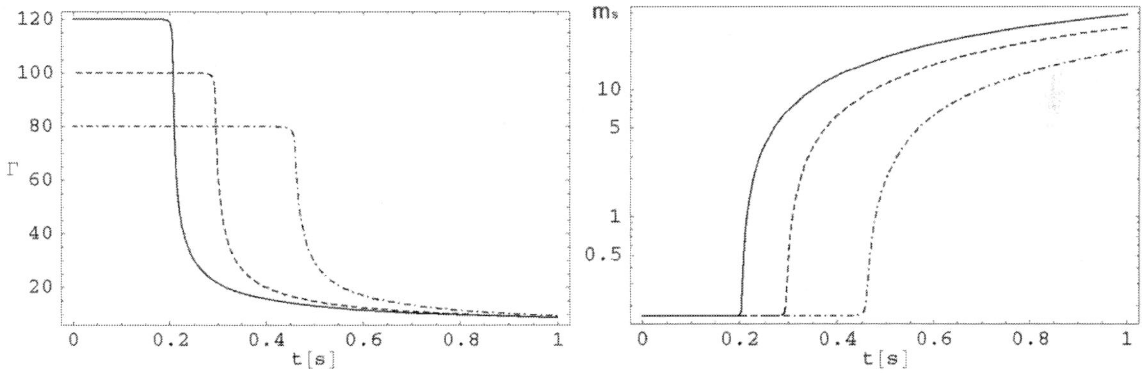

Figure 1. Evolution of Lorentz factor (left) and shock wave mass $m_s \times 10^{22}$ [gr.] (right) given in the logarithmic scale. We use three different values of $\Gamma_0 = 120$ - full line, $\Gamma_0 = 100$ - dashed line and $\Gamma_0 = 80$ - dash-doted line.

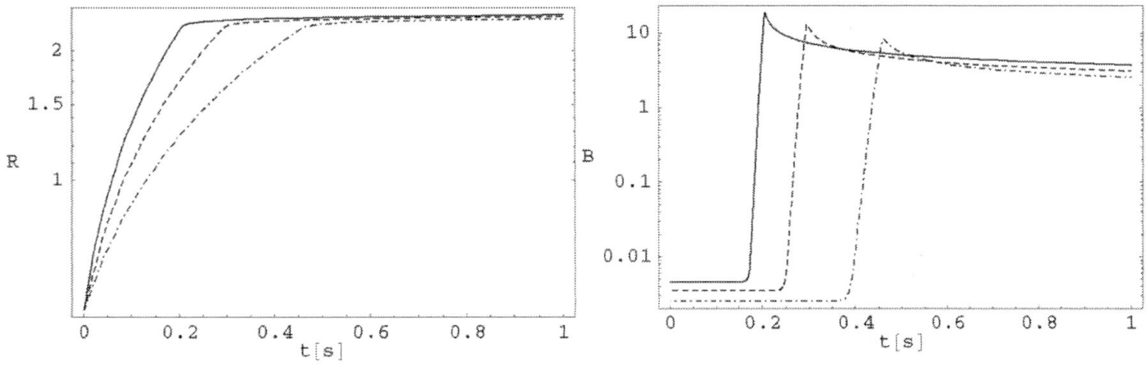

Figure 2. Evolution of distance from center of explosion $R \times 10^{14}$ [cm] (left) and vector of magnetic field $B \times 10^6$ Gauss, given in the logarithmic scale. We use three different values of $\Gamma_0 = 120$ - full line, $\Gamma_0 = 100$ - dashed line and $\Gamma_0 = 80$ - dash-doted line.

We can easily see sharp discontinuity in the moment of interaction, which causes decreasing of Lorentz factor for an order of magnitude, as well as rapid increase of shock wave mass. Evolution of distance show us what we initially propose that after a collision, resulting shock wave accumulate at some distance, which is not influenced by initial values of the Lorentz factor and ejected mass. Magnetic field in that moment sharply increases gaining the additional radiation power (see Eq. 7). That is expected if we know that equation for synchrotron radiation power is directly proportional to the intensity of the magnetic field.

3. Results and discussion

We select a sample of 30 recorded GRB events from a large BATSE database (3rd channel, $E = 100 - 300$ keV, for the light curve) which are suitable for light curve pulse extraction. The main criteria we used is that: (i) the light curves are well isolated with clear peak maximum (process of pulse creation is stochastic in nature and result in a wide variety of pulse shapes and combinations, e.g. it is often observed that two pulses are superposed one on another); (ii) we avoid small pulses, because of their low temporal resolution we could not achieve enough accuracy in a fitting process; (iii) the pulses are with different shapes as much as possible in order to include long and short duration pulses, as well as pulses of different intensity and symmetries in the shape.

During the fitting attempts we have done broad analysis of how parameters influence the shape of the light curve pulse. The one particulary interesting case is with the parameter θ_m, the opening angle of the shock wave cone. In the Figure 3 (left) we can see that for the bigger value of θ_m, resulting pulse is more wider and slow decay feature is much more expressed then in the case when the angle θ_m has lower values. In that case the generated pulse is more tinner and symmetrical in shape. That is in agrement with employed physical processes during shock expansion. Indeed, if we demand wider opening angle, we lower the energy density of the shock wave and the produced pulse is softer and more broadly disperse. On the contrary, if we narrow the jet, than the shock wave is more energetic and we have symmetrical but intensive light curve pulses.

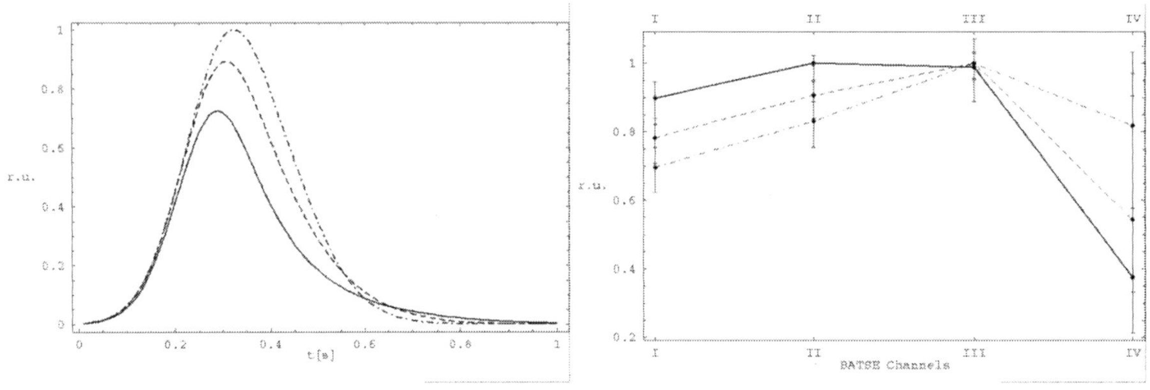

Figure 3. Shape of the light curve pulses (left) and its spectral curves (right) given in relative units compared to highest pulse. Three lines are appropriate to different values of the opening angle of ejected material $\theta_m = 0.09$ rad - full line, $\theta_m = 0.07$ rad - dashed line and $\theta_m = 0.05$ rad - dash-doted line.

Also, other parameters have substantial influence on the light curve pulse, see Figures 3. For example, if we demand higher values for the initial Lorentz factor, we expect to have shorter and more intensive pulses. Similar, with the increase of shock wave mass, pulses also increase

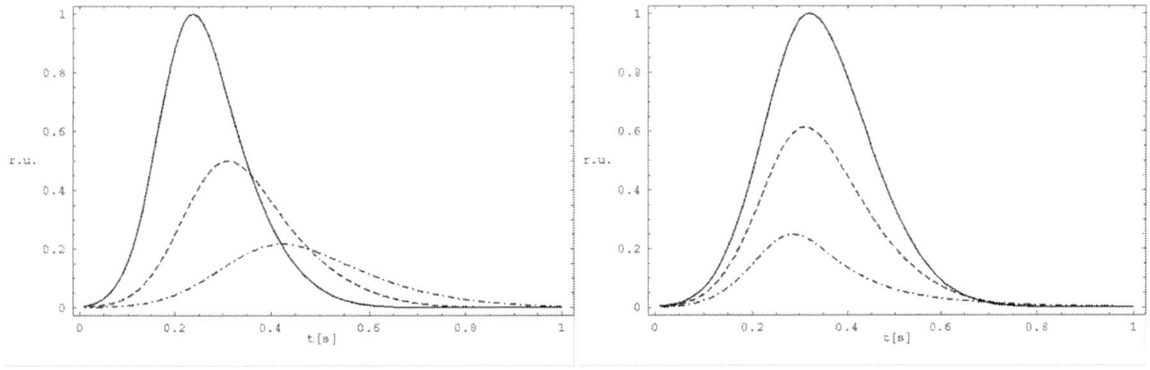

Figure 4. Shape of the light curve pulses for different values of Γ_0 (left) and M_{ej} (right) given in relative units compared to highest pulse. Three lines on the left and right Figures are appropriate to following set of parameter: $\Gamma_0 = 80, M_{ej} = 9 \times 10^{-11} \times M_{sun}$ - full line; $\Gamma_0 = 70, M_{ej} = 6 \times 10^{-11} \times M_{sun}$ - dashed line and $\Gamma_0 = 60, M_{ej} = 3 \times 10^{-11} \times M_{sun}$ - dash-doted line.

Table 1. Mean and most probable values taken from the large sample of fitted GRBs.

parameter GRB	n_0 [cm^{-3}]	Γ_0 -	M_{ej} x 10^{-11}[Msun]	θ_m [rad]	R_c x 10^{14}[cm]	ΔR x 10^{13}[cm]	n_b x 10^5[cm^{-3}]
Mean →	22.5	42.4	3.9	0.067	1.3	5.9	10.
Most probable →	16.	27	0.1	0.05	1.0	1.6	0.1

in its height. This behavior is connected with increase of initial energy of shock wave, when the interaction with the stationar barrier become quicker and more intense. In the opposite situation the produced pulses are smaller and less symmetrical in shape following the well known FRED (*fast rise slow decay*) law. In the case of density of surrounding media we have almost direct proportionality. That is well understanding if we know that the radiation power of the synchrotron emission is connected with the number of radiating particles. So, if we have denser barrier then the light curve pulse is more intensive and also broader.

Based on presented results we can make overall conclusion about parameters influence on the shape of the light curve pulses. Namely, in the interaction of the shock wave with the barrier, two cases can be achieved considering the parameters setup. First, the case when the shock wave is more energetic and compact compared with the barrier. In this case produced pulses are symmetrical, relatively shorter and intensive. Otherwise, when the incoming shock wave is broader and with lower energy then the barrier, pulses are more asymmetrical and broader in shape.

In the Figures 3 the results of fitting the light curve for particular GRBs are presented as well as theirs spectral curves. We select from our sample two opposite cases, first GRB911104 with the broad pulse duration and GRB000513 presenting the narrow pulses. With this we try to emphasize differences of shapes in the GRB pulses, and ability of our model to fit both mentioned types. With the similar results from others GRBs in the our sample we can group the data and statistically handle them to acquire most probable values and intervals. Example of basic parameters are presented in the Table 1 below.

At the end we try to synthesized overall l.c. that contains multiple pulses of different shapes and intensity. For this reason we propose that all pulses are created in the same manner, and that observed differences originate from the stochastic nature of basic parameters. In this case

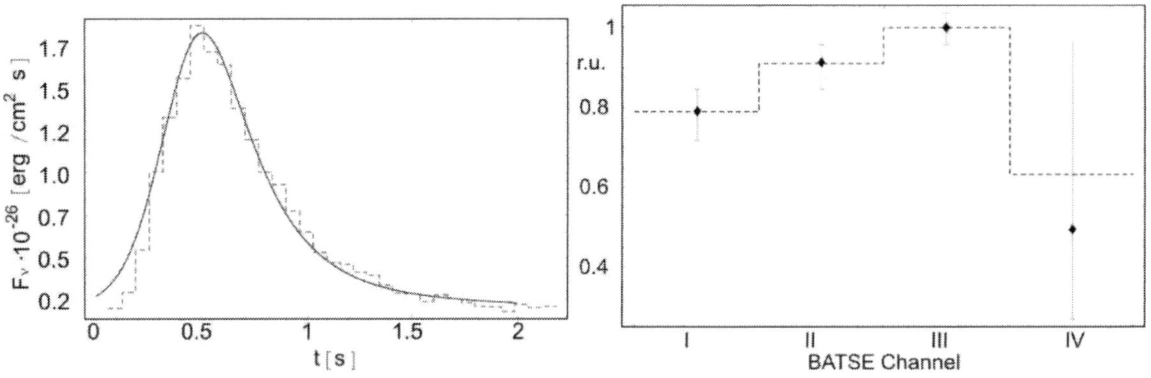

Figure 5. Left Figure presents the best fit (solid line) of pulse form GRB911104 (dashed line), with the abscissa axis designated as emitted flux F_ν and time t on the ordinate. On the right side we place the spectral fit for all four channels of BATSE instrument, where the values of the flux are put in the relative units (r.u.).

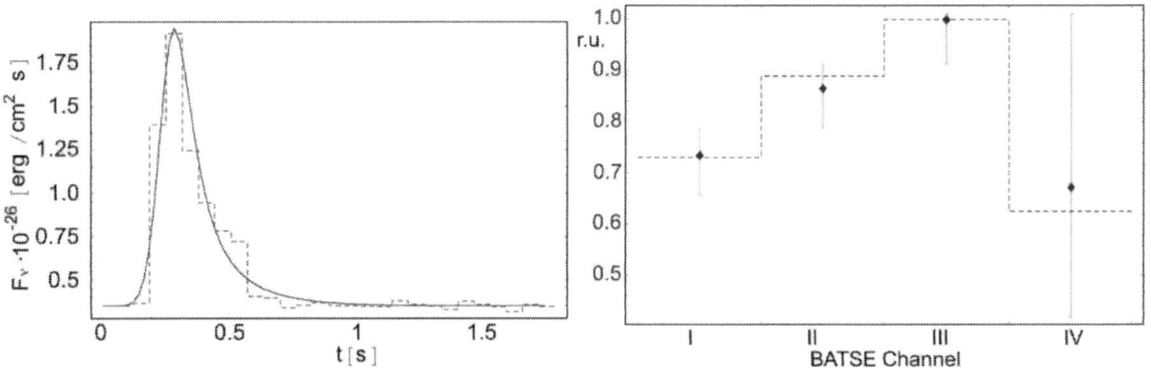

Figure 6. Same as in Fig. 3 but for GRB000513.

we account just a clear single pulses, which mean that all of them must have monotonic rise and decay. Pulses consisted from combination of two or more are avoided in this simulation. Another thing to consider is the time interval between pulse generation, which is tightly correlated with the dynamics of internal GRB engine. We have to notice that on this level of research we can take it as totaly random in some range which is acquired mostly from the observations.

Using this assumptions we can synthesize light curve by generating one collision for every separate pulse, and by superposition of them we get final overall l.c. For any particular interaction a new set of basic parameters are selected in the range specified with the use of Table 1. In Figure 7 the result of our simulation is presented.

4. Conclusion

In this paper we have develop a phenomenological model to simulate the evolution of shock waves expanding from the center of GRB, and colliding mutually producing the observed light curve pulses. As a result we acquire a range of basic parameters which describe the shock waves. We present mean and most probable values of parameters in the Table 1 in order to get clearer picture of physical processes which are behind the GRB event, and we reach following conclusions:

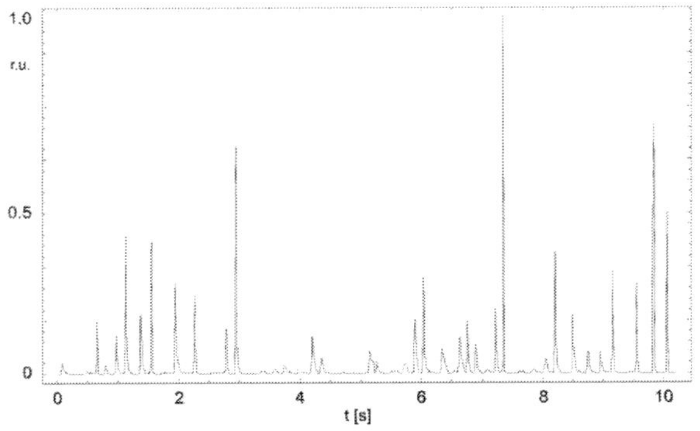

Figure 7. Synthetic light curve produced by simultaneous application of our model.

(i) The derived model is able to synthesize pulses similar to one observed in the l.c. of GRBs. The main advantage of this model is its simplicity, compared to the models which are based on the strict magneto-hydrodynamical approach. Nevertheless, using our simplified model we preserve accuracy and clear picture of physical processes employed in the central engine of GRBs.

(ii) The obtained values for internal shock parameters are extracted for different shapes of pulses in the GRB light curve. They are in a good agreement with estimations given by other authors earlier (see [4]). As we see in Table 1 there is some most probable values which show us that the physical processes responsible for GRB creation are similar, and that observed variation of pulse shapes are generated by stochastic variation in values of basic parameters.

(iii) Extracted parameters can present fundamental requirements for construction of broad models which describe GRB progenitor.

Acknowledgments

The work was supported by the Ministry of Science and Technological Development of Serbia through the project no. 146002, "Astrophysical Spectroscopy of Extragalactic Objects".

References

[1] Woosley S E 1993 *ApJ* **405** 273
[2] Narayan R, Paczynski B and Piran T 1992 *ApL* **395** 83
[3] Paczynski B 1991 *AcA* **41** 257
[4] Piran T 2005 *RvMP* **76** 1143
[5] Simić S 2004 *BaltA* **13** 297
[6] Simić S, Popović L Č, Andersen M I and Chrinstensen L 2007 *A&SS* **309** 173
[7] Kobayashi S, Piran T and Sari R 1997 *ApJ* **490** 92
[8] Blandford R D and McKee C F 1976 *Physics of Fluids* **19** 1130
[9] Huang Y F, Dai Z G and Lu T 1998 *A&A* **336** L69
[10] Rybicki G B and Lightman A P 1979 *Radiative Processes in Astrophysics* (New York: Wiley)
[11] Huang Y F, Gou L J, Dai Z G and Lu T 2000 *ApJ* **543** 90

Confinement and anisotropy of ultrahigh-energy cosmic rays in isotropic plasma wave turbulence

Miroslava Vukčević

Department of Physics, University of Montenegro, Podgorica, Montenegro

E-mail: `miras@cg.ac.yu`

Abstract. The mean free path and anisotropy of galactic cosmic rays is calculated in weak plasma wave turbulence that is isotropically distributed with respect to the ordered uniform magnetic field. The modifications on the value of the Hillas energy, above which cosmic rays are not confined to the Galaxy, are calculated too. The original determination of the Hillas limit has been based on the case of slab turbulence where only parallel propagating plasma waves are allowed. We use quasilinear cosmic ray Fokker-Planck coefficients to calculate the mean free path and the anisotropy in isotropic plasma wave turbulence. In isotropic plasma wave turbulence the Hillas limit is enhanced by about four orders of magnitude to $E_c = 2.03 \cdot 10^5 A n_e^{1/2} (L_{\rm max}/10\ {\rm pc})$ PeV resulting from the dominating influence of transit-time damping interactions of cosmic rays with obliquely propagating magnetosonic waves. Below the energy E_c the cosmic ray mean free path and the anisotropy exhibit the well known $E^{1/3}$ energy dependence. At energies higher than E_c both transport parameters steepen to a E^3-dependence. This implies that cosmic rays even with ultrahigh energies of several hundreds of EeV can be rapidly pitch-angle scattered by interstellar plasma turbulence, and are thus confined to the Galaxy.

1. Introduction

To unravel the nature of cosmic sources that accelerate cosmic rays to ultrahigh energies has been identified as one of the eleven fundamental science questions for the new century [1]. The most interesting are the cosmic rays with energies as large as $10^{20.5}$ eV. It has been argued that, due to the amplification of the magnetic field in the shock, the acceleration of cosmic rays in young supernova remnants is possible up to $\sim 10^{18}$ eV. This implies that such particles may have a Galactic origin. For ultrahigh-energy ($10^{18} - 10^{20.5}$ eV) cosmic rays an extragalatic origin is favored by many researchers. Extragalactic ultrahigh-energy cosmic rays (UHECRs) coming from cosmological distances ≥ 50 Mpc should interact with the universal cosmic microwave background radiation (CMBR) and produce pions. For an extragalactic origin of UHECRs the detection or non-direction of the Greisen-Kuzmin-Zatsepin cutoff [2] resulting from the photopion attenuation in the CMBR will have far-reaching consequences not only for astrophysics but also for fundamental particle physics as e.g. the breakup of Lorentz symmetry [3] or the non-commutative quantum picture of spacetime [4].

For many years the theoretical development of the resonant wave-particle interactions has mainly concentrated on the special case that the plasma waves propagate only parallel or antiparallel to the ordered magnetic field – the socalled slab turbulence. In this case only cosmic ray particles with gyroradii R_L smaller than the longest parallel wavelength $L_{\|,max}$ of the plasma waves can resonantly interact. Obviously this condition is equivalent to a limit on the maximum particle rigidity R:

$$R = \frac{p}{Z} \leq eB_0 \mathrm{L}_{\parallel,max}. \tag{1}$$

An alternative way to express the condition (1) is

$$E_{15}/Z \leq 40 \cdot \left(\frac{B_0}{4\mu\ \mathrm{G}}\right)\left(\frac{L_{\parallel,max}}{10\mathrm{pc}}\right), \tag{2}$$

where E_{15} denotes the cosmic ray particle energy in units of 10^{15} eV. The limit set by the right hand side of Eq. (2) is referred to as Hillas limit [5]. According to this limit, cosmic ray protons of energies larger than 40 PeV= $4 \cdot 10^{16}$ eV cannot be confined or accelerated in the Milky Way, and an extragalactic origin for this cosmic ray component has to be invoked. Moreover, as the cosmic ray mean free path in case of spatial gradients is closely related to the cosmic ray anisotropy ([6], Eq. (94)), the Hillas limit (2) implies strong anisotropies at energies above 40 PeV which have not been observed by the KASKADE experiment [7].

It is the purpose of this work to investigate how the Hillas limit (2) is affected if we discard the assumption of purely slab plasma waves, i.e. if we allow for oblique propagation angles θ of the plasma waves with respect to the ordered magnetic field component. In order to estimate confinement and relevant transport parameters we have to investigate also influence of strong coupling of cosmic ray particles with damped plasma waves. When discussing the nature of interstellar turbulence, it is necessary to consider the fact that the interstellar medium contains a number of plasmas of very diverse characteristics, not only cold plasma. In this work, we include the temperature effects to the first order [8], in deriving the relevant Fokker-Planck coefficients.

2. Relevant magnetohydrodynamic plasma modes without damping

Most cosmic plasmas have a small value of the plasma beta $\beta_P = c_S^2/V_A^2$, which is defined by the ratio of the ion sound c_S to Alfven speed V_A, and thus indicates the ratio of thermal to magnetic pressure. For low-beta plasmas the two relevant magnetohydrodynamic wave modes are the

(1) incompressional *shear Alfven waves*

(2) the *fast magnetosonic waves*

2.1. Resonant interactions of shear Alfven waves

For shear Alfven waves only interactions with $n \neq 0$ are possible. These are referred to as *gyroresonances* and corresponding the resonance parallel wavenumber is

$$k_{\parallel,A} = \frac{n\Omega_c}{\pm V_A - v\mu}, \tag{3}$$

which apart from very small values of $|\mu| \leq V_A/v$ typically equals the inverse of the cosmic ray particle's gyroradius, $k_{\parallel,A} \simeq n/R_L$ and higher harmonics.

2.2. Resonant interactions of fast magnetosonic waves

In contrast, for fast magnetosonic waves the $n = 0$ resonance is possible for oblique propagation due its compressive magnetic field component. The $n = 0$ interactions are referred to as *transit-time damping*, hereafter TTD

$$v\mu = \pm V_A / \cos\theta \tag{4}$$

as necessary condition which is independent from the wavenumber value k. Apparently all super-Alfvenic ($v \geq V_A$) cosmic ray particles are subject to TTD provided their parallel velocity $v\mu$ equals at least the wave speeds $\pm V_A$. Hence equation (4) is equivalent to the two conditions

$$|\mu| \geq V_A/v, \quad v \geq V_A. \tag{5}$$

Additionally, fast mode waves also allow gyroresonances ($n \neq 0$) at wavenumbers

$$k_F = \frac{n\Omega_c}{\pm V_A - v\mu \cos\theta}. \tag{6}$$

2.3. Implications for cosmic ray transport

The simple considerations of the last two subsections allow us the following immediate conclusions:

(1) The momentum diffusion coefficient

$$A_M = \frac{1}{2}\int_{-1}^{1} d\mu\,[D_{pp}(\mu) - \frac{D^2_{\mu p}(\mu)}{D_{\mu\mu}(\mu)}] = A_T + A_2 \tag{7}$$

has contributions both from transit-time damping of fast mode waves,

$$A_T \simeq \int_{V_A/v}^{1} d\mu D_{pp}^{\rm TTD}(\mu), \tag{8}$$

and from second-order Fermi gyroresonant acceleration by shear Alfven waves [6]

$$A_2 = \frac{1}{2}\int_{-1}^{1} d\mu\,[D^A_{pp}(\mu) - \frac{[D^A_{\mu p}(\mu)]^2}{D^A_{\mu\mu}(\mu)}]. \tag{9}$$

(2) On the other hand, the spatial diffusion coefficient

$$\kappa = \frac{v^2}{8}\int_{-1}^{1} d\mu (1-\mu^2)^2\, D^{-1}_{\mu\mu}(\mu) \tag{10}$$

is given by the integral over the *inverse* of the Fokker–Planck coefficient $D_{\mu\mu}$, so here the small values of $D_{\mu\mu}$ due to gyroresonant interactions in the interval $|\mu| < V_A/v$ determine the spatial diffusion coefficient and the corresponding parallel mean free path

$$\kappa = v\lambda/3 \simeq \frac{v^2}{8}\int_{-V_A/v}^{V_A/v} \frac{d\mu}{D^{\rm G}_{\mu\mu}(\mu)}. \tag{11}$$

The gyroresonances can be due to shear Alfven waves or fast magnetosonic waves. For relativistic cosmic rays the relevant range of pitch angle cosines $|\mu| \leq v_A/v$ is very small allowing us the approximation $D^{\rm G}_{\mu\mu}(\mu) \simeq D^{\rm G}_{\mu\mu}(0)$ so that

$$\kappa = v\lambda/3 \simeq \frac{v^2}{4}\frac{\epsilon}{D^{\rm G}_{\mu\mu}(0)} = \frac{vV_A}{4D^{\rm G}_{\mu\mu}(0)}. \tag{12}$$

(3) According to Eq. (90) of Schlickeiser [6] the streaming cosmic ray anisotropy due to spatial gradients in the cosmic ray density is given by

$$\delta = \frac{F_{\rm max} - F_{\rm min}}{F_{\rm max} + F_{\rm min}} = \frac{1}{2F}\frac{v}{4}\frac{\partial F}{\partial z}\int_{-1}^{1} d\mu (1-\mu^2)\, D^{-1}_{\mu\mu}(\mu), \tag{13}$$

which also is determined by the smallest value of $D_{\mu\mu}$ around $\mu = 0$. Approximating again $D_{\mu\mu}(\mu) \simeq D^{\mathrm{G}}_{\mu\mu}(0)$ for $|\mu| \leq \epsilon = V_A/v$ we derive the direct proportionality of the cosmic ray anisotropy with the parallel mean free path, i.e.

$$\delta \simeq \frac{v}{8}\frac{\partial F}{\partial \ln z}\frac{2V_A}{vD^{\mathrm{G}}_{\mu\mu}(0)} = \frac{V_A}{4}\frac{1}{D^{\mathrm{G}}_{\mu\mu}(0)}\frac{\partial F}{\partial \ln z} = \frac{1}{3}\lambda\frac{\partial F}{\partial \ln z}. \tag{14}$$

Introducing the characteristic spatial gradient of the cosmic ray density $< z >^{-1} \equiv (1/F)|\partial F/\partial z|$ Eq. (14) reads

$$\delta = \frac{\lambda}{3 < z >}. \tag{15}$$

Cosmic ray gradients derived from diffuse galactic GeV gamma-ray emissivities suggest a value of $< z > \simeq 2$ kpc.

2.4. Quasilinear cosmic ray mean free path and anisotropy isotropic plasma wave turbulence
Throughout this work we consider isotropic linearly polarized magnetohydrodynamic turbulence, and we adopt a Kolmogorov-like power law dependence (index $q > 1$) together with vanishing cross helicity of each plasma mode. According to [9] at particle pitch-angles outside the interval $|\mu| \geq \epsilon$ transit-time damping provides the dominant and overwhelming contribution to these Fokker-Planck coefficients. This justifies the approximations to derive Eqs. (12) and (14) for the cosmic ray mean free path and anisotropy, respectively. Both transport parameters are primarily fixed by the small but finite scattering due to gyroresonant interactions in the interval $|\mu| < \epsilon$. In the following, we consider both transport coefficients for positively charged cosmic ray particles with $\Omega > 0$ especially in the limit $k_{min}R_L >> 1$.

2.4.1. Gyroresonant Fokker-Planck coefficients at $\mu = 0$ Comparing the Fokker-Planck coefficients from fast mode waves and Alfven waves we have found that the latter one is always smaller by the small ratio $\epsilon = V_A/v$ than the first one:

$$D^A_{\mu\mu}(\mu = 0) \simeq \epsilon D^F_{\mu\mu}(\mu = 0), \tag{16}$$

so that the gyroresonant contribution from Alfven waves can be neglected in comparison to the gyroresonant contribution from fast mode waves.

2.4.2. Cosmic ray mean free path Neglecting $D^A_{\mu\mu}(\mu = 0)$ we obtain for the cosmic ray mean free path

$$\lambda(\gamma) \simeq \frac{3v\epsilon}{4D^F_{\mu\mu}(\mu = 0)} =$$

$$\frac{1}{\pi(q-1)}\frac{B_0^2}{(\delta B)^2_F}\frac{R_L(k_{min}R_L\epsilon)^{1-q}}{\sum_{n=1}^{\infty} n^{-(q+1)}H[n - \epsilon R_L k_{min}]}, \tag{17}$$

which exhibits the familiar Lorentz factor dependence $\propto \beta\gamma^{2-q} \simeq \gamma^{2-q}$ at Lorentz factors $\gamma \leq \gamma_c$ below a critical Lorentz factor defined by

$$\gamma_c = k_c/k_{\min} \tag{18}$$

with $k_c = \Omega_{0,p}/V_A = \omega_{p,i}/c$ being the inverse ion skin length. The Lorentz factor dependence $\lambda \propto \gamma^{2-q}$ especially holds at rigidities $1 \leq k_{\min}R_L \leq \epsilon = c/V_A$, in a rigidity range where the slab turbulence model would predict an infinitely large mean free path.

Expressing $k_{\min} = 2\pi/L_{\max}$ in terms of the longest wavelength of isotropic fast mode waves $L_{\max} = 10$ pc yields

$$\gamma_c = \frac{\omega_{p,i} L_{\max}}{2\pi c} = 2.16 \cdot 10^{11} n_e^{1/2} (\frac{L_{\max}}{10 \text{ pc}}). \tag{19}$$

The corresponding cosmic ray hadron energy is

$$E_c = A\gamma_c m_p c^2 = 2.03 \cdot 10^5 A n_e^{1/2} (\frac{L_{\max}}{10 \text{ pc}}) \text{ PeV}, \tag{20}$$

which is four orders of magnitude larger than the Hillas limit (2) for equal values of the maximum wavelength. This difference demonstrates the dramatic influence of the plasma turbulence geometry (slab versus isotropically distributed waves) on the confinement of cosmic rays in the Galaxy. With isotropically distributed fast mode waves, even ultrahigh energy cosmic rays obey the scaling $\lambda\gamma^{q-2} = const.$

Only, at ultrahigh Lorentz factors $\gamma > \gamma_c$ or energies $E > E_c$ the mean free path approaches the much steeper dependence

$$\lambda(\gamma > \gamma_c) \simeq \frac{1}{\pi(q-1)} \frac{B_0^2}{(\delta B)_F^2} R_L (k_{min} R_L \epsilon)^2$$

$$\propto \beta\gamma^3 \simeq \gamma^3. \tag{21}$$

independent from the turbulence spectral index q. Here the mean free path quickly attains very large values greater than the typical scales of the Galaxy.

2.4.3. Anisotropy Because of the direct proportionality between mean free path and anisotropy, the cosmic ray anisotropy (15) shows the same behaviour as a function of energy.

3. Relevant magnetohydrodynamic plasma modes for damped waves

It has been pointed out by Schlickeiser and Miller [9] that this transit-time damping (TTD) contribution provides the overwhelming contribution to particle scattering because in this interaction the cosmic ray partticle interacts with the whole wave spectrum, in contrast to gyroresonances that singles out individual resonant wave numbers. The inclusion of resonance broadening due to wave damping in the resonance function guarantees that this dominance also holds for cosmic ray particles at small pitch angle cosines $\mu \leq |V_a/v|$, unlike the case of neglible wave damping [10] .

The damping of magnetosonic waves is caused both by collisionless Landau damping and collisional viscous damping, Joule damping and ion-neutral friction. The dominating contribution is provided by viscous damping with the rate calculated for plasma parameters of the diffuse intercloud medium at very small propagation angles

$$\gamma_F \simeq 2.9 \cdot 10^5 \beta V_A^2 k^2 \sin^2\theta. \tag{22}$$

For the ratio of damping rate to real frequency we obtain

$$\frac{\gamma_F}{|\omega_R|} \simeq 2.9 \cdot 10^5 \beta V_A k \sin^2\theta = 2.9 \cdot 10^5 \beta \omega_R \sin^2\theta. \tag{23}$$

Because in the interstellar medium $\Omega_{0,p} = 3.6 \cdot 10^{-2} B(4\mu G)$ Hz, we see that in the far MHD-wave region $\omega_R \leq 10^{-4} \Omega_{0,p}/\beta$ the weak damping limit is fulfilled.

3.1. Fast magnetosonic plasma modes
In the MHD wave range that interests us here, the dispersion relation reads as

$$\omega_R \simeq jkV_A \tag{24}$$

describing forward ($j = 1$) and backward ($j = -1$) moving fast mode waves.
The mean free path:

$$\lambda^{0F}(T >> 1) = \frac{15V_A\alpha}{8}\frac{q}{q-1}(\frac{B_0}{\delta B})^2 10^{14} T^3, \tag{25}$$

where T=$k_{min}R_L = E$ and is normalized with respect to E_c (where E_c is defined as for undamped case (eq. 20)). At relativistic rigidities we find that $\lambda^0 \sim T^3$.

$$\lambda^{F0}(T << 1) = 36V_A\alpha(\frac{B_0}{\delta B})^2. \tag{26}$$

In this energy limit the mean free path is constant with respect to T.

3.2. Slow magnetosonic plasma modes
Dispersion relation for slow magnetosonic waves in low-β plasma reads [8]

$$\omega_R^2 \simeq k^2 V_A^2 (\frac{\eta\beta}{1+\beta} + \frac{\eta^2\beta^2}{(1+\beta)^3}) \tag{27}$$

with $\eta = \cos\theta$ and β is the ratio of thermal and magnetic pressure. In the last equation, in the first approximation, we neglect the second term in brackets since it is one order smaller then the first term.

The mean free path for slow mode waves:

$$\lambda^{0S}(T >> 1) = \frac{3V_A\alpha}{8\sqrt{2}}\frac{q}{q-1}(\frac{B_0}{\delta B})^2 \frac{\pi}{h} 10^{13} T^3, \tag{28}$$

where $T = k_{min}R_L$, $h = 2arctanh\sqrt{1-\delta} - 2\sqrt{1-\delta}$, with $\delta << 1$.
At relativistic rigidities we find that $\lambda^0 \sim T^3$.

$$\lambda^{0S}(T << 1) = 36V_A\alpha(\frac{B_0}{\delta B})^2. \tag{29}$$

In this energy limit the mean free path is constant with respect to T.

3.2.1. Anisotropy According to eq. (20, 21, 22) of VS, we can calculate anisotropy for damped waves using eq. (26) and eq. (29) (fast and slow, respectively). Anisotropy dependance on energy is the same as for mean free path.

4. Summary

We have investigated the implications of isotropically distributed interstellar magnetohydrodynamic plasma waves on the scattering mean free path and the spatial anisotropy of high-energy cosmic rays. We demonstrate a drastic modification of the energy dependence of both cosmic ray transport parameters compared to previous calculations that have assumed that the plasma waves propagate only parallel or antiparallel to the ordered magnetic field (slab turbulence). In case of slab turbulence cosmic rays with Larmor radius R_L resonantly interact with plasma waves with wave vectors at $k_{res} = R_L^{-1}$. If the slab wave turbulence power spectrum vanishes for wavenumbers less than $k_{\min}$, as a consequence then cosmic rays with Larmor radii larger than $k_{\min}^{-1}$ cannot be scattered in pitch-angle, causing the socalled Hillas limit for the maximum energy $E_{15}^H = 4Z \cdot (B_0/4\mu\text{G})(L_{\|,max}/\text{parsec})$ of cosmic rays being confined in the Galaxy. At about these energies this would imply a drastic increase in the spatial anisotropy of cosmic rays that has not been detected by KASKADE and other air shower experiments.

In case of isotropically distributed interstellar magnetohydrodynamic waves we demonstrated that the Hillas energy E^H is modified to a limiting total energy that is about 4 orders of magnitude larger $E_c = 2.03 \cdot 10^4 A n_e^{1/2}(L_{\max}/1 \text{ pc})$ PeV, where A denotes the mass number and $L_{\max}$ the maximum wavenumber of isotropic plasma waves. Below this energy the cosmic ray mean free path and the anisotropy exhibit the well known E^{2-q} energy dependence, where $q = 5/3$ denotes the spectral index of the Kolmogorov spectrum. At energies higher than E_c both transport parameters steepen to a E^3-dependence. This implies that cosmic rays even with ultrahigh energies of several tens of EeV can be rapidly pitch-angle scattered by interstellar plasma turbulence, and are thus confined to the Galaxy.

Considering damped fast and slow mode waves caused by dominate viscous damping, we have determined energy dependance of the mean free path. We have found that for small energies it is approximately constant and for high energies is proportional to the third power of energy of the particle. The analysis for the influence of different types of damping, as well as the influence of different geometries of turbulence, will be the subject of further research.

5. References

[1] Turner M S et al. 2002 *Report to the National Academy of Science*
[2] Zatsepin G T, Kuzmin V A 1966 *Soviet Phys. JETP Letters* **4** 78
[3] Coleman S abd Glashow S L 1997 *Phys. Lett. B* **405** 249
[4] Amelio-Camella G, Ellis J, Mavromatos N E, Nanopoulos D V and Srakar S 1997 *Nature* **393** 763
[5] Hillas A M 1984 *ARAA* **22** 425
[6] Schlickeiser R 1989 *ApJ* **336** 243
[7] Antoni T, Apel W D, Badea A F et al. 2004 *ApJ* **604** 687
[8] Dogan A, Spanier F, Vainio R and Schlickeiser R 2006 *J. Plasma Physics* **72** 419
[9] Schlickeiser R and Miller J A 1998 *ApJ* **492** 352
[10] Vukčević M, Schlickeiser R 2007 *Astron. Astroph.* **467** 15

γ-radiation from the Galactic Center: dark matter annihilation or more conservative astrophysical models?

Alexander F. Zakharov[1,2], **Achille A. Nucita**[3], **Francesco De Paolis**[4], **Gabriele Ingrosso**[4]

[1]Institute of Theoretical and Experimental Physics, Moscow, 117259, Russia
[2]Bogoliubov Laboratory of Theoretical Physics, Joint Institute for Nuclear Research, Dubna, Russia

E-mail: `zakharov@itep.ru`
[3]XMM-Newton Science Operations Centre, ESAC, ESA, PO Box 50727, 28080 Madrid, Spain

E-mail: `achille.nucita@le.infn.it`
[4]Department of Physics and *INFN*, University of Lecce, CP 193, I-73100 Lecce, Italy

E-mail: `Francesco.DePaolis,Gabriele.Ingrosso@le.infn.it`

Abstract. The existence of dark matter (DM) at scales of few pc down to $\simeq 10^{-5}$ pc around the centers of galaxies and in particular in the Galactic Center region has been considered in the literature. Under the assumption that such a DM clump, principally constituted by non-baryonic matter (like WIMPs) does exist at the center of our galaxy, the study of the γ-ray emission from the Galactic Center region allows us to constrain both the mass and the size of this DM sphere. Further constraints on the DM distribution parameters may be derived by observations of bright infrared stars around the Galactic Center. Here, we discuss the constraints that can be obtained with the orbit analysis of stars (as S2 and S16) moving inside the DM concentration with present and next generations of large telescopes. In particular, consideration of the S2 star apoastron shift may allow improving limits on the DM mass and size. Further technological progress in a star orbit reconstruction and apocenter shift could detect features of bulk matter distributions or put so strict constraints on bulk mass matter (including DM) distributions that it will be impossible to explain γ-flux with DM annihilation.

1. Introduction

For the black hole in the Galactic Center, Hall and Gondolo [1] used estimates of the enclosed mass obtained in various ways and tabulated by Ghez et al. [2, 3]. The black hole, stellar cluster and DM could contribute in the mass inside stellar orbits. Moreover, if a DM cusp does exist around the Galactic Center it could modify the trajectories of stars moving around it in a sensible way depending on the DM mass distribution.

In the last years intensive searches for dark matter (DM), especially its non-baryonic component, both in galactic halos and at galaxy centers have been undertaken (see for example [4, 5] for recent results). It is generally accepted that the most promising candidate for the DM non-baryonic component is neutralino. In this case, the γ-flux from galactic halos (and from our Galactic halo in particular) could be explained by neutralino annihilation

[6, 7, 8, 9, 10, 11, 12, 13]. Since γ-rays are detected not only from high galactic latitude, but also from the Galactic Center, there is a wide spread hypothesis (see [14] for a discussion) that a DM concentration might be present at the Galactic Center. In this case the Galactic Center could be a strong source of γ-rays and neutrinos [4, 7, 15, 16, 17, 18, 19, 20, 21, 22] due to DM annihilation. Since it is also expected that DM forms spikes at galaxy centers [23, 24, 25] the γ-ray flux from the Galactic Center should increase significantly in that case.

At the same time, progress in monitoring bright stars near the Galactic Center have been reached recently [2, 3, 26]. The astrometric limit for bright stellar sources near the Galactic Center with 10 meter telescopes is today $\delta\theta_{10} \sim 1$ mas and the Next Generation Large Telescope (NGLT) will be able to improve this number at least down to $\delta\theta_{30} \sim 0.5$ mas [28, 29] or even to $\delta\theta_{30} \sim 0.1$ mas [27, 28, 29] in the K-band. Therefore, it will be possible to measure the proper motion for about ~ 100 stars with astrometric errors several times smaller than errors in current observations.

Recently it was shown [30, 31] that it is possible to constrain the parameters of the DM distribution possible present around the Galactic Center by considering the induced apoastron shift due to the presence of this DM sphere and either available data obtained with the present generation of telescopes (the so called *conservative* limit) and also expectations from future NGLT observations or with other advanced observational facilities.

2. The mass concentration at the Galactic Center

Recent advancements in infrared astronomy are allowing to test the scale of the mass profile at the center of our galaxy down to tens of AU. With the Keck 10 m telescope, the proper motion of several stars orbiting the Galactic Center black hole have been monitored and almost entire orbits, as for example that of the S2 star, have been measured allowing an unprecedent description of the Galactic Center region. Measurements of the amount of mass $M(<r)$ contained within a distance r from the Galactic Center are continuously improved as more precise data are collected. Recent observations [2] extend down to the periastron distance ($\simeq 3 \times 10^{-4}$ pc) of the S16 star and they correspond to a value of the enclosed mass within $\simeq 3 \times 10^{-4}$ pc of $\simeq 3.67 \times 10^{6}$ $M_{\odot}$. Several authors have used these observations to model the Galactic Center mass concentration. Here and in the following, we use the three component model for the central region of our galaxy based on estimates of enclosed mass given by Ghez et al [2, 3] recently proposed [1]. This model is constituted by the central black hole, the central stellar cluster and the DM sphere (made of WIMPs), i.e.

$$M(<r) = M_{BH} + M_*(<r) + M_{DM}(<r) \ , \tag{1}$$

where M_{BH} is the mass of the central black hole Sagittarius A*. For the central stellar cluster, the empirical mass profile is

$$M_*(<r) = \begin{cases} M_*\left(\frac{r}{R_*}\right)^{1.6} \ , & r \le R_* \\ M_*\left(\frac{r}{R_*}\right)^{1.0} \ , & r > R_* \end{cases} \tag{2}$$

with a total stellar mass $M_* = 0.88 \times 10^{6}$ $M_{\odot}$ and a size $R_* = 0.3878$ pc.

As far as the mass profile of the DM concentration is concerned, Hall and Gondolo [1] have assumed a mass distribution of the form

$$M_{DM}(<r) = \begin{cases} M_{DM}\left(\frac{r}{R_{DM}}\right)^{3-\alpha} \ , & r \le R_{DM} \\ M_{DM}, & r > R_{DM} \end{cases} \tag{3}$$

M_{DM} and R_{DM} being the total amount of DM in the form of WIMPs and the radius of the spherical mass distribution, respectively.

Hall and Gondolo [1] discussed limits on DM mass around the black hole at the Galactic Center. It is clear that present observations of stars around the Galactic Center do not exclude the existence of a DM sphere with mass $\simeq 4 \times 10^6 M_\odot$, well contained within the orbits of the known stars, if its radius R_{DM} is $\lesssim 2\times10^{-4}$ pc (the periastron distance of the S16 star in the more recent analysis [3]). However, if one considers a DM sphere with larger radius, the corresponding upper value for M_{DM} decreases (although it tends again to increase for extremely extended DM configurations with $R_{DM} \gg 10$ pc). In the following, we will assume for definiteness a DM mass $M_{DM} \sim 2 \times 10^5 M_\odot$, that is the upper value for the DM sphere in [1] within an acceptable confidence level in the range $10^{-3} - 10^{-2}$ pc for R_{DM}. As it will be clear in the following, we emphasize that even a such small value for the DM mass (that is about only 5% of the standard estimate $3.67 \pm 0.19 \times 10^6$ $M_\odot$ for the dark mass at the Galactic Center [3]) may give some observational signatures.

Evaluating the S2 apoastron shift [1] as a function of R_{DM}, one can further constrain the DM sphere radius since even now we can say that there is no evidence for negative apoastron shift for the S2 star orbit at the level of about 10 mas. In addition, since at present the precision of the S2 orbit reconstruction is about 1 mas, we can say that even without future upgrades of the observational facilities and simply monitoring the S2 orbit, it will be possible within about 15 years to get much more severe constraints on R_{DM}.

Moreover, observational facilities will allow in the next future to monitor faint infrared objects at the astrometric precision of about 10 μas [32] and, in this case, previous estimates will be sensibly improved since it is naturally expected to monitor eccentric orbits for faint infrared stars closer to the Galactic Center with respect to the S2 star.

In the following section, we study the motion of stars as a consequence of the gravitational potential $\Phi(r)$ due the mass profile given in Eq. (1). As usual, the gravitational potential can be evaluated as

$$\Phi(r) = -G \int_r^\infty \frac{M(r')}{r'^2}\, dr' \ . \tag{4}$$

3. Apoastron Shift Constraints

According to GR, the motion of a test particle can be fully described by solving the geodesic equations. Under the assumption that the matter distribution is static and pressureless, the equations of motion in the PN-approximation become (see, for example, [33])

$$\frac{d\mathbf{v}}{dt} \simeq -\nabla(\Phi_N + 2\Phi_N^2) + 4\mathbf{v}(\mathbf{v}\cdot\nabla)\Phi_N - v^2\nabla\Phi_N \ . \tag{5}$$

We note that the PN-approximation is the first relativistic correction from which the apoastron advance phenomenon arises. In the case of the S2 star, the apoastron shift as seen from Earth (from Eq. (7)) due to the presence of a central black hole is about 1 mas, therefore not directly detectable at present since the available precision in the apoastron shift is about 10 mas (but it will become about 1 mas in 10–15 years even without considering possible technological improvements). It is also evident that higher order relativistic corrections to the S2 apoastron shift are even smaller and therefore may be neglected at present, although they may become important in the future.

As it will be discussed below, the Newtonian effect due to the existence of a sufficiently extended DM sphere around the black hole may cause an apoastron shift in the opposite direction

[1] We want to note that the periastron and apoastron shifts $\Delta\Phi$ as seen from the orbit center have the same value whereas they have different values as seen from Earth (see Eq. (7)). When we are comparing our results with orbit reconstruction from observations we refer to the apoastron shift as seen from Earth.

with respect to the relativistic advance due to the black hole. Therefore, we have considered the two effects comparing only the leading terms.

For the DM distribution at the Galactic Center we follow Eq. (3) as done in [1]. Clearly, if in the future faint infrared stars (or spots) closer to the black hole with respect to the S2 star will be monitored [32], this simplified model might well not hold and higher order relativistic corrections may become necessary.

For a spherically symmetric mass distribution (such as that described above) and for a gravitational potential given by Eq. (4), Eq. (5) may be rewritten in the form (see for details [34])

$$\frac{d\mathbf{v}}{dt} \simeq -\frac{GM(r)}{r^3}\left[\left(1+\frac{4\Phi_N}{c^2}+\frac{v^2}{c^2}\right)\mathbf{r}-\frac{4\mathbf{v}(\mathbf{v}\cdot\mathbf{r})}{c^2}\right], \tag{6}$$

$\mathbf{r}$ and $\mathbf{v}$ being the vector radius of the test particle with respect to the center of the stellar cluster and the velocity vector, respectively. Once the initial conditions for the star distance and velocity are given, the rosetta shaped orbit followed by a test particle can be found by numerically solving the set of ordinary differential equations in eq. (6).

In Fig. 1, as an example, assuming that the test particle orbiting the Galactic Center region is the S2 star, we show the Post-Newtonian orbits obtained by the black hole only and the black hole plus the stellar cluster plus the contribution of DM mass density with $R_{DM} = 10^{-3}$ pc. In each case the S2 orbit apoastron shift is given (for black hole (without stellar cluster or DM concentration) $\Delta\Phi$=580 arcsec, for black hole plus stellar cluster $\Delta\Phi$=460 arcsec, for black hole plus stellar cluster plus DM (with $R_{DM} = 10^{-3}$ pc) we have $\Delta\Phi$=–300 arcsec). As one can see, for selected parameters for DM and stellar cluster masses and radii the effect of the stellar cluster is almost negligible while the effect of the DM distribution is crucial since it enormously overcome the shift due to the black hole (for $R_{DM} = 10^{-3}$ pc). Moreover, as expected, its contribution is opposite in sign with respect to that of the black hole [35].

We note that the expected apoastron (or, equivalently, periastron) shifts (mas/revolution), $\Delta\Phi$ (as seen from the center) and the corresponding values $\Delta\phi_E^{\pm}$ as seen from Earth (at the distance $R_0 \simeq$ 8 kpc from the GC) are related by

$$\Delta\phi_E^{\pm} = \frac{d(1\pm e)}{R_0}\Delta\Phi, \tag{7}$$

where with the sign $\pm$ are indicated the shift angles of the apoastron (+) and periastron (-), respectively. The S2 star semi-major axis and eccentricity are $d = 919$ AU and $e = 0.87$ [3].

In Fig. 2, the S2 apoastron shift as a function of the DM distribution size R_{DM} is given for $\alpha = 0$ and $M_{DM} \simeq 2\times 10^5$ $M_\odot$. Taking into account that the present day precision for the apoastron shift measurements is of about 10 mas, one can say that the S2 apoastron shift cannot be larger than 10 mas. Therefore, any DM configuration that gives a total S2 apoastron shift larger than 10 mas (in the opposite direction due to the DM sphere) is excluded. The same analysis is done for two different values of the DM mass distribution slope, i.e. $\alpha = 1$ and $\alpha = 2$. In any case, we have calculated the apoastron shift for the S2 star orbit assuming a total DM mass $M_{DM} \simeq 2\times 10^5$ $M_\odot$. As one can see, the upper limit of about 10 mas on the S2 apoastron shift may allow to conclude that DM radii in the range about $10^{-3} - 10^{-2}$ pc are excluded by present observations for DM mass distribution slopes.

We notice that the results of the present analysis allows to further constrain the results of the Hall and Gondolo [1] who have concluded that if the DM sphere radius is in the range $10^{-3} - 1$ pc, configurations with DM mass up to $M_{DM} = 2\times 10^5$ $M_\odot$ are acceptable. The present analysis shows that DM configurations of the same mass are acceptable only for R_{DM} out the range between $10^{-3} - 10^{-2}$ pc, almost irrespectively of the α value.

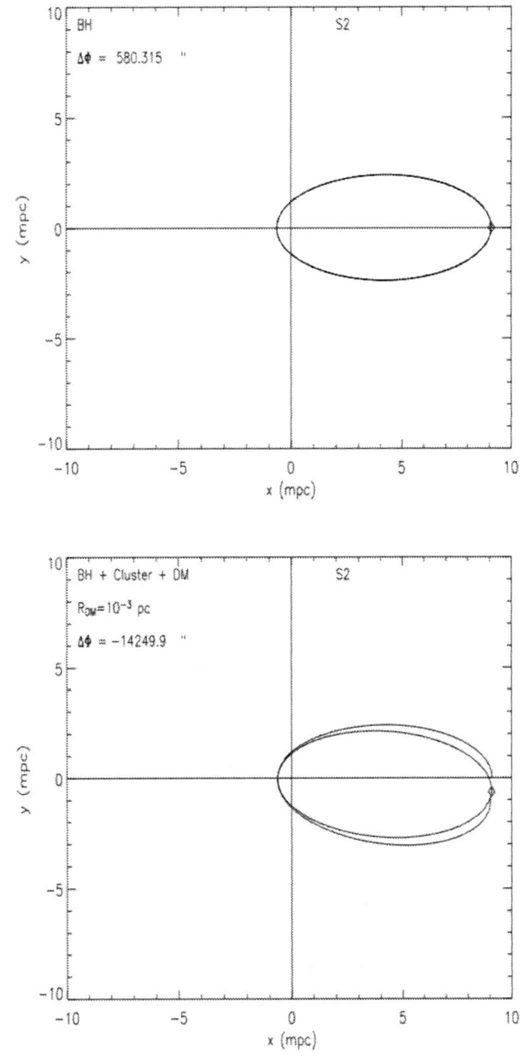

Figure 1. PN-orbits for different mass configurations at the Galactic Center. The S2 star has been considered as a test particle and its apoastron shift is indicated in each panel as $\Delta\Phi$ (in arcsec). The top panel shows the central black hole contribution to the S2 shift that amounts to about 580 arcsec (the classical GR apoastron shift is not distinctly visible in the orbit shape but it could be measured with the modern technology since $\Delta\phi_E^+$ is about 1 mas). In the bottom panel the contributions due to stellar cluster and DM mass are added (as derived in eq. (3)) and in the case, deviations from elliptical orbit are clearly seen (here we assume that DM mass $M_{DM} \simeq 2 \times 10^5$ $M_\odot$ and $R_{DM} = 10^{-3}$ pc).

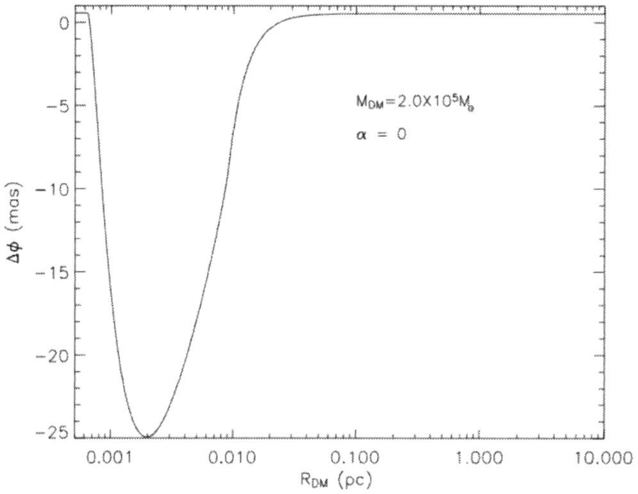

Figure 2. Apoastron shift as a function of the DM radius R_{DM} for $\alpha = 0$ and $M_{DM} \simeq 2 \times 10^5$ $M_\odot$. Taking into account present day precision for the apoastron shift measurements (about 10 mas) one can say that DM radii R_{DM} in the range $8 \times 10^{-4} - 10^{-2}$ pc are not acceptable.

4. Conclusions

In this paper we have considered the constraints that the upper limit (presently of about 10 mas) of the S2 apoastron shift may put on the DM configurations at the galactic center considered by Hall and Gondolo [1].

When (in about 10–15 years, even without considering improvements in observational facilities) the precision of S2 apoastron shift will be about 1 mas (that is equal to the present accuracy in the S2 orbit reconstruction) our analysis will allow to further constrain the DM distribution parameters. In particular, the asymmetric shape of the curves in Fig. 2 imply that any improvement in the apoastron shift measurements will allow to extend the forbidden region especially for the upper limit for R_{DM}. Quantitatively, we have a similar behavior curves for other choices of slope parameters α for DM concentrations.

In this context, future facilities for astrometric measurements at a level 10 μas of faint infrared stars will be extremely useful [32] and they give an opportunity to put even more severe constraints on DM distribution. In addition, it is also expected to detect faint infrared stars or even hot spots [36] orbiting the Galactic Center. In this case, consideration of higher order relativistic corrections for an adequate analysis of the stellar orbital motion have to be taken into account. Due to a great progress in precision of measurements, one could not exclude a possibility that matter density will be so low that alternative scenarios (to DM annihilation model) will be needed to explain γ-flux from the Galactic Center. Electromagnetic processes in plasma with a presence of a strong gravitational field near the Galactic Center may be important components of such alternative scenarios for the detected γ-flux.

In our considerations we adopted simple analytical expression and reliable values for R_{DM} and M_{DM} parameters following [1] just to illustrate the relevance of the apoastron shift phenomenon in constraining the DM mass distribution at the Galactic Center. If other models for the DM distributions are considered (see, for instance [37] and references therein) the qualitative aspects of the problem are preserved although, of course, quantitative results on apoastron shifts may be different.

Acknowledgments
AFZ is grateful to Dipartimento di Fisica Universita di Lecce and INFN, Sezione di Lecce where a part of this work was carried out and profs. M. Dimitrievič and L. Č. Popovič for their kind attention to our contribution and the hospitality in Serbia.

References

[1] Hall J and Gondolo P 2006, *Phys. Rev.* D **74** 063511
[2] Ghez A M et al. 2003 *Astron. Nachr.* **324** 527
[3] Ghez A M et al. 2005 *Astrophys. J.* **620** 744
[4] Bertone G., Hooper D. and Silk J. 2005 *Phys. Reports* **405** 279
[5] Bertone G. and Merritt D. 2005 *Modern Phys. Lett.* A, **20**, 1021
[6] Gurevich A V and Zybin K P 1997 *Phys. Lett.* A **225** 217
[7] Bergström L, Ullio P and Buckley J H 1998 *Astropart. Phys.* **94** 131301
[8] Tasitsiomi A and Olinto A V 2002 *Phys. Rev.* D **66** 023502
[9] Stoehr F, White S D M, Springel V et al. 2003 *Mon. Not. R. Astron. Soc.* **345** 1313
[10] Prada F, Klypin A, Flix J, Martinez M and Simonneau E 2004 Astrophysical inputs on the SUSY dark matter annihilation detectability (*Preprint* astro-ph/0401512)
[11] Prada F, Klypin A, Flix J, Martinez M and Simonneau E 2004 *Phys. Rev. Lett.* **93** 241301
[12] Profumo S 2005 *Phys. Rev.* D **72** 103521
[13] Mambrini Y, C. Munoz C, Nezri E and Prada F 2005 Gamma-Ray Excess from the Galactic Center and Supergravity Models (*Preprint* hep-ph/0509300)
[14] Evans N W, Ferrer F and Sarkar S 2004 *Phys. Rev.* D **69** 123501
[15] Bouquet A, Salati P and Silk J 1989 *Phys. Rev.* D, **40** 3168
[16] Stecker F W 1988 *Phys. Lett.* B **201** 529 (1988).
[17] Berezinsky V, Bottino A and Mignola G 1994 *Phys. Lett.* B **325** 136
[18] Bertone G, Nezri E, Orloff J and Silk J 2004 *Phys. Rev.* D **70** 063503
[19] O.Y. Gnedin and J.R. Primack, Phys. Rev. Lett., **93**, 061302 (2004).
[20] Bergström L, Bringmann T, Eriksson M and Gustafsson M 2005 *Phys. Rev. Lett.* **9** 138
[21] Horns D 2005 *Phys. Lett.* B **607** 225
[22] Bertone G and Merritt D. 2005 *Phys. Rev.* D, **72** 103502
[23] Gondolo P and Silk J 1999 *Phys. Rev. Lett.* **83** 1719
[24] Ullio P, Zhao H - S and Kamionkowski M 2001 *Phys. Rev.* D **64** 043504
[25] Merritt D 2003 *Proc.of the Fourth Int. Workshop on the Identification of Dark Matter*, York, UK, 2-6 September 2002, ed. N J C Spooner and V Kudryavtsev, World Scientific, ISBN 981-238-237-2, p. 96; (*Preprint* astro-ph/0301365)
[26] Genzel R et al. 2003 *Astrophys J.* **594** 812
[27] Ames G, Aubrun J, Bolte M et al. 2002 *California Extremely Large Telescope - Conceptual Design for a Thirty-Meter Telescope*, http://tmt.ucolick.org/reports_and_notes/reports/Web_final_Greenbook.pdf
[28] Weinberg N, Miloslavljević M and Ghez A M 2005 Astrometric Monitoring of Stellar Orbits at the Galactic Center with a Next Generation Large Telescope, (*Preprint* astro-ph/0512621)
[29] Weinberg N, Miloslavljević M and Ghez A M 2005 *Astrophys J.* **622** 878
[30] Zakharov A F, Nucita A A, De Paolis F , Ingrosso G 2007 *Phys. Rev.* D **76** 62001
[31] Ghez A M, Salim S, Weinberg et al. 2008 Probing the properties of the Milky Way's central supermassive black hole with stellar orbits (*Preprint* arXiv:0808.2870v1 [astro-ph])
[32] Eisenhauer F, Perrin G and Rabien S 2005 *Astron. Nachr.* **326** 561
[33] Weinberg S 1972 *Gravitation and Cosmology: Principles and Applications of the General Theory of Relativity*, (Wiley)
[34] Rubilar G G and Eckart A 2001 *Astron. & Astrophys.* **375** 95
[35] Nucita A A, De Paolis F , Ingrosso G et al. 2007 *Publ. Astron. Soc. Pacific* **119** 349
[36] Genzel R and V. Karas V 2007 The Galactic Center *Preprint* (arXiv:0704.1281v1[astro-ph])
[37] Merritt D, Harfst S and Bertone G 2007 *Phys. Rev.* D **75** 043517

AUTHOR INDEX

Aigner, F 124
Ambrico, P F 160
Antoniou, A 266
Baksht, E H....................191
Benedikt, J 106
Blanco, F 9
Bocvarski, Valerij 49
Bogaerts, A 212
Borge, M J G 9
Borisov, A G 73
Borka, D....................132
Brok, W J M....................212
Brunger, Michael J 1
Buckman, S J....................1, 33
Bundaleski, N 115, 140
Burgdörfer, J....................124
Cadež, Iztok....................274
Califano, F 220
Calzada, M D....................257
Caradonna, P....................1, 33
Cassimi, A....................140
Ciroi, S 243
Danezis, E 266
De Benedictis, S....................160
De Paolis, Francesco 299
Den, M....................231
Di Mille, F 243
Dilecce, G....................160
Dimitrijevic, M S....................266
Dujko, S 40
Eckert, J....................148
Efimova, V V 148
Faganello, M....................220
Flötgen, C....................106
Francis-Staite, Jessica....................1
Garcia, G 1
García, G....................9
Gauyacq, J-P....................73
Goeckner, M J 85
Gorfinkiel, J D 9
Hargreaves, Leigh 1
Hoffmann, V 148, 181
Horiuchi, R 231
Huerga, C....................9
Ingrosso, Gabriele 299
Ishiguro, S 231
Ishizaki, R 231
Ishizawa, A....................231
Ito, A....................231, 231
Jarvis, B....................85
Jindal, A K 85
Jones, A 1, 33
Joseph, E A 85
Kazansky, A K....................73
Khemliche, H 115, 140
Kobayashi, Akira....................170
Kostyrya, I D....................191
Kussel, G 106
Limão-Vieira, P 9
Lomaev, M I....................191
Lower, Julian C 1
Lyratzi, E 266
Mackochekanwa, Casten 1
Maddern, Todd....................1
Maira-Vidal, A....................9
Majstorovic, G Lj 203
Makochekanwa, C....................33
Markelj, Sabina 274
Martens, T....................212
Maunoury, L....................140
McEachran, Robert P 1
Miškovic, Z L....................95, 132

AUTHOR INDEX

Miura, H 231
Momeni, Anouchah 115
Mondal, Subhendu 1
Mowbray, D J 132
Muñoz, A 9
Nakajima, N 231
Nakamura, H 231
Nelson, C T 85
Neškovic, N 132
Nikitovic, Željka 22
Nucita, Achille A 299
Ohtani, H 231
Oiler, J C 9
Okamura, S 231
Overzet, L J 85
Padron-Wells, G 85
Pegoraro, F 220
Pelicon, Primoz 274
Petrovic, D 212
Petrovic, Ivan 49
Petrovic, S 132
Petrovic, Violeta 49
Petrovic, Zoran Lj 22
Pierce, R 85
Popovic, L C 266
Popovic, Luka C 284
Puac, N 57
Raball, V 106
Rafanelli, P 243
Roncin, P 115, 140
Rousseau, P 115, 140
Rupnik, Zdravko 274
Rybka, D V 191
Sakagami, H 231
Sant, S P 85
Sato, M 231
Simic, Saša 284
Simonovic, N 124
Škoric, M M 231
Slaughter, Daniel 1
Smid, P 148, 181
Solleder, B 124
Soulisse, Pierre 115
Steers, E B M 148, 181
Stojanovic, V 22, 65
Sudo, S 231
Sullivan, J P 1, 33
Takayama, A 231
Tarasenko, V F 191
Téllez, M 9
Tengblad, O 9
Todo, Y 231
Tomita, Y 231
Usami, S 231
Van Dijk, J 212
Vaona, L 243
Von Keudell, A 106
Voronov, M V 148
Vukcevic, Miroslava 292
Watanabe, T-H 231
Weiss, Z 181
White, R D 40
Wirtz, L 124
Yamamoto, T 231
Zakharov, Alexander F 299
Zhou, B S 85